3G Mobile Networks
Architecture, Protocols and Procedures

Based on 3GPP Specifications for UMTS WCDMA Networks

3G Mobile Networks

Architecture, Protocols and Procedures

Based on 3GPP Specifications for UMTS WCDMA Networks

Sumit Kasera

Senior Technical Leader, Hughes Software Systems
Gurgaon, India

Nishit Narang

Senior Technical Leader, Hughes Software Systems
Gurgaon, India

McGraw-Hill

New York   Chicago   San Francisco   Lisbon   London
Madrid   Mexico City   Milan   New Delhi   San Juan
Seoul   Singapore   Sydney   Toronto

The _McGraw·Hill_ Companies

Cataloging-in-Publication Data is on file with the Library of Congress

1 2 3 4 5 6 7 8 9 0   DOC/DOC   0 1 0 9 8 7 6 5 4

ISBN 0-07-145101-3

The sponsoring editor for this book was Stephen S. Chapman and the production supervisor was Sherri Souffrance. The art director for the cover was Handel Low.

Printed and bound by RR Donnelley.

This book was previously published by Tata McGraw-Hill Publishing Company Limited, New Delhi, India, copyright © 2004.

McGraw-Hill books are available at special quantity discounts to use as premiums and sales promotions, or for use in corporate training programs. For more information, please write to the Director of Special Sales, McGraw-Hill Professional, Two Penn Plaza, New York, NY 10121-2298. Or contact your local bookstore.

 This book is printed on recycled, acid-free paper containing a minimum of 50% recycled, de-inked fiber.

CONTENTS

Foreword   *xxi*

Preface   *xxiii*

Acknowledgements   *xxvii*

PART I—INTRODUCTION AND BACKGROUND

1. Introduction   **3**

    1.1 Introduction   *3*

    1.2 Second Generation Mobile Networks   *3*

        1.2.1 Limitations of 2G Networks   *4*

    1.3 2.5 Generation Mobile Networks   *5*

    1.4 International Mobile
Telecommunication-2000 (IMT-2000)   *6*

    1.5 Third Generation Partnership Program (3GPP)   *8*

        1.5.1 Interaction of 3GPP with other bodies   *8*

        1.5.2 Objective of 3GPP   *9*

        1.5.3 3GPP Technical Specifications Group (TSG)   *10*

        1.5.4 Stages of 3GPP Specifications   *12*

        1.5.5 Series of 3GPP Specifications   *13*

        1.5.6 Version Numbering of 3GPP Specifications   *13*

        1.5.7 Releases of 3GPP Specifications   *15*

        1.5.8 Evolution towards 3GPP Networks   *16*

    Summary   *17*

2. Principles Of WCDMA   **19**

    2.1 Introduction   *19*

2.2 Requirements for Third Generation Air Interface   *19*

2.3 Schemes for Radio Access   *21*

2.4 WCDMA Overview   *22*

    2.4.1 Direct-Sequence CDMA   *22*

    2.4.2 Wideband   *23*

    2.4.3 Synchronization Aspects   *23*

    2.4.4 Modes of Operation   *23*

2.5 Spreading and De-spreading   *24*

    2.5.1 Autocorrelation and Cross Correlation   *26*

    2.5.2 Benefits of Spreading   *27*

2.6 Scrambling   *28*

2.7 Rake Receiver   *30*

2.8 Multipath Diversity and Macrodiversity   *30*

2.9 Power Control Mechanisms   *32*

2.10 Soft and Softer Handover   *36*

2.11 SRNS Relocation   *39*

Summary   41

PART II—UMTS NETWORK ARCHITECTURE AND PROTOCOLS

3. UMTS Network Architecture   45

3.1 Introduction   *45*

3.2 Basic Structure of UMTS Network   *45*

    3.2.1 User Equipment (UE)   *47*

    3.2.2 Access Network (AN)   *48*

    3.2.3 Core Network (CN)   *48*

3.3 Access Stratum and Non-access Stratum   *53*

    3.3.1 Access Stratum (AS)   *54*

    3.3.2 Non-Access Stratum (NAS)   *54*

3.4 Hierarchical Network Organization   *56*

    3.4.1 Public Land Mobile Network (PLMN)   *56*

    3.4.2 Location Area (LA)   *57*

    3.4.3 Routing Area (RA)   *58*

    3.4.4 UTRAN Registration Area (URA)   *59*

    3.4.5 Cell Global Identity (CGI)   *59*

3.5 Addresses and Identifiers   *60*

    3.5.1 Subscriber Identity   *60*

3.5.2 Service Identity *60*

3.5.3 Temporary Identities *62*

3.5.4 PDP Address *64*

3.5.5 Equipment Identity *64*

3.5.6 Location Number *64*

3.5.7 Identifying Network Entities *65*

3.6 Service Aspects *66*

3.7 Service Classification *67*

3.7.1 Bearer Services *68*

3.7.2 Other Bearer Service *69*

3.7.3 Teleservices *70*

3.7.4 Supplementary Services *71*

3.7.5 Other Services *71*

3.7.6 Toolkits *71*

3.8 Quality of Service (QoS) Architecture *71*

3.9 UMTS QoS Classes *73*

3.9.1 Conversational Class *73*

3.9.2 Streaming Class *75*

3.9.3 Interactive Class *75*

3.9.4 Background Class *75*

Summary 76

4. User Equipment   **77**

4.1 Introduction *77*

4.2 Components of User Equipment *78*

4.2.1 Universal Integrated Circuit Card (UICC) *78*

4.2.2 Mobile Equipment (ME) *79*

4.2.3 User Equipment Combination *80*

4.3 Interfaces of User Equipment *81*

4.3.1 External Interfaces *82*

4.3.2 Internal Interfaces *83*

4.4 UE Functions *84*

4.4.1 Mobile Termination Functions *84*

4.4.2 Terminal Equipment Functions *85*

4.4.3 Terminal Adaptation Functions *85*

4.4.4 USIM Functions *85*

4.5 UE Protocols *85*

    4.5.1 Access Stratum Protocols *86*

    4.5.2 Non-Access Stratum Protocols *86*

4.6 Classification of UE *87*

Summary 88

5. Access Network   **89**

5.1 Introduction *89*

5.2 Access Network Entities *89*

    5.2.1 Base Station Sub-system (BSS) *89*

    5.2.2 Radio Network Sub-system (RNS) *90*

5.3 Network Interfaces *91*

    5.3.1 Abis Interface between BSC and BTS *92*

    5.3.2 Iub Interface between RNC and Node B *92*

    5.3.3 Iur Interface between RNCs *93*

    5.3.4 A Interface between MSC/VLR and BSS *94*

    5.3.5 Gb Interface between SGSN and BSS *94*

    5.3.6 Iu Interface between CN and RNS *94*

5.4 Radio Interface Protocol Architecture *95*

5.5 UTRAN Protocol Architecture *97*

    5.5.1 Iu_CS Protocol Architecture *98*

    5.5.2 Iu_PS Protocol Architecture *100*

    5.5.3 Iur Protocol Architecture *101*

    5.5.4 Iub Protocol Architecture *103*

    5.5.5 Iu_BC Protocol Architecture *104*

5.6 Functions *105*

    5.6.1 Transfer of User Data *106*

    5.6.2 System Access Control *106*

    5.6.3 Security Functions *106*

    5.6.4 Mobility Management *107*

    5.6.5 Radio Resource Management *107*

    5.6.6 Broadcast and Multicast Services *109*

    5.6.7 Other Functions *110*

5.7 Radio Interface Protocols *110*

    5.7.1 Physical Layer *110*

    5.7.2 Medium Access Control (MAC) *113*

5.7.3 Radio Link Control (RLC) *117*

5.7.4 Packet Data Convergence Protocol (PDCP) *121*

5.7.5 Broadcast/Multicast Control (BMC) *123*

5.7.6 Radio Resource Control (RRC) *124*

5.8 ATM-based Transport Network Protocols *130*

5.8.1 Asynchronous Transfer Mode (ATM) *131*

5.8.2 ATM Adaptation Layer 2 (AAL2) *135*

5.8.3 ATM Adaptation Layer 5 (AAL5) *137*

5.8.4 Service Specific Connection Oriented Protocol (SSCOP) *139*

5.8.5 Service-Specific Co-ordination Function for NNI (SSCF-NNI) *140*

5.8.6 Service-Specific Co-ordination Function for UNI (SSCF-UNI) *141*

5.8.7 Message Transfer Part 3 for Broadband (MTP3b) *142*

5.8.8 Signaling Transport Converter (STC) *143*

5.8.9 AAL2 Signaling *144*

5.9 Application Layer Protocols *147*

5.9.1 Radio Access Network Application Part (RANAP) *148*

5.9.2 Radio Network Sub-system Application Part (RNSAP) *150*

5.9.3 NBAP *153*

5.9.4 Service Area Broadcast Protocol (SABP) *157*

5.9.5 Iu User Plane (UP) Protocol *158*

5.9.6 Framing Protocols for Iub and Iur Interface *161*

Summary *163*

6. Core Network **164**

6.1 Introduction *164*

6.2 Entities Common to CS and PS Domain *165*

6.2.1 Home Location Register (HLR) *165*

6.2.2 Authentication Center (AuC) *166*

6.2.3 Equipment Identity Register (EIR) *167*

6.2.4 Short Message Service (SMS) Entities *167*

6.3 Entities Specific to the CS Domain *167*

6.3.1 Visitor Location Register (VLR) *168*

6.3.2 Mobile Switching Center (MSC) *168*

6.3.3 Gateway Mobile Switching Center (GMSC) *169*

6.4 Entities Specific to the PS Domain   *169*

   6.4.1 Serving GPRS Support Node (SGSN)   *169*

   6.4.2 Gateway GPRS Support Node (GGSN)   *170*

   6.4.3 Border Gateway (BG)   *170*

6.5 Service-specific Entities of the Core Network   *170*

   6.5.1 Gateway Mobile Location Center (GMLC)   *171*

   6.5.2 CAMEL Entities   *171*

   6.5.3 Cell Broadcast Center (CBC)   *171*

6.6 Network Interfaces of CS Domain   *171*

   6.6.1 B Interface between MSC and VLR   *172*

   6.6.2 C Interface between GMSC and HLR   *173*

   6.6.3 D Interface between VLR and HLR   *174*

   6.6.4 E Interface between MSC and MSC   *174*

   6.6.5 F Interface between MSC and EIR   *174*

   6.6.6 G Interface between VLR and VLR   *174*

   6.6.7 Nb/Nc Interface between MSC and GMSC   *175*

   6.6.8 Interface between VLR and SMS-MSC   *175*

6.7 Interfaces of PS Domain   *175*

   6.7.1 Gn/Gp Interface between two GSNs   *176*

   6.7.2 Gi Interface between GGSN and PDN   *178*

   6.7.3 Gr Interface between SGSN and HLR   *178*

   6.7.4 Gs Interface between SGSN and MSC/VLR   *179*

   6.7.5 Gf Interface between SGSN and EIR   *179*

   6.7.6 Gc Interface between GGSN and HLR   *180*

   6.7.7 Gd Interface between SGSN and SMS-MSC   *180*

6.8 CS Domain Protocol Architecture   *180*

   6.8.1 User Plane   *180*

   6.8.2 Control Plane   *181*

6.9 PS Domain Protocol Architecture   *184*

   6.9.1 User Plane   *184*

   6.9.2 Control Plane   *185*

6.10 Core Network Functions   *189*

   6.10.1 Mobility Management   *189*

   6.10.2 Call Handling   *193*

   6.10.3 Session Management   *193*

   6.10.4 Supplementary Services   *194*

6.10.5 Short Message Service   *195*

6.10.6 Security Functions   *195*

6.11 Subscriber Data   *196*

6.12 SS7 Protocols   *200*

    6.12.1 Message Transfer Part (MTP)   *200*

    6.12.2 Signaling Connection Control Part (SCCP)   *202*

    6.12.3 ISDN User Part (ISUP)   *203*

6.13 Application Protocols   *204*

    6.13.1 Transaction Capabilities (TCAP)   *204*

    6.13.2 Mobile Application Part (MAP)   *209*

    6.13.3 NAS Signaling   *220*

    6.13.4 GPRS Tunneling Protocol (GTP)   *224*

    6.13.5 Base Station Sub-system Application Part + (BSSAP+)   *226*

Summary   *229*

PART III—PROCEDURES IN UMTS NETWORK

7. Radio Resource Control Procedures   **233**

7.1 Introduction   *233*

7.2 RRC Protocol States   *233*

7.3 RRC Connection Management Procedures   *236*

    7.3.1 Broadcast of System Information   *237*

    7.3.2 Paging   *239*

    7.3.3 UE Dedicated Paging   *240*

    7.3.4 RRC Connection Establishment   *240*

    7.3.5 RRC Connection Release   *243*

    7.3.6 Signaling Connection Release Procedure   *244*

    7.3.7 Transmission of UE Capability Information   *244*

    7.3.8 Direct Transfer of NAS Messages   *246*

    7.3.9 Security Functions   *247*

7.4 Radio Bearer Control Procedures   *249*

    7.4.1 Radio Bearer Establishment   *250*

    7.4.2 Radio Bearer Reconfiguration   *251*

    7.4.3 Radio Bearer Release   *252*

    7.4.4 Transport Channel Reconfiguration   *253*

    7.4.5 Physical Channel Reconfiguration   *253*

7.5 RRC Connection Mobility Procedures *254*

    7.5.1 Cell Update Procedure *255*

    7.5.2 UTRN Registration Area (URA) Update Procedure *256*

    7.5.3 UTRAN Mobility Information *257*

    7.5.4 Soft Handover and Active Set Update *257*

    7.5.5 Hard Handover *258*

    7.5.6 Inter-system Handover *259*

7.6 Measurement Procedures *260*

    7.6.1 Measurement Control *260*

    7.6.2 Measurement Report *261*

Summary 262

8. UTRAN Signaling Procedures   **263**

8.1 Introduction *263*

8.2 UTRAN Global Signaling Procedures *264*

    8.2.1 System Information Broadcasting *264*

    8.2.2 Service Area Broadcast *265*

8.3 UTRAN Signaling Procedures for a Specific UE *265*

    8.3.1 Paging *266*

    8.3.2 NAS Signaling Connection Establishment *267*

    8.3.3 RRC Connection Establishment *268*

    8.3.4 RRC Connection Release *272*

    8.3.5 Radio Access Bearer Establishment *273*

    8.3.6 Radio Access Bearer Release *278*

    8.3.7 Physical Channel Reconfiguration *278*

    8.3.8 Transport Channel Reconfiguration *280*

    8.3.9 Soft Handover *282*

    8.3.10 SRNC Relocation *285*

    8.3.11 Cell Update *287*

    8.3.12 URA Update *289*

    8.3.13 Direct Transfer *290*

Summary 291

9. Mobility Management   **292**

9.1 Introduction *292*

9.2 State Model for Mobility Management *293*

9.3 Hierarchical Management of Location Information *295*

9.4 Paging *296*

9.5 MM/GMM Procedures Overview *296*

9.6 MM Procedures in the Mobile Station *298*

  9.6.1 MS 'Idle Mode' Procedures *299*

  9.6.2 MS 'Connected Mode' Procedures *304*

9.7 MM Procedures in the Access Network *304*

9.8 MM Procedures in the Core Network *305*

  9.8.1 MM Procedures in CS Domain *305*

  9.8.2 MM Procedures in PS Domain *313*

  9.8.3 Super-Charger Functionality *320*

Summary 323

10. Call Handling   **324**

10.1 Introduction *324*

10.2 Architecture of MO and MT Calls *325*

  10.2.1 Architecture of Mobile-Originated Call *325*

  10.2.2 Architecture of Mobile-Terminated Call *326*

  10.2.3 Architecture of a Basic Mobile-to-Mobile Call *328*

10.3 Mobile-originated Call Handling *330*

10.4 Mobile-terminated Call Handling *333*

  10.4.1 Retrieval of Routing Information *333*

  10.4.2 MT Call Handling at VPLMN *335*

10.5 Interaction of CF and CB Services With
Call Handling Procedures *338*

  10.5.1 Interaction of CF and CB Services with MT Calls *339*

  10.5.2 Interaction of CB Service with MO Calls *340*

10.6 Support for Optimal Routing *341*

  10.6.1 Conditions for Optimal Routing *342*

  10.6.2 Information Flows for Optimal Routing *343*

10.7 Immediate Service Termination (IST) *350*

  10.7.1 IST Alert Service *350*

  10.7.2 IST Command Service *352*

Summary 353

11. Session Management   **354**

11.1 Introduction *354*

11.2 Session Management Concepts *355*

  11.2.1 Addressing *355*

11.2.2 PDP Context Activation and Deactivation *358*

11.2.3 Packet Routing *358*

11.2.4 Encapsulation and Tunneling *359*

11.2.5 Packet Filtering *361*

11.3 PDP Protocol States *361*

11.4 PDP Context Activation Procedures *362*

11.4.1 PDP Context Activation Procedure *363*

11.4.2 Secondary PDP Context Activation Procedure *364*

11.4.3 Network-requested PDP Context Activation Procedure *365*

11.5 PDP Context Modification Procedures *367*

11.5.1 MS-Initiated PDP Context Modification Procedure *368*

11.5.2 SGSN-Initiated PDP Context Modification Procedure *369*

11.5.3 Other PDP Context Modification Procedures *370*

11.6 PDP Context Deactivation Procedures *370*

11.6.1 MS-Initiated PDP Context Deactivation Procedure *370*

11.6.2 SGSN-Initiated PDP Context Deactivation Procedure *371*

11.6.3 Other PDP Context Deactivation Procedure *372*

Summary *372*

12. Supplementary Services **373**

12.1 Introduction *373*

12.2 Supplementary Service Concepts *374*

12.2.1 Association with Basic Services *374*

12.2.2 SS Operations *377*

12.2.3 SS State Information *378*

12.3 Call Independent SS Management *379*

12.3.1 Call Independent SS Management Procedures *380*

12.3.2 Man-Machine Interface for SS Management *382*

12.4 Supplementary Services in UMTS *385*

12.4.1 Enhanced Multilevel Precedence and
Pre-emption (eMLPP) *387*

12.4.2 Call Deflection (CD) *388*

12.4.3 Line Identification *389*

12.4.4 Call Forwarding (CF) *392*

12.4.5 Call Barring *393*

12.4.6 Call Waiting (CW) and Call Hold (CH) *394*

12.4.7 Multiparty *397*

12.4.8 Closed User Group (CUG) *398*

12.4.9 Advice of Charge (AoC) *402*

12.4.10 User-to-User Signaling (UUS) *404*

12.4.11 Explicit Call Transfer (ECT) *406*

12.4.12 Multi-Call (MC) *408*

12.4.13 Other Supplementary Services *410*

12.5 Unstructured Supplementary Service Data *411*

12.5.1 USSD Architecture *411*

12.5.2 USSD Message Flows *413*

Summary 415

13. Value-added Services   **416**

13.1 Introduction *416*

13.2 Short Message Service *416*

13.2.1 SMS Network Architecture *416*

13.2.2 Mobile-Originated SMS Procedures *418*

13.2.3 Mobile-Terminated SMS Procedures *419*

13.3 Cell Broadcast Service *420*

13.3.1 CBS Network Architecture *421*

13.3.2 CBS Message Transfer Procedures *422*

13.4 Multimedia Messaging Service *422*

13.4.1 MMS Reference Architecture *422*

13.4.2 MMS Protocol Framework *424*

13.4.3 MMS Message Transfer Procedures *427*

13.5 Location Services *428*

13.5.1 LCS Logical Reference Model *428*

13.5.2 LCS Control Procedures *429*

13.5.3 LCS Network Architecture *430*

13.5.4 Mechanisms for Determination of Location Information *431*

13.5.5 Location-based Services *433*

13.6 Service Capability Features *435*

13.6.1 SCF Types *436*

13.6.2 SCF Toolkits *436*

Summary 437

14. Security Management   **438**

14.1 Introduction *438*

14.2 User Domain Security *440*

14.3 Network Access Security *441*

    14.3.1 Mutual Authentication *442*

    14.3.2 Data Confidentiality *451*

    14.3.3 Data Integrity *453*

    14.3.4 User Identity Confidentiality *454*

    14.3.5 Access Security Flow Diagram *455*

14.4 Network Domain Security Using MAPsec *457*

    14.4.1 Protection Modes and Message Formats *457*

    14.4.2 Components of MAPsec Protocol *459*

    14.4.3 Operations of MAPsec *461*

    14.4.4 Key Distribution in MAPsec *462*

14.5 Network Domain Security Using IP Security *463*

    14.5.1 Architecture for NDS/IP *464*

    14.5.2 Encapsulating Security Payload (ESP) *465*

    14.5.3 Internet Key Exchange (IKE) *468*

Summary *470*

PART IV—IP INITIATIVES IN UMTS NETWORK

15. IP-based Signaling Transport   **473**

15.1 Introduction *473*

15.2 IP-based Signaling Transport
From SIGTRAN *474*

    15.2.1 Requirements *474*

    15.2.2 SIGTRAN Protocol Layering *476*

15.3 Stream Control Transmission Protocol (SCTP) *477*

    15.3.1 Functions of SCTP *478*

15.4 SS7 MTP3 User Adaptation Layer (M3UA) *479*

    15.4.1 Functions of M3UA *479*

    15.4.2 Scenarios for Deployment of M3UA in
UMTS Network *480*

15.5 SCCP User Adaptation Layer (SUA) *481*

    15.5.1 Functions of SUA *482*

    15.5.2 Scenarios for Deployment of SUA in UMTS Network *482*

15.6 Comparison Between M3UA and SUA *483*

Summary *484*

16. IP Multimedia Subsystem   **486**

16.1 Introduction *486*

16.2 Entities of IP Multimedia Subsystem *488*

    16.2.1 Home Subscriber Server (HSS) *490*

    16.2.2 Call Session Control Function (CSCF) *491*

    16.2.3 Server Locator Function (SLF) *492*

    16.2.4 Application Server (AS) *492*

    16.2.5 Entities used for Interworking *493*

    16.2.6 Signaling Gateway Function (SGW) *493*

16.3 Network Interfaces of IP Multimedia Subsystem *494*

    16.3.1 Cx Interface between HSS – CSCF *496*

    16.3.2 Dx Interface between CSCF and SLF *496*

    16.3.3 Sh Interface between HSS and AS *497*

    16.3.4 Si interface between HSS – CAMEL *498*

    16.3.5 ISC Interface between S-CSCF and AS *498*

    16.3.6 Gm Interface between UE and CSCF *498*

    16.3.7 Mc Interface between MGCF and MGW *499*

    16.3.8 Mg Interface between MGCF and S-CSCF *499*

    16.3.9 Mw Interface between x-CSCF and y-CSCF *499*

    16.3.10 Mi Interface between S-CSCF and BGCF *499*

    16.3.11 Mj Interface between BGCF and MGCF *499*

    16.3.12 Mk Interface between BGCF and BGCF *499*

16.4 IMS Addressing *499*

    16.4.1 IMS Private User Identity (IMPI) *500*

    16.4.2 IMS Public User Identity (IMPU) *500*

    16.4.3 Relationship of IMPI and IMPU *500*

16.5 Subscriber Data *501*

16.6 Session-unrelated Procedures *503*

    16.6.1 Establishing IMS Transport *503*

    16.6.2 Registration *504*

    16.6.3 De-registration *507*

    16.6.4 Profile Update *510*

16.7 Session-related Procedures *511*

    16.7.1 Service Control *512*

    16.7.2 Session Origination *513*

    16.7.3 Interworking Procedure *518*

16.7.4 Session Termination *518*

16.8 IMS Protocols *520*

16.8.1 Session Initiation Protocol (SIP) *521*

16.8.2 Diameter *521*

16.9 Security in IP Multimedia Subsystem *522*

16.9.1 Access Security *523*

Summary 524

APPENDICES

Appendix A: Deployment of 3G Networks   **526**

A.1 Japan *526*

A.2 UK *527*

A.3 Italy *528*

Summary 528

Appendix B: Fourth Generation (4G) Mobile Networks   **529**

B.1 Why 4G? *529*

B.2 What is 4G? *530*

B.3 How to Achieve 4G? *532*

B.4 When Should We Expect 4G? *534*

References   **535**

Abbreviations   **550**

Index   **560**

FOREWORD

It gives me great pleasure to write the foreword of the book *3G Mobile Networks: Architecture, Protocols and Procedures*, a comprehensive text for engineers and wireless networking professionals. While considerable system development efforts are currently underway in the 3G arena, especially for the UMTS networks based on 3GPP specifications, the available literature is inadequate in terms of ease of understanding, comprehensiveness and presentation. As a leader in creating communication software for wireless networks, we at Hughes Software Systems (HSS) are constantly ramping up engineers to work on this emerging technology. We have always felt the need for a greater variety of introductory and reference reading material. This book fills both these requirements very well. I am sure that this will find a place on the bookshelves of many professionals.

The strength of this book is its organization, procedural descriptions and detailed illustrations. Spanning across topics such as UMTS Network Architecture (including Access Network and Core Network), Protocols (including RRC, NBAP, RANAP, MM/GMM, MAP and GTP), Procedures (including UTRAN procedures, Mobility Management, Call/Session Handling and Security Management), and Services (including Supplementary Services and Value-Added Services), the book touches almost every imaginable aspect of a 3G Network. The coverage of contemporary topics such as IP Multimedia Sub-system (IMS) and SIGTRAN are also noteworthy. Appendix A providing the status of deployment of a 3G UMTS Network, complements the theory and procedural description. Similarly, a brief introduction of 4G Networks in Appendix B sets the tone for the tomorrow that we are all working towards. Possibly, with a greater emphasis on 3G services in the next edition, this book will be titled *3G Mobile Networks: Architecture, Protocols, Procedures and Services*.

I congratulate my colleagues at HSS, who through their deep understanding and a comprehensive treatment of the subject, have given the book its unique flavor. I am quite hopeful that this book will be a valuable guide to people working on 3G UMTS Networks, and anyone who wants a view of what the future holds.

VINOD SOOD

Vice President & Head of Engineering
Hughes Software Systems, India

PREFACE

Raison d'être

The mobile telecommunications industry is witnessing a period of robust growth. The net result of this high-growth rate is that the mobile subscriber base has reached 1.3 billion users, surpassing the seemingly invincible fixed-line user base of 1.2 billion. It is expected that by 2007, the number of mobile users will be a staggering two billion; by that time, one in every three persons will have a mobile phone. Even now, around 450 million new handsets are being purchased every year. In certain parts of Europe and Asia around 80% of the population has a mobile phone. While developed countries are getting saturated, developing countries like China and India, where the mobile-density is low, are offering exciting opportunities to expand the market.

Given this scenario, certain trends emerge. Firstly, as the subscriber base reaches saturation levels, growth needs to be sustained from new services and not from new subscribers. Also integration of voice and data will become quite critical. Secondly, as bandwidth limitations of existing wireless networks become a stumbling block, operators will need to devise novel means to overcome the limitation. This will entail deployment of new generation of networks that provide greater bandwidth, allowing application developers to popularize bandwidth-intensive multimedia applications. Lastly, in developing countries where mobile penetration is low, even simple wireless services will act as engines of growth for the next five to seven years.

Among the current wireless networks, the Global System for Mobile communications (GSM), which is categorized as a Second Generation (2G) network, is the most popular. All 2G networks including GSM suffer from low data rates; the pre-dominance of voice results in low-efficiency for packet-switched services. 2.5G networks is an advancement over 2G networks which includes High-Speed Circuit Switched Data (HSCSD), General Packet Radio Services (GPRS) and Enhanced Data

Rates for Global Evolution (EDGE). While the 2.5G networks provide improve-ments over 2G networks, there is a pressing need for greater bandwidth and service capabilities. This need led ITU-T to form a vision of Third Generation (3G) networks with a single radio interface (with speeds of up to 2 Mbps) providing global roaming. The vision remained unfulfilled and a set of five technologies for the radio interface emerged. These technologies were under the umbrella of the International Mobile Telecommunications 2000 (IMT-2000) standards. Among these, the European standardization body ETSI proposed Wideband CDMA (WCDMA) system that supported two modes of operation, one for paired spectrum referred to as Frequency Division Duplex (FDD) mode and the other for unpaired spectrum referred to as Time Division Duplex (TDD) mode. Apart from Europe, other countries like Japan and Korea also supported systems similar to those selected by ETSI. To co-ordinate the initiatives undertaken by different bodies, a group was formed by the standardization bodies of these countries. This group was called the Third Generation Partnership Program (3GPP) which came into existence in 1998. The 3GPP standardized the 3G networks based on WCDMA. These networks are now also referred to as Universal Mobile Telecommunications System (UMTS) or UMTS WCDMA network.

With the GSM's massive subscriber base of nearly 1 billion and its ability to offer a convenient platform for mobile operators to evolve from 2G to 3G, the ensuing years will see rapid deployments in the 3G arena, especially for the UMTS WCDMA, which will reuse significant parts of the core network.

While UMTS WCDMA technology has attracted a lot of interest in various quarters including mobile equipment vendors, software firms, system developers and network operators, we felt that the existing literature was inadequate for the stakeholders to grasp the technology. Our first hand experience in developing systems for UMTS WCDMA networks, as we have and felt the need in the industry, we embarked upon the task of writing this book.

The Book

We took a lot of time organizing the book to provide a clearly defined structure to the book. After considerable thought, we organized the book into four parts as depicted in Figure P.1.

Part I provides an introduction to WCDMA networks based on 3GPP specification. Chapter 1 provides an overview of the Third Generation Partnership Program (3GPP). Apart from details of 3GPP, it also gives a background and limitations of 2G and 2.5G networks. Chapter 2 provides an overview of the WCDMA principles.

Part II covers UMTS architecture and the accompanying protocols. The first chapter in this part, Chapter 3, provides an overview of UMTS network architecture. According to this architecture, the UMTS network is divided into three logical parts—User

Equipment (UE), Access Network (AN) and Core Network (CN). These logical parts are explained in Chapters 4–6. An overview of various protocols used in UMTS network is provided in Chapters 5 and 6.

In **Part III,** important procedures for both Access Network and Core Network are covered. Chapters 7–14, cover a gamut of topics including RRC procedures, UTRAN signaling procedures, Mobility Management, Call Handling, Session Management, Supplementary Services, Value-Added Services and Security Management.

Part IV looks at the increasing importance of Internet Protocol (IP) in UMTS Network. In particular, two important developments are discussed including the introduction of IP as a carrier of signaling data (Chapter 15) and the introduction of an overlay IP Multimedia Sub-system (IMS) over the packet-switched domain (Chapter 16).

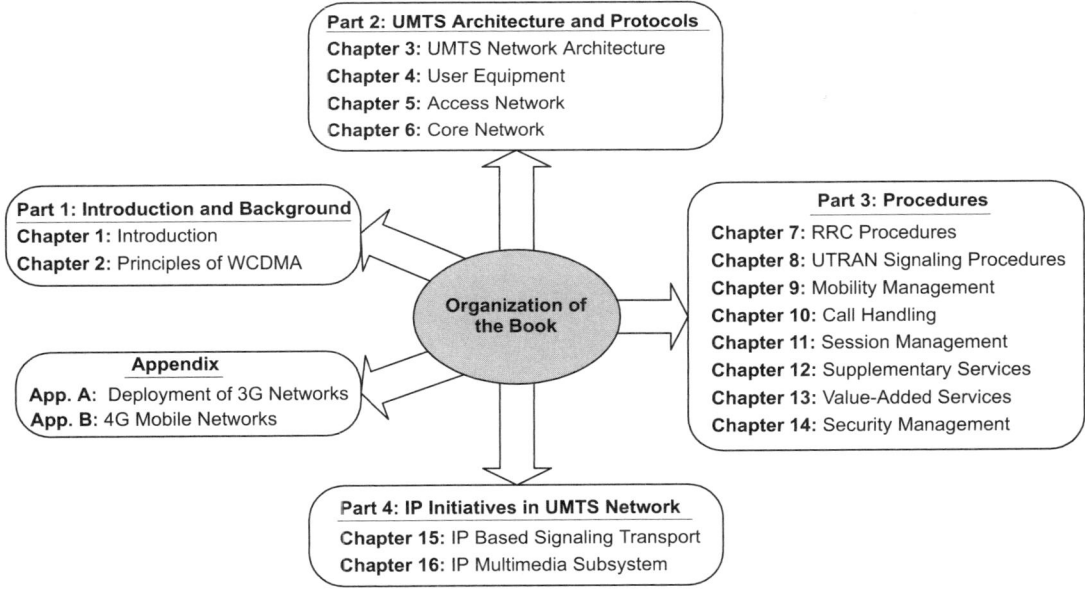

Fig. P.1 *Organization of this book*

Apart from these four parts, **Appendix A** of this book looks at the 3G deployments taking place in various parts of the world. The book concludes with a brief mention of Fourth Generation (4G) mobile networks in **Appendix B**.

Website

To have a greater interaction with the readers after the publication of the book, the following website have been created—**www.tatamcgrawhill.com/digital_solutions/ SumitNishit (***http://3gbook.tripod.com/***).**

This website offers the following:

- Preface
- Table of Contents
- Errata
- Feedback and Review Comments
- References
- Other Related Material

Readers are encouraged to visit the sites and use the available material.

Suggestions

Your comments, feedback and constructive criticism are valuable to us. Please feel free to drop a mail at *umts3gbook@yahoo.com*. We would be glad to incorporate your comments in the subsequent editions of the book.

SUMIT KASERA
NISHIT NARANG

ACKNOWLEDGEMENTS

We would first like to thank our organization *Hughes Software Systems, Gurgaon, India* where we have spent nearly five years of our professional life. It goes without saying that Hughes Software Systems is one of the best telecom software companies in India. In particular, we would like to thank two senior executives of Hughes Software Systems, Mr Vinod Sood and Mr Gautam Brahma for being role models.

We would also like to thank our technical reviewer Pradeep Kirnapure who painstakingly reviewed the whole manuscript and provided valuable comments. The comments helped us in improving the content as well as the technical accuracy of the book.

We would also thank the Senior Consultants in our company—Prabir Datta and Dinesh Singh and our managers Sunil Godse, Tarun Singhal, Vibha Bansal and Sudatta Kar.

We would like to thank Rajiv Gupta and Prashant Vashisht for being good friends and something more.

We would like to thank the entire team at Tata McGraw-Hill for publishing this book.

We wish to also thank our alma mater Indian Institute of Technology (IIT) and all its professors for providing us the necessary technological foundation to write a book.

In addition, Sumit would like to acknowledge the role of his family in making this book a reality. Among the family members, his father Jagdish Prasad Kasera's contribution stands out. He would also like to thank his mother S. L. Kasera, wife Manisha, sister Smita, brother-in-law Gaurav, brother Rajiv, and aunt Manju Banka for their support. Sumit would also like to thank his friends and colleagues Nishit Narang, Alhad Wakankar, Anjali Gupta, Ritesh Singh, Narendra Singhal, Saket Saraf, Dipak Kumar Singh, Vivek Singh and many others for continuous encouragement and support.

In addition, Nishit would like to acknowledge the role of his father Ramesh Narang, a professor in Mathematics at the Delhi University, who painstakingly coached him throughout his student life. He would also like to thank his mother (Dr.) Saroj Narang (professor in Sanskrit at Delhi University), wife Sumita Narang (Senior Software Engineer at Hughes Software Systems) and other family members and friends.

SUMIT KASERA
NISHIT NARANG

3G Mobile Networks

Architecture, Protocols and Procedures

Based on 3GPP Specifications for UMTS WCDMA Networks

Part

1

Introduction
and
Background

In the early 1990s, before the commercial success of GSM and other Second Generation (2G) mobile networks, it was felt that the existing standards were inadequate in terms of bandwidth and service capabilities. Thus, research work was started by various organizations to move towards the next generation mobile network. Since the main bottleneck to providing greater bandwidth was the 'air interface', much of the research focus was on this area. Five different radio interface systems were proposed by various standardization bodies. The telecommunication body, International Telecommunication Union (ITU) wanted a single radio interface that would provide global roaming. However, due to various reasons, technical and other, a set of five standards was adopted for the radio interface of the Third Generation (3G) networks. It was felt that the adoption of multiple standards would foster competition and also lead to migration of the installed base of 2G networks to 3G networks. Among the various standardized radio interface systems, the Wideband Code Division Multiple Access (WCDMA) system adopted by the Third Generation Partnership Program (3GPP) has the largest backing from the industry and the standardization bodies. This book discusses the 3G WCDMA networks based on 3GPP specification.

Part 1 includes two chapters which provide an introduction to WCDMA networks based on 3GPP specification. Chapter 1 provides an overview of the Third Generation Partnership Program (3GPP) and also provides a background of 2G and 2.5G networks. The limitations of the 2G/2.5G networks are also explained in Chapter 1.

Chapter 2 provides an overview of the WCDMA principles. The WCDMA concepts are central to understanding the air interface used in the WCDMA network. The topics covered in this chapter include Code Division Multiple Access (CDMA) and Wideband-CDMA (WCDMA), spreading and scrambling, rake receivers, multipath diversity, and power control.

□□

CHAPTER

1

INTRODUCTION

1.1 INTRODUCTION

The first commercial mobile networks were launched in the mid-1980s. The mobile communication world has since then been witnessing rapid changes marked by significant improvement in the services offered by mobile networks. The transition from First Generation (1G) networks to the Third Generation (3G) networks being deployed today is a clear indication that satisfying consumer demands for better and improved services, and generating more revenue for the operator have been the main areas of focus. No wonder, by the year 2002, the number of mobile users in the world had already exceeded the 1 billion mark. In many countries, the rate of growth of mobile subscribers has far outstripped the growth in users of fixed wireline networks. These facts indicate the high acceptability that mobile services have found with consumers.

The initial wireless networks of the mid-1980s, referred to as the First Generation (1G) standards, were based on analog communication in the radio path. These networks had limited regional scope, mostly confined to national boundaries, and were an outcome of agreements between the regional telecom operator and the domestic industry. The early nineties saw the replacement of these analog networks with the digital Second Generation (2G) networks. Global System for Mobile communications (GSM) became the most popular of the 2G standards. By the end of 2002, GSM accounted for over 66 per cent of the world's total 2G market. The success of 2G technologies provided the necessary thrust to mobile wireless communications and paved the way for enhanced networks in the future.

1.2 SECOND GENERATION MOBILE NETWORKS

There were a couple of important changes that marked the shift from 1G to 2G standards. Firstly, unlike the analog communication in 1G, the 2G standards were based on digital communication, both in the radio path and between network entities. Secondly, 2G standardization processes were aimed at making the notion of global roaming more realistic. Standardization in 1G was never elaborate, leading to national standards which offered no roaming beyond national boundaries. However, 2G standards brought about semi-global acceptance, and consequently, roaming to larger regions. Primarily, there are four competing 2G technologies, as explained here:

- **Global System for Mobile communication (GSM):** Among the 2G technologies, the GSM is the most popular. The European Telecommunications Standards Institute (ETSI) developed the GSM specifications in 1989. The early GSM systems used a 25 MHz frequency spectrum in the 900 MHz band. This 25 MHz frequency spectrum was then divided into 124 carrier frequencies of 200 kHz each. A single 200 kHz radio channel was shared between eight users, by allocating a unique time slot to each one of them. The GSM is therefore viewed as a combination of two multiplexing techniques, namely Frequency Division Multiple Access (FDMA) and Time Division Multiple Access (TDMA). Of late, there are two other variants of GSM operating at frequency of 1800 MHz (1.8 GHz) and 1900 MHz (1.9 GHz).

- **Personal Digital Communications (PDC):** While a majority of countries adopted the GSM technology, the PDC gained immense popularity in Japan. Outside Japan, however, PDC which operates at 800 MHz and 1500 MHz frequencies, is not very popular. Since the success of PDC is limited to Japan, the Japanese telecommunication companies are trying to work towards 3G technologies that have global acceptance. Another reason for Japan to show great interest in 3G is that the 2G capacity is fast running out of bandwidth and facing scalability issues. This is possibly why Japan is one of the first countries where the 3G network is commercially deployed.

- **IS-95:** While most of the 2G networks were based on FDMA and TDMA, a company by the name Qualcomm designed a Code Division Multiple Access (CDMA) scheme. This scheme uses separate codes to distinguish between data transmitted by different users on the same frequency. The IS-95 is popular in South Korea and the United States apart from same other countries.

- **US-TDMA (or D-AMPS):** Another 2G system popular in North America is the Digital-Advanced Mobile Phone Services (D-AMPS). It is a backward compatible with AMPS and is a digital version of First Generation AMPS technology.

1.2.1   Limitations of 2G Networks

Though the 2G networks brought about a major change in the way mobile networks were built, they had their limitations, some of which are as follows:

- **Low transfer rates:** The 2G networks are primarily designed to offer voice services to the subscribers. Thus, the transfer rate offered by these networks was low.

Though the rates vary across technologies, the average rate is of the order of tens of kilobits per second.

- **Low efficiency for packet-switched services:** With the rising popularity of the Internet, there is a growing demand among customers for access to the Internet not just at home or the office, but also when they are on the move. Wireless Internet access with the 2G networks is not efficiently implemented.

- **Multiple standards:** With a multitude of competing standards in place, a wireless user can roam in only those networks that support the same standard. This allows the user only limited roaming. Though the 2G standards were an improvement over their 1G predecessors, they still lacked the ability to offer complete global roaming, and were semi-global in this respect.

1.3   2.5 GENERATION MOBILE NETWORKS

Efforts were made to enhance 2G networks by removing these major impediments to their growth. The underlying objective was to make minimum changes in the existing network architecture. The outcome was the 2.5 Generation Networks.

The different categories of 2.5G networks are explained as follows:

- **High-Speed Circuit-Switched Data (HSCSD):** To circumvent the drawback of low data rates, a simple solution is to use multiple time slots instead of one. Given that a GSM channel provides speeds of 9.6Kbps or 14.4Kbps, by using up to four channels, a speed of 57.6Kbps can be obtained. This is the principle of HSCSD. The advantage of this scheme is that it requires minimum changes in the network architecture; its flip side is that it uses circuit-switching, which is considered inefficient in terms of resource usage. For better resource utilization, packet-switching is used. GPRS, which is explained next, uses packet switching.

- **General Packet Radio Services (GPRS):** GPRS offers data services (e.g. Internet access) by using a packet-switching domain (i.e. GPRS reserves radio resources only when there is data to send; this provides efficient use of network resources). GPRS uses 1 to 8 radio channels in the 200 kHz frequency to offer speeds of up to 115Kbps. An important advantage of GPRS is that it provides a means to migrate towards 3G networks. This is because the core network components of GPRS (e.g. SGSN and GGSN) are an integral part of 3G Core Network. 'Though the software components in entities of 2.5 network are different from the software components in entities of 3G network, the basic network architecture and network interfaces remain the same."

- **Enhanced Data Rates for Global Evolution (EDGE):** By using better modulation techniques, the data rates of GSM and GPRS could be increased up to three times.

This improvement is called EDGE for GSM and Enhanced GPRS (EGPRS) for GPRS. Using EGPRS, a speed of up to 384Kbps can be obtained.

1.4 INTERNATIONAL MOBILE TELECOMMUNICATION-2000 (IMT-2000)

Even before the commercial success of GSM and other 2G mobile networks, it was felt that the then existing standards were inadequate in terms of bandwidth and service capabilities. Thus, research work was started by various organizations to move towards the next generation mobile network. Since the main bottleneck in providing greater bandwidth and enhanced multimedia services was the air interface, much of the research focus was on this area.

As different standardization bodies worked in parallel, each came up with its own air interface. By 1997, the Special Mobile Group (SMG) of European Tele-communication Standards Institute (ETSI) came up with five candidate schemes. These schemes were as follows:

- Wideband Code Division Multiple Access (WCDMA)
- Wideband Time Division Multiple Access (WTDMA)
- Time Division Multiple Access (TDMA) / Frequency Division Multiple Access (FDMA)
- Orthogonal Frequency Division Multiple Access (OFDMA)
- Opportunity Driven Multiple Access (ODMA)

Out of these, ETSI selected two systems; one for *paired spectrum* and another for *unpaired spectrum*. In paired spectrum, frequencies reside in paired bands that typically are of equal bandwidth. The paired bands are required for uplink and downlink directions. Thus, the paired spectrum permits simultaneous communication in both directions. In unpaired spectrum, a single frequency band is used for both forward and reverse directions. Unlike the paired spectrum, the bandwidth in unpaired spectrum can be altered in uplink and downlink direction. This allows asymmetric communication (e.g. downloads from the Internet) where downlink bandwidth is much more than the uplink bandwidth.

The Wideband CDMA (WCDMA) was adopted for paired frequency bands for Frequency Division Duplex (FDD) operation. Wideband TDMA (WTDMA) was adopted for unpaired frequency bands for Time Division Duplex (TDD) operation. Apart from Europe, other countries like Japan and Korea also supported systems similar to those selected by ETSI. To co-ordinate the work done by different bodies, a group was formed by the standardization bodies of these countries. This group was called the Third Generation Partnership Program (3GPP) and it came into existence in 1998. The 3GPP adopted the Frequency Division Duplex (FDD) and Time Division Duplex (TDD) modes of WCDMA as its air interface.

In North America, the cellular operators formed another group by the name 3GPP2. This group chose the multi-carrier cdma2000 radio system.

Then, the Universal Wireless Communications Consortium (UWCC) chose the UWC136 (also known as the IMT-Single Carrier or IMT-SC) system.

Yet another system, called Digital Enhanced Cordless Telecommunications (DECT) system, was chosen by the ETSI DECT project.

Overall, five different radio interface systems were thus proposed. The original goal, as envisaged by ITU, was to have a single radio interface that provided global roaming. However, due to various reasons, technical and other, a set of five standards was adopted for Third Generation (3G) networks. It was felt that adoption of multiple standards fosters competition and also caters to the migration of the installed base of 2G networks. The standards adopted for 3G networks fall under the umbrella of International Mobile Telecommunications 2000 (IMT-2000) systems.

Here, it is important to clarify the terms Third Generation (3G) and IMT-2000. Third Generation (3G) is defined as 'a term coined by the global cellular community to indicate the next generation of mobile service capabilities in terms of bandwidth and network functions. These service capabilities in turn allow advanced services and applications, including multimedia'. The term IMT-2000 refers to a set of radio interface standards that fulfill the requirements from 3G networks.

Figure 1.1 depicts the various radio interface technologies that fall under IMT-2000. As shown in the figure, different standardization bodies are working on standards for 3G networks based on different radio interface technologies. However, the most prominent among these is the work done by 3GPP, which has been focused upon in this book. *As this book is completely based on the 3GPP specification, it is important to look at the various aspects of 3GPP. The following sections highlight different aspects of 3GPP in detail.*

Note that there is another related term—Universal Mobile Telecommunications System (UMTS)—that is quite often used. UMTS is a term coined by ETSI and as defined by it, is synonymous with 3G networks. Since ETSI forms an integral part of

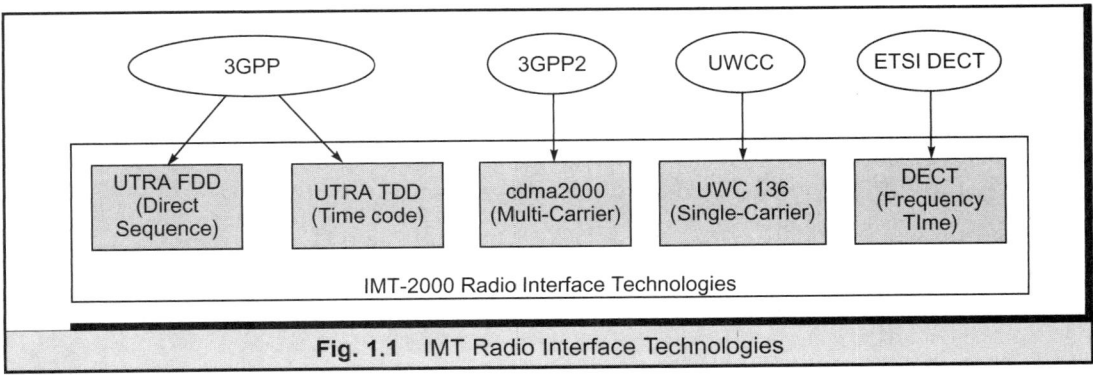

Fig. 1.1 IMT Radio Interface Technologies

the 3GPP body, the 3G networks, as standardized by the 3GPP, can be seen as a UMTS network. It is common to associate the term WCDMA with UMTS and refer to the network as 'UMTS WCDMA network'.

1.5   THIRD GENERATION PARTNERSHIP PROGRAM (3GPP)

As different countries started developing their own 3G radio interface systems, it became evident that a certain degree of coordination was required between different standardization bodies working on the 3G systems. This was particularly important because one of the stated goals of IMT-2000 was 'Global Roaming' whereby the same mobile handset could be used in different parts of the world. Since the solutions adopted by some of these standardization bodies were quite similar, initiatives were taken to create a single forum for the development of a common specification based on Universal Terrestrial Radio Access (UTRA) radio interface. The result of this initiative was the Third Generation Partnership Program (3GPP), which formally came into existence in December 1998 when the 'The 3rd Generation Partnership Project Agreement' was signed.

1.5.1   Interaction of 3GPP with other Bodies

In order to work efficiently, 3GPP interacts with various bodies. These bodies fall under the following categories:

- **Organizational Partners:** The 3GPP brings together a number of telecommunications standards bodies. These bodies, formally referred to as Organizational Partners, are as follows:
    1. Association of Radio Industries and Businesses (ARIB) of Japan
    2. Telecommunication Technology Committee (TTC) of Japan
    3. European Telecommunication Standards Institute (ETSI) of Europe
    4. T1 of USA
    5. Telecommunications Technology Association (TTA) of Korea
    6. China Wireless Telecommunication Standards Group (CWTS) of China

- **Market Representation Partners:** Apart from the standardization bodies, 3GPP works in association with the following 'Market Representation Partners', bodies that provide a consolidated view of market requirements:
    1. UMTS Forum
    2. Global mobile Suppliers Association (GSA)
    3. GSM Association
    4. 3G.IP
    5. 3G Americas

6. Mobile Wireless Internet Forum (MWIF)
7. IPv6 Forum

- **Observers:** These are some of the standards development organizations that may become organizational partners in future:

    1. Telecommunications Industry Association (TIA) of North America
    2. Telecommunications Standards Advisory Council of Canada (TSACC) of Canada
    3. Australian Communications Industry Forum (ACIF) of Australia

- **Mobile Competence Centre (MCC):** This has been established to ensure efficient day-to-day running of the 3GPP. The MCC is based at the ETSI headquarters in Sophia Antipolis, France.

- **Individual Members:** These are members of the organizational partners who are actively engaged in 3GPP work (e.g. development of technical standards).

Apart from the above, the 3GPP also interacts with other standardization bodies. For example, it works with the Internet Engineering Task Force (IETF) so that IETF standards can be used/customized in the 3G environment.

1.5.2 Objective of 3GPP

The 3GPP can be defined as a collaborative agreement between the different standardization bodies that it interacts with. The objectives of 3GPP are summarized as follows:

- To produce globally applicable technical specifications and technical reports for a Third Generation Mobile System. The 3G Mobile System so designed comprises of:

    - **The Access Network** which is based on the Universal Terrestrial Radio Access (UTRA) radio interface. The UTRA includes two modes of operation: Frequency Division Duplex (FDD) and Time Division Duplex (TDD).
    - **The Core Network** for the 3GPP Third Generation Mobile System is evolved from the GSM core network. Some new architectural entities have been added (e.g. Gateway Mobile Location Center (GMLC))

- Apart from producing specifications for Third Generation Mobile Systems, the 3GPP is also responsible for maintenance and development of GSM Technical Specifications and Technical Reports (which were developed by ETSI). This includes the GSM standards as well as the standards for evolved networks like General Packet Radio Service (GPRS) and Enhanced Data rates for Global Evolution (EDGE).

1.5.3 3GPP Technical Specifications Group (TSG)

The detailed technical and standardization work in 3GPP is carried out by the Technical Specifications Group (TSG). The TSG develops, approves and maintains the technical specifications and technical reports. The difference between a technical specification and a technical report is that while the former is a technical standard, the latter is generally for informational purposes. It is common for a standard to be first published as a technical report and then proceed to become a technical specification. Not all technical reports become a technical specification.

To cater to different areas in which standardization activity is carried out, there are five different TSGs:

- **Core Network (CN) TSG:** This TSG deals with specifications of the Core Network part of the 3GPP systems, in particular its following aspects:
    - Call control, session management and mobility management between User Equipment and Core Network.
    - Signaling between the Core Network nodes. The signaling is used to exchange subscriber information and to control network services.
    - Interconnection of the Core Network with external networks (e.g. PSTN network and the Internet).
    - Interworking with 2G networks (e.g. handover to/from GSM).
    - Core Network aspects of the Iu interface.
    - Packet related matters such as mapping the Quality of Service.
    - Operations and Maintenance requirements of the Core Network.
    - It is customary to use of O and M or Operation and Maintenance

- **GSM/EDGE Radio Access Network (GERAN) TSG:** This TSG was created to maintain consistency between GSM and 3G specifications. It is responsible for the specification of the Radio Access part of GSM/EDGE, in particular, its following aspects.
    - Layer 1, Layer 2 and Layer 3 specifications for the GERAN radio interface.
    - RF aspects of GERAN.
    - Abis, A and Gb interface specifications.
    - Operations and Maintenance requirements of GERAN.
    - Conformance test specifications for testing of all aspects of GERAN base stations.
    - Conformance test specifications for testing of all aspects of GERAN terminals.

- **Radio Access Network (RAN) TSG:** This TSG deals with the definition of the functions, requirements and interfaces of the UTRA network in its two modes, namely the FDD mode and the TDD mode. In particular, it deals with the following aspects of the RAN:

- Layer 1, Layer 2 and Layer 3 specifications for the radio interface.
- Iub, Iur and Iu specification.
- Operations and Maintenance requirements of UTRAN.
- Specifications for radio performance.
- Conformance test specifications for testing of all aspects of base stations.

- **Service and System Aspects (SA) TSG:** While other TSGs work in their individual domains, the Service and System Aspects (SA) TSG is responsible for the overall architecture and service capabilities of systems based on 3GPP specifications. Thus, SA TSG coordinates the work of different TSGs, in particular, the following aspects:

  - Definition, evolution and maintenance of the overall system architecture, which includes all the subsystems of the 3GPP system (i.e. the mobile terminal, SIM/USIM, GERAN/UTRAN and Core Network).
  - Development of a framework for services, service capabilities and service architecture.
  - Definition of bearer capabilities offered by different subsystems, including Quality of Service requirements for access to both packet-switched and circuit-switched networks.
  - Formulating charging and accounting definition.
  - Definition of a security framework.
  - Providing network management.
  - CODEC aspects including principles for definition of end-to-end transmission.

- **Terminal TSG:** It deals with the specifications of the Terminal Equipment (TE) interfaces, ensuring that the terminals are based on the relevant 3GPP specifications. In particular, the Terminal TSG deals with the following aspects of TE:

  - UTRAN-based Terminal Equipment performance specifications.
  - USIM and its interface with Mobile Terminal.
  - Service capability protocols.
  - Messaging.
  - Services end-to-end interworking.
  - Framework for terminal interfaces and service execution.
  - Conformance test specifications of terminals, including radio aspects.
  - Multi-mode terminals

Figure 1.2 depicts the TSG Organization in 3GPP. As shown in the figure, there is a project coordination group that controls the overall operations of different TSGs.

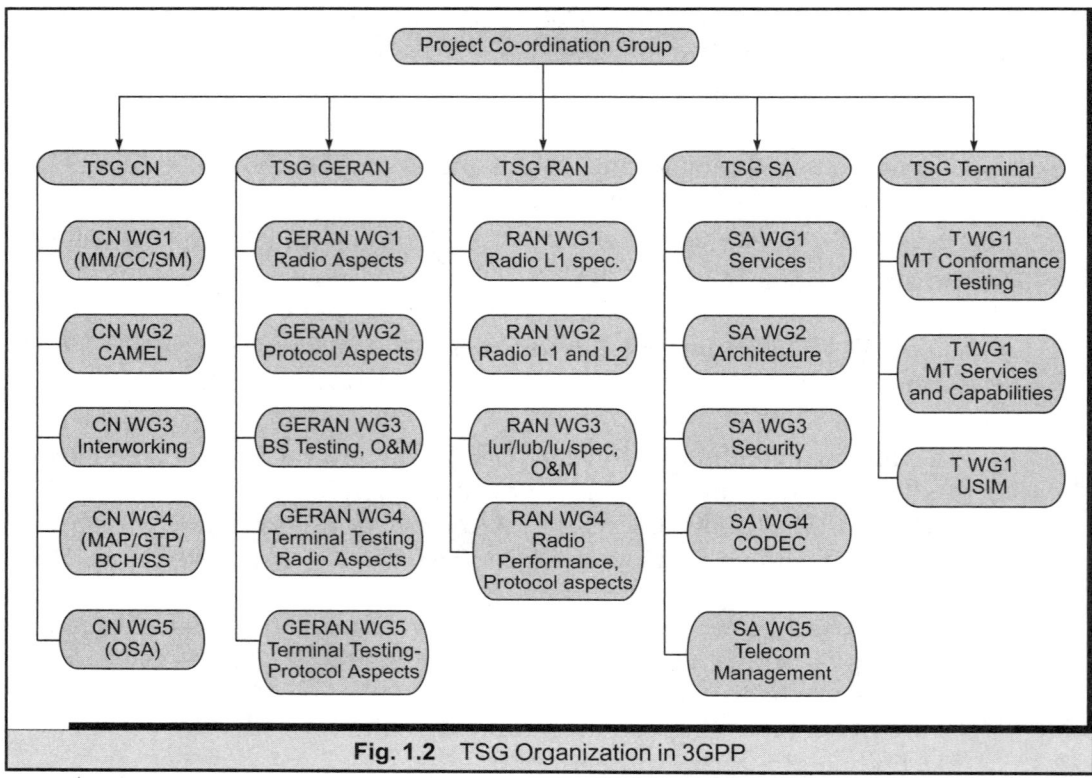

Fig. 1.2   TSG Organization in 3GPP

Among the TSGs, the Service and System Aspects (SA) TSG co-ordinates the work of different TSGs. Each TSG is further divided into different working groups. For details of the scope of work of each Working Group (WG) in a TSG, the reader is referred to the 3GPP website (www.3gpp.org).

1.5.4   Stages of 3GPP Specifications

Standardization of a service in 3GPP goes through three distinct stages: Stage 1, Stage 2 and Stage 3. As a specification proceeds from Stage 1 to Stage 3, a greater level of detail is added.

In *Stage 1*, a service is defined in terms of the functionality it offers to the end-user. This can be viewed as the requirements from the service. These requirements then define how the service is exactly implemented.

In *Stage 2*, the requirements available from Stage 1 are used to define broad functional blocks that will be used to provide the service. Using these functional blocks, an abstract architecture is defined. The information flow between the functional blocks is also defined at this stage. However, the exact protocol used is left for Stage 3.

In *Stage 3,* the functional blocks are assigned to actual physical entities. Further, the protocols used between these entities are also defined at this stage.

The various stages in the development of a specification follow a waterfall model whereby the Stage 2 specifications require Stage 1 specifications, and Stage 3 specifications require Stage 2 specifications. There is, however, a feedback mechanism so that Stage 2 and 3 work can appropriately modify the Stage 1 and/or Stage 2 specifications.

1.5.5 Series of 3GPP Specifications

The 3GPP specifications are numbered as 3GPP TS *ab.def* (for example 3GPP TS 29.002). The first part of the number ('ab') defines the series of 3GPP specifications. The latter part of the specification number ('def') defines the document number under the particular series.

Table 1.1 depicts the various series of 3GPP specifications. Apart from these series, there are other series for GSM specifications (1 to 13 and 41 to 55 series). For details of these series, the reader is referred to the 3GPP website and 3GPP TR 21.900.

1.5.6 Version Numbering of 3GPP Specifications

Apart from the series and the specifications number, a 3GPP specification is also associated with a version number of the form *x.y.z*. A 3GPP specification is uniquely referred to by its series number, the specification number and the version number.

In the version number, the digit 'x' identifies the release number. The different values of release number are as follows:

- **0:** Implies that the specification is still a draft.
- **1:** Implies that the specification is presented to TSG for information. The specification is assumed to be 60 per cent complete.
- **2:** Implies that the specification is presented to TSG for approval, and is assumed to be 80 per cent complete.
- **3:** Implies that the specification is approved by TSG for Release 99. Changes can be carried out using the Change Control mechanism.
- **4:** Implies that the specification is approved by TSG for Release 4. Changes can be carried out using the Change Control mechanism.
- **5:** Implies that the specification is approved by TSG for Release 5. Changes can be carried out using the Change Control mechanism.
- **6:** Implies that the specification is approved by TSG for Release 6. Changes can be carried out using the Change Control mechanism.

Table 1.1   Series of 3GPP Specifications

Series number	Subject of specification series	Description
21 series	Requirements	These are generally temporary specifications that graduate to other specifications.
22 series	Service aspects	These are typically Stage 1 specifications that define the service and service features.
23 series	Technical realization	These are typically Stage 2 specifications that define functional blocks and the information flow between these blocks.
24 series	Signaling protocols (user equipment to network)	These are Stage 3 specifications of protocols between User Equipment and the Core Network.
25 series	Radio aspects	These are specifications for the radio aspects. The specifications are further classified into the following sub-series:· • 25.1bb: UTRAN radio performance • 25.2bb: UTRA layer 1 • 25.3bb: UTRA layers 2 & 3 • 25.4bb: UTRAN Iub, Iur & Iu interfaces
26 series	CODECs	Include specifications for codecs for speech and video.
27 series	Data	Includes specifications for defining functions necessary to support data applications.
28 series	Signalling protocols	These are Stage 3 specifications of protocols between radio subsystem (e.g. BSS) and edge of Core Network (e.g. MSC).
29 series	Signalling protocols	These are Stage 3 specifications of signalling protocols between elements of the Core Network.
30 series	Program management	Includes specifications for project plans, project work program and stand-alone documents for major work items.
31 series	User Identity Module (SIM/USIM)	Includes specifications for Subscriber Identity Module (SIM) and Universal Subscriber Identity Module (USIM) and its interfaces with other entities.
32 series	Operations and Maintenance	Defines specifications for application of TMN. Also includes specifications for operation, administration and maintenance (OAM).
33 series	Security aspects	Defines specifications for security aspects.
34 series	Test specifications	These define test specifications.
35 series	Security algorithms	Defines specifications for encryption algorithms that provide confidentiality and authentication.

At the time of writing of this book, the Release 6 specifications of 3GPP were being developed. The first formal release of 3GPP was in Release 99. From then on, two more releases, namely Release 4 and Release 5 have been developed by 3GPP. The functional division of various 3GPP releases is explained in the next sub-section.

1.5.7 Releases of 3GPP Specifications

In the previous subsection, it was mentioned that the 3GPP has gone through four releases. The first release, called Release 99 (or Rel99 in short), was frozen in March 2000. The main development over GSM was the development of WCDMA-based Universal Terrestrial Radio Access (UTRA). The UTRA was developed for FDD as well as for the 3.84 Mcps TDD mode. Apart from this, other features of this release were:

- CAMEL phase 3
- Basic version of Open Service Architecture (OSA)
- Location Services (LCS)
- Narrowband AMR (a new codec)

Apart from this, there were other improvements like multi-call and High-Speed Circuit Speech Data (HSCSD), among others.

After Release 99, the numbering convention was altered and it was de-coupled from naming conventions based on calendar years. The next release, which was frozen in March 2001, was called the Release 4 (or Rel4 for short). The main features of this release are summarized as follows:

- Low chip rate TDD mode (1.28Mcps).
- The MSC was split into a Media Gateway and an MSC server to provide bearer independent circuit-switched network architecture.
- GERAN concept was established (i.e. EDGE/GPRS Iu interface).
- Streaming service so that real-time video could be retrieved.
- Messaging services, which included enhanced messaging (i.e. rich text and still image) and multimedia messaging (i.e. multimedia attachments).
- Introduction of SIGTRAN in Core Network.

Apart from this, there were other improvements like real-time facsimile, transcoder free operation, emergency calls in CS domain, security aspects and Mobile Execution Environment (MExE), among others.

Release 4 was followed by Release 5 (or Rel5 for short). The biggest development in Release 5 was the introduction of IP Multimedia Subsystem (IMS). Before Release 5, the Core Network consisted of a CS and PS domain. Release 5 saw a new overlay sub-system based on Session Initiation Protocol (SIP). The IMS was developed for introduction of multimedia services based on SIP protocol. Apart from this, other features of this release were:

- Introduction of High-Speed Downlink Packet Access (HSDPA), which has the ability to provide throughput of the order of 10Mbps.
- CAMEL phase 4
- Wideband AMR (a new 16 kHz codec)
- End-to-end QoS in PS domain
- Global Text Telephony (GTT) for real-time text.

Apart from this, there were improvements in radio interface. SIGTRAN was introduced in UTRAN. There were other enhancements in GERAN, LCS, OSA and MExE, among others. Release 5 also enabled intra-domain connection of RAN nodes to multiple CN nodes.

Release 5 was followed by Release 6 (or Rel6). At the time of writing of this book, work on Release 6 was underway and is expected to be completed in mid-2004. The features slated for Release 6 are as follows:

- Improvements in IMS (through introduction of IMS messaging and IMS group support).
- Multimedia Broadcast/Multicast Service (MBMS)
- Push services
- Digital Rights Management
- Speech enabled services
- Identity (or number) portability
- Presence Service
- Generic User Profile
- Wireless LAN interworking
- Radio optimization
- Priority service
- Other enhancements to LCS, OSA, MExE and emergency calls, among others

Figure 1.3 depicts the chronology of 3GPP releases. Work on first release started at the beginning of 1999 and ended by early 2000. Around that time, work on Release 4 had also started. Since then, a new release is coming in every year or so. Prior to 1999, there were GSM releases; these are not shown in the figure.

1.5.8 Evolution towards 3GPP Networks

While developing a new technology, it is important to provide an evolution path to it. Given the huge costs involved in the deployment of Second Generation networks, it was important to provide easy migration to Third Generation networks. To achieve this end, 3GPP adopted a two-pronged approach. Firstly, it designed a completely

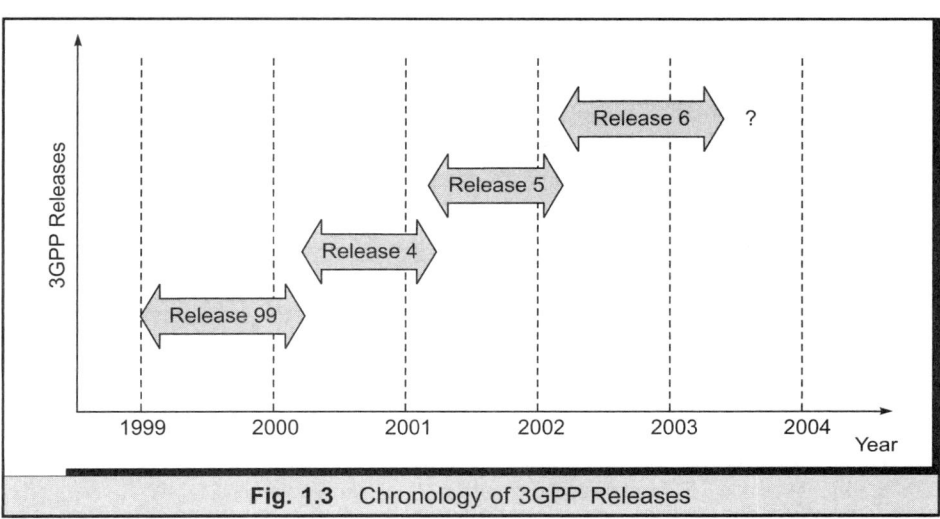

Fig. 1.3 Chronology of 3GPP Releases

new Radio Access technology based on Wideband CDMA (WCDMA) providing FDD and TDD modes of operation. This was necessary to provide higher levels of bandwidth. Secondly, it derived network elements from 2G and 2.5G networks. These elements include Home Location Register (HLR), Visitors Location Register (VLR), Serving GPRS Support Node (SGSN), Gateway GPRS Support Node (GGSN), Authentication Center (AuC) and Equipment Register (EIR). The net result was that a new handset was required along with software/hardware upgrades in network elements. This strategy ensured that high-bandwidth 3G services could be provided with minimum cost for upgrade.

Figure 1.4 provides an evolution path towards 3G network based on W-CDMA (note that the figure provides some of the possible evolution paths and is in no way exhaustive). A GSM operator may directly move from GSM to W-CDMA. Alternatively the GSM operator may use GPRS as a stepping-stone towards W-CDMA. Moving further from GPRS, EGPRS may also be used as an intermediate step. The exact path taken depends upon various factors including the existing infrastructure, availability of 3G licenses, capital employed and market demand, among others.

SUMMARY

The 2G standards for mobile communication revolutionized the way mobile communication networks were built. The GSM, PDC, IS-95 and D-AMPS were the important 2G mobile networks. Among these, the GSM was, and still is, the most popular standard. However, the progress of 2G networks is hindered by serious limitations, including lower bit rates, inefficient circuit-switching and a semi-global roaming because of a plethora of standards.

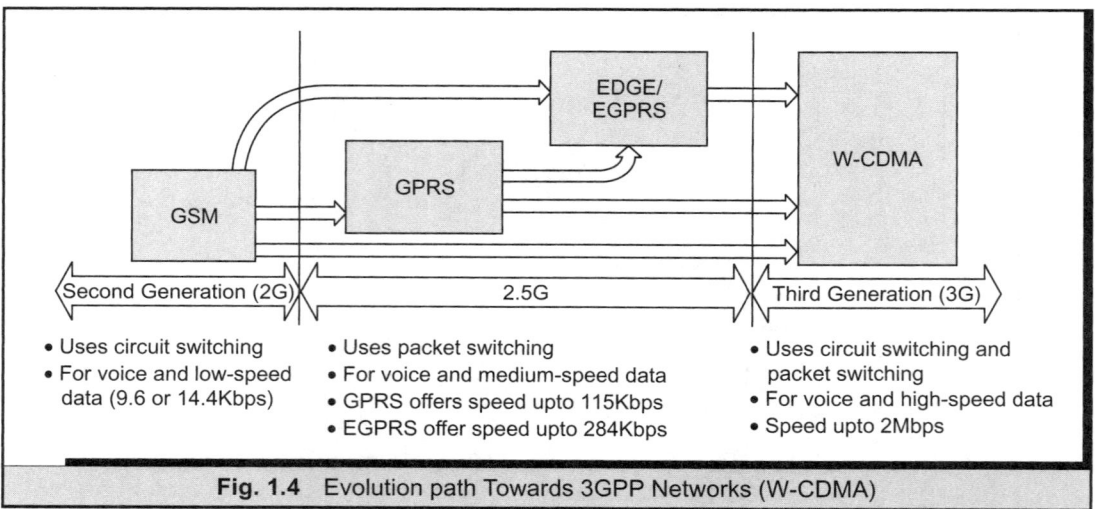

Fig. 1.4   Evolution path Towards 3GPP Networks (W-CDMA)

The 2.5G generations like HSCSD, GPRS and EDGE were built over the existing 2G networks to provide greater bandwidth. However, these networks still fall short of providing high-speed wireless communication.

At present, 3G standards are being developed to overcome the limitations of 2G/2.5G networks. The goal of 3G standardization is to offer greater bandwidths, improved services, and true global roaming. Various standardization bodies are working on the development of 3G standards. The main area of focus is the air interface that is capable of providing bandwidths up to 2Mbps. Initially, the goal, as envisaged by ITU, was to have a single radio interface that provided global roaming. However, due to various reasons, a set of five standards was adopted for Third Generation (3G) networks. These standards fall under the umbrella of International Mobile Telecommunications 2000 (IMT-2000) systems.

On reviewing the efforts of various standardization bodies, we find that the work done by 3GPP is most prominent.

Moreover, even as the skeptics wonder about the commercial viability of 3G networks, these are already being deployed in Japan, with Europe following closely and customers are beginning to get the feel of the improvements. NTT DoCoMo launched the world's first commercial 3G service—FOMA—on October, 2001. 'Appendix A provides the status of deployment of 3G networks in various parts of the world.'

Third Generation standards are still evolving, and future releases of these standards propose an all-IP network. This would aid faster development and deployment of new services, and would benefit both the customer as well as the operator. The mobile industry and academia is abuzz with the sounds of 3G, and it would be reasonable to say that all this hype is not without reason.

PRINCIPLES OF WCDMA

2.1 INTRODUCTION

Wideband Code Division Multiple Access (WCDMA) is the radio access scheme used within the UMTS Third Generation mobile networks as defined by the 3GPP. Evolved from the Code Division Multiple Access (CDMA) scheme, WCDMA offers high bandwidth (upto 2Mbps) to users of the mobile network. This chapter explains some of the basic principles of WCDMA.

It begins by describing the requirements for the air interface of the Third Generation mobile networks, and the available choice of radio access schemes for the air interface. It then provides an overview of WCDMA, which is followed by a discussion on WCDMA-related concepts. The focus of this chapter is on concepts that are specific to WCDMA networks. These include spreading, scrambling, use of RAKE receivers, multipath diversity and macrodiversity, power control mechanisms, handovers and relocation procedures.

2.2 REQUIREMENTS FOR THIRD GENERATION AIR INTERFACE

The Second Generation mobile networks were built mainly to provide voice services to subscribers. As support for voice calls was the primary requirement, the air interface was developed accordingly to achieve this goal with the maximum possible efficiency. However, a lot has changed since the time the Second Generation systems were first introduced. Subscribers are no longer content with just voice-based services. Value-added services are being offered to subscribers of the Second Generation systems by enhancing these networks with additional network elements and resources. Already, a need is being felt to offer multimedia services over the mobile network. Support for multimedia services, at higher bandwidths, places the first requirement on the air interface of a Third Generation system.

Besides the ever-growing needs of subscribers, several technological advances have taken place, which make the existing technology, used in the air interface of Second Generation systems, obsolete. This same air interface technology, which at the time of introduction of Second Generation systems was seen as providing the maximum possible efficiency, is no longer considered efficient. Several advances have been made on the air interface, which increase its efficiency many folds. Thus, it was evident that the air interface of the Third Generation systems would require a major overhaul from that of the Second Generation systems.

Before proceeding on a discussion of the radio access technology used for the air interface of Third Generation systems, it is instructive to take a look at the requirements for Third Generation systems. Some of the primary requirements for the Third Generation mobile networks, which are relevant from the air interface perspective, are as follows:

- Support of bit rates of up to 2Mbps.
- Mechanisms to efficiently support 'Bandwidth on Demand' and Variable Bit Rate (VBR) services.
- Support for services with different quality requirements (e.g. speech, video, packet data, etc.). Quality requirements for services differ in terms of
  - **Delay Requirements:** Refers to support for both real-time and non-real time services.
  - **Error Requirements:** Refers to support for bit error rates as low as 10^{-6}.
  - **Bandwidth Requirements:** Refers to support for services requiring variable bandwidth. Also includes support for asymmetric uplink and downlink bandwidths for applications like web-browsing.
- Simultaneous co-existence with 2G systems with support for inter-system handover.
- Higher spectrum efficiency.

In a nutshell, the air interface for 2G systems was required to support voice calls at fixed bit rates. Bit rate requirements were as low as 9.6Kbps. Services considered were not as diverse as is required of a Third Generation system. Only circuit-switched services, with similar requirements, were offered.

Simply put, the aim of the above discussion is to bring forward the point that the air interface for a 3G system requires a complete overhaul from what was being offered by its predecessors. Also, looking at the requirements, it is evident that the choice of the radio access technology for the air interface is of prime importance when it comes to meeting these requirements. The following section discusses the available choices for the radio access technology. The subsequent sections then explain the features of the WCDMA technology in detail.

2.3 SCHEMES FOR RADIO ACCESS

Radio access schemes provide techniques that allow multiple users to simultaneously access the radio spectrum, in an efficient manner. Radio spectrum being a scarce resource, it is essential to utilize it efficiently. This section discusses four of the most popular mechanisms used for radio access, namely:

- Time Division Multiple Access (TDMA)
- Frequency Division Multiple Access (FDMA)
- Hybrid FDMA/TDMA Scheme
- Code Division Multiple Access (CDMA)

In the Time Division Multiple Access (TDMA) scheme, the entire frequency spectrum is available to each user. Time-sharing is used to provide access to multiple users. In technical terms, the frequency spectrum is divided into multiple time slots, and each user is allotted a particular time slot, which gives him access to the entire frequency spectrum. Each subscriber is allocated one time slot for each direction, uplink and downlink.

In the Frequency Division Multiple Access (FDMA) scheme, the entire frequency spectrum is divided into smaller frequency channels. Each user accessing the radio interface is allocated a particular frequency channel, which is a portion of the entire frequency spectrum. The frequency channel is available to the user for the entire call duration. Thus, unlike the TDMA scheme, where the sharing of resources is done based on time slots allocated to users, the FDMA scheme shares resources by allocating portions of frequency spectrum to users. Each subscriber is allocated one frequency channel in each direction, uplink and downlink.

For a comparison of the TDMA and the FDMA schemes, let us consider the entire available 'radio resource' as a product of the total frequency spectrum and the total time. Then, as already discussed, while the TDMA scheme divides the total available 'radio resource' on time basis (time-sharing), the FDMA scheme divides the resource on frequency channel basis.

The GSM system, which is the most popular Second Generation mobile system, uses a scheme that derives from a combination of the FDMA and the TDMA schemes. In this, the entire frequency spectrum is divided into multiple frequency channels, similar to the FDMA scheme. Further, each frequency channel is divided into time slots and is time-shared between multiple users, like in the TDMA system. Thus, each user is allocated a time slot within a particular frequency channel (TDMA), which itself is a part of the total frequency spectrum (FDMA). Hence, this scheme has aptly been named the Hybrid FDMA/TDMA scheme.

The Code Division Multiple Access (CDMA) scheme offers the entire frequency spectrum to multiple users simultaneously—for the entire duration of time. In other words, the entire radio resource, as defined above, is allocated to one particular user. This raises the question of how multiple access is provided to multiple users simultaneously. CDMA introduces the concept of spreading codes, which are applied as part of a secondary modulation of user signals. These codes are used to transform the user signals into a spread-spectrum-coded version of the original signal, before transmission. The receiver follows the reverse process to recover the original signal from the coded signal. Different users are allocated different spreading codes, and the signals of multiple users are thus differentiated on the basis of the spreading code. Spreading codes can therefore be considered as the third possible parameter on which sharing of resources can be achieved—besides time slots and frequency channels. CDMA systems use this spreading code to provide multiple access to users.

Just like it is possible to apply the TDMA and the FDMA schemes in a hybrid fashion, it is also possible to apply the CDMA scheme in combination with the TDMA and the FDMA schemes. This is what is actually used within the UMTS Radio Interface. The latter part of the following section discusses this hybrid scheme in detail.

2.4 WCDMA OVERVIEW

WCDMA has emerged as the most widely adopted air interface technology for Third Generation systems. The 3GPP consortium has created specifications for the use of WCDMA technology in its networks, where it is referred to as the 'UTRA' (Universal Terrestrial Radio Access). Evolved from the CDMA radio access scheme, WCDMA is a Wideband Direct-Sequence CDMA (DS-CDMA) system. To understand this definition, it is imperative to understand some of the terms/concepts used with respect to WCDMA:

- Direct-Sequence CDMA
- Wideband
- Synchronization Aspects
- Modes of Operation

2.4.1 Direct-Sequence CDMA

The term 'Direct-Sequence CDMA (DS-CDMA)' stems from the 'Direct-Sequence Spread Spectrum (DSSS)' technique used in DS-CDMA. Recollect from the previous section that the CDMA scheme uses the concept of spreading codes to transform the user signal into a spread-spectrum-coded signal. These spreading codes are used to provide access to multiple users simultaneously. Multiple spreading techniques exist; notable among them are the DSSS and the 'Frequency Hopping Spread Spectrum (FHSS)'.

The DSSS uses a carrier that remains fixed to a specific frequency band. The data signal is spread onto a much larger range of frequencies using a specific encoding scheme, rather than being transmitted on a narrow band. This encoding scheme is known as Pseudo-Noise sequence (or PN sequence). Frequency Hopping Spread Spectrum, on the other hand, attempts to achieve the same result by sending its transmissions over a different carrier frequency at different times. The DSSS is the simpler of the spreading techniques, whereby the original signal is directly multiplied by a faster-rate spreading code. Section 2.5 describes the DSSS spreading technique in more detail.

2.4.2 Wideband

In WCDMA, the term *Wideband* refers to the higher bandwidth carrier signal of approximately 5 MHz. Original DS-CDMA systems, like the IS-95, used a carrier bandwidth of about 1 MHz. These systems are referred to as *Narrowband CDMA* systems. In contrast, WCDMA is a CDMA based access scheme, which uses the DSSS spreading technique with a carrier bandwidth of around 5 MHz. The higher carrier bandwidth makes WCDMA a wideband system.

2.4.3 Synchronization Aspects

The proposals for WCDMA radio interface are broadly divided into two categories, *network synchronous* and *network asynchronous*, depending upon whether the base stations within the network are synchronized in time or not. A network where the base stations are time synchronized with each other are categorized as synchronous networks. If synchronization between base stations is not required, then such a network is termed as an asynchronous network. Second Generation IS-95 systems are based on the synchronous network concept. Though synchronized networks provide more efficient utilization of the radio interface, it requires a lot of functionality within the base stations, and hence more costly hardware. Additional techniques, like Global Positioning System (GPS), would be required to time synchronize the base stations. This further places a restriction on deployment of base stations—indoor and micro base stations can only be deployed if the GPS signals can be received indoors. While Third Generation systems based on the 3GPP specifications utilize the asynchronous network based scheme, CDMA 2000 based proposals use schemes that are synchronous network based.

2.4.4 Modes of Operation

WCDMA supports two basic modes of operation: one for paired spectrum and the other for the unpaired spectrum. Here, pairing refers to the frequency bands available for communication. The Frequency Division Duplex (FDD) mode is used for the paired spectrum, while for unpaired spectrum, it is the Time Division Duplex (TDD) mode.

In the FDD mode of operation, separate 5 MHz carrier frequencies are used in the uplink and the downlink direction. Since two frequency bands of equal bandwidths are available, one is used for uplink direction (i.e. from mobile station to base station) and the other for downlink direction (i.e. from base station to mobile station). Thus, the information transfer in FDD mode is symmetric. Further, data can be exchanged in both directions simultaneously. The traditional GSM uses the FDD mode.

In the TDD mode of operation, only one 5 MHz carrier is time-shared between the uplink and the downlink. Since only one frequency band is available, the TDD is said to use the unpaired spectrum. The data is alternated in the uplink and downlink direction. The main benefit of TDD mode is that the bandwidths in forward and backward direction can be altered. Thus, it is possible that the downlink bandwidth is much more than the uplink bandwidth. This is helpful in certain applications (e.g. downloads from the Internet) where a small request is followed by large amounts of information. The unpaired nature of TDD makes it use the spectrum more efficiently. Further, as the spectrum becomes scarce, getting an unpaired spectrum will be easier as compared to obtaining a paired spectrum. This will provide TDD an edge over FDD. It is presumed that TDD will be used in hot spots (e.g. airports) to provide high data rate connectivity efficiently.

In a certain sense, the FDD mode of operation can be considered as a hybrid scheme that uses CDMA and FDMA schemes, since the CDMA scheme is applied within each frequency channel. On the other hand, the TDD mode of operation uses a hybrid scheme based on CDMA, FDMA and TDMA, as the same carrier frequency is further time-shared for the uplink and the downlink direction.

With this introduction to WCDMA forming the basis of discussion, the subsequent sections cover some of the important concepts associated with WCDMA in greater detail.

2.5 SPREADING AND DE-SPREADING

Spreading is a technique used to transform the user's original signal into a signal form, that is 'spread' over a larger bandwidth than what is required for the original signal. Codes used for this transformation of the signal are called Spreading Codes (also known as Channelization Codes). The concept of codes and their use in providing multi-access was briefly discussed in Section 2.3. This section explains the process of spreading and de-spreading.

The process of spreading involves transformation of the user signal by multiplication (XOR) with bits in a spreading code. Each user is assigned a unique spreading code.

The bits in the spreading code are termed as 'chips', while the bits within the user signal are called symbols. Figure 2.1 depicts the process of spreading. The transformation involves bit-wise XOR of the user signal with the bits in the spreading code to form the spread signal.

The process of spreading can also be seen as using the chips in the spreading code to 'chop' the user signal into smaller parts. The spread signal is actually spread over a larger bandwidth as a result of the spreading process. The ratio between the bandwidth required to transmit the spread signal to the original bandwidth requirements is called the Spreading Factor. Values of the spreading factor within the UTRAN lie between 4 and 512. In Figure 2.1, a spreading factor of 4 is depicted.

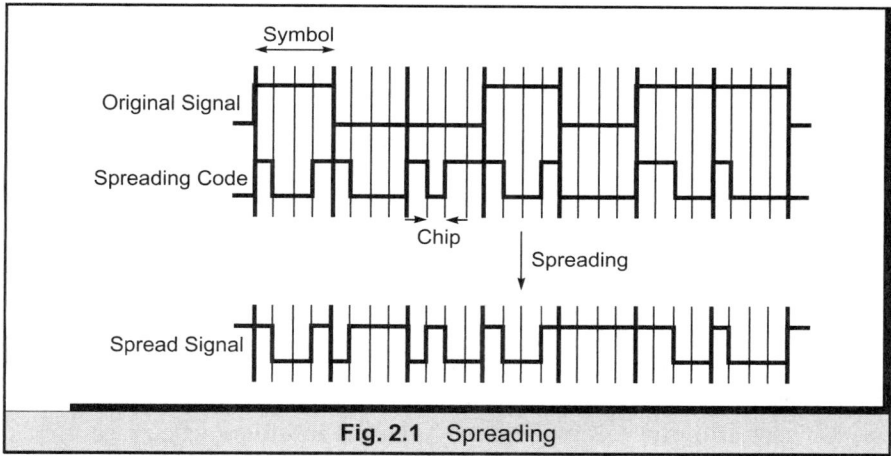

Fig. 2.1   Spreading

De-spreading is the reverse process of spreading. It involves recovering of the original signal at the receiving end from the spread signal. The same sequence of spreading code is used in the de-spreading process. The process involves bit-wise XOR of the spread signal with the spreading code to recover the original signal. Figure 2.2 depicts the process of de-spreading.

Having gone through the description of Spreading and Despreading, at first thought, the entire process seems to be a waste of bandwidth. Available bandwidth on the air interface is a scarce resource, and must be efficiently utilized amongst multiple users. Why would WCDMA then use spreading to transform the original signal into a signal that would require more bandwidth than the original signal? There are many reasons why spreading is so important for WCDMA. However, before delving into the benefits of spreading, it is important to understand two key concepts: Autocorrelation and Cross Correlation.

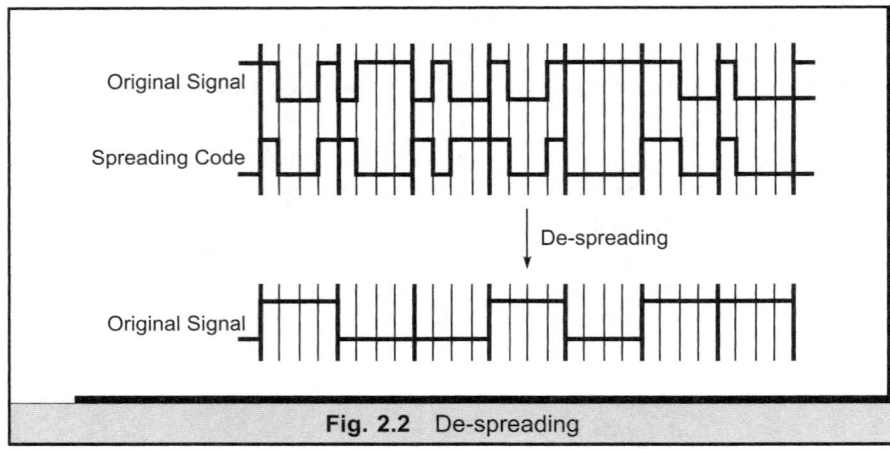

Fig. 2.2   De-spreading

2.5.1   Autocorrelation and Cross Correlation

Propagation of the radio signal over the radio interface is generally characterized by multiple reflections and diffractions, which are a result of obstacles within the path from the mobile station to the base station (or vice versa). These obstacles could be tall buildings, hills, etc., which reflect/diffract the signal, resulting in a concept known as *Multipath Propagation* (refer to Section 2.8 for details of multipath propagation). As a result of this multipath propagation, the same signal is received by the mobile/base station more than once—at different time intervals and with different power levels. Thus, in such a scenario, the receiver has to have the intelligence of receiving the same signal multiple times, and then using these signals to obtain the original signal. For this purpose, a rake receiver is used, which is discussed in Section 2.7.

The Rake Receiver consists of multiple rake fingers, each receiving a multipath signal. The receiver itself acts as a correlator, correlating the signal received by each rake finger. It is here that the concept of autocorrelation finds its significance. Autocorrelation measures the amount of correlation between the received signal and a delayed version of the same signal received later in time. The higher the autocorrelation, the easier it is to receive a multipath signal. Since a signal is received through multiple paths as various components, these (components) may result in interference at the receiver. A spreading code that provides good autocorrelation properties of the spread signal can resist this interference.

Cross correlation, on the other hand, measures the correlation between a signal spread using a particular spreading code, to the same signal spread using some other pseudo-random code. Spreading codes should have a low cross correlation with other spreading codes. A low cross correlation between spreading codes results in lower interference between signals received at a receiver. Thus, a good spreading code should have high autocorrelation properties, but lower cross correlation with other spreading codes. However, it is generally not possible to have both high autocorrelation and lower cross correlation at the same time. A trade-off between the two properties is normally required.

2.5.2 Benefits of Spreading

The use of spreading techniques provides many benefits. Firstly, by using spreading (codes), it is possible to provide a multi-access environment. Spreading codes are unique to each user. As a result of this, each user's signal is transformed differently before transmission on the air interface. This allows multiple users to use the same frequency channel simultaneously by transmitting the signal after transformation with different spreading codes. This is what forms the basis of CDMA—a code-based distinction between multiple simultaneous users.

Spreading codes have low cross correlation between them. Hence, when the spread signals of different users are received at the receiver, and these have been spread by different spreading codes, then they (the signals) can be easily separated from each other. Since the receiver knows the spreading code used by the sender, the original signal can be recovered by using this spreading code. Any other noise received along with the spread signal cannot be recovered using this spreading code, and hence gets separated from the original signal as noise.

Figure 2.3 depicts the concept of recovery of the original signal by separating it from interference/noise. An interfering signal present in the same band typically appears as a higher power, narrow band signal (see part (a) of figure). At the receiving end, the

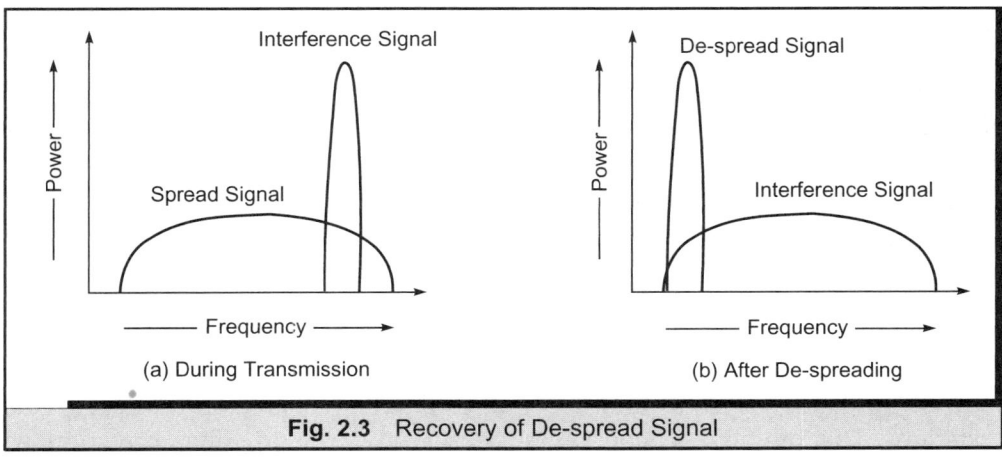

Fig. 2.3 Recovery of De-spread Signal

de-spreading process recovers the original signal (De-spread signal) and spreads out the interference instead (see part (b) of figure).

Second, the spreading of the signal over a larger bandwidth results in a frequency reuse factor of one. This means that the same frequency can be reused in adjacent cells, resulting in high spectral efficiency. On the other hand, the FDMA/TDMA hybrid scheme of GSM results in a frequency reuse factor of at least four, implying that the same frequency can at best be reused in every fourth cell.

Third, as a result of higher autocorrelation between spreading codes, the problem of multipath interference is tackled. Using a rake receiver can constructively correlate signals received from multiple paths. Resolving multipath interference was not so simple in GSM systems, which did not use the concept of spreading.

Fourth, the spreading of user signals before transmission results in improved security of the transmitted signal. Since the spreading code used to spread the signal is only known to the sender and the intended receiver, it is not possible for any other receiver in-between to capture this signal. However, this only provides security from non-resourceful hackers. A resourceful hacker can always capture a user signal and use brute-force mechanism to derive the original signal using all spreading codes, especially since the number of spreading codes is fixed.

Another significant property of spread signals is that unlike the original signal, they cannot be jammed. This property stems from the fact that while jamming a particular frequency could jam the original signal, the spread signal is spread over a much larger frequency band, requiring the jamming of the entire frequency band. This is more difficult than jamming the transmission at a particular frequency. Hence, spreading also finds a great deal of importance in military communications.

2.6   SCRAMBLING

The previous section discusses the benefits of using spreading before transmission of signals over the radio interface. However, spreading alone is not sufficient to provide an efficient solution to the transmission problem. The following issues still remain to be addressed:

- The spreading codes are orthogonal in nature. This means that two spreading codes would have a negligible cross correlation between them, provided they are synchronized in time. However, time synchronization between signals cannot be guaranteed in the uplink direction, where multiple mobile stations can be communicating asynchronously with the base station. In such a case, the signals from multiple mobile stations may interfere with each other, leading to difficulty in de-spreading and separating the original signals from different mobile stations. This problem is generally not observed in the downlink direction, where one base station is coordinating transmission to multiple mobile stations. Here, the base station can ensure that the timing synchronization is maintained. Also, the TDD mode of transmission in the uplink direction can actually address this problem easily, but the FDD mode needs some other mechanism to tackle it.

- Second, as a result of reflections and diffractions, multipath components of signals can be received at the receiver. However, the orthogonality of these multipath components cannot be guaranteed. Again, this results in higher cross correlation between signals received at the base station, and hence, difficulties in separating the user signals.
- Also, WCDMA based systems normally have a frequency reuse factor of one. This means that adjacent cells can be using the same frequency. This leads to a problem in transmission on the downlink direction. The problem stems from the fact that the number of spreading codes is finite, and two different mobile stations in adjacent cells can be allocated the same spreading codes. In such a case, a mobile station on the border of two cells (where it can receive transmission from both base stations) cannot figure out if the transmission is for it, or for some other mobile station in the adjacent cell, which uses the same spreading code.

To solve these problems, the process of scrambling is used, which provides means to distinguish between signals from multiple mobile stations in the uplink, and helps reduce inter-base station interference in the downlink. Scrambling follows the process of spreading at the transmitting end, using pseudo-random codes called Scrambling Codes (Figure 2.4).

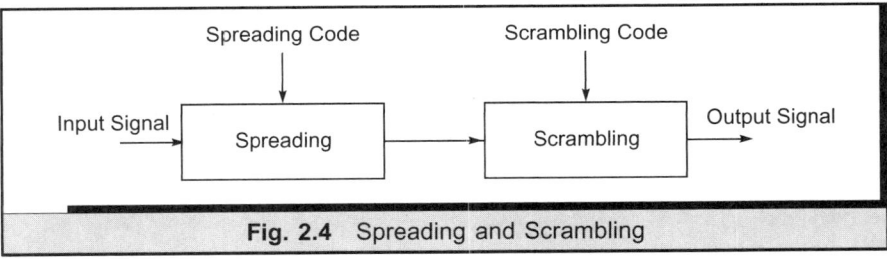

Fig. 2.4 Spreading and Scrambling

Each mobile station and base station is assigned a unique Scrambling Code; the former in the uplink and the latter in the downlink direction. Like in the case of Spreading Codes, Scrambling Codes too have high autocorrelation properties. The process of scrambling a signal using scrambling codes is similar to the process of spreading, as shown in Figure 2.1. Thus, scrambling is used in addition to spreading, to solve the problems arising due to lack of synchronization between mobile stations, multipath reception of signals, and reuse of spreading codes in adjacent cells.

2.7   RAKE RECEIVER

Reflections and diffractions of signals from obstacles in the radio path lead to multipath propagation, which resuls in the same signal being received by the mobile station/base station more than once, at different time intervals, and with different power levels. Section 2.8 describes the concept of multipath propagation in detail. In such a scenario, the receiver has to be intelligent enough to receive the same signal multiple times, and then use these signals to obtain the original signal. For this purpose, a rake receiver is used. The Rake receiver resembles a garden rake, and hence the name. The Rake receiver is also pictorially represented in the form of a garden rake (Figure 2.5).

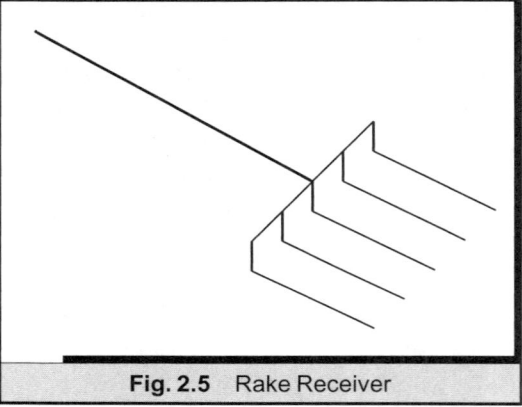

Fig. 2.5   Rake Receiver

Each finger of the Rake receiver receives a copy of the original signal, which reaches its destination by following a particular path. The fingers de-spread the received signal, and these de-spread signals from each finger are combined into the original signal. Thus, the Rake receiver can broadly be seen as comprising of two main components:

- **Receiving Fingers:** Each receiving finger includes the functionality to de-spread the received multipath signal using the spreading codes.

- **Combiner:** It combines the de-spread signal components of the original signal to obtain the signal transmitted by the sender.

The components of the Rake receiver are depicted in Figure 2.6.

2.8   MULTIPATH DIVERSITY AND MACRODIVERSITY

Multipath Diversity is a term used for a phenomenon where a signal is received by the receiver multiple times via different paths. This happens as a result of the signal being reflected and diffracted due to high-rise buildings, mountains, or other such obstacles in its path (Figure 2.7).

Each component of a multipath signal, from the transmitting end to the receiving end, can experience varying delays, based on the path taken by the signal. At the

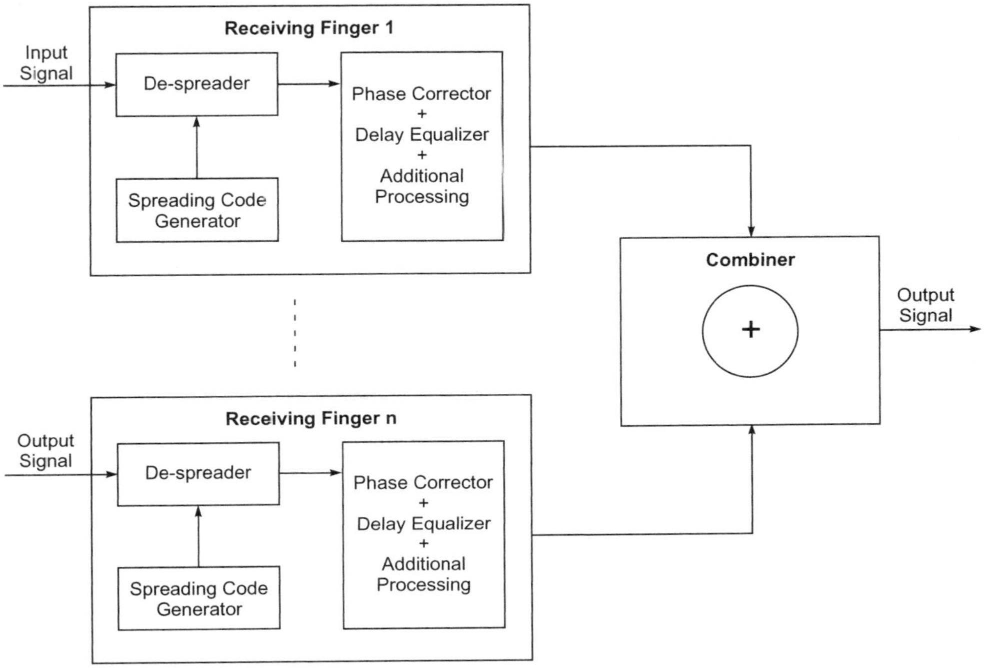

Receiving Finger 1

Input
Signal → De-spreader

Spreading Code
Generator

Phase Corrector
+
Delay Equalizer
+
Additional
Processing

Receiving Finger n

Output
Signal → De-spreader

Spreading Code
Generator

Phase Corrector
+
Delay Equalizer
+
Additional
Processing

Combiner

+

Output
Signal

Fig. 2.6 Rake Receiver Components

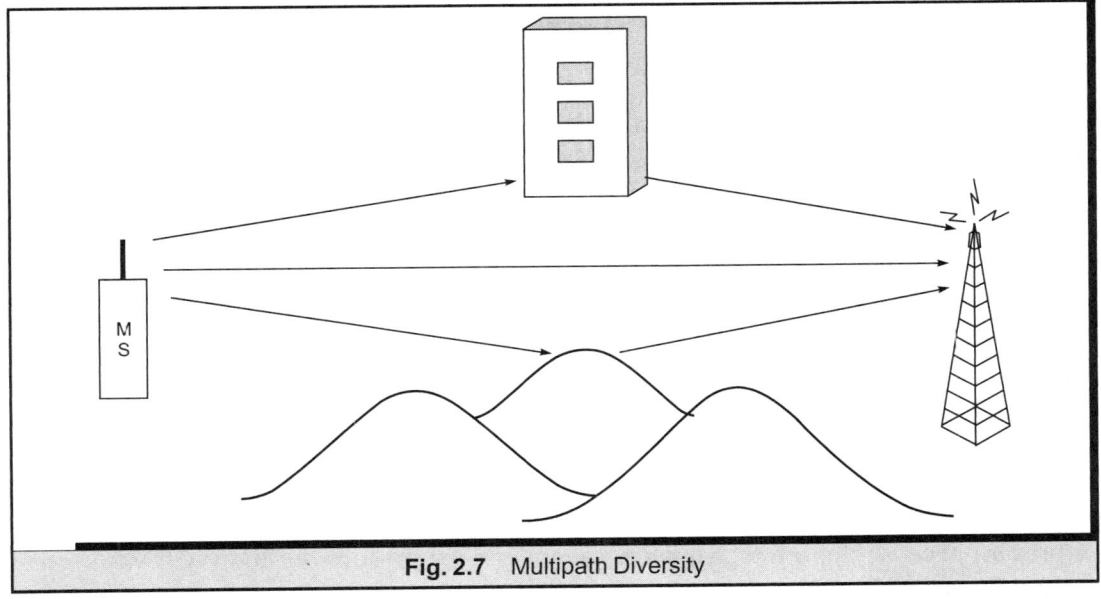

Fig. 2.7 Multipath Diversity

receiving end, it should be possible for the receiver to receive the same signal multiple times, each at different times and with different power levels. For this purpose, a Rake receiver—which is capable of receiving multiple multipath components and then combining them into a composite signal—is used.

Macrodiversity is a term used in WCDMA networks to denote a scenario where the same signal can be transmitted to a mobile station via multiple base stations. This generally happens when the mobile station is in a region which falls in the coverage area of more than one base station. In such a situation, the UTRAN utilizes the services of two or more base stations to transmit and receive the signals to/from the mobile station.

In a macrodiversity situation, the bandwidth used by the mobile station is more than what is normally used. This results from multiple base stations participating in the transmission and reception of signals to/from the mobile station; hence, the duplication of signal transmission/reception. However, the gain from macro-diversity is that it allows the signal to travel from more than one base station, thereby compensating the losses in transmission by combining the signals received from multiple base stations. Macrodiversity also provides protection from *shadowing*, which is a situation where the mobile station gets shadowed behind an obstacle. In such a scenario, while the signal from the mobile station to one base station would get blocked, the other base stations would still be able to service the mobile, thus preventing complete service outage for the mobile station.

However, despite its benefits, a macrodiversity situation is not always recommended, for besides using more bandwidth, it also increases the overall interference level in the system. Hence, the UTRAN makes a decision as to when it is beneficial to use macro-diversity, and when it is not. In case the UTRAN decides to go in for macrodiversity, a Soft Handover (SHO) procedure is used to achieve the macrodiversity situation. The SHO procedure is described in Section 2.10.

A special case of Macrodiversity is the Site-Selection Diversity Transmit (SSDT) scheme (Figure 2.8). In the SSDT scheme, while the transmission from the mobile station is received by multiple base stations in the uplink direction, only one base station transmits to the mobile in the downlink direction. The base station that is best suited for data transmission is assigned the role of transmitting data in the downlink to the mobile station. However, the control information to the mobile station continues to be transmitted from all the base stations that are involved. The control information is discussed in detail in Chapter 7.

2.9 POWER CONTROL MECHANISMS

Power control mechanisms define the means to control the power at which the base station and the mobile station transmit signals to each other. Power control is important because the same frequency carrier is used throughout the WCDMA system

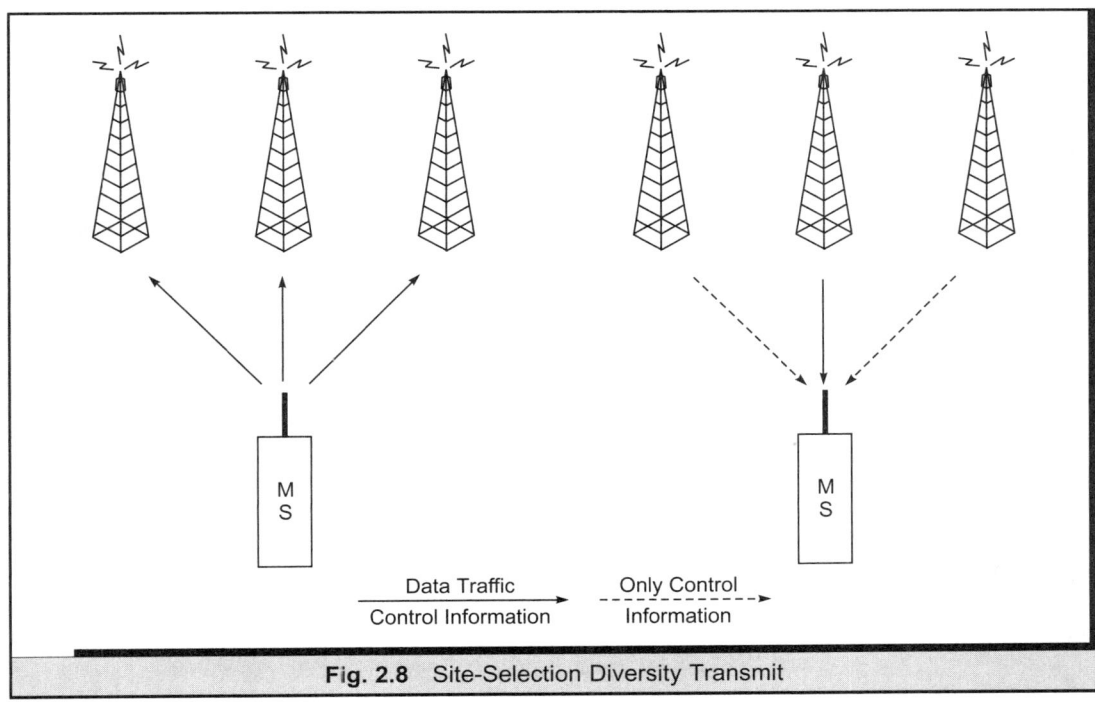

Fig. 2.8 Site-Selection Diversity Transmit

for communication; hence, interference resulting from simultaneously communicating entities needs to be minimized. The need for power control can be attributed to the following factors:

- To prevent the 'Near-Far Effect' in radio communication: The near-far effect is a term used to denote a situation wherein a mobile station further away from the base station cannot be serviced because its transmission is overshadowed by a mobile station that is closer to the base station. Consider a scenario where all mobile stations within a cell use constant power levels to communicate with the base station. In this case, since the mobile stations are not equidistant from the base station, the signal from a mobile station further away from the base station would suffer greater attenuation. As a result, the power levels of signals received at the base station from multiple mobile stations would be inversely proportional to the distance between the mobile and the base station. Clearly, the signals with higher power levels (from near distance mobile stations) would overshadow the signals with lower power levels (from far-off mobile stations), thus preventing the latter from being serviced. Hence, power control mechanisms are required to monitor and control the power level at which each mobile station should communicate.

- To transmit signals to mobile stations on cell boundary: Mobile stations on cell boundaries are more prone to interference from signals being transmitted in adjacent cells. To prevent this, it is desirable for the base station to transmit signals to such mobile stations using marginally increased power. This marginal additional power would neutralize the interference from neighbouring cells.
- To reduce interference in neighbouring cells: WCDMA systems have a frequency reuse factor of one, leading to adjacent cells using the same frequency carrier. If no power control mechanisms are employed, high power transmissions in one cell could easily interfere with communications in adjacent cells. Hence, the power used in transmission within one cell should be bounded to prevent such interference in communication in neighbouring cells.

Two types of power control mechanisms are used within a WCDMA network: *Open Loop Power Control* and *Closed Loop Power Control*. An open loop power control mechanism (Figure 2.9) uses the measurements made on the signal received in the downlink direction to decide on the power level for transmission in the uplink direction. The logic for the open loop power control is simple: the lower the received signal power, the farther away is the Mobile Station (MS) from the base station; the farther the MS, the higher the power level it will have to use for transmission to the base station.

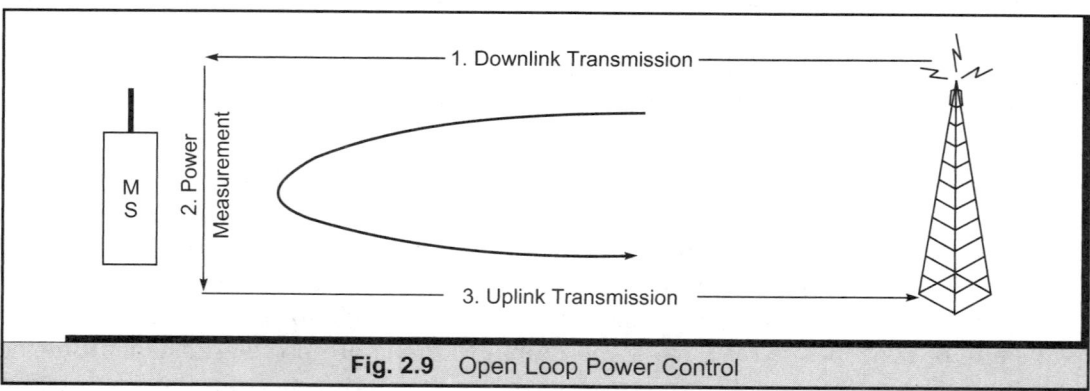

Fig. 2.9 Open Loop Power Control

The Open Loop Power Control mechanism has the advantage that it leads to a quick power adjustment. The process works extremely fast; the signal received in one direction is evaluated in terms of its quality, and the transmission in the other direction is done on the basis of this measurement. However, the Outer Loop Power Control mechanism has a major problem; it assumes that the fading of the signal in the downlink direction can give a fair idea of the fading of the uplink signal. But this is generally not the case with the WCDMA FDD mode, where the uplink and the downlink frequencies used for transmission are different. This difference in frequencies in opposite directions leads to a difference in the fading of signals. Hence, the Open Loop Power Control is generally used in the following situations:

- In the UTRAN TDD mode, where the frequency used in the uplink and downlink direction is the same.
- To provide a coarse initial power setting to be used by the mobile station at the beginning of a connection.

The other type of power control mechanism is known as the Closed Loop Power Control mechanism. Unlike the open loop mechanism, where the measurements made on signals received in one direction are used when transmitting signals in the other direction, the closed loop power control uses measurements made on signals transmitted in one direction for further transmission in the same direction. This mechanism uses a feedback loop, where the receiving entity performs the quality measurements on the received signal, and sends back power adjustment commands to the sending end. If the received quality of the signal is low, the receiving end would inform the sending end to increase the power level for transmission. The reverse is true in case the quality of the received signal is higher than required quality.

For the closed loop power control mechanism, the Signal-to-Interference Ratio (SIR) is used as the basis for measurement of quality of the received signal. A 'SIR^{Target}' value, stored at the receiving end, defines the required quality of the signal by defining the target value for the SIR. If the measured SIR for the received signal falls below the 'SIR^{Target}' value, the sender is notified to increase its power level. Similarly, if the SIR measured for the received signal is higher than the 'SIR^{Target}' value, the sender is notified to 'power down'. The closed loop power control mechanism is depicted in Figure 2.10. It shows the procedure used to control the power in the uplink direction. A similar process can be used to control the power in the downlink direction.

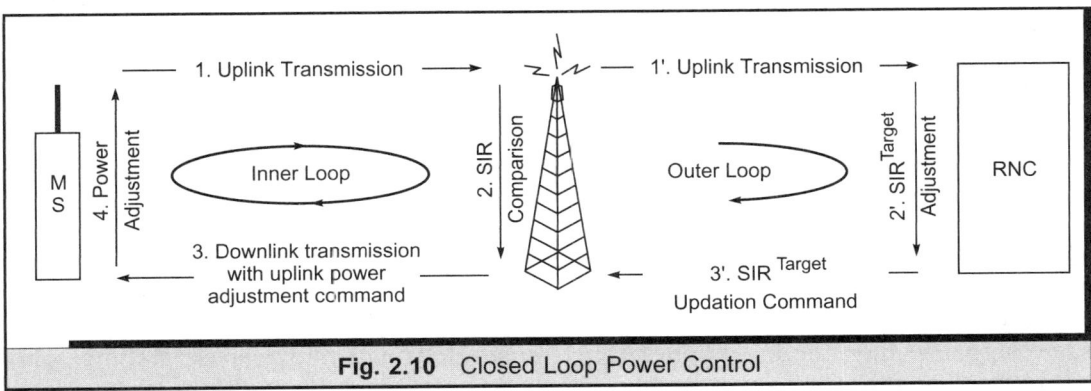

Fig. 2.10 Closed Loop Power Control

Though the closed loop power control mechanism provides more accurate information on power adjustment, its inherent drawback lies in the fact that it uses a feedback mechanism, which is normally slow. To remedy this shortcoming, the process of quality measurement and the sending of feedback power control commands is done for every time slot (of 667 microseconds). Thus, the delay caused in a feedback mechanism is partially overcome by maintaining a faster rate of measurement and feedback. The power control feedback commands in each time slot, in the reverse direction, are sent along with the data transmitted in that direction, as part of the Transmit Power Control (TPC) bit.

The closed loop power control internally consists of two loops: an inner loop and an outer loop. The inner loop power control mechanism (Steps 1 to 4 in Figure 2.10) is responsible for quality measurement of received signals and power control feedback commands, as already discussed. The outer loop power control mechanism is used by the RNC to update the 'SIR^{Target}' value, if required. The base station measures the quality of the received signal as a SIR value. However, a better measurement of the signal quality would be in terms of the Bit Error Rate (BER) or the Block Error Rate (BLER). The job of the RNC is to maintain the BER/BLER at a constant value, so as to maintain the quality. Thus, the RNC calculates the BER/BLER of the signal on the basis of the received signal. In case the RNC determines that the BER/BLER value has increased, thus lowering the quality of transmission, it increases the value of the 'SIR^{Target}', and accordingly notifies the base station, (Steps 1' to 3' in Figure 2.10).

To summarize, the closed loop power control consists of two loops, namely:

- **Outer Loop:** The Outer Loop Power Control mechanism is used by the RNC to update the 'SIR^{Target}' value in the base station. The updation of the 'SIR^{Target}' value is based on the quality measurement performed at RNC using the BER/BLER as the parameter for judgement. A higher BER/BLER value leads to an increase in the 'SIR^{Target}' value, and vice versa.

- **Inner Loop:** The inner loop power control mechanism is used by the base station to send power control feedback commands to the mobile station by comparing the SIR of the received signal with the 'SIR^{Target}' value available with it. If the SIR of the received signal is less than the current 'SIR^{Target}' value, then 'power-up' commands are sent to the mobile station, and vice versa. The Inner Loop Power Control is also called the Fast Closed Loop Power Control mechanism.

2.10 SOFT AND SOFTER HANDOVER

Section 2.8 discussed the concept of macrodiversity, wherein it was stated that the same signal to the mobile station can be transmitted via multiple base stations. This happens when the mobile station is in a region which falls in the coverage area of more than one base station. In this situation, the UTRAN utilizes the services of both the base stations to transmit and receive the signals to/from the mobile station. The state of macrodiversity is achieved as a result of the Soft Handover (SHO) procedure, which is discussed in this section.

Figure 2.11 depicts a scenario where Soft Handover (SHO) procedure is carried out. SHO is typically employed at the boundaries of cells, where the regions covered under two or more cells overlap. In this state, the mobile station maintains simultaneous radio links with more than one Node-B. The Node-Bs to which the mobile station is thus connected is called its 'Active Set'. The transmission to/from the mobile station takes place using both the radio links simultaneously. Within the mobile station, a Rake receiver is used to combine the signal received from the two links. Note that this case is different from the case where multiple signals are received as a result of Multipath propagation. In multipath propagation, the multiple signals received are time-delayed versions of the original signal, and are spread using the same spreading code. However, in the SHO scenario, the signals are received over different links, from different Node-Bs, and are spread using different spreading codes. In the uplink direction, the base stations forward the signals received over the radio links to the RNC. These signals are then combined at the RNC.

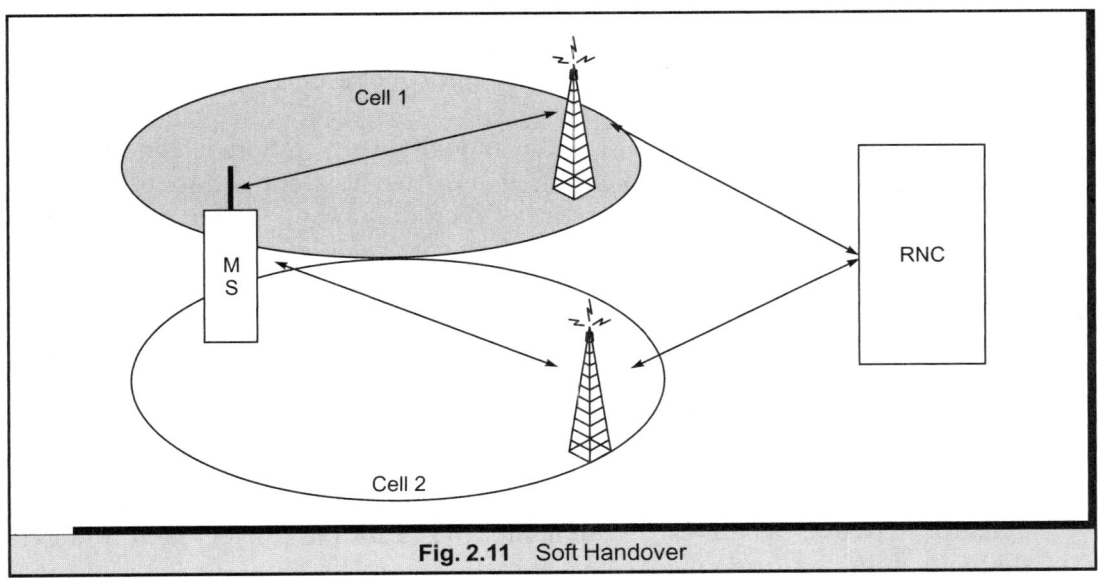

Fig. 2.11 Soft Handover

Soft Handover is carried out for reasons similar to those given for power control. Just before the procedure starts, the mobile station is placed in the region of two base stations. However, it has a radio link with only one of these base stations. If only one base station were to power control the mobile station, then being on the cell boundary, the base station would direct the mobile station to increase its power level. This would result in interference with the transmissions in the other base station's cell. Practically though, transmissions at such high power levels are not required, provided the services of both the base stations can be used. If transmission to/from the mobile station can be made through both the base stations, then low power transmissions through these base stations can be combined at the mobile station/RNC to produce a good quality signal. Thus, the advantage of using SHO lies in the fact that by using multiple radio links, the transmission power to/from the mobile station can be kept in control. Thus, after the Soft Handover is carried out one power control loop is kept active for each base station to which the mobile station is connected. Each base station can then power control the mobile station, and ensure that the latter's transmission does not interfere with other transmissions in either cell.

The flip side of SHO is that in this state, a UE consumes more network resources than required with a single connection to the network. This is because the same bits of information transmitted uplink and downlink between the UE and the RNC would go through multiple paths (from different base stations) in duplicate, thus requiring more RAN resources. The signals received from multiple paths would then be combined in the RNC/mobile station to produce a good quality signal. It is this trade-off between network resources and signal quality that is to be considered before the SHO is carried out. The UTRAN must decide when the UE should enter the SHO state in order to receive the additional signal gain at the cost of network resources. The UE must not enter the SHO state independently.

A concept similar to SHO is that of a 'Softer Handover'. In Softer Handover, the mobile station is located in the overlapping area of two adjacent sectors of the same base station (Figure 2.12). Communication between the mobile station and the base station takes place via two radio links, one for each sector. The combining of the signals takes place internally within the base station. The RNC is not affected as a result of this condition. From the mobile station's perspective, the process is similar to that followed in the SHO scenario. However, unlike SHO, only one power control loop is maintained between the mobile station and the base station in the case of a Softer Handover.

Besides SHO and Softer Handover, the WCDMA networks also provide support for other handovers. These include:

- **Inter-frequency Hard Handovers (HHO):** This type of handover is used in operator networks where each base station has multiple carriers, and a mobile station is migrated from one frequency carrier to another.

- **Inter-system Handovers:** These handovers take place between the WCDMA networks and the GSM based networks. They may also take place between WCDMA FDD and TDD networks, or between WCDMA and GSM networks.

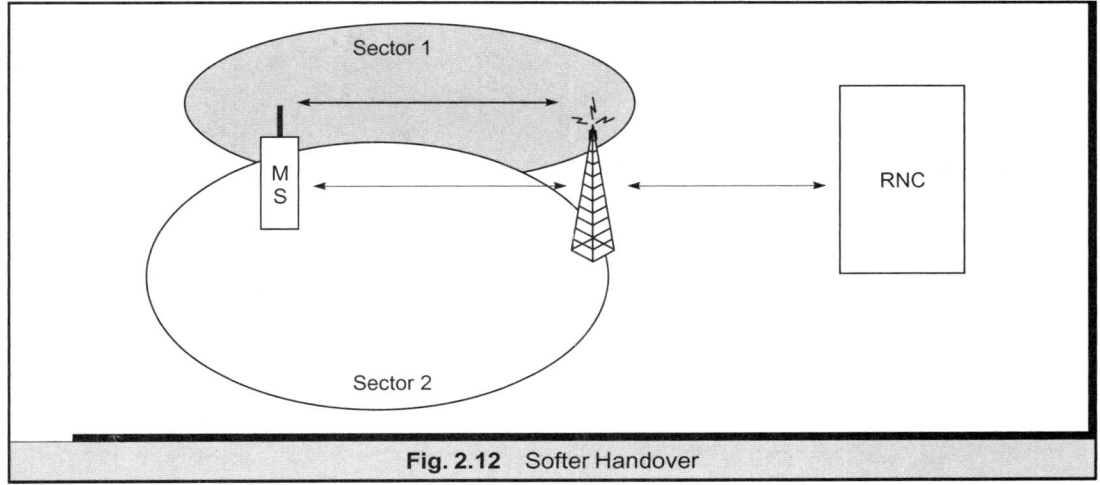

Fig. 2.12 Softer Handover

Procedures for carrying out the handovers within the UTRAN are discussed in detail in Chapter 8.

2.11 SRNS RELOCATION

Another important concept in WCDMA networks is that of SRNS Relocation. In UMTS terminology, the RNC and its associated Node-Bs (base stations) are collectively called a Radio Network Subsystem (RNS). At any point in time, one RNC entity is responsible for maintaining the connection of the mobile station with the UTRAN and the Core Network. This RNC is called the Serving RNC (SRNC) for the mobile station. The RNS to which this RNC belongs is called the Serving RNS (SRNS). This section discusses the procedure involved in the relocation of SRNS when the mobile station is on the move (Figure 2.13).

The following steps are involved in this relocation process:

(a) Initially, the mobile station is assumed to have carried out the SHO procedure; it is simultaneously serviced by base stations of cells 1 and 2. Both base stations are associated with RNC-1, which is also the serving RNC for the mobile station.

(b) As the mobile station moves, it enters into a region that is an overlap of cells 2 and 3. The base stations serving these cells, and the mobile station, are connected to different RNCs: RNC-1 (also the SRNC for the mobile station) and RNC-2. In

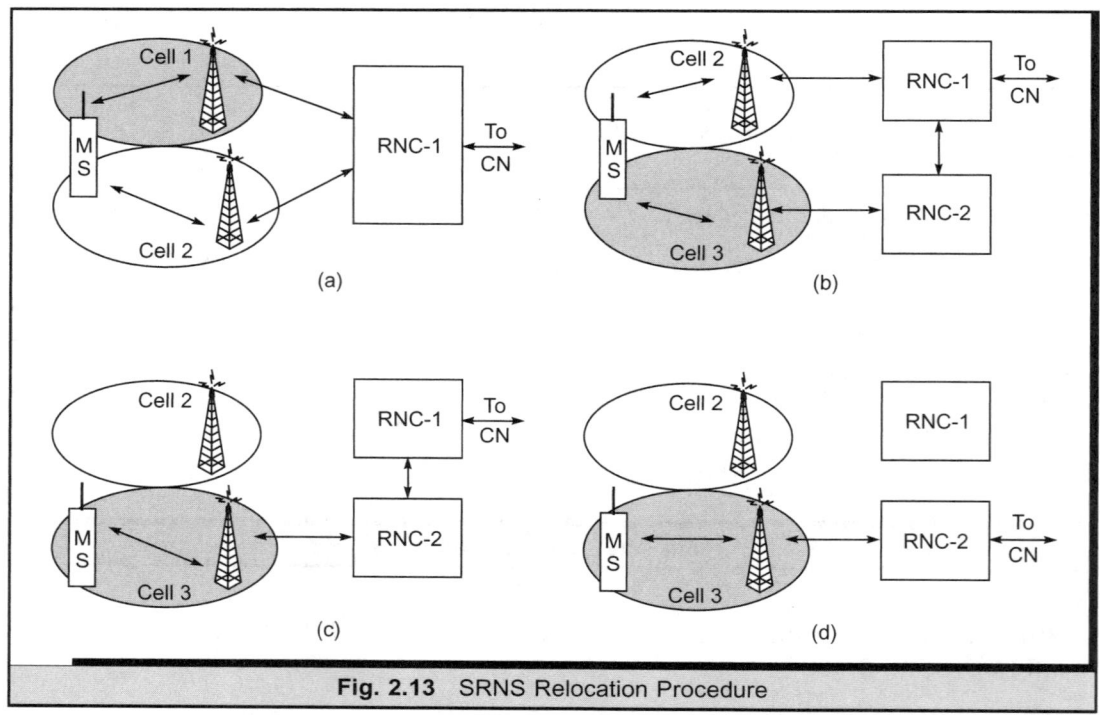

Fig. 2.13 SRNS Relocation Procedure

UMTS terminology, RNC-2 is called the Drift RNC (DRNC) for the mobile station. Traffic received by RNC-2 is forwarded to the SRNC. The SRNC is responsible for combining the traffic received from its base station and the DRNC, and then forwarding it to the CN.

(c) As the mobile station continues to move, it comes entirely within the region of cell 3. All traffic to/from the mobile station goes through the base station of cell 3. However, instead of following a direct path, the traffic to/from the mobile station to the Core Network goes through the DRNC and the SRNC. Since the mobile station is no longer in the region of the SRNS, it is advisable to relocate the SRNC (and SRNS) of the mobile station to RNC-2. This would reduce the network resource requirements for the mobile station, since the bandwidth used in forwarding the traffic between the DRNC and the SRNC would no longer be required.

(d) The SRNS Relocation procedure is carried out. RNC-2 is made the new SRNC for the mobile station. All traffic from the mobile station to the Core Network takes the direct path via the base station of cell 3 and RNC-2. This completes the relocation of the SRNS.

A detailed description of the procedure followed within the UTRAN for the SRNS relocation is provided in Chapter 8.

SUMMARY

The Second Generation (2G) mobile networks were developed to provide voice services to mobile subscribers. However, a need to provide multi-media services and higher bandwidths to the mobile subscribers has currently risen. To fulfil these demands, it requires a migration from the existing 2G radio access scheme towards a scheme that can offer higher bandwidths. 3GPP has defined WCDMA as the radio access scheme to be used within the Third Generation mobile network. Two modes of operations are supported in WCDMA.: FDD mode for the paired spectrum and TDD mode for the unpaired spectrum. It is presumed that the initial deployments of WCDMA will use the FDD mode of operation. On the other hand, TDD mode implementations will be used in hot spots like airports, to provide high data rate connectivity at these spots.

To support multi-user access to the radio resource, WCDMA uses a process called 'Spreading'. Spreading involves transformation of the user signal by multiplication with bits of the Spreading Code, which is unique for each user in a cell. Besides Spreading, WCDMA-based networks also use the process of Scrambling, to reduce the interference between different signals in both uplink and downlink direction. Scrambling is carried out using codes known as Scrambling Codes. Other concepts related to WCDMA networks include Multipath Diversity and Macrodiversity, Power Control, Handovers and Relocation, which have been already discussed in the chapter.

Part

2

UMTS Network Architecture and Protocols

This part of the book discusses the UMTS network architecture and the associated protocols.

The first chapter in this part, Chapter 3, provides an overview of the subject; its objective is to familiarize the reader with some of the basic concepts of the UMTS network, as subsequent chapters are built upon these concepts. The chapter starts with a discussion on the basic structure of the UMTS network, which comprises of three logical parts: the User Equipment (UE), the Access Network (AN) and the Core Network (CN). This discussion is followed by the stratification of the UMTS network into Access Stratum (AS) and Non-Access Stratum (NAS). The concept of AS and NAS is important in understanding the message flows between UE and the Access Network and between UE and the Core Network. The hierarchical organization of UMTS network—which divides the UMTS PLMN into Location Area, Routing Area, UTRAN Registration Area and Cells—is discussed next. After this discussion, the various addresses/identifiers used in UMTS are elaborated upon. The next topic is service aspects and service classification in the UMTS network. A brief reference to Bearer service, Teleservice and Supplementary Service is also provided. Then, the UMTS QoS Architecture is discussed. Finally, the four QoS classes in UMTS, namely the Conversational class, the

Streaming class, Interactive class and Background class are elaborated upon.

Chapter 4 discusses User Equipment (UE). It begins with a description of the various components of UE. These components include the Universal Subscriber Identity Module (USIM), Mobile Termination (MT) and Terminal Equipment (TE). This is followed by a discussion on the various interfaces of UE, i.e. the interface of UE with the Access Network, and the internal interfaces between the various UE components. Next, the functions of UE are explained. Finally, the User Equipment protocols, which comprise of Access Stratum and Non-Access Stratum protocols, are discussed.

Chapter 5 discusses the Access Network and also how it interfaces with the UE on the air interface side and the Core Network on the network side. To explain the various aspects of Access Network, the chapter first details the various entities that form the Access Network. These entities are: Base Station Controller (BSC) and Base Transceiver Station (BTS) in the Base Station Sub-system (BSS); and Radio Network Controller (RNC) and Node B in the Radio Network Sub-system (RNS). The chapter then explains the various interfaces. The focus is on RNS, which is relatively new as compared to the BSS. The protocol architecture applicable over these interfaces is explained followed by the functions of

Access Network. The functions include user data transfer, system access control, security management, mobility management and radio resource management. Finally, the Access Network protocols are explained. The protocols used in UTRAN can be classified into three distinct categories, namely the 'radio interface' protocols, 'transport network layer' protocols and 'application layer' protocols. The important radio interface protocols are Radio Link Control (RLC), Medium Access Control (MAC) and Radio Resource Control (RRC). The ATM-based transport network protocols include the ATM layer, adaptation layers and the ATM signaling protocols residing over it. The application layer protocols include Radio Access Network Application Part (RANAP), Radio Network Sub-system Application Part (RNSAP) and Node B Application Part (NBAP), among others. The focus of this chapter is on RNS, which was introduced in Rel99 for 3G Universal Terrestrial RAN (UTRAN).

Chapter 6 explains the various aspects of the Core Network. It first focuses on its various entities, which fall under four categories. The first category includes entities that are common to the CS and the PS domain (these include HLR, AuC, EIR and SMS-entities). Apart from these, there are entities like MSC and VLR that are specific to the CS domain. The third category includes entities like SGSN, GGSN and Border Gateway that are specific to the PS domain. Besides the common, CS-specific and PS-specific entities, there are some service-specific entities. The entities included in this category are Gateway Mobile Location Center (GMLC), CAMEL entities and Cell Broadcast Center (CBC). The various interfaces between these Core Network entities are explained next. To make presentation easy, the interfaces of CS and PS domains are explained separately. Thereafter, the chapter covers the protocol architecture applicable over these interfaces. The Core Network functions are taken up next. These include mobility management, call handling, security functions and message service functions, among others. Finally, the Core Network protocols are briefly explained. They are divided into two broad categories: First, the SS7 protocols that include MTP, SCCP and ISUP; second, the application protocols that include TCAP/MAP, GTP and BSSAP+. The NAS Signaling protocol between UE and Core Network is also covered under the application protocols.

❏❏

3

UMTS NETWORK ARCHITECTURE

3.1 INTRODUCTION

This chapter provides an overview of the UMTS network architecture. Its objective is to familiarize the reader with some of the basic concepts of UMTS network, as subsequent chapters are built upon these concepts.

The chapter starts with a discussion on the basic structure of UMTS network, which divides the network into three logical parts: User Equipment (UE), Access Network (AN) and the Core Network (CN). Each of these parts is discussed in detail in the next three chapters, i.e. Chapters 4, 5, and 6 respectively.

The discussion on the logical structure of UMTS network is followed by the stratification of UMTS network into Access Stratum (AS) and Non-Access Stratum (NAS). The concept of AS and NAS is important in understanding the message flows between UE and Access Network as well as between UE and the Core Network.

The hierarchical organization of UMTS network, which divides the UMTS PLMN into Location Area, Routing Area, UTRAN Registration Area and Cells, is discussed next. Thereafter, the various addresses/identifiers used in UMTS are elaborated, upon. The next topic is service aspects and service classification in the UMTS network, with brief mention of Bearer service, Teleservice and Supplementary Service. Then, the UMTS QoS Architecture is discussed. Finally, the four QoS classes in UMTS, namely the Conversational class, Streaming class, Interactive class and Background class are elaborated upon.

3.2 BASIC STRUCTURE OF UMTS NETWORK

A typical UMTS network can be modeled on its three basic parts or *sub-systems*, namely User Equipment (UE), Access Network (AN) and the Core Network (CN). This basic model of the UMTS network is depicted in Figure 3.1.

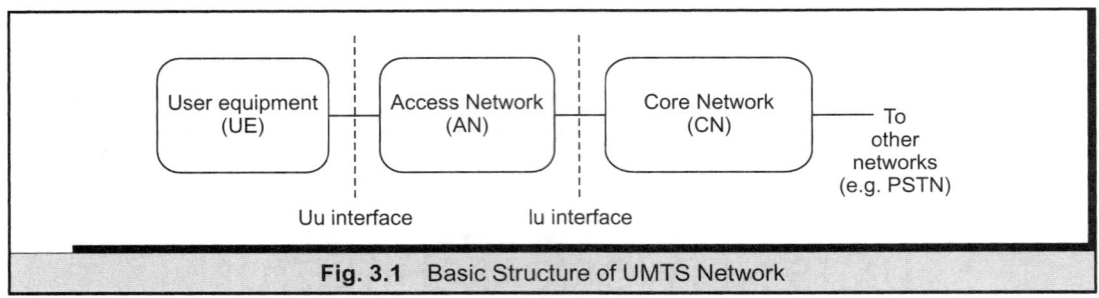

Fig. 3.1   Basic Structure of UMTS Network

The User Equipment (UE) is used by a subscriber/user to access the services provided by the network. To connect to the network, a UE interfaces with the Access Network using the WCDMA air interface, which is referred to as the Uu interface. Two modes of operation are used over the Uu interface: the Frequency Division Duplex (FDD) mode for the *paired spectrum* and the Time Division Duplex (TDD) mode for the *unpaired spectrum*. These modes of operation were discussed in Chapter 2.

The Access Network (AN) performs functions specific to the radio access technique. In case of UMTS, the Access Network performs functions specific to the WCDMA air interface. The Access Network has two different types of entities: the Base Transceiver Station (BTS) that terminates the radio connection with the UE, and a Base Station Controller (BSC) that controls the resources of the BTS. BSC and one or more BTS collectively form the Access Network. The BSC interfaces with the Core Network over the Iu interface.

The Core Network (CN) performs the core functions of the network, which include mobility management, call control, switching and routing. The Core Network also manages the subscription information of a subscriber and provides services based on this information.

The basic structure of the UMTS network is similar to that used in any wireless network. In particular, it is modeled on the lines of GSM/GPRS network architecture. Thus, at the architectural level there are many similarities between the two. However, the actual protocols residing on these entities are quite different. This difference is created by the introduction of WCDMA-based air interface in the UMTS Access Network, leading to significant changes in the protocols residing at the User Equipment and the Access Network. Thus, the GSM/GPRS mobile handsets are rendered useless in a UMTS environment (unless they are backward compatible). In contrast, the Core Network of GSM/GPRS is almost entirely reused in the UMTS. Even though there are changes and enhancements in Core Network protocols, the main network entities (e.g. HLR, VLR, SGSN and GGSN) and the important Core Network protocols (e.g. MAP, GTP and ISUP) exist in the UMTS as well. This implies that upgrading the Core Network to make it compliant with UMTS standards is easier as compared to upgradation of the UE or the Access Network.

The following sub-sections provide the details of the User Equipment (UE), the Access Network (AN) and the Core Network (CN).

3.2.1 User Equipment (UE)

The User Equipment (UE) is a device used by a subscriber/user to access network services. To make its design modular, the UE is divided into two logical parts: the Mobile Equipment (ME) and the Universal Subscriber Identity Module (USIM). The logical structure of UE is depicted in Figure 3.2.

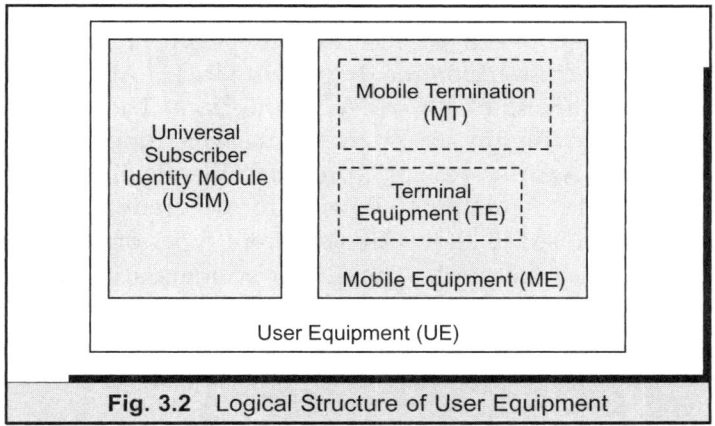

Fig. 3.2 Logical Structure of User Equipment

The Mobile Equipment (ME), or the Mobile handset, is manufactured by equipment vendors. The ME is further divided into two distinct functional groups, namely, Mobile Termination (MT) and Terminal Equipment (TE). The MT performs functions like radio transmission termination, authentication, and mobility management. The TE manages the hardware (e.g. speaker, microphones, video cameras, and user display) and hosts user applications (e.g. Web browser). The division of Mobile Equipment into MT and TE is also referred to as *MT-TE functionality split*. As an example, a Mobile Termination (MT) unit may be physically connected to a Laptop (which acts as a TE). The same MT may also provide services to other Terminal Equipment like a camera, using a Bluetooth interface.

Besides the Mobile Equipment (containing the MT and the TE), the User Equipment also contains a Universal Subscriber Identity Module (USIM) application. The USIM contains the logic required to unambiguously and securely identify the user. In particular, it contains the permanent identity of the user (called the IMSI), the shared secret key (used for authentication), phone book and a host of other information. The USIM application resides on a Smart Card that can be inserted or removed from the ME. The smart card is called the UMTS Integrated Circuit Card (UICC). The USIM on the UICC card is provided by the service provider. Hence, even if the UICC card is moved from one ME to another, the service provider and the service configuration remains the same.

Chapter 4 provides the details of the User Equipment (UE). The topics covered in the chapter include the logical structure of UE (especially the MT-TE functionality split) and details of the USIM application.

Note that in this book, the User Equipment (UE) is also referred to as Mobile Station (MS). Thus, the terms UE and MS are used interchangeably.

3.2.2 Access Network (AN)

Access Network resides between the UE and the Core Network. It performs the functions specific to the access technique. In case of UMTS, Access Network performs functions specific to the access of WCDMA air interface. The Core Network, on the other hand, may be used with any access technique. This functional split between the Core Network and the Access Network provides the flexibility to keep the Core Network fixed, while at the same time allowing for different access techniques.

The Access Network in UMTS allows two different types of access network systems to interface with the Core Network. These two systems are the Base Station Sub-system (BSS) and the Radio Network Sub-system (RNS). While BSS is the legacy of the GSM era, the RNS is the newly standardized access network for UMTS networks following Rel99 onwards 3GPP specifications. The Core Network can connect to one or both of these Access Network types.

Both types of access systems (i.e. the BSS and the RNS) have a similar structure. They comprise of a Base Station Controller (BSC) and one or more Base Transceiver Station (BTS). The BTS terminates the radio connection with the UE; the BSC controls one or more BTS. The nomenclature of BTS and BSC is specific to Base Station Sub-system (BSS). In Radio Network Sub-system (RNS), the BTS is referred to as Node B, while the BSC is referred to as Radio Network Controller (RNC). The Radio Network Sub-system is also known as the universal Terrestrial Radio Access Network (UTRAN). The Access Network comprising of BSS and RNS is depicted in Figure 3.3.

Chapter 5 provides details of the Access Network, especially the UTRAN. The topics covered in the chapter include UTRAN network architecture, the network entities (e.g. Node B and RNC), the interfaces (e.g. Iu/Iur/Iub interface), the functions performed by and the protocols used in Access Network (e.g. RRC and RANAP).

3.2.3 Core Network (CN)

In the preceding sections, the functions and structure of User Equipment (UE) and Access Network (AN) were discussed. The UE provides an interface to the end user.

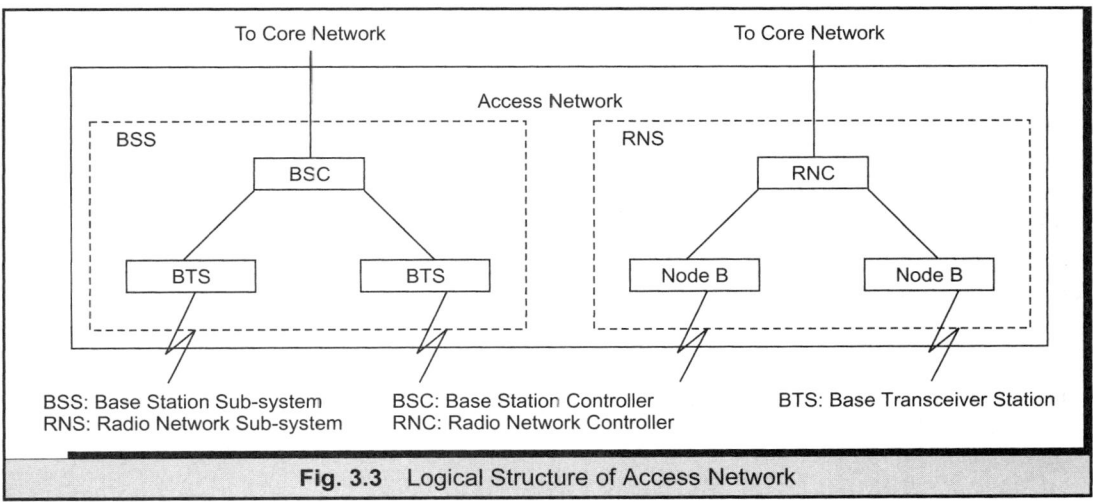

Fig. 3.3 Logical Structure of Access Network

Thus, its functions are limited to terminating the radio interface and hosting user applications. The Access Network also provides limited functions; its scope is restricted to managing radio connection with UE and associated radio resources. Thus, apart from these two entities, there is a clear need for a sub-system that would perform the following functions:

- **Mobility Management:** This refers to tracking the location of the UE. In a mobile network, where the position of the UE is not fixed, this is a very important function.

- **Call Control:** This refers to establishment and release of voice call between the UE and an end-point. Here, the end-point may be another UE or even a point outside the mobile network (e.g. a fixed telephone of a PSTN). Whatever be the case, a call control function is required to establish/release a voice connection between UE and an end-point.

- **Switching:** This refers to switching a voice call between the UE and an end-point. The switching function is performed after a voice connection is established.

- **Session Management:** This refers to establishment and releases of sessions for data transfer between UE and an end-point. The end-point is typically outside the mobile network (e.g. an Internet Server).

- **Routing:** This refers to routing of data packets between the UE and an end-point.

- **Authentication:** This esures that the user availing the service is authenticated.

- **Equipment Identification:** This ensures that the handset through which services are availed is genuine (and not stolen).

The above are just some of the functions that are performed by the Core Network (CN). In simple terms, the Core Network consists of the entities that provide support for various network features and services and performs functions like mobility management, call control, switching, session management, routing, authentication and equipment identification.

It is evident from above that there are two classes of traffic handled by the Core Network, namely, *voice and data*. The 2G mobile networks, like GSM networks, were designed primarily for voice. The GPRS networks provided capability for data transfer. Based on the fact that the UMTS Core Network is an evolved GSM/GPRS core network, the former is divided into two domains: the Circuit Switched (CS) domain and the Packet Switched (PS) domain. The CS domain provides services related to voice transfer, the PS domain to those related to data transfer. The entities of the Core Network and its decomposition into the CS and PS domain is depicted in Figure 3.4.

The CS domain uses Circuit-Switched (CS) connections for communication between UE and the destination. A CS connection is defined as *a connection for which dedicated network resources are allocated at the time the connection is established and are freed when the connection is released.* An example of CS connection is the connection established in the PSTN network during a telephonic conversation.

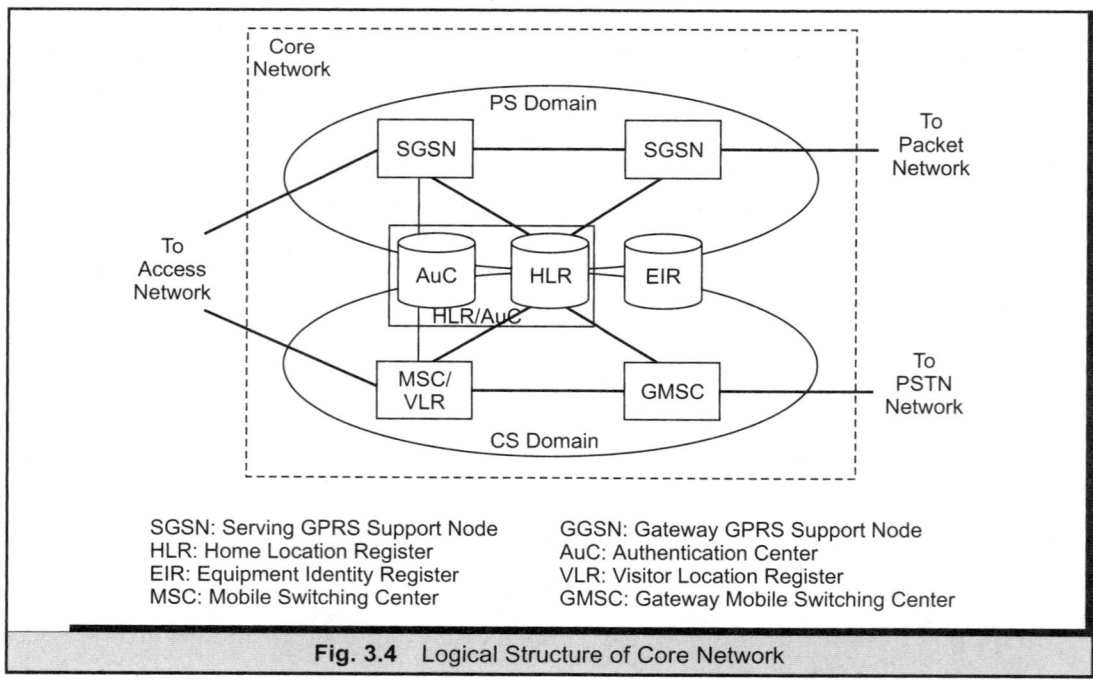

SGSN: Serving GPRS Support Node   GGSN: Gateway GPRS Support Node
HLR: Home Location Register   AuC: Authentication Center
EIR: Equipment Identity Register   VLR: Visitor Location Register
MSC: Mobile Switching Center   GMSC: Gateway Mobile Switching Center

Fig. 3.4 Logical Structure of Core Network

To establish/release CS connections and to switch voice streams, a switching entity is required. For this, the CS domain has the Mobile Switching Center (MSC). Alongside the MSC, there is another entity in the CS domain, called the Visitor Location Register (VLR). The VLR contains the subscriber profile obtained from the Home Location Register (HLR). MSC queries the VLR for subscriber information and provides services to the subscriber based on the queried information. It is customary to represent MSC and VLR as one entity: MSC/VLR. Apart from MSC/VLR, the CS domain has the Gateway Mobile Switching Center (GMSC). The GMSC provides connectivity to external CS networks (including the CS domain of other UMTS networks and the PSTN networks).

The PS domain uses Packet-Switched (PS) connections for communication between UE and the destination. A PS connection is defined as *a connection that transports the user information using autonomous concatenation of bits called packets; each packet is routed independently from the previous one.* An important aspect of PS connection is that resources are not reserved for a connection; rather, they are shared between various communicating entities. This sharing of resources results in better resource utilization. An example of PS connection is the connectionless transfer of IP datagrams in the Internet.

To route packets in the PS domain, a routing entity is required. For this, the PS domain has the Serving GPRS Support Node (SGSN). Unlike in the CS domain, where one entity holds the database (VLR) and another switches CS the connections (MSC), in PS domain, the SGSN performs both the functions. Apart from SGSN, the PS domain has the Gateway GPRS Support Node (GGSN) which performs functions similar to those of GMSC (i.e. GGSN provides connectivity to external PS networks).

Apart from entities belonging to the CS and PS domain, there are entities that are common to both the domains. Important among these is the Home Location Register (HLR) that is located in the subscriber's home network. The HLR holds the permanent and subscribed information of the subscriber. The permanent information includes the permanent identity of the user (called the IMSI). The subscribed information includes information about the services that are provisioned in the HLR, based on the services subscribed to by the subscriber.

Then, there is the Authentication Center (AuC), which holds authentication information. This information is used for authentication and other security-related functions. It is customary to represent the AuC as a part of HLR. Thus, the term AuC/HLR is used to represent the entity that performs the functions of HLR and AuC.

The entities common to CS and PS domain also include Equipment Identity Register (EIR). The EIR monitors the legitimacy of a User Equipment (UE) used in the UMTS network.

Apart from HLR, AuC and EIR, there are few other common entities like SMS Gateway MSC (SMS-GMSC) and SMS Interworking MSC (SMS-IWMSC). There are also service specific entities like Gateway Mobile Location Center (GMLC), Camel entities and Cell Broadcast Center (CBC). For sake of simplicity, these entities are not depicted in Figure 3.4.

Chapter 6 provides details of the Core Network. The topics covered include Core Network architecture, its entities (including the common entities, the CS and PS domain entities and service-specific entities), interfaces (for both CS and PS domain), the functions performed by Core Network and the protocols used in it (e.g. MAP, GTP and ISUP).

Apart from the CS and PS domain, another sub-system, called the IP Multimedia Sub-system (IMS), is introduced in Rel5 specifications. The IMS uses the services of PS domain for providing IP based multimedia services. To avoid complication, the IMS too is not depicted in Figure 3.4. Chapter 16 provides the details of this sub-system.

3.2.3.1 Domain Split in Core Network

The CS and PS domain divide the Core Network on the basis of its functionality. Another way to divided CN is based on its position *with respect to the user*. This classification divides the Core Network into *Serving Network Domain, Home Network Domain and Transit Network Domain* (see Figure 3.5).

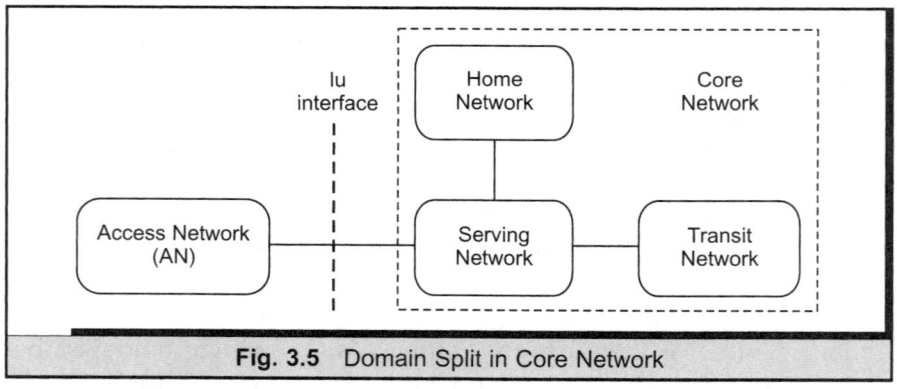

Fig. 3.5 Domain Split in Core Network

The serving network domain is defined as that part of the Core Network which is connected to the access network currently providing access to a user. Thus, the serving network is defined in the context of a particular user. It is responsible for switching and routing calls/packets (i.e. transfer of user information from source to destination), for which it interacts with the home network to obtain subscriber information. It is possible that the serving network is the home network, in which case it already has the subscriber information. However, it is clear that the serving network need not necessarily be the home network. The serving network changes with the change in location of the user.

The home network, in contrast, performs functions independent of the location of the user. The home network contains the permanent data of the user (e.g. permanent subscriber identity) and subscriber information. Thus, it is responsible for the management of subscription information of the user.

The transit network domain is an optional part in communication between source and destination. It is required when the destination party is outside the serving network. Thus, the transit network lies between the serving network and the destination network.

3.3   ACCESS STRATUM AND NON-ACCESS STRATUM

One way of modeling the UMTS network is to divide it into UE, Access Network and Core Network. A different way of modeling it is by dividing it into Access Stratum (AS) and Non-Access Stratum (NAS). The AS protocols provide the means to carry information over the air interface as well as to manage the resources of the air interface. In contrast, the NAS protocols are those that apply between UE and the Core Network, for which the access stratum acts as a relay (see Figure 3.6).

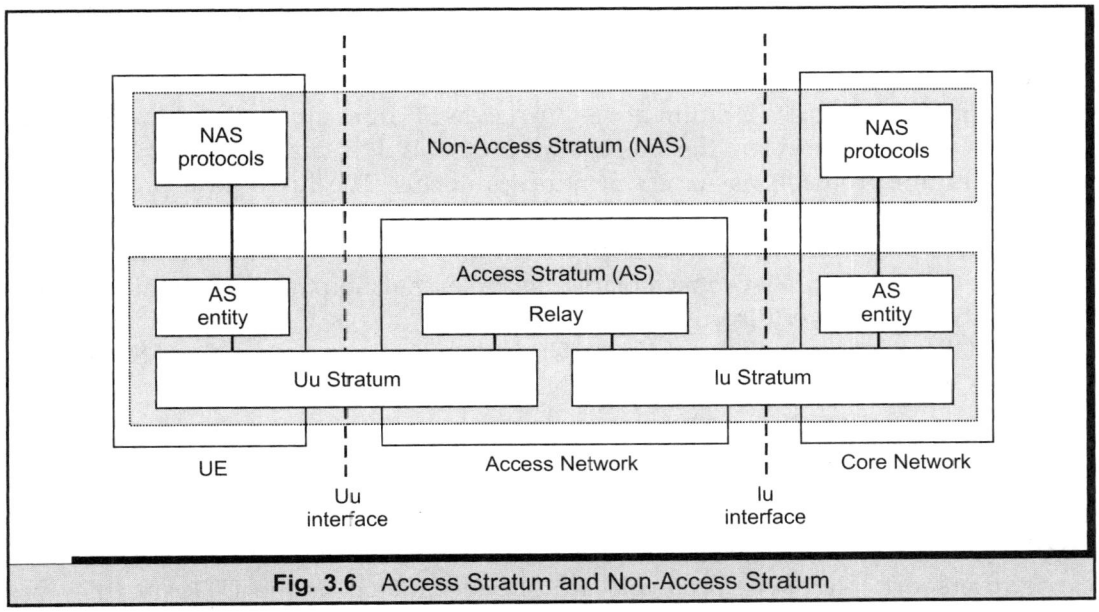

Fig. 3.6   Access Stratum and Non-Access Stratum

3.3.1 Access Stratum (AS)

The Access Stratum (AS) provides the means to carry information over the air interface and also to manage the resources of the air interface. It contains parts of the UE and parts of the Access and Core Network.

The Access Stratum is further divided into two distinct components: the *Uu stratum* and the *Iu stratum*.

The AS uses the Uu stratum for communication between the UE and the Access Network. The Uu stratum is used to manage the radio resources between the UE and the Access Network. The Uu stratum protocols include Medium Access Control (MAC), Radio Link Control (RLC), Broadcast/Multicast Control (BMC), Packet Data Convergence Protocol (PDCP) and Radio Resource Control (RRC) protocol. Among these, the RRC is the main signaling protocol between UE and Access Network (in particular, between the UE and the RNC).

Chapter 5 provides the details of the Uu stratum protocols and Chapter 7 the details of the Uu stratum procedures (specifically, the RRC procedures).

The AS uses the Iu stratum (AN-CN interface) for communication between the Access Network and the Core Network. The Iu stratum is used by the Core Network to manage the resources provided by the Access Network to the UE. The Radio Access Network Application Part (RANAP) is the main Iu stratum protocol used between RNC and MSC/VLR and between RNC and SGSN. Chapter 5 provides details of the RANAP protocol and Chapter 8 the details of the Iu stratum procedures.

In simple terms, the Access Stratum (AS) provides services to the Non-Access Stratum (NAS). One of the important services provided by AS is to transport NAS messages between NAS entities.

The protocols in Access Stratum and Non-Access Stratum are depicted in Figure 3.7. The boxes shaded in gray are the Uu stratum protocols. It is customary to refer to only the Uu stratum protocols as Access Stratum protocols. The Iu stratum protocols are seldom referred to as part of Access Stratum protocols. This can best be explained if the AS and NAS protocols are viewed from the UE point of view. For a UE, the Iu interface is not visible because it can be replaced by a relay function that delivers all messages received by the AS layer directly to the NAS layer protocols. Given this, the protocols between UE and Access Network (over Uu interface) form part of the AS protocols and those between the UE and the Core Network form part of the NAS protocols.

3.3.2 Non-Access Stratum (NAS)

The Non-Access Stratum (NAS) protocols are those that apply between UE and the Core Network. For these protocols, the access stratum (i.e. the UTRAN) acts as a carrier/transport. The NAS protocols are not terminated at the UTRAN (they are terminated at the Core Network).

The NAS protocols are depicted in Figure 3.7. There is a Mobility Management (MM) layer for CS domain and GPRS Mobility Management (GMM) layer for PS domain. The MM/GMM procedures enable mobility of user terminals, such as keeping track of the subsciber's present location. During mobility management, it is also ensured that the identity of the user is kept confidential and that only authenticated users can avail network services. Chapter 9 provides details of the Mobility Management (MM) and GPRS Mobility Management (GMM) procedures.

A number of protocols reside over the MM/GMM layer. This includes the Call Control (CC), Session Management (SM), Supplementary Service (SS) and Short Message Service (SMS) protocols.

The Call Control procedures involve handling of the mobile-originated (MO) and mobile-terminated (MT) calls. These procedures, also referred to as Call Handling (CH) procedures, are detailed in Chapter 10.

In the PS domain, there is no concept of calls. Hence, the call handling procedures are not applicable here. However, there is an analogous concept, which is termed as *sessions*. A session can be viewed as a context maintained by the UE and the SGSN/ SGSN for information exchange in the PS domain. Chapter 11 provides details of the Session Management (SM) procedures.

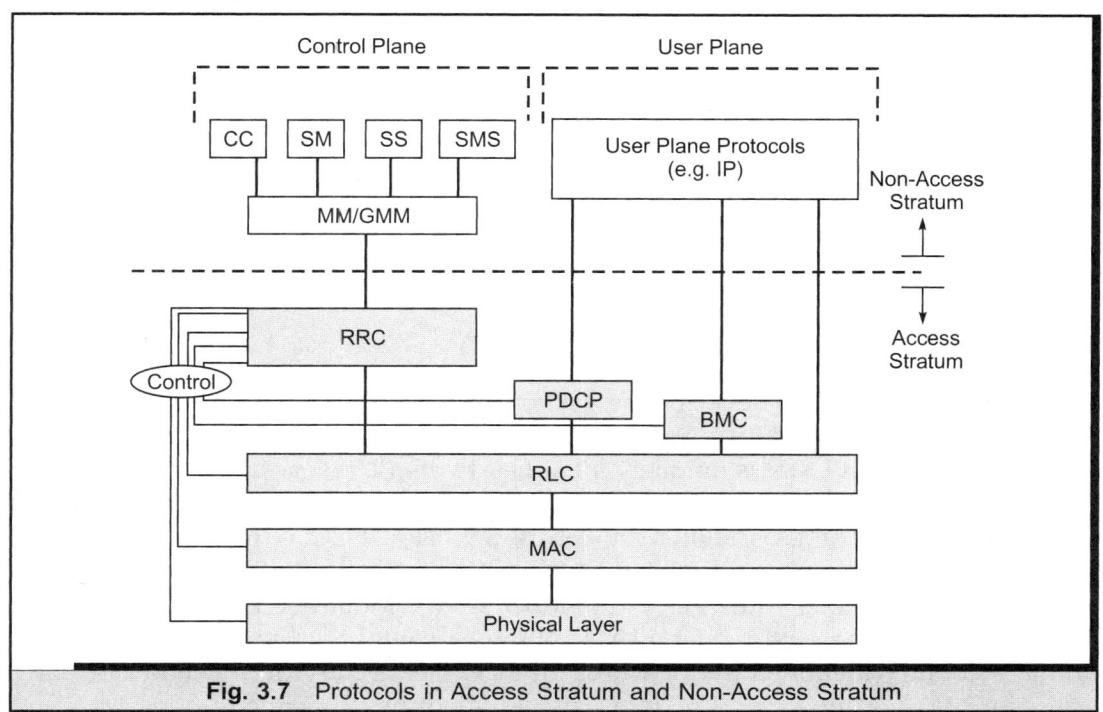

Fig. 3.7 Protocols in Access Stratum and Non-Access Stratum

Supplementary Services (SS) are modification or supplement to the basic services. Examples of SS are Call Forwarding and Call Barring. Chapter 12 provides details of the SS procedures.

Short Message Service (SMS) is the means by which a short text (of up to 160 characters) can be exchanged between the UE and a Short Message Service Center (SMSC). Chapter 13 provides details of the SMS procedures.

3.4 HIERARCHICAL NETWORK ORGANIZATION

To support mobility of a subscriber from one location to another, the UMTS network architecture is organized as a multi-tier hierarchical structure. This hierarchical structure enables a particular network entity to have only that much information as is required for its functioning. For example, a VLR has information only of the Location Area of an MS but does not know the exact cell location. Despite this, the exact cell location can be determined through paging when required. The hierarchical division is done to reduce the storage/processing load on the network entities (like the VLR), save radio resources and battery consumption of the MS.

The hierarchical structure of UMTS network is depicted in Figure 3.8. As shown in the figure, at the lowest level of UMTS hierarchy is the cell. At the next level is the UTRAN Registration Area (URA), which is a collection of cells. Then comes the Location Area (LA) and the Routing Area (RA). A collection of one or more cells forms a Location Area (LA) for the CS domain, and a Routing Area (RA) for the PS domain. Further, a Location Area contains one or more Routing Areas and a Routing Area contains one or more UTRAN Registration Areas. At the highest level of the hierarchy is a Public Land Mobile Network (PLMN), which is not depicted in the figure.

The hierarchical structure of UMTS network is explained in greater detail in the following sub-sections.

3.4.1 Public Land Mobile Network (PLMN)

In UMTS, at the highest level of the hierarchy is a Public Land Mobile Network (PLMN). A PLMN is defined as a telecommunications network providing mobile cellular services. A PLMN is uniquely identified by its *PLMN identifier*.

The PLMN identifier comprises of Mobile Country Code (MCC) and Mobile Network Code (MNC), as shown in Figure 3.9. The MCC is of three digits and identifies the country to which the Public Land Mobile Network (PLMN) belongs. The next two or three digits of the PLMN identifier are the Mobile Network Code (MNC). The MNC identifies a particular PLMN within a country. It is recommended that within a country identified by the MCC, all PLMN either use only two or else three digits for MNC. A mixture of the two schemes is not recommended.

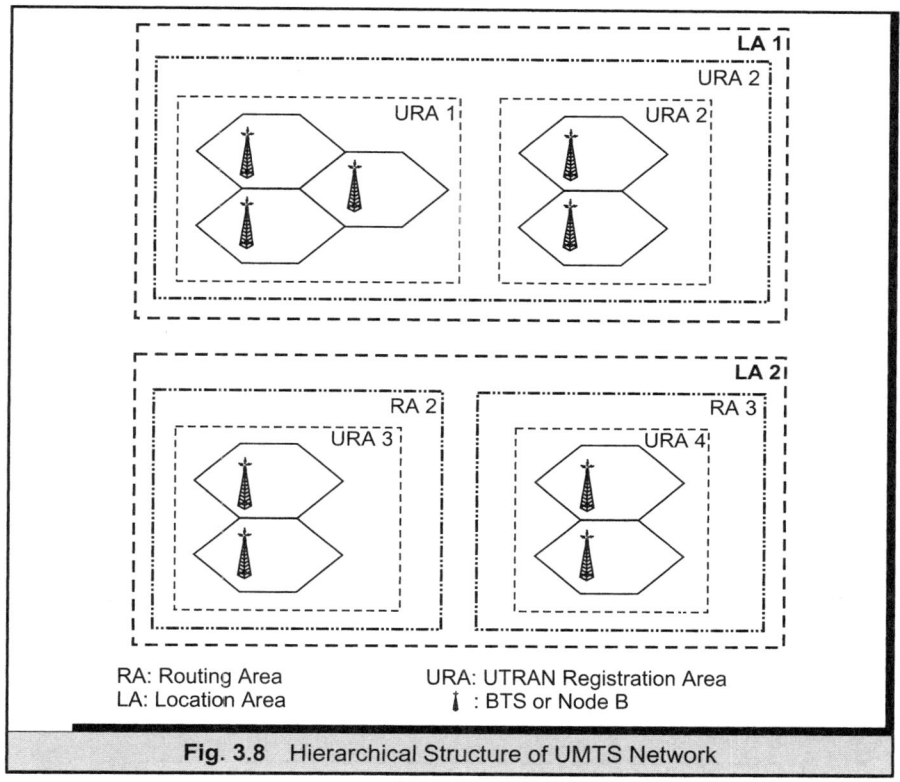

Fig. 3.8 Hierarchical Structure of UMTS Network

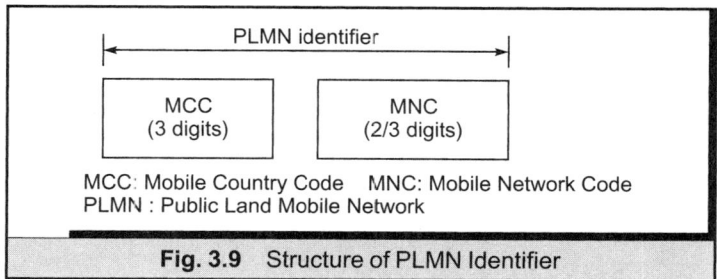

Fig. 3.9 Structure of PLMN Identifier

3.4.2 Location Area (LA)

Location Area (LA) is defined as an area in which an MS may move freely without updating its current location at the VLR. In case an MS moves outside its location area, it informs the VLR of its current location through the location update procedure.

A location area includes one or more cells. The reason for grouping of cells into location area is to facilitate efficient location management. To understand this, note that location management requires tracking the current location of the MS so that a terminating call can be delivered to the MS. Since the MS updates its location only at the change of location area, the VLR has accurate information on this. When a terminating call for an MS arrives, the VLR pages the MS to seek the exact location of the MS (in terms of its current cell location). Upon receiving the paging request, the MS responds with information on its current cell location. This information is used to set up a connection with the MS.

Hence, it is evident that if the location area is as small as a cell, there is no need to page the MS. However, this would require constant activity between the MS and VLR whenever the MS moves to a new cell, resulting in consumption of radio resources and battery power. If the location area is very large, the paging has to be performed in a very large area, which is undesirable. Thus, grouping of cells into location area allows us to achieve a balance between the accuracy of information maintained at the VLR as against the uplink radio capacity and the battery power consumed in the process.

Each location area is uniquely identified by a Location Area Identity (LAI). The structure of LAI is shown in Figure 3.10. The MCC and MNC are the same as Mobile Country Code and Mobile Network Code of the PLMN to which the LA belongs. The last two octets of the LAI are the Location Area Code (LAC) that identifies a location with a PLMN. Collectively, the LAI forms a unique identifier for a location area across all PLMNs.

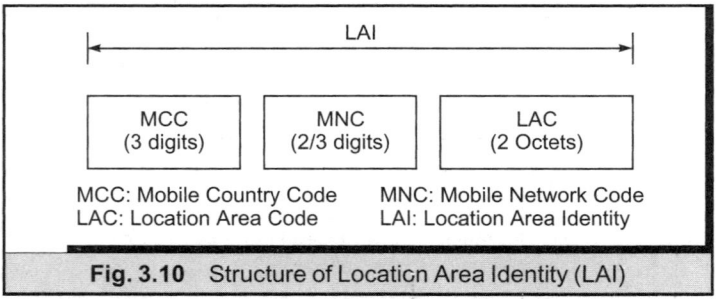

Fig. 3.10 Structure of Location Area Identity (LAI)

3.4.3 Routing Area (RA)

The Routing Area (RA) for PS domain is analogous to the location area for the CS domain. Routing area is defined as an area in which an MS may move freely without updating its current location at the SGSN. In case an MS moves outside its routing area, it informs the SGSN of its current location through the routing area update procedure.

Like a location area, a routing area too may include one or more cells. Grouping of cells into a Routing Area facilitates efficient location management, like in the case of a location area, where balance is achieved between the frequency of location updates and the area in which paging is done for mobile-terminated sessions.

One important difference between a routing area and location area is that the former is always contained within a location area. In other words, a location area may contain one or more routing areas.

Each routing area is uniquely identified by a Routing Area Identity (RAI). Since a routing area is a subset of a location area, the Routing Area Identity (RAI) is derived from Location Area Identity (LAI). In fact, the RAI is LAI plus a Routing Area Code (RAC) of 1-octet. The RAC uniquely identifies a routing area in a location area. Putting it simply, RAI= LAI + RAC.

3.4.4 UTRAN Registration Area (URA)

A UTRAN Registration Area (URA) is defined as an area covered by a number of cells. It is only internally known in the UTRAN. The URA is used to provide a layer of abstraction between cells and the routing area. A URA contains one or more cells and a routing area contains one or more URA. The URA is used to track the location of an MS within the UTRAN. The use of URA for mobility management is explained in Chapter 9. A URA is uniquely identified using the URA identity.

3.4.5 Cell Global Identity (CGI)

At the lowest level of UMTS hierarchy is the cell. Each cell is identified by the Cell Identity (CI). A CI is unique within a location area. To identify a cell uniquely across PLMNs, an identity called the Cell Global Identity (CGI) is defined. CGI is obtained by the concatenation of Location Area Identity and the Cell Identity. The structure of CGI is depicted in Figure 3.11.

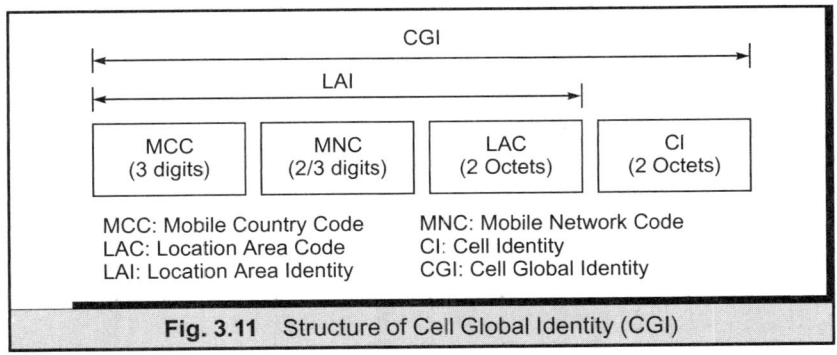

Fig. 3.11 Structure of Cell Global Identity (CGI)

3.5 ADDRESSES AND IDENTIFIERS

In UMTS, a number of identifiers are used for the purpose of addressing and identification. Each identifier serves a specific purpose. First comes the International Mobile Subscriber Identity (IMSI) that uniquely identifies a subscriber. An IMSI may be associated with multiple Mobile Subscriber ISDN (MSISDN) numbers. The MSISDN can be viewed as the mobile phone numbers or the service identity. Apart from these two identifiers, there is a temporary identifier, TMSI, which is used to hide the IMSI. There are other temporary identities as well, the need and functions of which are detailed later in this section.

While MSISDN is relevant in the CS domain, the equivalent identity in the PS domain is the Packet Data Protocol (PDP) address. In simple terms, the PDP address identifies the network address using which entities outside the PS domain communicate with the MS.

Apart from these, there is the International Mobile Equipment Identity (IMEI), which uniquely identifies a MS.

Then there are the E.164 addresses, used to identify network entities.

All these identifiers and addresses are explained in the following sub-sections (also see Table 3.1). The reader is referred to 3GPP TS 23.003 for complete information on numbering, addressing and identification schemes used in the UMTS network.

3.5.1 Subscriber Identity

A subscriber is uniquely identified by its International Mobile Subscriber Identity (IMSI). The IMSI is stored in the USIM and kept hidden from ordinary access. As shown in Figure 3.12, the IMSI is divided into three distinct parts. The first three digits of the IMSI form the Mobile Country Code (MCC). The MCC identifies the country of domicile of the mobile subscriber. The next two or three digits form the Mobile Network Code (MNC). The MNC identifies the home PLMN of the subscriber. The last field of IMSI is the Mobile Subscriber Identification Number (MSIN). The MSIN uniquely identifies a subscriber within a PLMN. The combination of MNC and MSIN is called the National Mobile Subscriber Identity (NMSI).

3.5.2 Service Identity

The mobile number used to contact a person is the Mobile Subscriber ISDN (MSISDN) number and not the IMSI. Thus, an MSISDN can be viewed as a service identity because a subscriber may have multiple MSISDN, where each MSISDN identifies a particular service.

Table 3.1 UMTS Addresses and Identifiers

Identity	Description	Composition
IMSI	Permanent identity that uniquely identifies a subscriber.	MCC + MNC + MSIN
MSISDN	Service identity that is used for communication with a subscriber.	CC + NDC+ SN
TMSI	Temporary identity that is used to hide the permanent identity IMSI of a subscriber.	Four octets (chosen by operator)
LMSI	Temporary identity that is used by VLR to optimize database search.	Four octets (allocated by VLR)
MSRN	Temporary identity that is allocated by VLR and is used to route calls directed to a MS.	CC + NDC+ SN
RNTI	Temporary identity used as UE identifiers to exchange signalling messages between UE and UTRAN.	Refer 3GPP TS 25.401
PDP Address	Static or dynamic network address used to communicate with other entities of a Packet Data Network (PDN).	Typically IPv4 or IPv6 address
IMEI	Permanent identity that uniquely identifies an MS.	TAC + SNR
Location Number	Refers to the geographical position of the MS in terms of standardized co-ordinates.	CC + NDC+ LSP
E.164 address	Used by MSC, GMSC, SGSN, GGSN, EIR, HLR and VLR for the purpose of signaling.	CC + NDC+ SN
GSN address	Used by the GSNs to communicate with each other over IP backbone.	IPv4 or IPv6 address
RNC identifier	Used to uniquely identify an RNC.	MCC + MNC + RNC-id

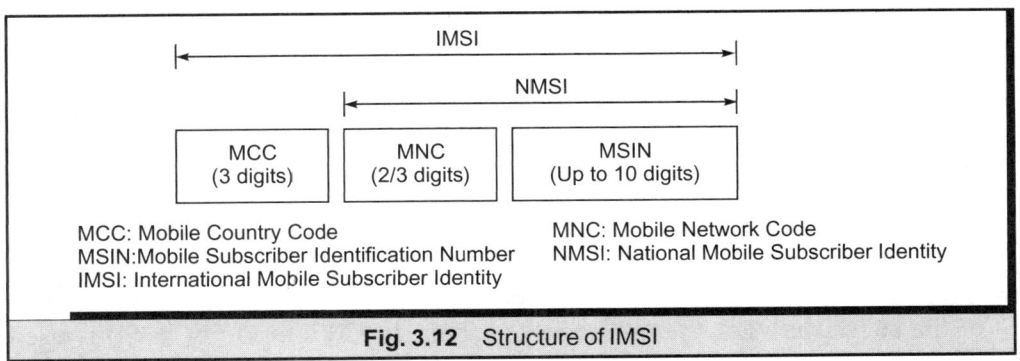

Fig. 3.12 Structure of IMSI

The MSISDN numbers are based on the ISDN numbering plan and are allocated in such a manner that fixed line ISDN or PSTN subscribers can call any mobile subscriber. The ISDN numbering plan is based on ITU-T specification E.164.

Figure 3.13 shows the structure of MSISDN. Like IMSI, an MSISDN number is made up of three distinct parts: a Country Code (CC), a National Destination Code (NDC) and a Subscriber Number (SN). There is a one-to-one analogy between the elements of IMSI and those of MSISDN. The basic difference between the two is the number of digits allocated to individual elements. The CC is from one to three digits.

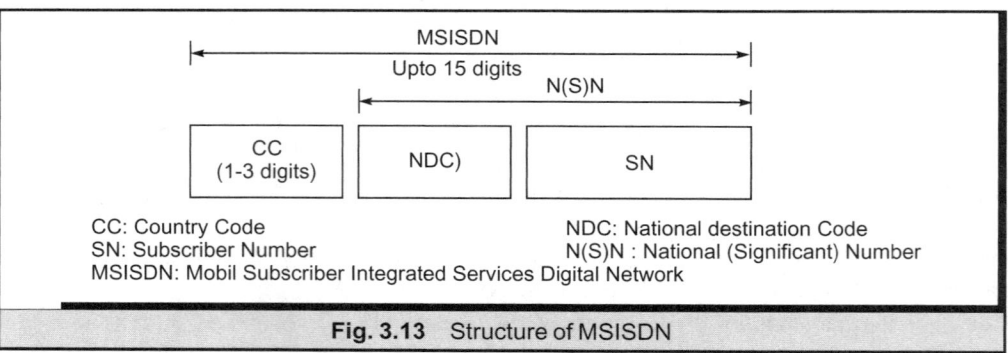

Fig. 3.13 Structure of MSISDN

The MSISDN can have a maximum of 15 digits. The size of National (Significant) Number depends upon the size of the CC and can be of a maximum of 14 digits (when CC is of one digit).

3.5.3 Temporary Identities

Apart from IMSI and MSISDN, there are temporary identifiers used for specific purposes. These temporary identifiers are as follows:

- Temporary Mobile Subscriber Identity (TMSI)
- Local Mobile Station Identity (LMSI)
- Mobile Station Roaming Number (MSRN)
- Radio Network Temporary Identity (RNTI)

3.5.3.1 *Temporary Mobile Subscriber Identity (TMSI)*

From security point of view, there is a requirement to hide the permanent identity IMSI of the subscriber. For this, a temporary identity TMSI (and not IMSI) is used on the air interface. The TMSI, or Temporary Mobile Subscriber Identity, is allocated by the VLR or SGSN. It is also possible that two temporary identities are used, one for CS and another for PS domain. Under such circumstances, the identities are called TMSI and P-TMSI for CS and PS domain respectively. The TMSI has only local significance and is applicable within the area controlled by VLR (or SGSN).

The TMSI consists of four octets. The exact encoding of TMSI is chosen by agreement between the network operator and equipment manufacturer to suit local needs.

3.5.3.2 *Local Mobile Station Identity (LMSI)*

For the purpose of optimizing database search, a VLR may use a local identifier called the Local Mobile Station Identity (LMSI). The VLR sends the LMSI to the HLR during message exchange along with IMSI/MSISDN. The HLR does not use the LMSI but keeps it along with IMSI/MSISDN in its database. In all further correspondence with the VLR, the HLR includes the LMSI sent earlier by the VLR. The VLR then uses the LMSI to optimize database search.

The LMSI consists of four octets and is allocated by the VLR.

3.5.3.3 *Mobile Station Roaming Number (MSRN)*

To facilitate roaming, a VLR allocates a roaming number called the Mobile Station Roaming Number (MSRN). The MSRN is used to route calls directed to a MS. When a mobile-terminated call is received by GMSC, it queries the VLR (via HLR) for a number, using which it can route the call. The VLR allocates a MSRN for the MS and passes it to HLR, which in turn forwards it to GMSC. The GMSC then uses the MSRN to route the call to the MS via MSC/VLR.

The MSRN is of the same format as the MSISDN, but is not the same as the MSISDN. The MSRN is allocated by the visited network according to the numbering plan of the visited PLMN. In certain cases, the MSRN may be the same as the MSISDN (for example, when the subscriber is in the home network).

3.5.3.4 *Radio Network Temporary Identity (RNTI)*

While TMSI, LMSI and MSRN are allocated by Core Network, there are temporary identities allocated by the UTRAN. One such temporary identity is the Radio Network Temporary Identity (RNTI). The RNTI is used as a UE identifier to exchange signalling messages between UE and UTRAN. There are various types of RNTI, out of which only the s-RNTI is discussed here.

The s-RNTI is allocated by the *Serving RNC*, which is in charge of the radio connections between the UE and UTRAN. The Serving RNC allocates the s-RNTI for those UE that have a RRC connection. The s-RNTI is used by the UE to identify itself to the Serving RNC and is used by the Serving RNC, which in turn uses it (s-RNTI) to address a particular UE.

Apart from s-RNTI, there are d-RNTI, c-RNTI, u-RNTI and few other identifiers. For details of these, the reader is referred to 3GPP TS 25.401.

3.5.4 PDP Address

For an MS to communicate with entities of a Packet Data Network (PDN), it must have an address applicable in the PDN. Note that the PDN lies outside the PLMN, which implies that the addresses in the PLMN (like IMSI) are alone not sufficient for communication with PDN entities. Since the most common PDN is based on the Internet Protocol (IP), a MS must have an IP address for communicating with other entities in the PDN. The IP address may be an IPv4 or an IPv6 address. In either case, the MS must have an address, called the Packet Data Protocol (PDP) address, to communicate with entities in a PDN.

The PDP address is assigned either *statically* or allocated *dynamically* by the GGSN. A static PDP address is allocated by the network operator of the home PLMN. Since the allocation is static, it is of permanent nature.

However, network addresses are a scarce resource and it does not make sense to allocate them on a permanent basis, more so because a subscriber may not need to use one all the time. Hence, the addresses are generally allocated dynamically, so that a small set of these may be shared between a large number of subscribers.

A dynamic PDP address is allocated during the *activation* of *PDP context*. A PDP context can be viewed as a set of information maintained by UE, SGSN and GGSN. It contains a PDP type (that identifies the type of PDN, for example IPv4); the PDP address (say a dynamically allocated IPv4 address); QoS information; and other session information. Activating a PDP context refers to creating the PDP context at the UE, SGSN and GGSN so that the UE can communicate with an entity in PDN using the PDP address maintained in the PDP context. After the communication is over, the PDP context is deactivated.

Chapter 11 provides details of PDP context and also the procedures related to its activation and deactivation.

3.5.5 Equipment Identity

A MS is identified by its International Mobile Equipment Identity (IMEI). The IMEI is a 15-digit identifier (its structure is shown in Figure 3.14). The first eight-digits are the Type Allocation Code (TAC). The next six-digits from the Serial Number (SNR). The last digit is spare and is set at 0.

The IMEI is used to track stolen a MS. The IMEI of a handset can be known by typing the string *#06# (star hash 0 6 hash) on the MS.

3.5.6 Location Number

In section 3.4, the location of a subscriber was defined in terms of Location Area, Routing Area, UTRAN Registration Area and Cells. This location referred to the

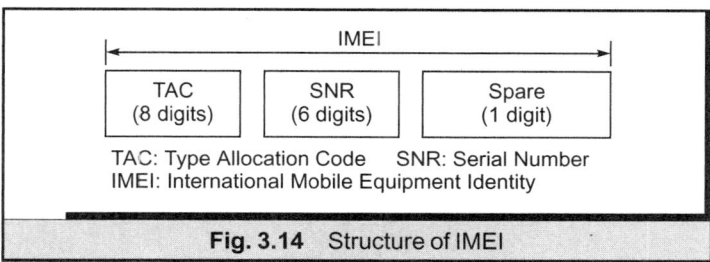

Fig. 3.14   Structure of IMEI

location of the MS in the hierarchical structure of the network. There is yet another notion of *location* (or *position*) of the MS. According to this notion, the location (or position) refers to the geographical position of the MS in terms of standardized co-ordinates. This information is used by application service providers to provide specialized services, also referred to as Location Services (LCS). An example of LCS is providing the list of restaurants around a given geographical location to the MS. Chapter 13 gives the details of Location Services (LCS).

To provide Location Services, a location number is required. The location number defines a specific location within a PLMN. It contains the Country Code (CC), National Destination Code (NDC) and a Locally Significant Part (LSP). The structure of the location number is depicted in Figure 3.15. The exact structure of LSP is a matter of agreement between the PLMN operator and the national numbering authority in the country containing the PLMN.

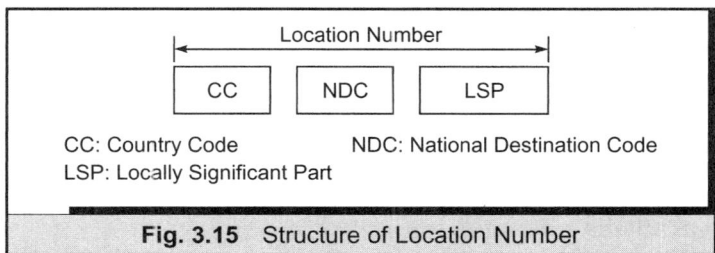

Fig. 3.15   Structure of Location Number

3.5.7 Identifying Network Entities

The preceding sections elaborated upon the identifiers used for subscriber identification, service identification, and equipment identification. Apart from these, there are identifiers used for addressing network entities.

The core network entities, including MSC, GMSC, SGSN, GGSN, EIR, HLR and VLR, are identified using the E.164 numbers. Recall from Section 3.5.2 that the MSISDN number is based on E.164 format.

Apart from the E.164, the SGSN and GGSN also require a GSN address (i.e. an IP address). This is because the GSNs communicate with each other using the IP protocol. The structure of the GSN address is depicted in Figure 3.16. The format of the address depends upon whether the protocol used is IPv4 or IPv6. Depending upon this, the GSN address part is of 4 bytes of 16 bytes.

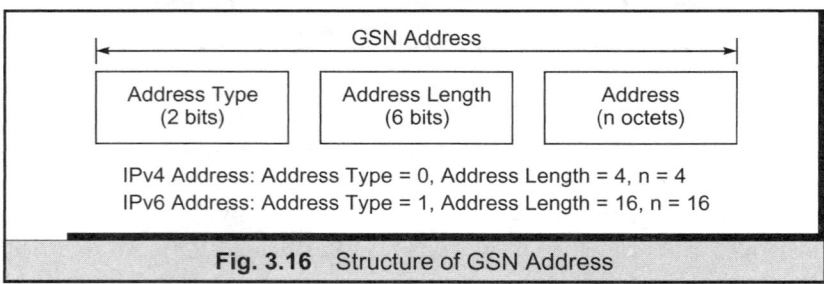

Fig. 3.16 Structure of GSN Address

In the UTRAN, the RNC is uniquely identified by the RNC identifier. The globally unique RNC identifier comprises of the PLMN identifier (MCC + MNC) and a RNC identifier.

3.6 SERVICE ASPECTS

One of the important developments in standardization of services in UMTS is the realization that instead of standardizing the services, it is much better to standardize service capabilities. In 2G and 2.5G networks, complete sets of teleservices, applications and supplementary services have been standardized. The net result of this is that a new service or changes in existing service requires considerable effort. Further, standardization of services prevents operators from offering distinct or specialized services. In such an environment, the time it takes for the service to become commercially available is also quite long.

To change this trend, instead of standardizing the services, the service capabilities are standardized. For this, two things are required. The first requirement is the presence of a wide variety of bearers with their specified QoS parameters. These bearers help in providing lower layer capabilities to satisfy the QoS requirements of a wide variety of services. This is different from providing standardized services along with their standardized bearer capabilities. In the first case, technological improvements in service realization (e.g. a better codec for voice transfer) would result in use of a bearer with lower bandwidth. In the second case, however, as the service is tied to the bearer, it is very difficult to change the applications or the bearers. In simple terms, the idea is to decouple a service from its underlying bearer and to provide a set of bearers so that the QoS requirements of a wide variety of services can be satisfied.

The second requirement is to have the necessary mechanisms in place to realize a service. For this, certain aspects like the functionality provided by various network elements, the communication between these elements and the data stored in them must be defined. This implies that an application developer must have standardized mechanisms (which is different from a standardized service) to provide different applications to the subscribers.

As an example, the IP Multimedia Subsystem (IMS) described in Chapter 16 primarily talks about a framework for realizing the services. Thus, instead of standardizing a host of teleservices or supplementary services, a set of bare minimum services is standardized. For the rest, a framework is provided so that an array of services can be provided through Applications Servers. However, there is no indication of what these services are; only the means to provide them have been defined.

3.7 SERVICE CLASSIFICATION

Services in UMTS (Figure 3.17) are classified into various categories, which are as follows:

- Bearer Services (includes circuit bearer services and GPRS-based packet bearer services)

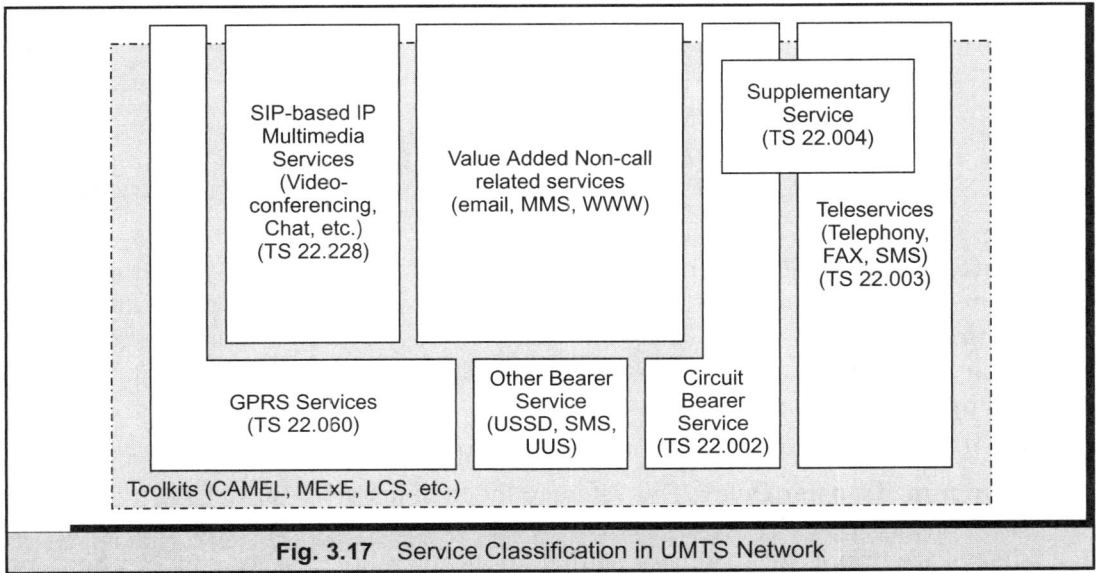

Fig. 3.17 Service Classification in UMTS Network

- Other Bearer Service (e.g. SMS)
- Teleservices
- Supplementary Services
- Value Added Service
- IP Multimedia Service

For details of above services, the reader is referred to 3GPP TS 22.101.

3.7.1 Bearer Services

Bearer Services refer to services that *provide the capability of transmission of signals between two communicating entities.* As the name suggests, the bearer service defines the lower layer capabilities as seen in the context of OSI reference model used in communication. Thus, the bearer service provides a communication link between two entities for information transport. There is the freedom to use any higher layer protocol over the bearer services.

A particular bearer service is defined by a set of characteristics that distinguishes it from other bearer service. Each characteristic of the set has a particular value such that the collection of the values of a particular bearer service uniquely defines the service.

The different characteristics associated with a bearer service are broadly classified as *information transfer* and *information quality*. The parameters defining the nature of *information transfer* are as follows:

- **Nature of Service:** This refers to whether the service is connection-oriented or connectionless.

- **Traffic Type:** It refers to the basic nature of traffic. The different types of traffic are:
    - Guaranteed (Constant Bit Rate)
    - Non-Guaranteed (Dynamically Variable Bit Rate)
    - Real-Time Dynamically Variable Bit Rate with minimum guaranteed bit rate

- **Traffic Characteristics:** These indicate whether the service is point-to-point or point-to-multipoint. Point-to-point traffic is further classified into uni-directional, bi-directional symmetric and bi-directional asymmetric. Point-to-multipoint traffic is further classified as multicast and broadcast. Note that point-to-multipoint by nature is uni-directional.

The parameters defining the nature of *information quality* are as follows:

- **Minimum Transfer Delay:** This refers to the time taken to transfer the information from one access point to its delivery at the other access point. Minimum transfer delays are important in time-sensitive applications (e.g., voice conversation and video conferencing) because after inordinate delay, data becomes useless for that application. For example, if a movie frame reaches its destination after its succeeding frames have been viewed, it is of no use and has to be discarded.

- **Delay Variation:** This parameter refers to variation in delay over time. The transient changes in network load cause jittery transmission, affecting real-time services.

- **Bit Error Ratio:** This is the ratio of undetected bit error versus the total transferred information bits. While some data loss is permissible in voice conversation, it is strictly prohibited in data applications.

- **Data Rate:** This has to do with the amount of data transferred between the two access points in a given period of time. The typical rates supported in UMTS are 384Kbps in outdoors and up to 2Mbps indoors.

Specific values of aforementioned characteristics define a particular bearer service. There are two basic categories of bearer services, namely *Circuit Bearer Services* and *Packet Bearer Services*. The Circuit bearer services are defined in 3GPP TS 22.002. The services include *synchronous bearer service* and *asynchronous bearer service*. While the synchronous mode supports the Transparent (T) data service, the asynchronous mode supports both Transparent (T) data service and Non-Transparent (NT) data service.

The Packet bearer services (also referred to as GPRS bearer services) are defined in 3GPP TS 22.060. These services provide means to transmit data between user-network access points and are primarily used to carry IP packets.

3.7.2 Other Bearer Service

As mentioned in Section 3.7.1, there are two types of bearer services, namely the circuit bearer services and packet bearer services. Apart from these bearers, there are other services that can be used as bearer services. These include Short Message Service (SMS), Unstructured Supplementary Service Data (USSD) and User-to-User Signaling (UUS).

The SMS is actually a Teleservice (as explained in Section 3.7.3). However, SMS can be used as a bearer over which applications can be built (e.g. a dating application). The details of Short Message Services are given in Chapter 13.

The Unstructured Supplementary Service Data (USSD) provides a mechanism whereby mobile users and PLMN operators can communicate with each other using means that are transparent to the MS and the intermediate network entities. This mechanism allows the development of services that are operator-specific. The Unstructured Supplementary Service Data (USSD) service is explained in Chapter 12.

The User-to-User Signaling (UUS) is actually a Supplementary Service but can act as a bearer. The UUS allows a subscriber to send (or receive) a limited amount of subscriber generated information to (or from) another user in the call. This information is passed transparently through the network without any modification of the contents. The network does not try to interpret or act upon this information. The User-to-User Signaling (UUS) service is also explained in Chapter 12.

3.7.3 Teleservices

Teleservices are defined as *services that provide the full capabilities for communication by means of terminal equipment and network functions.* In simple terms, the scope of a Teleservice is not restricted to merely providing transport of user information, like the bearer service. A Teleservice provides complete services as seen from user's point of view.

A bearer capability is associated with every Teleservice, and it defines the technical characteristics of a Teleservice. Here, bearer capability merely refers to the lower layer capabilities (i.e. bearer capabilities) of the Teleservice. Figure 3.18 depicts the difference between Bearer service and Teleservice. As shown in the figure, while the former extends till the mobile termination, the latter extends till the terminal equipment, providing complete services to the user.

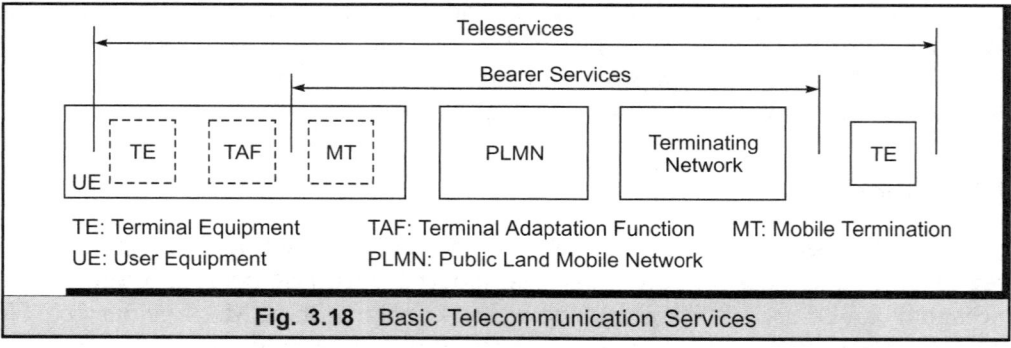

Fig. 3.18 Basic Telecommunication Services

The different groups of Teleservices, along with the individual Teleservice as defined in 3GPP TS 22.003, are as follows:

- **Speech Service:** It includes plain speech service and emergency calls.

- **Short Message Service:** It includes mobile-originated and mobile-terminated Short Message Service (SMS). It also includes the Cell Broadcast Service (CBS).

- **Facsimile Transmission:** It includes alternate speech and facsimile group 3 and automatic facsimile group 3.

- **Voice Group Service:** It includes Voice Group Call Service (VGCS) and Voice Broadcast Service (VBS). The VGCS enables a calling subscriber to establish a

Voice Group Call to destination subscribers belonging to a predefined Group Call Area and Group identity. The Group Call Area and Group identity collectively identify a Voice Group Call. On similar lines, VBS enables a calling subscriber to distribute speech into a predefined geographical area to reach all or a group of service subscribers located in this area.

3.7.4 Supplementary Services

Supplementary services are those that modify or supplement the basic services (i.e. bearer service and teleservice). Unlike the basic services that are independent in themselves, supplementary services do not have any independent existence. Examples of supplementary services are Call Forwarding and Call Barring. A network operator may or may not offer supplementary services to the subscribers. Supplementary services are explained in detail in Chapter 12.

3.7.5 Other Services

Apart from Bearer and Teleservice, there are value-added non-call related services like email, MMS and WWW. These services can run over various types of bearers.

Then there are the SIP based IP Multimedia Services that use GPRS as bearers. Chapter 16 provides details of IP Multimedia Services.

3.7.6 Toolkits

In order to create or modify the various services, there are standardized toolkits available by 3GPP such as CAMEL, Location Service (LCS) or external solutions (e.g. Internet mechanisms). Pre-paid is an example of an application created with toolkits that may apply to all of these services categories. Chapter 13 provides details of CAMEL and LCS.

3.8 QUALITY OF SERVICE (QoS) ARCHITECTURE

The Quality of Service (QoS) Architecture in UMTS defines how various entities interact (at various levels of abstraction) to provide end-to-end QoS. The QoS Architecture in UMTS is depicted in Figure 3.19.

To realize a certain QoS, a Bearer Service with clearly defined characteristics and functionality is set up from the source to the destination of the service. This bearer service includes all aspects necessary to provision and get a QoS as described by a contract. These aspects include the control signaling, user plane transport and QoS management functionality.

At the highest level, the end-to-end QoS applies between two Terminal Equipments (TEs). The scope of UMTS, however, is limited to providing QoS guarantees from the Mobile Termination (MT) till the edge of the Core Network. The guarantees cease to apply outside the PLMN (corresponding to external bearer), as what happens outside cannot be ascertained by the UMTS network. Similarly, the local bearer service between the MT and the TE is also outside the scope of UMTS bearer service. This is because the interface between MT and TE is not standardized. Nor is it necessary for such a distinction to always exist.

The UMTS bearer is provided by using the Radio Access Bearer (RAB) Service and the Core Network Bearer Service. The RAB is defined as *the service that the access stratum provides to the non-access stratum for transfer of user data between MT and CN*. As depicted in Figure 3.19, the RAB Service extends between the MT and the edge of the Core Network. It provides confidential transport of signaling data and user data between MT and CN Edge Node with the QoS necessary for the negotiated UMTS Bearer Service or with the default QoS for signaling. This service is based on the characteristics of the radio interface and is maintained for a moving MT. The RAB Service itself is realized using Radio Bearer Service and the Iu Bearer Service.

The role of the Radio Bearer Service is to cover all aspects of the radio interface transport. The Radio Bearer is defined as *the service provided by the Layer 2 of the access*

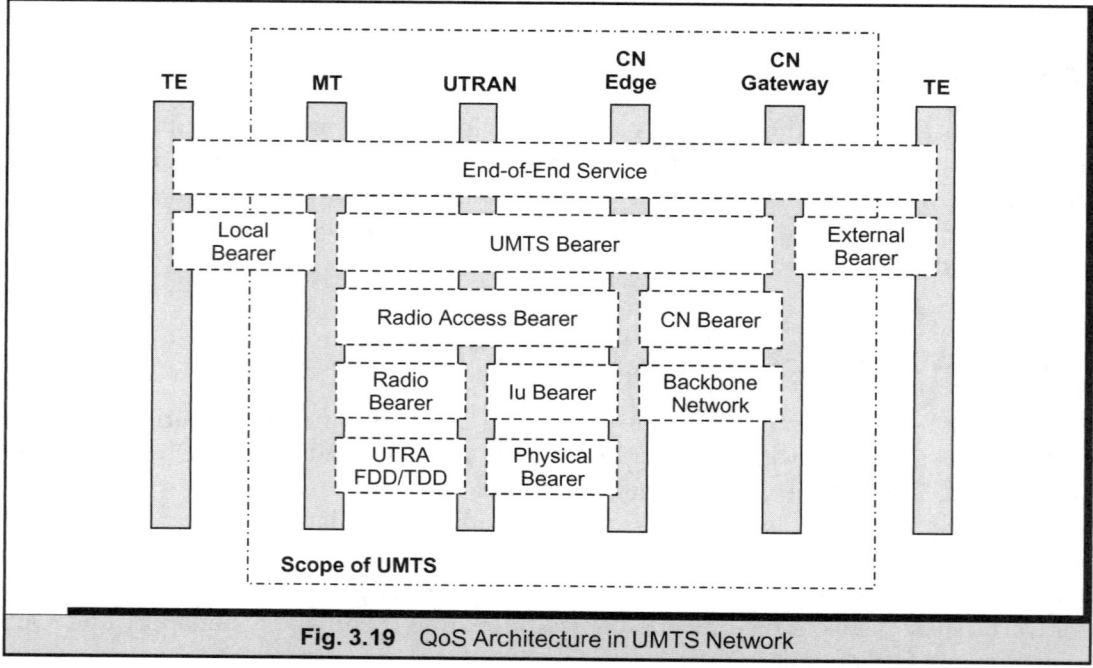

Fig. 3.19 QoS Architecture in UMTS Network

stratum for transfer of user data between MT and UTRAN. The Radio Bearer Service uses the UTRA FDD/TDD.

The Iu-Bearer Service, together with the Physical Bearer Service, provides the transport between UTRAN and CN. Iu bearer services for packet traffic provide different bearer services for a variety of QoS.

The Core Network Bearer Service of the UMTS core network connects the UMTS CN Edge Node with the CN Gateway to the external network. The role of this service is to efficiently control and utilize the backbone network in order to provide the contracted UMTS bearer service. The UMTS packet core network supports different backbone bearer services for a variety of QoS.

3.9   UMTS QoS CLASSES

With respect to Quality of Service, the UMTS is quite different from 2G or 2.5G network. The 2G network was designed to carry voice. Thus, the QoS was characterized by busy hour call blocking probability, call drop rate and voice quality. In GPRS (2.5G), even though primitive QoS features were introduced, the service remained a best effort. In contrast, the UMTS provides traffic with different bandwidth and QoS requirements. The QoS requirements mainly relate to QoS parameters like *delay, delay variation* and *bit-error*. Based on these QoS parameters, four different QoS classes are defined. These classes are as follows:

- Conversational class
- Streaming class
- Interactive class
- Background class

Table 3.2 provides a brief description of each QoS class and also the QoS requirements for each class. For example, Conversational Class has stringent requirements for delay and delay variation. This implies that for this class, delay and delay variation should be as low as possible so as to provide the desired service level. Each of the above QoS classes is further explained in the sections that follow.

For details of QoS classes, the reader is referred to 3GPP TS 23.107.

3.9.1   Conversational Class

Among all the QoS classes, the technical requirement of conversational class is most stringent. The stringency arises from strict upper bounds on transfer delay and delay variation. This implies that not only must the data streams reach the destination within a specified time period, the variation in time taken by different packets/streams to reach the destination must also be minimal. The bounds on delay and delay variation

Table 3.2 UMTS QoS Classes

| QoS class | Description | Requirements on QoS parameters | | | Applications |
		Delay	Delay Variation	Bit-error	
Conversational class	Characterized by strict upper bounds on the transfer delay and delay variation.	Stringent	Stringent	No	Voice and video conferencing
Streaming class	Characterized by real-time one-way data flow aimed at a human user.	Constrained	Constrained	No	Video-on-demand
Interactive class	Characterized by a request-response protocol where a request is followed by download of requested data.	Loose	No	Stringent	Web-browsing and Internet-based email
Background class	Characterized by no requirements on delay or delay variation so that applications using it can run in background.	No	No	Stringent	E-mail, SMS and FAX

are governed by human perception of video and audio conversation. This QoS class, however, permits some data loss.

Given its conversational nature, the conversational class is also referred to as *real-time service*. Typical applications of conversational class include voice conversation and video conferencing.

3.9.2 Streaming Class

In terms of requirements for real-time transfer, this class comes after the conversational class. It is characterized by real-time data flow aimed at a human user. The flow is one-way and thus the name *streaming class* or *streaming service*.

Given the streaming and non-interactive nature of the application, the requirements for transfer delay are much less as compared to the conversational class. The delay variation requirement is similar to that of the conversational class, though somewhat less stringent. What this implies is that due to the non-interactive nature of the application, some delay is permitted in the time taken for the data stream to reach its destination. Further, with the use of buffering, minor variations in delay can also be tackled. However, if the delay variation exceeds a value beyond which buffering does not solve the problem, there is degradation in audio/video quality.

The streaming class is a relatively new concept. New 3G applications like video-on-demand belong to the streaming class.

3.9.3 Interactive Class

The interactive class is used by applications that adopt a request-response protocol where a request is followed by download of requested data. Given the request-response nature of those applications, there is an upper limit on the transfer delay. However, given the non real-time nature of communication, the requirement on the transfer delay is much less stringent as compared to the conversational class. Further, there is no requirement on delay variation. However, there are stringent requirements on data loss.

The applications using interactive class include Web-browsing, Internet-based email and data base queries.

3.9.4 Background Class

The background class is used by applications that do not have any delay or delay variation requirements. The only requirement is on data loss, which is not acceptable. This QoS class is named so because applications using it can run in the background (i.e. as a low priority task). This implies that given the real-time nature of other QoS classes, this class gets lower priority. Thus, this QoS class is served as a background activity as and when bandwidth is available in the network.

The applications using background class include email, SMS and FAX.

SUMMARY

This chapter provided an overview of the UMTS network architecture. The UMTS network is divided into three parts: UE, Access Network and Core Network. The next three chapters describe these three standard components.

CHAPTER

4

USER EQUIPMENT

4.1 INTRODUCTION

Chapter 3 presented an overview of the UMTS Network architecture and discussed its three basic components: the User Equipment, Access Network and the Core Network. This chapter explains the various aspects of User Equipment (UE) and discusses its components, functions, protocols and interfaces; all relevant information on UE has been organised as per the outline provided in Figure 4.1.

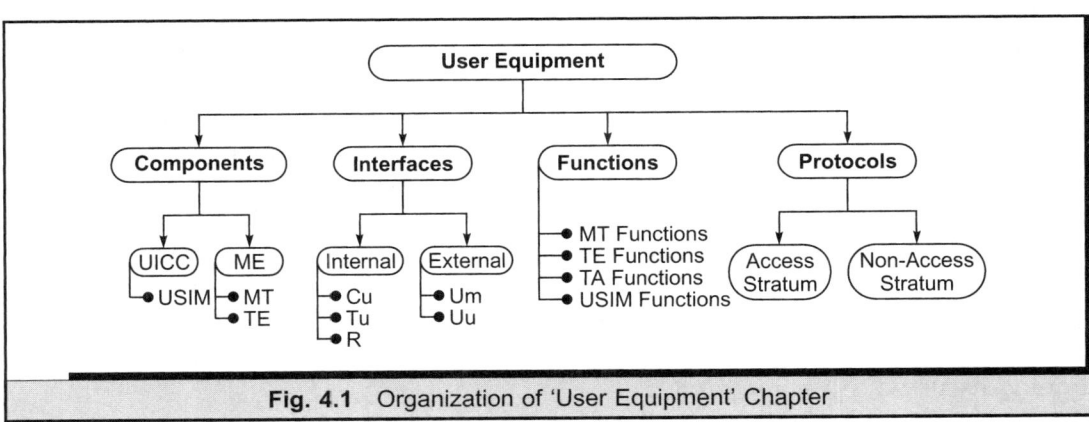

Fig. 4.1 Organization of 'User Equipment' Chapter

In GSM networks, the term *Mobile Station* was used instead of *User Equipment*. The term User Equipment has been newly introduced in UMTS. In the literature, however, User Equipment (UE) is also referred to as Mobile Station (MS); hence, this book uses the terms UE and MS interchangeably.

4.2 COMPONENTS OF USER EQUIPMENT

User Equipment (UE) is a device used by a subscriber/user to access network services. To give it a modular design, this device is divided into two components: Mobile Equipment (ME) and Universal Integrated Circuit Card (UICC) (Figure 4.2).

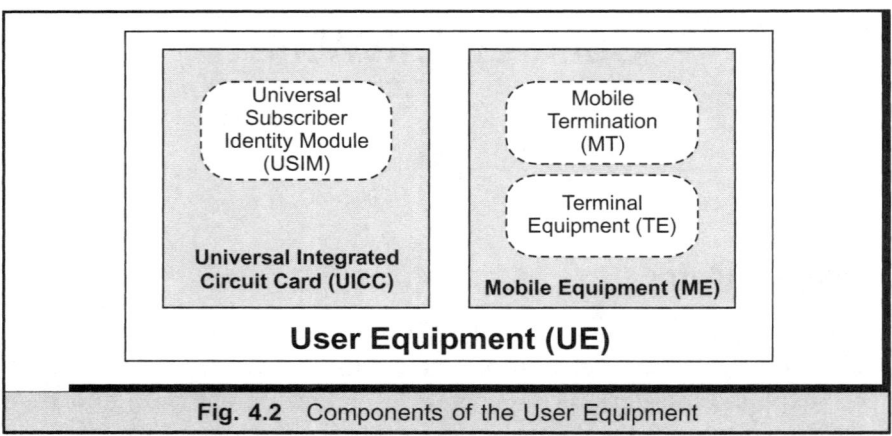

Fig. 4.2 Components of the User Equipment

The UICC is a smart card that contains an application called the Universal Subscriber Identity Module (USIM). The USIM contains the logic required to unambiguously and securely identify the user. Thus, the USIM is the user-dependent part of the UE and is provided by the service provider.

The ME, on the other hand, is the user-independent part of UE. It is manufactured by an equipment manufacturer, who is normally an entity distinct from the service provider. The ME is further divided into two distinct logical components, namely Mobile Termination (MT) and Terminal Equipment (TE).

The next two sections provide a detailed description of the components of the User Equipment. This division of the UE allows the equipment manufacturer and service provider to independently manufacture/provide the ME and UICC respectively. Further, it enables the introduction of the *User Equipment Combination* concept, which is briefly discussed in Section 4.2.3.

4.2.1 Universal Integrated Circuit Card (UICC)

The Universal Integrated Circuit Card (UICC) is a smart card of defined electro-mechanical specifications. It can easily be inserted into an ME or removed from it. The UICC contains application software modules and includes an application called the

Universal Subscriber Identity Module (USIM). The USIM application holds the logic required to identify the user.

Multiple USIMs can optionally be located on a UICC. However, only a single USIM can be active on a UICC at any given time. The USIM application resident in the UICC is explained in the following section.

4.2.1.1 *Universal Subscriber Identity Module (USIM)*

The Universal Subscriber Identity Module (USIM) is the user-dependent part of the UE. It is a UMTS application resident on a UICC card, which inter-operates with a UMTS terminal to provide the mobile user access to the UMTS services. In other words, a USIM application resident on a UICC is required to access a 3GPP-based UMTS mobile network.

The USIM contains the logic required to unambiguously and securely identify the mobile user. In particular, the USIM contains the permanent identity of the user (called the IMSI (International Mobile Subscriber Identity)), the shared secret key (used for authentication), the user phone book and a host of other information. The service provider of the mobile user provides the user-dependent information (IMSI, authentication keys, etc.) contained in the USIM.

4.2.2 Mobile Equipment (ME)

The Mobile Equipment (ME) or the mobile handset is the user-independent part of the UE. Equipment vendors, who may be distinct from the service provider, manufacture the ME. The ME contains a slot to hold the UICC, which is required to access the UMTS network.

The ME itself is divided into two logical components: *Mobile Termination (MT)* and *Terminal Equipment (TE)*. Each of these is explained in the following sections. The MT and TE interface with each other via a Terminal Adaptation (TA) function, which is explained in Section 4.2.2.3. The division of Mobile Equipment into Mobile Termination (MT) and Terminal Equipment (TE) is also referred to as *MT-TE functionality split*.

4.2.2.1 *Mobile Termination (MT)*

The Mobile Termination (MT) is that part of the ME that performs functions like radio transmission termination, authentication and mobility management. It consists of the following functional groups:

- **Radio Termination (RT):** This functional group of the MT is related to the radio access network, and contains functions specific to the radio access technology. The RT implements the Access Stratum protocols mentioned in Section 4.5.1.

These protocols include the Radio Resource Control (RRC), Radio Link Control (RLC) and Medium Access Control (MAC) protocols. The access stratum protocols are used on the radio interface, and are discussed in detail in Chapter 5.

- **Network Termination (NT):** This functional group, on the other hand, is the Core-Network (CN) dependent part of the MT. The NT implements the Non-Access Stratum protocols mentioned in Section 4.5.2, which are used between the UE and the CN. These protocols include the Mobility Management (MM), GPRS Mobility Management (GMM), Call Control (CC), and Session Management (SM) protocols, besides others. The Non-Access Stratum protocols are discussed in detail in Chapter 6.

4.2.2.2 Terminal Equipment (TE)

The Terminal Equipment (TE) component of ME manages the hardware (e.g. speaker, microphones, video cameras and user display) and end-user applications (e.g. Web browser). It may also control the mobile termination by using the modem control command set (AT commands). The 3GPP specification 27.007 defines the AT command set for the User Equipment.

An example of a Terminal Equipment could be a laptop, which is to be physically connected to a Mobile Termination (MT) unit. The same MT might also be providing services to another Terminal Equipment, for example a Camera, via a Bluetooth interface.

4.2.2.3 Terminal Adaptation (TA)

The Terminal Equipment (TE) interacts with the Mobile Termination (MT) via a Terminal Adaptation (TA) function. The TA function provides facilities to allow manual or automatic call control of circuit-switched services. The functions of the TA are discussed in detail in Section 4.4.3.

4.2.3 User Equipment Combination

A User Equipment Combination is defined as the set of user equipment components that are connected and used together in a particular scenario. An example of a user equipment combination may consist of an MT and all the TEs that are connected to it (Figure 4.3).

As shown in the Figure, the television set and video equipment act as the Terminal Equipment and are connected to the MT via a Bluetooth network. The laptop, which is also acting as Terminal Equipment, is connected to the MT via a physical connection.

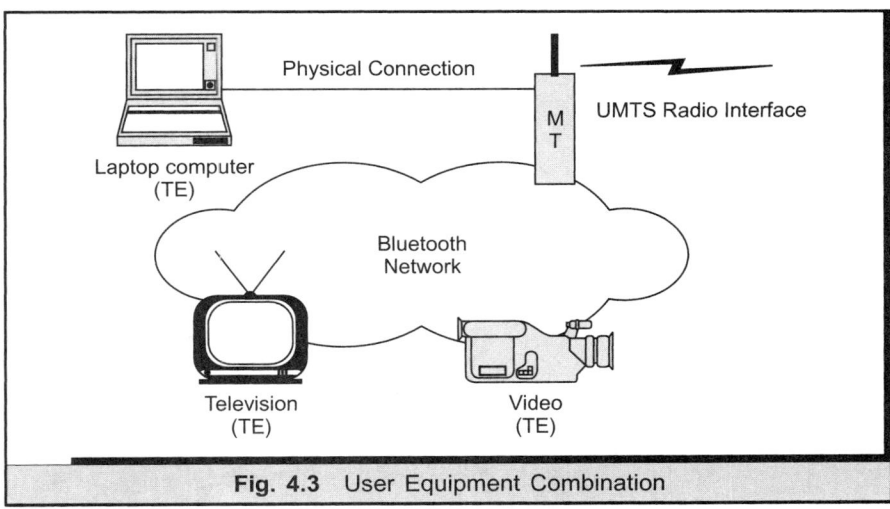

Fig. 4.3 User Equipment Combination

In case of a User Equipment Combination, multiple independent applications residing on different TEs can be simultaneously connected to a single MT. In this case, independent users are possibly using the different TEs. However, all users of TEs employ only one subscription to the mobile network, and this subscription information is stored in one USIM of the UE. Hence, in this case, the user(s) identity is possibly different from the mobile subscriber identity.

4.3 INTERFACES OF USER EQUIPMENT

The User Equipment interfaces with the Access Network over the radio interface. It may interface with both the 2G-based GSM access networks, and the UTRAN. Further, the User Equipment is logically divided into multiple components, as already discussed in Section 4.2. The internal interfaces of the UE between its various components, as well as its external interfaces with the access networks are depicted in Figure 4.4.

The UE interfaces with a GSM/GPRS BSS using the Um interface. Further, it interfaces with the RNS in the UTRAN using the Uu interface. The external interfaces of the UE with the access networks are discussed in Section 4.3.1.

Within the UE, the UICC interfaces with the Mobile Equipment (ME) using the Cu interface. Within the ME, the R interface is used between the Terminal Equipment (TE) and the Mobile Termination (MT) via the Termination Adaptation (TA). Further, within the MT, the Radio Termination (RT) and the Network Termination (NT) themselves interact with each other over the Tu interface. The internal interfaces of the UE, between its various components, are discussed in Section 4.3.2.

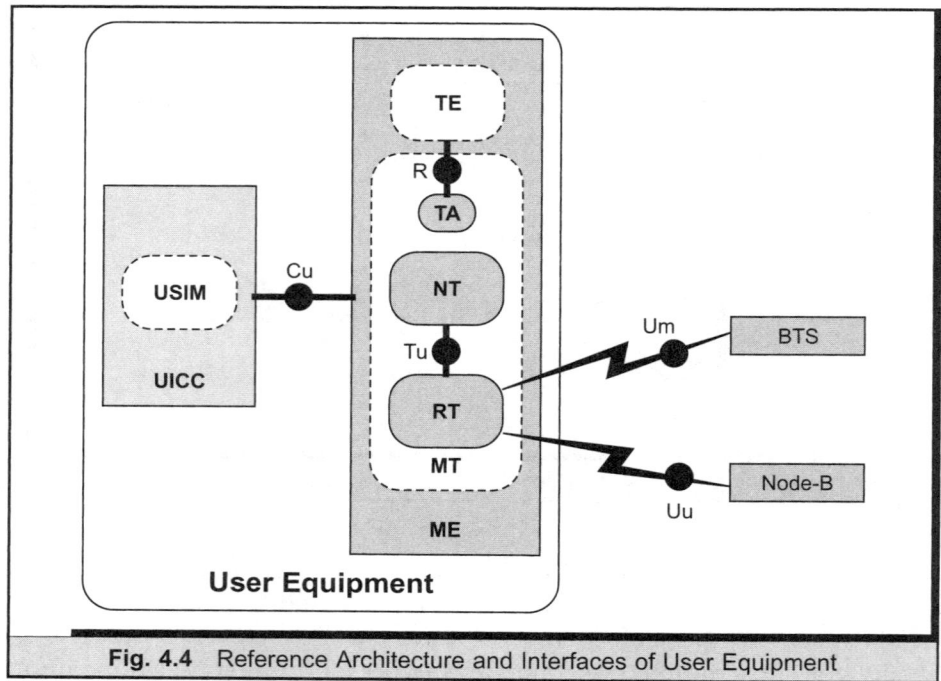

Fig. 4.4   Reference Architecture and Interfaces of User Equipment

4.3.1 | External Interfaces

The external interfaces of User Equipment are listed in Table 4.1. These interfaces are further elaborated in the following subsections. Note that the GSM legacy interface (Um interface) and the associated protocols are not discussed in this chapter in detail. For more information on these legacy interfaces, the reader is referred to the specifications as given in Table 4.1.

Table 4.1 | External Interfaces of UE

Interface	Between	Description	Air Interface	Specification
Um	MS – BSS	Radio interface between MS and BSS.	Hybrid TDMA/FDMA	44.xxx/45.xxx
Uu	MS – RNS	Radio interface between MS and RNS.	WCDMA (FDD and TDD)	24.xxx/25.xxx

4.3.1.1 *Um Interface between MS and BSS*

The User Equipment interfaces with the GSM/GPRS Base Station Subsystem using the Um interface. In other words, the Um interface is the GSM/GPRS network interface for providing circuit and packet data services over the radio interface to the UE. The Um interface is based on a Hybrid TDMA/FDMA radio access technology, which was briefly discussed in Chapter 2. The details of the protocols used over the Um interface are beyond the scope of this book. For more details on the Um interface, the reader is referred to the relevant GSM/GPRS specifications.

4.3.1.2 *Uu Interface between MS and RNS*

The User Equipment interfaces with the RNS in the UTRAN using the Uu interface. This interface is used by the UE to send/receive data and control information to/from the RNS. The Uu interface is based on the WCDMA radio access technology, which was discussed in Chapter 2. The radio interface protocols on the Uu interface between the UE and the RNS are called the Access Stratum protocols for the *Uu Stratum*. These protocols are mentioned in Section 4.5.1.

4.3.2 Internal Interfaces

The internal interfaces of the User Equipment are listed in Table 4.2. These interfaces are further elaborated in the following subsections.

Table 4.2 Internal Interfaces of UE

Interface	Between	Description	Protocol	Specification
Cu	UICC – ME	Internal interface between the UICC and the ME components of the UE.	UICC-Terminal Transmission Protocols	31.101 or ETSI TS 102 221
R	TE – MT	Internal interface between the TE and MT components of the ME.	AT Command Set	27.007 and 27.005
Tu	NT – RT	Internal interface between the NT and RT components of the MT.	Proprietary	Not applicable

4.3.2.1 *Cu Interface*

The Cu interface is the interface between the UICC and the Mobile Equipment (ME) components of User Equipment. This interface is standardized in 3GPP TS 31.101. Standardization of the Cu interface allows the network operator and mobile equipment vendor to independently manufacture the UICC and ME respectively, while taking into account the interface requirements between the two. The Cu interface standards define the transmission protocols for the physical layer, data link layer, transport layer and the application layer of the communication stack between the UICC and the ME.

4.3.2.2   R Interface

The R interface is the interface between the Terminal Equipment (TE) and the Mobile Termination (MT) components of the Mobile Equipment. The R interface between TE and MT uses the AT command set defined in 3GPP specifications TS 27.007 and 27.005. Standardization of the R interface allows multiple Terminal Equipments (TEs) to interface with a single MT. The case of a single TE interfacing with a single MT is considered as a special case of this scenario, which was discussed in Section 4.2.3.

4.3.2.3   Tu Interface

The Tu interface connects the UTRAN and the CN specific parts together in the mobile terminal. It is used to interface between the Network Termination (NT) and the Radio Termination (RT) components of the Mobile Termination (MT). Due to performance requirements, the Tu interface is embedded within the UE hardware, and is left open as a proprietary interface, to be implemented by the mobile equipment vendor.

4.4   UE FUNCTIONS

The functions of the UE consist of the set of functions performed by each component of the UE. These can be categorized as follows:
- Mobile Termination Functions
- Terminal Equipment Functions
- Terminal Adaptation Functions
- USIM Functions

4.4.1   Mobile Termination Functions

The Mobile Termination (MT) performs the following functions:
- Provides radio attachment to the mobile network.
- Radio transmission channel management.
- Presentation of a Man-Machine Interface (MMI) to a user (refer to Chapter 12 for discussion on MMI).
- Speech encoding/decoding.
- Error protection for all information sent across the radio path.
- Flow control of signaling and user data.
- Rate adaptation of user data.
- Support for multiple terminal equipment.
- Mobility Management functions (refer to Chapter 9).

4.4.2 Terminal Equipment Functions

These are as follows:

- Control of hardware in the TE, which may include speakers, microphone, video cameras, displays, etc.
- Providing access to services and capabilities provided by the MT.

4.4.3 Terminal Adaptation Functions

The Terminal Adaptation (TA) provides facilities to allow manual or automatic call control functions associated with circuit-switched services. The TA also performs the following functions:

- Inter-operability between the ITU-T V series/ISDN type interfaces and the PLMN interface. This includes performing the electrical, mechanical, functional and procedural conversions between the two interfaces.
- Bit rate adaptation between the ITU-T V series/ISDN type interfaces and the PLMN interface.
- Flow control.
- Synchronization procedure for the data transfer phase between two user terminals. This is described in 3GPP TS 27.001.
- Filtering of channel control information. This function is described in 3GPP TS 27.001.
- Terminal compatibility checking.
- Splitting and combining of data flow in case of multiple sub-stream data configurations.

4.4.4 USIM Functions

The USIM performs the following functions:

- Unambiguously identifies a subscriber.
- Provides storage for subscription and subscriber related information.
- May contain other user applications that use the USIM Application Toolkit (USAT). The USAT is defined in 3GPP specification TS 31.111.

4.5 UE PROTOCOLS

The Network Termination (NT) and Radio Termination (RT) components of Mobile Termination (MT) contain the UE protocols required for accessing the services offered by the mobile network. The UE protocols are divided into two broad categories:

Access Stratum protocols and *Non-Access Stratum* protocols. Each of these is defined in the following sections.

4.5.1 Access Stratum Protocols

Access Stratum protocols are the protocols used on the radio interface between the UE and UTRAN. These protocols are used for the transfer of user and control data between the UE and UTRAN. The Access Stratum protocols of the UE are implemented in the Radio Termination (RT) component of Mobile Termination (MT). These protocols include the following:

- **Medium Access Control (MAC) protocol:** The main functionality of the MAC layer is to map higher layer data on to appropriate *transport channels* of the physical layer. It is the only layer that needs the intelligence to manipulate the physical layer. The data transfer service provided by MAC is an unacknowledged service.

- **Radio Link Control (RLC) protocol:** RLC provides various link layer services, which include reliable data transfer, segmentation and re-assembly functions, flow control, error control and sequence numbering.

- **Broadcast/Multicast Control (BMC) protocol:** The Broadcast/Multicast Control (BMC) layer is used to carry user-plane information in the downlink direction. The information is broadcast or multicast in nature and is sent in an unacknowledged mode.

- **Packet Data Convergence Protocol (PDCP):** The Packet Data Convergence Protocol (PDCP) layer is used to carry user-plane information for the PS-domain. It carries data protocols like IP and PPP. Compression of redundant header information of the data protocols forms one of the most important functions of the PDCP protocol.

- **Radio Resource Control (RRC) protocol:** The RRC is the most important of the Radio Interface protocols. It is used for signaling between UTRAN and UE. RRC controls the lower layers including PDCP, BMC, RLC, MAC and the Physical layer.

The radio interface protocol architecture and the above mentioned protocols between the UE and UTRAN are discussed in detail in Chapter 5.

4.5.2 Non-Access Stratum Protocols

The Non-Access Stratum protocols are used between UE and the Core Network (CN). These protocols are used for transparent transfer of user and control data between the

UE and CN. The Non-Access Stratum protocols of the UE are implemented in the Network Termination (NT) component of Mobile Termination (MT). They include the following:

- **Mobility Management (MM)/GPRS Mobility Management (GMM) protocol:** The objective of MM/GMM protocol is to support mobility of user terminals, such as informing the network of the present location. During mobility management, it is also ensured that the identity of the user is confidential and that only authenticated users can avail the network services.

- **Call Control (CC) protocol:** The objective of the CC protocol is to support delivery of mobile-originated and mobile-terminated calls for a mobile user.

- **Session Management (SM) protocol:** The SM protocol is analogous to the CC protocol. However, while the CC protocol works for the CS domain, the SM protocol is applicable to the PS domain.

- **Supplementary Service (SS) protocol:** The SS protocol is used to provide the mobile user the ability to control the supplementary services offered by the mobile network.

- **Security-related protocols:** A security-related protocol (e.g. Authentication and Key Agreement (AKA) protocol) is used to provide user and network authentication, as well as confidential/secured transfer of information between the UE and the mobile network.

The non-access stratum protocols between the UE and CN are discussed in detail in Chapter 6.

4.6 CLASSIFICATION OF UE

Based on its capabilities, the UE can be classified into various categories. This classification is done mainly according to the Mobile Termination's (MT's) capability to support several access and core network technologies by its RT and NT components. The various categories of UE can be defined as follows:

- **Single Radio Mode UE:** In this UE, the MT can support only one radio access technology. For example, a UMTS-only UE would support only the UMTS radio access technology, and would not support roaming into a GSM-based access network.

- **Multi-Radio Mode UE:** This is a UE in which the MT can support more than one radio access technology. An example of a dual-radio mode UE (i.e. a UE which supports two radio access technologies) is a GSM-UMTS UE. This dual-mode UE can support roaming into both UMTS and GSM-based access networks.

- **Single Network Mode UE:** In this case, the MT can support only one Core Network technology. For example, a UMTS-only UE would support only the UMTS CS-domain and/or PS-domain, and would not support the GSM Network Sub-System (NSS). A UMTS-only single network mode UE may further be classified into the following categories:
  - **CS Operation Mode UE:** A UE that can only access services offered by the CS-domain of the UMTS Core Network.
  - **PS Operation Mode UE:** A UE that can only access services offered by the PS domain of the UMTS Core Network.
  - **CS/PS Operation Mode UE:** A UE that can access services offered by both the CS domain and the PS domain of the UMTS Core Network.

- **Multi-Network Mode UE:** This is a UE in which the MT can support more than one Core Network technology. An example of a multi-network mode UE can be a UE that can support both the UMTS Core Network (CN) and the GSM Network Sub-System (NSS).

These categories of UE will depend upon the capabilities of the RT and NT components of the MT. However, it is expected that the initial user equipment manufactured for use in UMTS networks would at least be dual-radio mode UEs, which will also support roaming into the existing GSM-based access networks.

SUMMARY

The UMTS network architecture consists of three basic components: User Equipment (UE), Access Network and Core Network . Of these, the User Equipment is used by the mobile subscriber to access the services provided by the network. The User Equipment consists of three components: Universal Subscriber Mobile Identity (USIM), Mobile Termination (MT) and Terminal Equipment (TE). The USIM is a user-dependent part of the UE, which contains logic to unambiguously identify the mobile subscriber. In comparison, the MT and the TE and the user-independent parts of UE. The MT performs functions like radio transmission termination, authentication and mobility management, whereas the TE manages the hardware (e.g. speaker, microphones, video cameras and user display.)

3GPP specifications define the interfaces between the various components of the UE. These include Cu interface, R interface and TU interface which have been discussed in details in the chapter. The functions of the UE are also discussed including the functions of MT, TE, USIM and the Terminal Adaptation functions. Based on the capabilities of UE, it can be classified into various categories (for e.g. Single Radio Mode UE or Multi-Radio Mode UE) which have been described in the chapter.

ACCESS NETWORK

5.1 INTRODUCTION

In Chapter 3, an overview of the UMTS Network architecture was presented, the main focus being on its three basic components: the User Equipment, Access Network and the Core Network. Chapter 4 elaborated upon the User Equipment.

This chapter discusses the Access Network and its various aspects. The information has been organised as per the outline provided in Figure 5.1.

5.2 ACCESS NETWORK ENTITIES

The 3GPP standards allow two different types of Access Network systems to interface with the Core Network. These two systems are the Base Station Sub-system (BSS) and the Radio Network Sub-system (RNS). While BSS is a legacy of the GSM era, the RNS is the newly standardized access network for Rel99 onwards UTRAN. The MSC (or SGSN) can connect to one or both of these Access Network types. The Access Network Architecture is depicted in Figure 5.2.

5.2.1 Base Station Sub-system (BSS)

The Access Network in the GSM is called the Base Station Sub-system (BSS). Since there is a large number of GSM users, it is necessary to allow them to avail services from a UMTS Core Network. Thus, BSS also forms part of the UMTS Access Network.

The BSS comprises of one Base Station Controller (BSC) and one or more Base Transceiver Station (BTS).

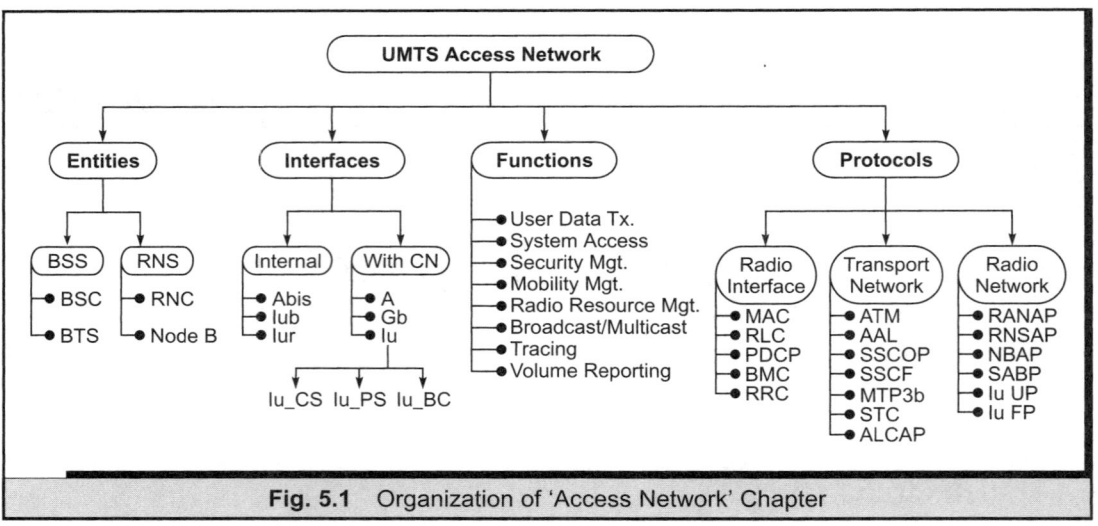

Fig. 5.1   Organization of 'Access Network' Chapter

5.2.1.1   *Base Station Controller (BSC)*

As the name suggests, the Base Station Controller (BSC) controls one or more BTS. The primary task of BSC is to control the radio resources of GSM Radio Access Network.

The important functions of BSC include radio resource management, control of BTS, inter-cell handovers and power control.

5.2.1.2   *Base Transceiver Station (BTS)*

The Base Transceiver Station (BTS) provides services in a cell. A BTS works on the instructions of BSC. The interface between BTS and BSC is based primarily on proprietary solutions.

The important functions of BTS include channel coding, encryption decryption, and transcoding and rate adaptation.

5.2.2   Radio Network Sub-system (RNS)

One of the most important standardization accomplishments in the UMTS network is the standardization of fully open Radio Network Sub-system (RNS). This is unlike BSS, which is based primarily on proprietary solutions. The RNS, also known as Universal Terrestrial Radio Access Network (UTRAN), comprises of one Radio Network Controller (RNC) and one or more Node B. The RNC is similar to BSC while Node B is similar to BTS. The open interfaces of RNS enable RNC and Node B of different vendors to interoperate with each other.

5.2.2.1 Radio Network Controller (RNC)

The Radio Network Controller (RNC) controls one or more Node B. The important functions of RNC include radio resource management, control of Node B, encryption decryption, admission control and power control (downlink power control and uplink outer loop power control).

Depending upon its function, the RNC can assume various roles. These roles are: Controlling RNC, Drift RNC and Serving RNC. A *Controlling RNC* has overall control of the logical resources of a Node B. Thus, the term Controlling RNC is always used in the context of the Node B that it controls.

From the point of view of a UE, RNC can assume two different roles: the Serving RNC and the Drift RNC. The *Serving RNC* is the one that is in charge of the radio connections between the UE and UTRAN. It is not necessary that there be a direct connection between the Serving RNC and the UE. Instead, there can be an intermediate RNC, called the *Drift RNC*, which works on behalf of the Serving RNC. More formally, the Serving RNC terminates the Radio Access Network Application Part (RANAP) signaling with the Core Network over the Iu interface. The Serving RNC also terminates the Radio Resource Control (RRC) signaling with the UE over the air interface. The Drift RNC and Serving RNC communicate using the Radio Network Sub-system Application Part (RNSAP).

5.2.2.2 Node B

The Node B works as per the instructions of RNC using the Node B Application Part (NBAP) protocol. The primary task of Node B is to interface with the UE over the air interface. A Node B serves one or more cells.

The important functions of Node B include channel coding, rate matching, spreading/despreading and inner-loop power control. Note that functions like spreading/despreading and inner-loop power control are new to UTRAN and are not performed by BTS.

5.3 NETWORK INTERFACES

The Access Network lies between the UE and the Core Network. Its components are the BSC and BTS for the Base Station Sub-system (BSS) and the RNC and Node B for the Radio Network Sub-system (RNS) (Figure 5.2).

The RNC interfaces with the CS and PS domain using the Iu interface. The Iu_CS and Iu_PS interfaces are used for the CS and PS domains, respectively. The equivalent for the BSS is the A and Gb interface.

Within the RNS, a RNC interfaces with the Node B over the Iub interface. The RNC can also interface with another RNC over the Iur interface. The Iub interface is

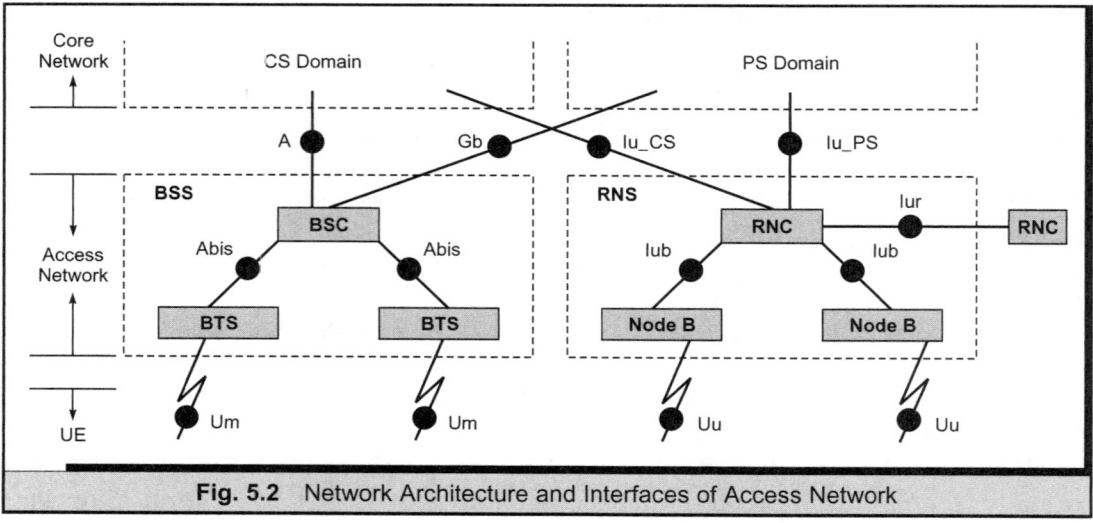

Fig. 5.2 Network Architecture and Interfaces of Access Network

equivalent to Abis interface in the BSS. However, there is no equivalent for the Iur interface in the BSS.

Over the air interface, the RNS communicates with the UE over the Uu interface. For the BSS, the interface with UE is called the Um.

All the interfaces of Access Network are listed in Table 5.1. These are further elaborated in the following sub-sections. Note that the GSM legacy interface (that includes Abis, A, and Gb interface) and the associated protocols applicable over this interface are covered very briefly in this chapter. For more information on these legacy interfaces, the reader is referred to the specifications mentioned in Table 5.1.

5.3.1 Abis Interface between BSC and BTS

The interface between Base Station Controller (BSC) and Base Transceiver Stations (BTS) is referred to as the Abis interface. It used to support services offered to GSM subscribers. This interface also allows control of the radio equipment and radio frequency allocation in the BTS.

Though Abis interface is standardized in 3GPP TS 48.05x specifications as mentioned in Table 5.1, the solutions are typically proprietary. Being a GSM legacy interface, the Abis interface and the associated protocols applicable over this are not discussed any further in this chapter.

5.3.2 Iub Interface between RNC and Node B

On the lines of Abis, the interface between RNC and Node B is referred to as the Iub interface. Unlike Abis, one of the objectives of standardizing this interface is to allow

| Table 5.1 | Access Network Interfaces |

Interface	*Between*	*Description*	*Protocol*	*Specification*
Abis	BSC – BTS	Used between BSC and BTS to support services offered to GSM subscribers.	No specific name.	48.051/48. 052/ 48. 054/48. 056/ 48. 058
Iub	RNC – Node B	Used between RNC and Node B to support services offered to UMTS subscribers.	NBAP	25.430/25.431/ 25.432/25.433/ 25.434/25.435
Iur	RNC – RNC	Used between two RNCs to support handover, radio resource handling and synchronization.	RNSAP	25.420/25.421/ 25.422/25.423/ 25.424/25.425
A	MSC/VLR – BSS	Used between MSC/VLR and BSS for BSS management, call handling and mobility management.	BSSMAP	48.004/48.006/ 48.008
Gb	SGSN – BSS	Used between SGSN and BSS for data transfer and mobility management.	BSSGP	48.014/48.016/ 48.018
Iu	MSC/VLR – RNS SGSN – RNS	Used between RNS and CN for mobility management, call handling, and packet data transfer.	RANAP	25.410/25.411/ 25.412/25.413/ 25.414/25.415

inter-connection of RNCs and Node Bs from different manufacturers. This can be seen as a significant step towards developing open and well-defined standards.

The Iub interface is used to support the services offered to the UMTS subscribers. This interface allows control of radio equipment and radio frequency allocation in the Node B.

The Iub interface provides the means of user data transport between UE and RNC via Node B. For this, a signaling protocol also exists so that user data can be handled.

The Iub interface specifications are given in the 3GPP TS 25.43x series (Table 5.1). In particular, the Node B Application Part (NBAP) protocol used between RNC and Node B is defined in 3GPP TS 25.433. The NBAP protocol is used by the RNC to control the resources of Node B. This protocol, is explained in detail in Section 5.9.3 and also used by Node B to send measurement reports to RNC.

Other Iub interface specifications in the 25.43x series provide information on its general aspects and on the transport protocols.

5.3.3 Iur Interface between RNCs

This is an open interface that allows RNCs supplied by different manufacturers to inter-operate. The Iur interface provides the means for a UE, that has a connection with UTRAN, to move between RNSs. For this, there are functions to support handover, radio resource handling and synchronization between RNSs. Besides, there are means to transfer user data and signaling information between Serving RNC (SRNC) and Node B via the Drift RNC (DRNC).

The Iur interface specifications are given in the 3GPP TS 25.42x series (Table 5.1). In particular, the Radio Network Sub-system Application Part (RNSAP) protocol, used between two RNCs is defined in 3GPP TS 25.423. The RNSAP protocol, used primarily for inter-RNC soft handover, is explained in detail in Section 5.9.2.

Other Iur interface specifications in the 25.42x series provide information on its general its aspects and on the transport protocols.

5.3.4 A Interface between MSC/VLR and BSS

The A interface between MSC/VLR and BSS is used to interface between the Core network and the Access Network. This interface provides the means for BSS management, call handling and mobility management, and is based on BSS Management Application Part (BSSMAP) as defined in 3GPP TS 48.008. Being a GSM legacy interface, the A interface and the associated protocols applicable over it are not elaborted upon in this chapter.

5.3.5 Gb Interface between SGSN and BSS

While the BSS interfaces with the MSC/VLR using the A interface in the CS domain, it interfaces with the SGSN through the Gb interface in the PS domain. The Gb interface is used for packet data transmission between UE and the Core Network via BSS. Apart from this, it is also used for mobility management. The Gb interface is based on the BSS GPRS Protocol (BSSGP) specified in 3GPP TS 48.018. Being a GSM legacy interface, the Gb interface and the associated protocols applicable over it have not been discussed in detail in this chapter.

5.3.6 Iu Interface between CN and RNS

The Interface between the Core Network and RNS is called the Iu interface. For the CS domain, this interface is refered to as the Iu_CS interface while for the PS domain it is referred to as the Iu_PS interface.

The important functions applicable over this interface relate to RNS management, mobility management, call handling, and packet data transfer. These functions are summarized as follows:

- The establishment, maintenance and release of Radio Access Bearers (RABs).
- Procedures to perform intra-system handover, inter-system handover and SRNS relocation.

- Procedures for transfer of NAS signaling messages between UE and Core Network.
- Simultaneous access to multiple Core Network for a single UE.
- Procedures for resource reservation for packet data transfer.

The Iu interface is defined in 3GPP TS 25.41x specifications (Table 5.1). In particular, the Radio Access Network Application Part (RANAP) protocol used between MSC/VLR and RNC or between SGSN and RNC is defined in 3GPP TS 25.413. The RANAP protocol is explained in detail in Section 5.9.1.

Other Iu specifications in the 25.41x series provide information on its general aspects and on the transport protocols.

While designing the Iu interface, it was endeavoured to maximize the commonality of the various protocols that apply over it. Hence, a single RANAP protocol is used for both CS and PS domain. However, the transport protocols carrying the RANAP messages had to be different, for keeping them similar would have led to severe inefficiencies.

5.4 RADIO INTERFACE PROTOCOL ARCHITECTURE

The UMTS Radio Interface is used for the establishment, reconfiguration, and release of Radio Bearers. These bearers refer to the radio interface transport based on the UTRA FDD/TDD, and are used to exchange messages with the UE over the radio interface, which consists of multiple protocol layers. The lower-most of these layers is the physical layer (also called layer 1), over which there are two protocol layers: the Medium Access Control (MAC) protocol layer, and the Radio Link Control (RLC) protocol layer. The RLC and the MAC sub-layers jointly form the layer 2 of the Radio Interface Protocol Stack. The protocol layers over the RLC layer differ for the user plane and the control plane. In the user plane, Packet Data Convergence Protocol (PDCP), and Broadcast/Multicast Control Protocol (BMC) form the layer 3 of the protocol stack. The Radio Resource Control (RRC) protocol is the layer above the RLC layer in the control plane of the protocol stack. Note that all the radio interface protocols (including RRC, PDCP, BMC, RLC and MAC) form part of the Access Stratum (AS) protocols. The Core Network protocols that use AS to communicate with UE form the Non-Access Stratum (AS) protocols (e.g. Mobility Management protocol).

Figure 5.3 depicts the UMTS Radio Interface Protocol Stack. The physical layer of the protocol stack is responsible for transporting data received from higher layers over the physical channels. It hides all details of the underlying physical media, and provides transport channels to the MAC layer, which are independent of the physical layer. Transport channels provided by the physical layer to the MAC layer are categorized as 'Dedicated Channels' and 'Common Channels', depending upon whether the

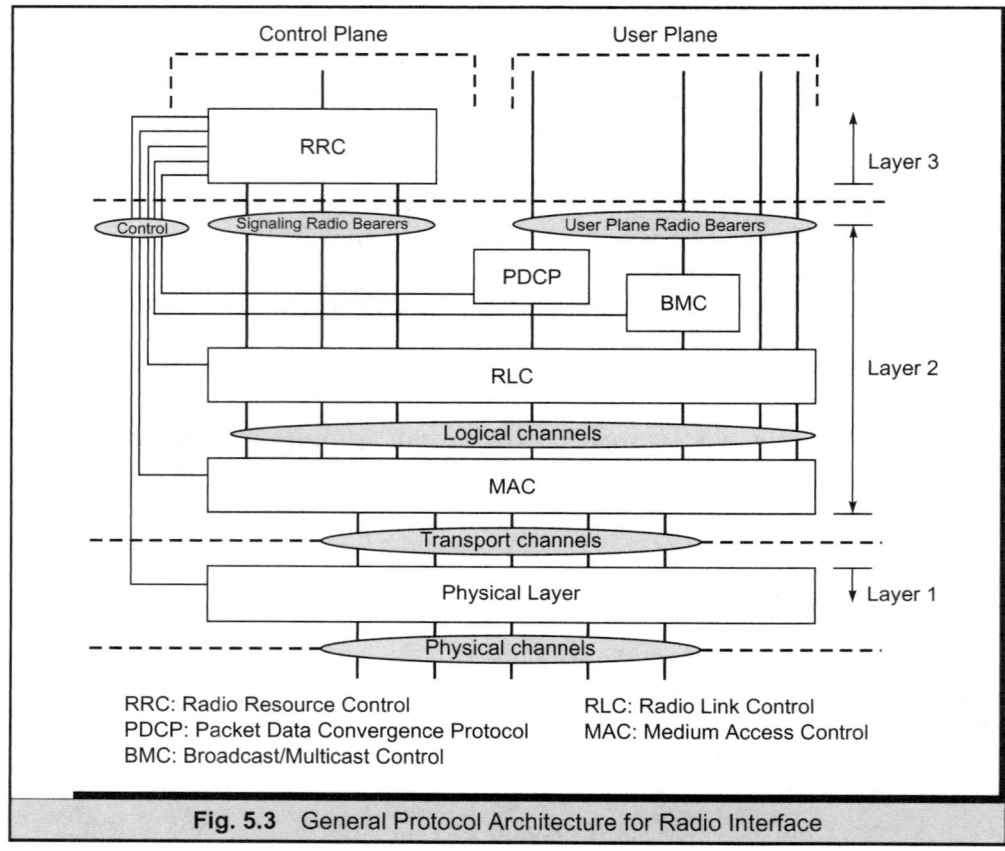

Fig. 5.3   General Protocol Architecture for Radio Interface

channel is dedicated to a particular UE, or shared between multiple UEs. The functionality of the physical layer is detailed in Section 5.7.1.

MAC, in turn, provides services to the RLC layer by means of logical channels. Logical channels are categorized as per the type of data they carry, and can either be 'Control Channels', or 'Traffic Channels', depending upon whether they are carrying control information or user traffic. The MAC layer internally maps the data received over the logical channels to the transport channels. The complete functionality of this layer is detailed in Section 5.7.2.

The RLC layer of the protocol stack provides services to higher layers in both the control plane and the user plane. It provides Service Access Points (commonly referred to as SAPs) to the higher layers, to invoke some service of the RLC layer. There are multiple types of SAPs provided by RLC, and each category of SAP defines the nature of service that RLC provides to the higher layer. The different types of SAPs and the services provided by the RLC are covered in greater detail in Section 5.7.3.

In the control plane of the protocol stack, RRC provides Service Access Points (SAPs) to the higher layers of the Non-Access Stratum (NAS). All higher layer signaling messages (for mobility management, session management, call control, etc.) are transmitted over the radio interface encapsulated into RRC messages. The architecture and functionality of the RRC protocol stack is covered in detail in Section 5.7.6.

In the user plane of the stack, PDCP and BMC are used. PDCP is used only for the PS domain services, and its primary functionality is header compression. BMC is used to transfer messages originating from the Cell Broadcast Center, over the radio interface. The details of PDCP and BMC are provided in Sections 5.7.4 and 5.7.5 respectively. Note that it is also possible for user plane information to be carried directly over RLC. In such a scenario, the PDCP and BMC do not come in the picture.

5.5 UTRAN PROTOCOL ARCHITECTURE

The Radio interface (discussed in the previous section) is used between the RNS and UE. This section discusses the protocol architecture for various UTRAN interfaces (including Iu_CS, Iu_PS, Iur, Iub and Iu_BC), which exists within the RNS and between the RNS and Core Network.

The 3GPP TS 25.401 defines a generic model for UTRAN protocols (Figure 5.4). The salient point of this model is that it is divided into two horizontal parts, namely the *Radio Network Layer* and the *Transport Network Layer*. Further, these layers are kept separate from each other. This enables the altering of one set of protocols without altering the other set.

The radio network layer has a *control plane* and a *user plane*. The control plane includes application layer protocols (like RANAP, NBAP, and RNSAP), which are used to manage Radio Access Bearers (RABs). In the user plane, the radio network layer includes User Plane (UP) protocols. These protocols facilitate transfer of user data. Examples of these protocols are, Iu User plane Protocol and Iu Framing Protocol.

In the transport network layer, there is again a *control plane* and a *user plane*. The control plane, also referred to as Access Link Control Application Part (ALCAP), is used to set bearers for user plane. For certain type of bearers, the ALCAP does not exist. This happens when the bearers are established through configuration.

The user plane in the transport network layer is used to carry data bearers and signaling bearers. As shown in Figure 5.4, there are three types of bearers in this layer. There are signaling bearers that carry the control plane information of radio network layer. Then there are the signaling bearers that carry control plane information of the transport network layer (i.e. they carry ALCAP signaling messages). Both these types of signaling bearers are pre-configured. The third types of bearers, called the data bearers, are used to carry user plane information of the radio network layer. These bearers are established using ALCAP. In cases where ALCAP does not exist, the data bearer is also pre-configured (i.e. established through configuration).

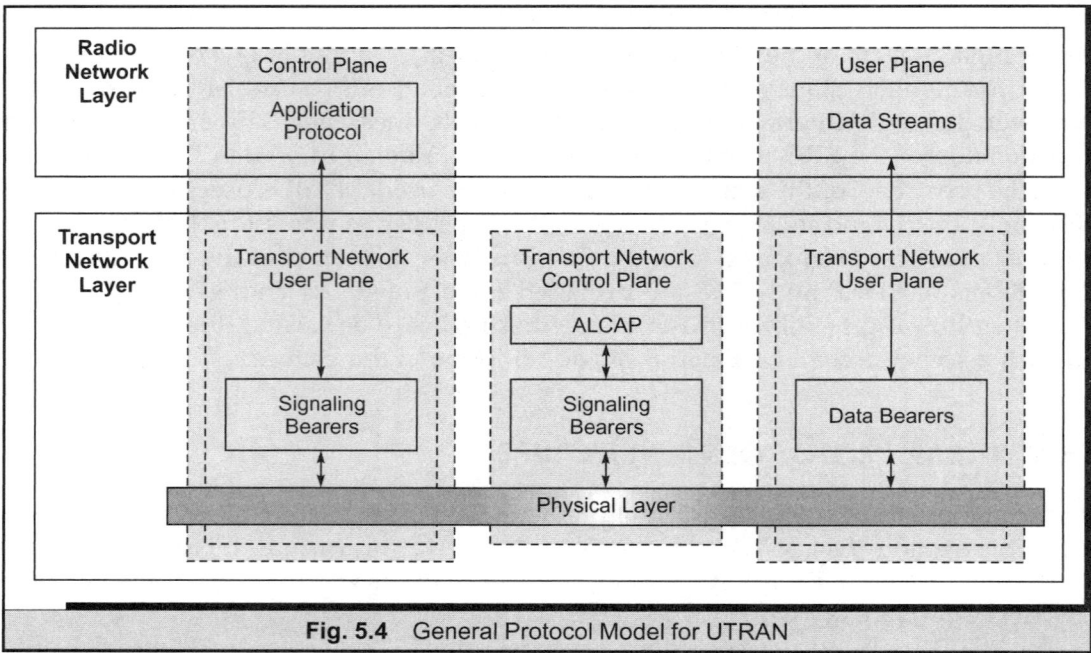

Fig. 5.4   General Protocol Model for UTRAN

The transport protocols used in the transport network layer are primarily based on ATM. The ATM protocol uses different adaptation layers for different services. In the context of the user plane, AAL2 for CS domain and AAL5 for the PS domain is used. Both AAL2 and AAL5 are explained later in this chapter. The option of IP as the transport layer is also available. Over the years, there has been a migration of pure ATM-based UTRAN to an all-IP UMTS architecture. The role of IP in UTRAN is discussed in detail in Chapter 15 (Note that even an in all-IP UMTS architecture, the IP protocol needs a link layer technology, which can be ATM).

The UTRAN Interface Protocol Architecture for each of the following interface is explained through actual examples:

- Iu_CS
- Iu_PS
- Iur
- Iub
- Iu_BC

5.5.1   Iu_CS Protocol Architecture

The Iu_CS interface exists between MSC/VLR and RNC. The protocol architecture for this interface (Figure 5.5) is explained in the following sub-sections.

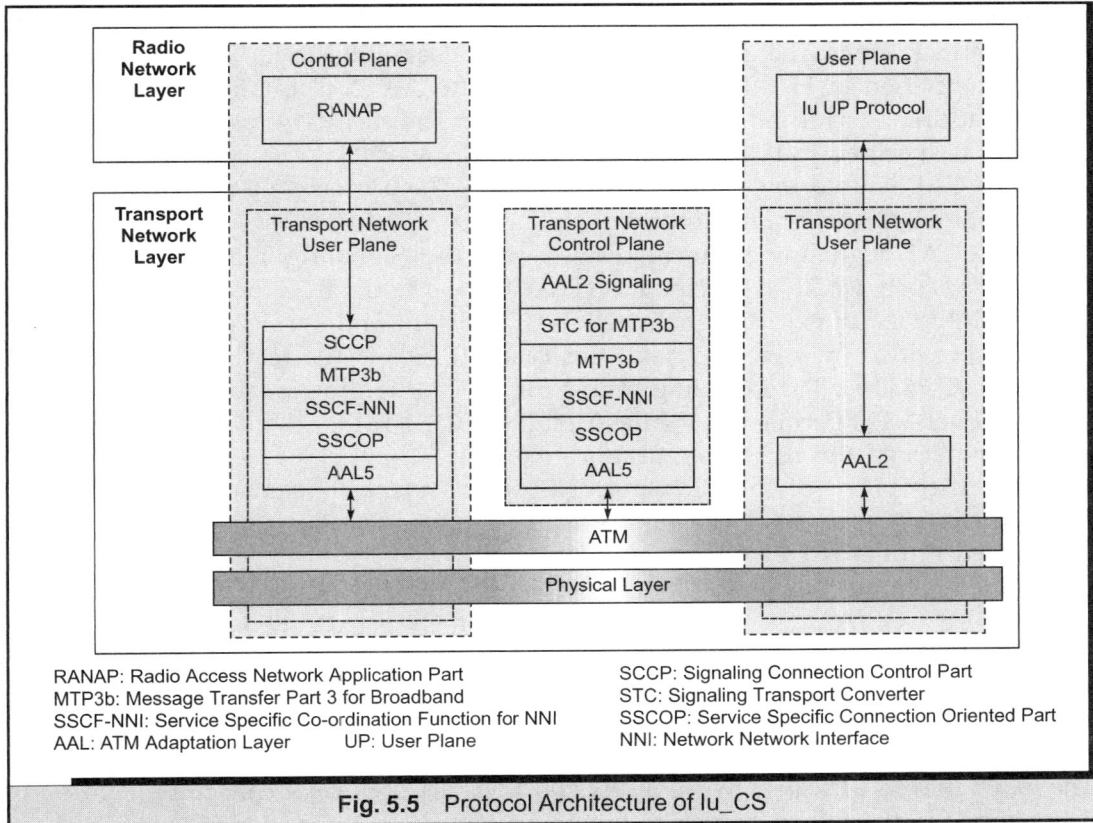

RANAP: Radio Access Network Application Part	SCCP: Signaling Connection Control Part
MTP3b: Message Transfer Part 3 for Broadband	STC: Signaling Transport Converter
SSCF-NNI: Service Specific Co-ordination Function for NNI	SSCOP: Service Specific Connection Oriented Part
AAL: ATM Adaptation Layer UP: User Plane	NNI: Network Network Interface

Fig. 5.5 Protocol Architecture of Iu_CS

5.5.1.1 Radio Network Layer

In the control plane of the radio network layer, the Radio Access Network Application Part (RANAP) protocol is used for signaling between MSC/VLR and RNC. Its function involve the establishment, maintenance and release of RABs; support of intra-system handover, inter-system handover and SRNS relocation; and transfer of NAS signaling messages between UE and the Core Network. The RANAP protocol is explained in greater detail in Section 5.9.1.

In the user plane of the radio network layer, the Iu User Plane (UP) protocol is used for transport of user data between the UTRAN and the CN over the Iu interface. Iu UP protocol is explained in Section 5.9.5.

5.5.1.2 Transport Network Layer

In the control plane of the transport network layer, the Access Link Control Application Part (ALCAP) is used. In the context of Iu_CS, the ALCAP is implemented using the AAL2 signaling protocol, as defined in ITU-T Q.2630.1. Apart from Q.2630.1, which defines the AAL2 signaling protocol for Capability Set 1, there is another protocol specified in ITU-T Q.2630.2 for Capability Set 2. The latter provides some optional additions over the former. For the sake of simplicity it is assumed that AAL2 signaling is specified in ITU-T Q.2630.1.

The ALCAP is used to establish AAL2 bearers. The bearers so established are used to carry the Iu_CS user plane information.

The ALCAP Signaling protocol resides over the Signaling Transport Converter (STC) for MTP3b, which is specified in ITU-T Q.2150.1. The STC for MTP3b is used so that the interface of AAL2 signaling is kept constant and a converter converts the interface provided by lower layer (in this case MTP3b) to the requirements of AAL2 signaling (Q.2630.1). The signaling bearer for ALCAP is pre-configured. It comprises of AAL5 over ATM Permanent Virtual Circuit (PVC). ATM PVC is a virtual connection that is established through configuration.

In the user plane of the Transport Network Layer, there are the data bearers and the signaling bearers. The data bearers are the AAL2 connections, which are established through ALCAP.

For the signaling bearers, a suite of protocols comprising of SCCP over MTP3b/SSCF-NNI/SSCOP/AAL5 are used. The SSCP protocol is explained in Chapter 6. The protocols following the SCCP are explained in different sub-sections of Section 5.8.

5.5.2   Iu_PS Protocol Architecture

The Iu_PS interface exists between SGSN and RNC. Its protocol architecture (Figure 5.6) is explained in the following sub-sections.

5.5.2.1   Radio Network Layer

The Radio Network Layer for Iu_PS is similar to the one defined for Iu_CS. Thus, RANAP protocol is used in the control plane and Iu User Plane (UP) protocol is used in the user plane.

5.5.2.2   Transport Network Layer

The control plane in the transport network layer does not exist. This implies that the data bearers required to carry user plane information of the radio network layer do not need to be established. Rather, they are pre-configured (i.e. AAL5 Permanent Virtual Connections). This remains a dominant feature of stacks using AAL5, which does not require the ALCAP signaling protocol for its establishment or release.

In the user plane, the data bearers comprise of GTP-U/UDP/IP/AAL5. For this stack, the GTP-U is the application layer, UDP is the transport layer, IP the network layer, and AAL5 is the link layer.

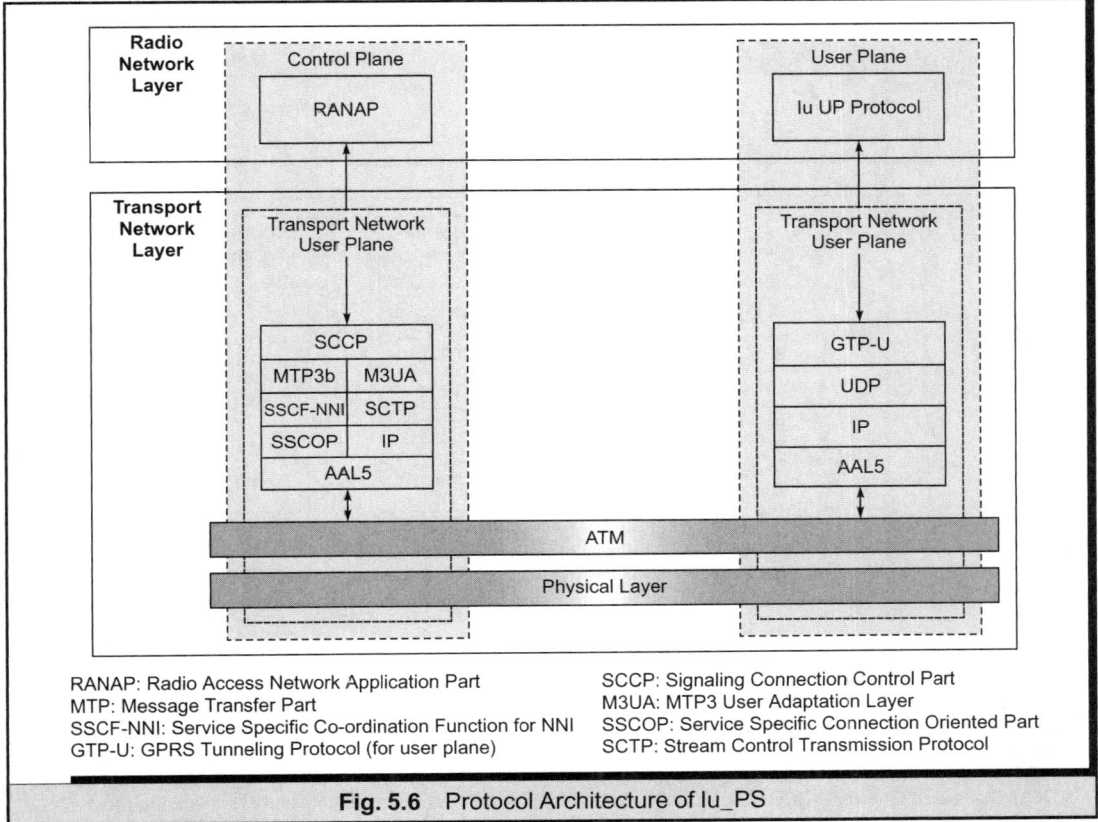

Fig. 5.6 Protocol Architecture of Iu_PS

RANAP: Radio Access Network Application Part
MTP: Message Transfer Part
SSCF-NNI: Service Specific Co-ordination Function for NNI
GTP-U: GPRS Tunneling Protocol (for user plane)

SCCP: Signaling Connection Control Part
M3UA: MTP3 User Adaptation Layer
SSCOP: Service Specific Connection Oriented Part
SCTP: Stream Control Transmission Protocol

Like the CS domain, the signaling bearers for Iu_PS comprise of SCCP over MTP3b/ SSCF-NNI/SSCOP/AAL5. However, standards specify one more configuration, which is based on IP (and not on ATM). It uses SCCP over M3UA/SCTP/IP. This configuration, also referred to as Signaling Transport, (or SIGTRAN) is explained in detail in Chapter 15.

5.5.3 Iur Protocol Architecture

The Iur interface exists between two RNCs. Its protocol architecture (Figure 5.7) is explained in the following sub-sections.

5.5.3.1 Radio Network Layer

In the control plane of the radio network layer, the Radio Network Sub-system Application Part (RNSAP) protocol is used for signaling between two RNCs over the

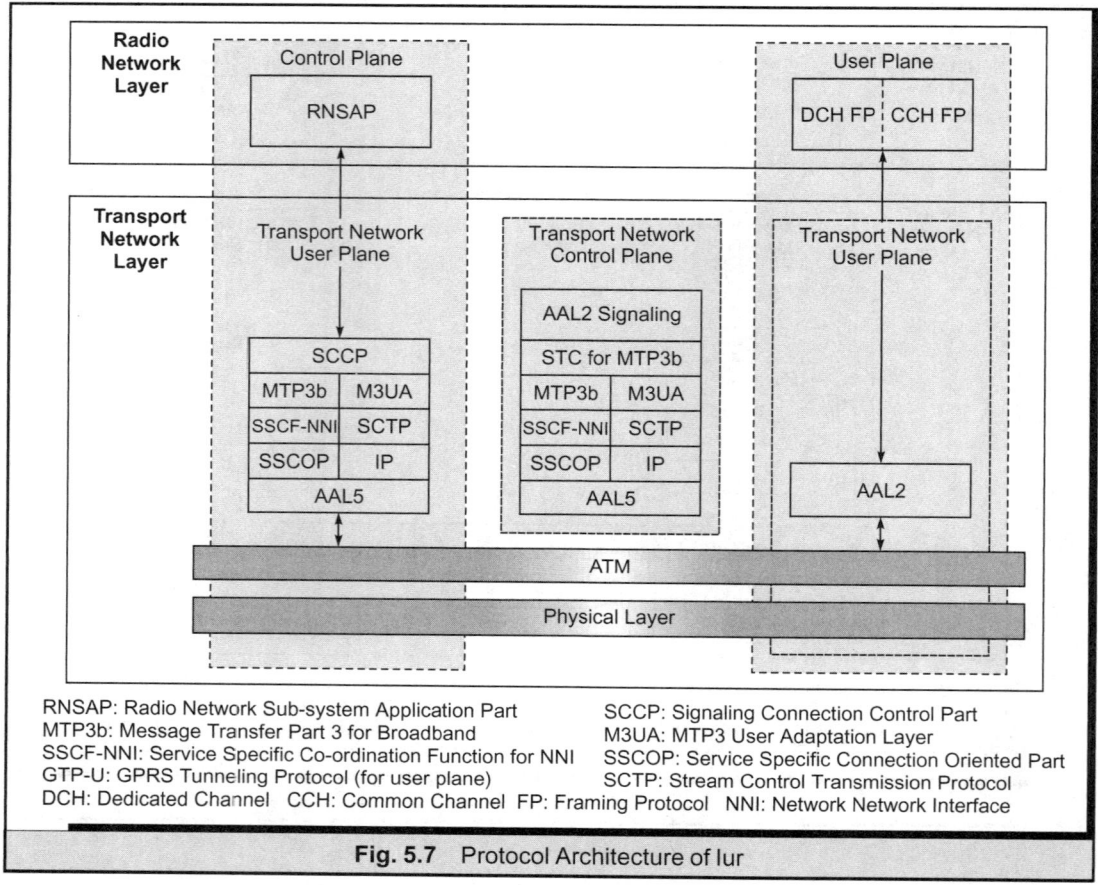

Fig. 5.7 Protocol Architecture of Iur

Iur interface. The important function of RNSAP is inter-RNC soft handover. The RNSAP protocol is explained in Section 5.9.2.

Framing protocols are used in the user plane of the radio network layer. These protocols define frame structures for carrying user data over the Iur interface. There are two types of framing protocols: one for Dedicated Channel and another for Common Channel. More information on framing protocols is provided in Section 5.9.6.

5.5.3.2 Transport Network Layer

In the control plane of the transport network layer, the Access Link Control Application Part (ALCAP) is used. The protocol stack for ALCAP in Iur is similar to that used for ALCAP in Iu_CS interface (as discussed in Section 5.5.1.2). However, in Iur interface, there is a major addition for the underlying transport, as shown in Figure 5.7. The enhancement is that below STC for MTP3b apart from MTP3b/SSCF-NNI/SSCOP, the IP option is also available (M3UA/SCTP/IP).

In the user plane of the transport network layer, the data bearers are similar to those of Iu_CS, comprising of AAL2 connections. The specifications also define an IP-based data bearer using UDP/IP. This option is not depicted in the figure.

For the signaling bearers, there are two options similar to those used for Iu_PS. One option is SCCP over MTP3b/SSCF-NNI/SSCOP/AAL5; another is SCCP over M3UA/SCTP/IP (the SIGTRAN option).

5.5.4 Iub Protocol Architecture

The Iub interface exists between Node B and RNC. Its protocol architecture (Figure 5.8) is explained in the following sub-sections.

5.5.4.1 Radio Network Layer

In the control plane of the radio network layer, the Node B Application Part (NBAP) protocol is used for signaling between RNC and Node B over the Iub interface. The important function of NBAP is to provide RNC the means to control the resources of Node B. The NBAP protocol is explained in Section 5.9.3.

Framing protocols are used in the user plane of the radio network layer. These protocols define frame structures for carrying user data over the Iub interface. There are different framing protocols for different types of channels. For more information on these, the reader is referred to Section 5.9.6.

5.5.4.2 Transport Network Layer

In the control plane of the transport network layer, the Access Link Control Application Part (ALCAP) is used. While the AAL2 signaling protocol for Iub is the same as that used for Iu_CS and Iur interface, and is based on ITU-T Q.2630.1, the underlying transport is different in this case. Here, a different Signaling Transport Converter (STC) is used, which resides over SSCF-UNI (and not MTP3b). This STC for SSCF-UNI is defined in ITU-T Q.2150.2.

In the user plane of the transport network layer, the data bearers are similar to those of Iu_CS and Iur, comprising of AAL2 connections.

The signaling bearers comprise of SCCP over MTP3b/SSCF-UNI/SSCOP/AAL5. Note that for Iub, the SSCF-UNI is used instead of SSCF-NNI. This is because the Iub interface between Node B and RNC can be viewed as a User-Network Interface (UNI), in contrast to other interfaces (e.g. Iu) which are essentially Network-Node Interfaces (NNIs). Network-Node Interface is also referred to as Network-Network Interface.

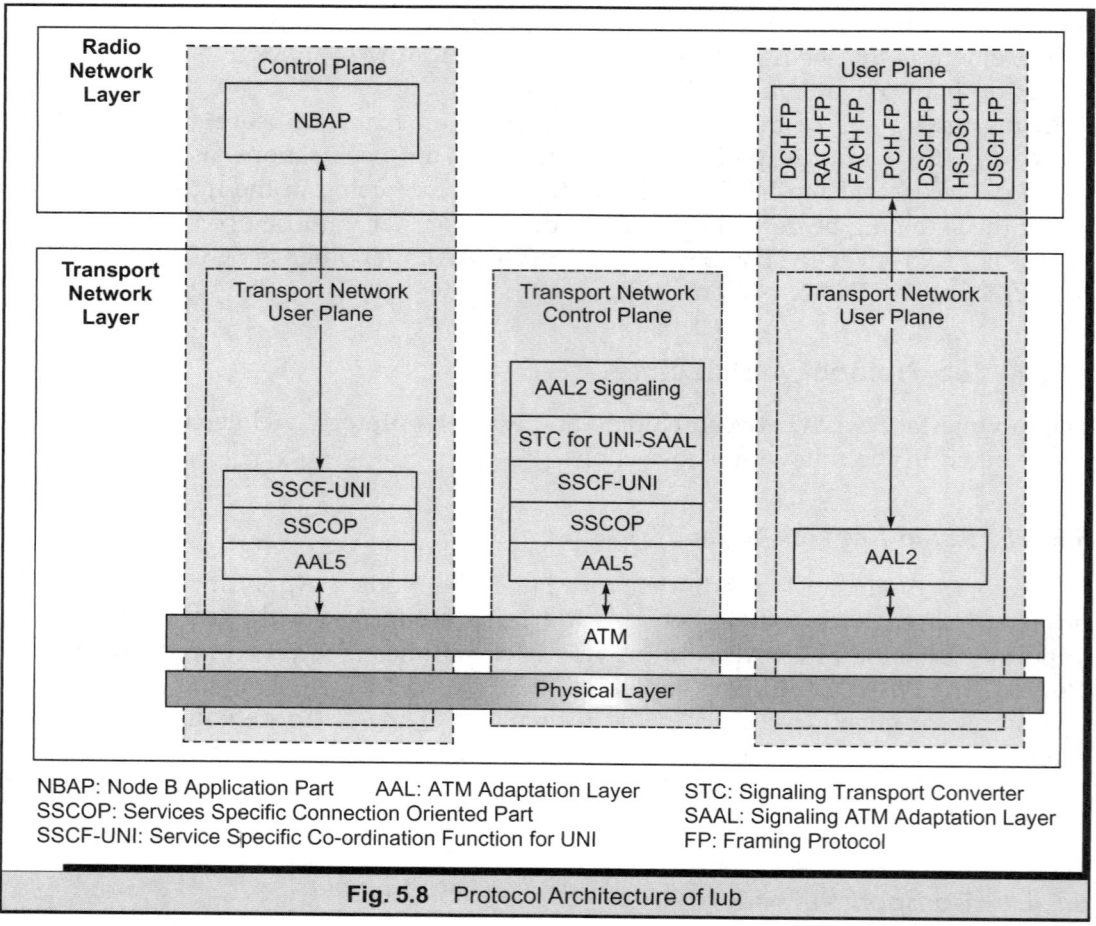

Fig. 5.8   Protocol Architecture of Iub

5.5.5 Iu_BC Protocol Architecture

The Iu_BC interface exists between Cell Broadcast Center (CBC) and RNC. Its protocol architecture (Figure 5.9) is explained in the following sub-sections.

5.5.5.1 Radio Network Layer

In the Radio Network Layer, there is only one layer called the Service Area Broadcast plan. This contains the Service Area Broadcast Protocol (SABP), which is used to control the information broadcasted by an RNC in a Service Area. The SABP protocol is explained in Section 5.9.4.

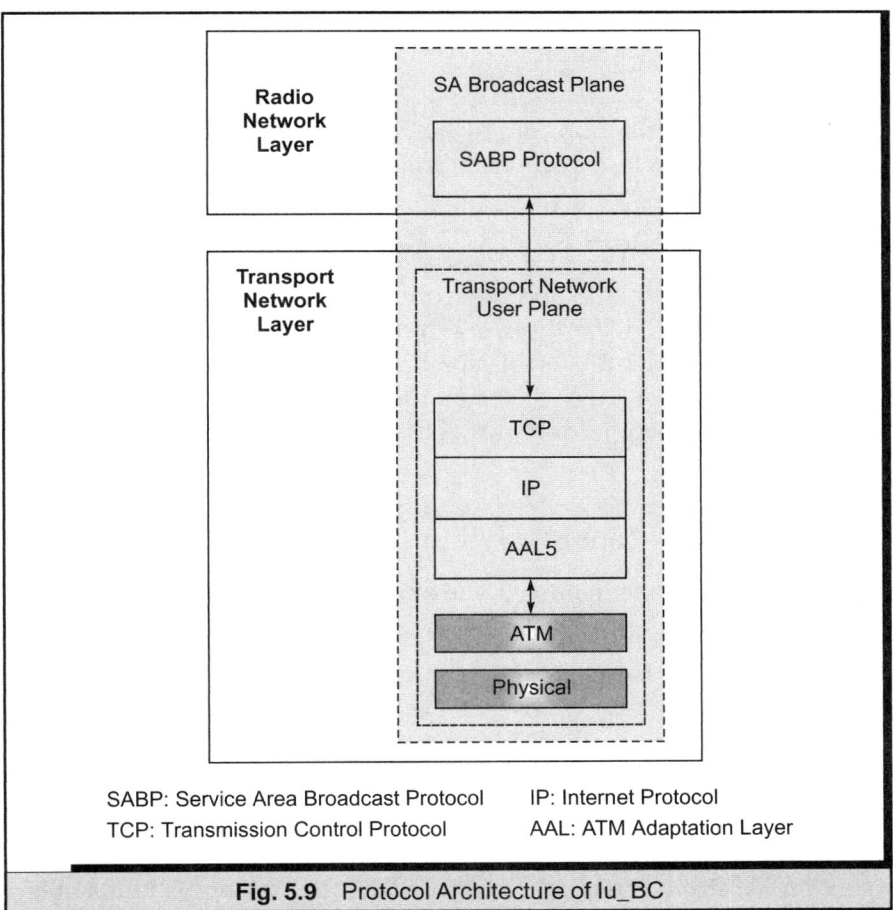

Fig. 5.9 Protocol Architecture of Iu_BC

5.5.5.2 *Transport Network Layer*

Like the radio network layer, there is only one vertical layer in the transport network layer. This is based on TCP/IP over ATM using AAL5.

5.6 FUNCTIONS

The UTRAN performs a host of functions. These are as follows:
- Transfer of User Data
- System Access Control
- Security Functions
- Mobility Management

- Radio Resource Management
- Broadcast and Multicast Services
- Other Functions

The details of UTRAN functions are available in 3GPP TS 25.401.
There functions are briefly explained in following sub-sections.

5.6.1 Transfer of User Data

The UTRAN lies between the MS and the Core Network. It communicates with the MS over the Uu interface and with the Core Network over the Iu interface. Thus, the UTRAN acts as a relay for transfer of signaling data (NAS signaling) and user data between MS and CN. Note that the signaling between MS and UTRAN (i.e. AS signaling using RRC protocol) terminates at the UTRAN and is not relayed to the Core Network.

5.6.2 System Access Control

The System Access Control encompasses those functions that enable a UMTS user to avail of UMTS services. Some of these functions are:

- **Admission Control:** Here, the goal is to ensure that the load on the system is below the permissible limits. For this, the admission control, based on the current load of the system and the possibility of it interfering with other existing connections, admits or rejects requests for new Radio Access Bearers (RABs). The admission control is typically performed during initial UE access, RAB assignment/reconfiguration and at handover.

- **Congestion Control:** The aim is to ensure that the system is in a stable state. For this, the congestion control function monitors, detects and handles situations wherein the system is reaching a near overload or an overload situation with the already connected users.

- **System Information Broadcasting:** This function is used by the UTRAN to provide system level information (for both the Core Network and the Access Network) to the MS. The System Information is broadcasted by the RRC protocol.

In the existing literature, Admission Control and Congestion Control are often categorized under the heading Radio Resource Management (see Section 5.6.5). However, the above classification is based on 3GPP TS 25.401.

5.6.3 Security Functions

The UTRAN performs two important security functions, namely *data confidentiality* and *data integrity*. Confidentiality refers to the property that does not allow information to be made available to unauthorized entities. For this, the data is encrypted and sent over the air interface. Encryption can be applied to both the user and the signaling data.

Ensuring data integrity implies that the contents of a packet are not altered in an unauthorized manner. The data integrity function is applied only to signaling data.

The UTRAN controls the security mode (i.e. whether or not to use ciphering and/or data integrity). The RRC protocol controls the security mode.

Moreover, the UTRAN also participates in the ciphering and data integrity (provided that these are applied). The ciphering function is applied at the RLC layer (for acknowledged and un-acknowledged mode) or at the MAC layer (in case the transparent mode of RLC is used). The data integrity function is applied at the RRC layer.

The details of security functions are provided in Chapter 14.

5.6.4 Mobility Management

The UTRAN performs a host of mobility management functions, which involve tracking the current location of an MS. Some of these functions are as follows:

- **Handover:** This is to ensure that the motion of the MS does not affect the state of its established connections. Handover comes in the picture in various situations, e.g. at the boundaries of two cells or of two systems (e.g. GSM and UMTS). As briefly discussed in Chapter 2, there may be various types of handovers: Soft Handover, Softer Handover, and Hard Handover. The UTRAN (in particular the RRC protocol) manages various types of handovers to provide seamless mobility.

- **SRNS Relocation:** This refers to the procedure through which the SRNS role is taken over by another RNS. The SRNS relocation function manages the Iu interface connection mobility from one RNS to another. The SRNS relocation was briefly discussed in Chapter 2.

- **Paging support:** Paging is used by the Core Network to request MS to establish a signaling connection with the Core Network (e.g. when there is an incoming call for the MS). UTRAN supports the broadcast of a page message in an area where the MS is known to be present.

- **Positioning:** This function is used to determine the geographic position of a UE. It may be used for location services (discussed in Chapter 13) and other similar services.

5.6.5 Radio Resource Management

Radio resource management is concerned with the allocation and maintenance of radio communication resources. This includes the following:

- **Radio resource configuration and operation:** This is the primary goal of radio resource management. It involves configuring the radio network resources (like cells and common transport channels) and allocating/de-allocating them into or out of operation.

- **Radio environment survey:** In order to understand the behavior of the radio interface and to use radio resources effectively, a radio environment survey is performed. This survey involves measurement of various parameters (e.g. received signal strengths, estimated bit error ratios, transmission range, synchronization status, received interference level, etc.).

- **Combining/splitting control:** This function controls the combining/splitting of information streams to receive/ transmit the same information through multiple physical channels (possibly in different cells) from/towards a single MS.

- **Connection set-up and release:** In an end-to-end connection between MS and another entity, the UTRAN is involved in MS-UTRAN connection (over Uu interface), intra-UTRAN connection (between Node B and RNC), and the UTRAN-CN connection. As such, the UTRAN facilitates setup and release of these connections and modification of parameters of these connections.

- **Allocation and de-allocation of Radio Bearers:** A Radio Access Bearer (RAB) extends from the MS to the edge of the Core Network. A RAB is associated with various QoS parameters. The function of UTRAN is to map the QoS of a RAB to the physical radio channels.

- **Dynamic Channel Allocation (DCA) for TDD:** The Dynamic Channel Allocation (DCA) is applied to the TDD mode. There are two variants of DCA: *Fast DCA* and the *Slow DCA*. The Slow DCA is the process of assigning radio resources, including time slots, to different TDD cells according to the varying cell load. Fast DCA is the process of assigning resources to radio bearers, and is related to admission control.

- **Radio protocols function:** As discussed earlier, the radio interface protocols include RLC and MAC, among others. These protocols carry out various radio protocols functions like segmentation and reassembly, acknowledged/unac-knowledged delivery and multiplexing of radio bearers on transport channels.

- **RF power control:** As discussed in Chapter 2, power control defines the means to control the power at which the MS and Base Station transmit signals to each other. This is required to minimize interference between two users who are communicating simultaneously as well as to maintain the quality of existing connections. The power control functions include *open loop* and *closed loop* power control for uplink and downlink directions.

- **Radio channel coding/decoding:** In order to allow detection or correction of signal errors introduced by the transmission medium, UTRAN introduces redundant information in the source data flow. The coding algorithm and the extent of redundancy information present in the data differs for different types of logical channels and data. At the receiving end, the source information is reconstructed using the redundant information added by the channel coding function. Note that the decoding function detects and corrects possible errors received in the data.

- **Channel coding control:** This function generates control information required by the channel coding/decoding execution functions. Control information includes channel coding scheme, code rate, etc.

- **CN Distribution function for NAS messages:** The UTRAN allows a MS to communicate with multiple CN at the same time. For this, a distribution function is used to route the NAS message to the appropriate CN domain. In the downlink direction (i.e. from CN to MS), the UTRAN provides the information to MS so that the latter can determine the originating CN. In the uplink direction, the UE inserts appropriate values for the CN domain indicator in the AS message. The RNC then uses this CN domain indicator to distribute the NAS message to the corresponding RANAP instance for transfer over Iu interface.

- **Uplink Synchronization:** This is used to synchronize the uplink radio signals from the MS to the UTRAN. When Node B detects an uplink burst, it evaluates the received power level and timing. Node B then replies by sending the adjustment function to MS so that the MS can modify its timing and power level for the next transmission.

5.6.6 Broadcast and Multicast services

UTRAN provides various Broadcast and Multicast services, which are as follows:

- **Broadcast/Multicast Information Distribution:** The RNC uses Broadcast/ Multicast Control (BMC) protocol to distribute broadcast/multicast information (like CBS messages) to the BMC entities configured per cell for further processing.

- **Broadcast/Multicast Flow Control:** In order to ensure that broadcast/multicast information does not cause congestion, the source of information is informed so that the situation can be tackled easily.

- **Cell Broadcast Service (CBS) Status Reporting:** The RNC collects various statistics (e.g. No-of-Broadcast-Completed-List, Radio-Resource-Loading-List). This information is transmitted to the Cell Broadcast Center (CBC) on receipt of a query from the latter.

5.6.7 Other Functions

UTRAN performs certain other functions. One of them is tracing, which allows tracking various events associated with a MS and its activities.

Apart from tracing, UTRAN also collects reports for the volume of unacknowledged data. This report is sent to the CN for accounting purposes.

5.7   RADIO INTERFACE PROTOCOLS

In Section 5.4, the radio interface protocol architecture was discussed, wherein, various protocols used in the radio interface were briefly mentioned. This section elaborates upon these protocols particularly the following ones:

- Physical Layer
- Medium Access Control (MAC)
- Radio Link Control (RLC)
- Packet Data Convergence Protocol (PDCP)
- Broadcast/Multicast Control (BMC)
- Radio Resource Control (RRC)

5.7.1 Physical Layer

The physical layer is the lower-most layer of the UMTS radio interface stack. It is the layer that is responsible for the actual transmission of higher layer data over the physical channels. The physical layer has to perform various functions, which are in one way or another related to the process of data transmission. The functions performed by the physical layer are summarized as follows:

- **Data Transfer over Physical Channels:** This is the primary function of the physical layer. The transmission of higher layer data takes place using physical channels.

- **FEC Encoding/Decoding:** Besides transmission of user data over the physical channels, the physical layer is also responsible for correction of errors introduced during transmission. To reduce transmission errors, the physical layer uses Forward Error Correction (FEC) schemes. Three different schemes are available to the physical layer, namely *Convolutional Coding, Turbo Coding,* and *No FEC Coding.* The main principle behind FEC coding is to add redundancy to the transmitted bit stream, so that the receiver can correct the occasional bit errors during transmission. Normally, Convolutional Coding is considered more suitable for low data rates, while Turbo Coding is used for higher rates. In cases where error correction is not needed (e.g. when the physical link is very reliable), no FEC coding is required. FEC Coding is also referred to as *Channel Coding.*

- **Error Detection:** Error detection is concerned with the detection of errors within received blocks of data. A Cyclic Redundancy Check (CRC) method is used for detecting errors in transport blocks. In case the physical layer detects a CRC error, it indicates this to the higher layer (layer 2), where the re-transmission function can take the required action. In the UTRAN physical layer, the channel coding functionality is combined with the error detection function to form a 'Hybrid ARQ' scheme. While the channel coding function aims at fixing as many errors as possible, the error detection function checks that there are no further errors. If there are, the error detection function informs the higher layers of the same.

- **Radio Measurements:** The RRC layer requests measurement collection from the lower layers. While the radio measurements are made in the physical layer, the traffic volume measurements are carried out in the MAC. The 3GPP TS 25.215 and 3GPP TS 25.225 define radio measurements to be carried out on the air interface for the FDD and the TDD modes respectively. The measurements to be performed at the physical layer include 'Signal-to-Interference Ratio (SIR)', 'Block Error Rate (BLER)', 'Bit Error Rate (BER)', 'Round Trip Time (RTT)' and 'Transmitted (Code/Carrier) Power'.

- **Multiplexing/De-multiplexing Transport Channels to CCTrCHs and mapping of CCTrCHs on physical channels:** A UE can use several transport channels simultaneously, which are multiplexed by the physical layer into a Coded Composite Transport CHannel (CCTrCH). A CCTrCH consists of transport channels of the same type. At the transmitting side, the physical layer multiplexes the transport channels into CCTrCHs and then maps each CCTrCH onto physical channels. Bits in a CCTrCH can be divided over more than one physical channel, depending upon the configuration. At the receiving side, the physical layer extracts the CCTrCH from the physical channel(s), and then de-multiplexes the transport channels from the CCTrCHs. Data on these transport channels is then forwarded to layer 2. Figure 5.10 depicts the multiplexing and mapping process in the physical layer at the transmitting end. Table 5.2 describes the transport channels and their mapping onto physical channels.

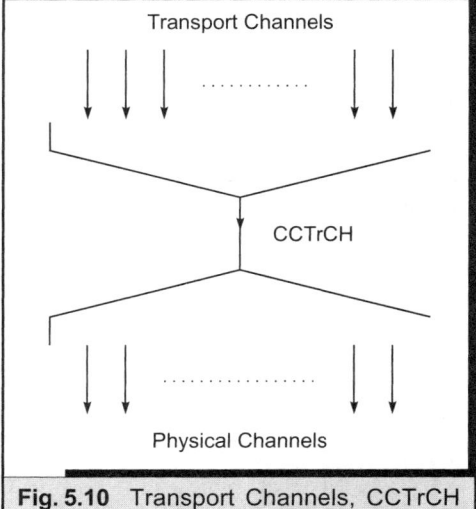

Fig. 5.10 Transport Channels, CCTrCH and Physical Channels

Table 5.2   Transport Channels and Mapping with Physical Channels***

Type	Name of Transport Channel	Abbr.	Description	Mapping to Physical Channel	
				Uplink	Downlink
Common Transport Channel	Random Access Channel	RACH	A contention based uplink channel used for transmission of relatively small amounts of data, e.g. initial access or non-real-time dedicated control or traffic data.	PRACH	N/A
	Common Packet Channel	CPCH	A contention based channel used for transmission of bursty data traffic. This channel only exists in FDD mode and only in the uplink direction. The common packet channel is shared by the UEs in a cell and therefore, it is a common resource. The CPCH is fast power controlled.	PCPCH (FDD)	N/A
	Forward Access Channel	FACH	Common downlink channel without closed-loop power control used for transmission of relatively small amounts of data.	N/A	SCCPCH
	Downlink Shared Channel	DSCH	A downlink channel shared by several UEs carrying dedicated control or traffic data.	N/A	PDSCH
	Uplink Shared Channel	USCH	An uplink channel shared by several UEs carrying dedicated control or traffic data, used in TDD mode only.	PUSCH (TDD)	N/A
	Broadcast Channel	BCH	A downlink channel used for broadcast of system information into an entire cell.	N/A	PCCPCH
	Paging Channel	PCH	A downlink channel used for broadcast of control information into an entire cell allowing efficient UE sleep mode procedures. Currently identified information types are paging and notification. Another use could be UTRAN notification of change of BCCH information.	N/A	SCCPCH
Dedicated Tr. Channel	High-Speed Downlink Shared Channel	HS-DSCH	A downlink channel shared between UEs by allocation of individual codes, from a common pool of codes assigned for the channel.	N/A	HS-PDSCH
	Dedicated Channel	DCH	A channel dedicated to one UE used in uplink or downlink.	DPCCH/DPDCH (FDD), DPCH (TDD)	DPCCH/DPDCH (FDD), DPCH (TDD)

PRACH: Physical Random Access Channel
PCPCH: Physical Common Packet Channel
PCCPCH: Primary Common Control Physical Channel
SCCPCH: Secondary Common Control Physical Channel
PDSCH: Physical Downlink Shared Channel

PUSCH: Physical Uplink Shared Channel
HS-PDSCH: High-Speed Physical Downlink Shared Channel
DPCCH: Dedicated Physical Control Channel
DPDCH: Dedicated Physical Data Channel
DPCH: Dedicated Physical Channel

- **Rate Matching:** Data available over transport channels is transferred over physical channels in a periodic fashion. The period of transmission is called the Transmission Time Interval (TTI). A TTI of 10msec is used in UTRAN for most transport channels. The number of bits available on transport channels can vary with each TTI. However, the physical channels carry data at a fixed rate, and hence, rate matching has to be done by the physical layer to ensure that the radio frames transferred over the physical links are completely filled. The rate matching function is performed by the physical layer in both the uplink and downlink direction to adjust the rates between the transport and physical channel(s). In the uplink direction, if the number of bits available in the transport channels is less than the bits carried by the radio frame, the physical layer repeats the bits so as to fill the complete radio frame. In the downlink direction, if the bits are not sufficient in the transport channels, then transmission is interrupted till the next TTI. This is called Discontinuous Transmission (DTX).

- **Spreading and Scrambling:** As discussed in Chapter 2, spreading and scrambling is used in WCDMA to offer a multi-access environment. The spreading and scrambling operation is carried out within the physical layer, using the spreading and scrambling codes.

- **Inner Loop Power Control:** The power control mechanism in WCDMA consists of an inner loop power control and an outer loop power control mechanism. As discussed in Chapter 2, the inner loop power control calculates the SIR of the received signals and sends Transmit Power Control (TPC) bits to the sending side. The outer loop power control calculates the SIR_{target}. While the outer loop power control is handled by the RRC, the inner loop power control is carried out entirely in Layer 1. This makes the inner loop power control mechanism very fast; hence, it is also known as fast power control mechanism.

5.7.2 Medium Access Control (MAC)

The Medium Access Control (MAC) layer is the lower sub-layer of layer 2 of the protocol stack. The MAC communicates with the physical layer using transport channels. It is the only layer that needs the intelligence to manipulate the physical layer. The main functionality of the MAC layer is to map data received on *logical channels* from RLC to appropriate *transport channels* of the physical layer. The MAC layer uses the services of the physical layer and in turn provides service to the RLC layer. The most important service provided by MAC layer is data transfer. The data transfer service is unacknowledged; if required, reliable acknowledge-mode service is provided by the layer above the MAC Layer (i.e. RLC layer).

The MAC layer provides various logical channels to RLC layer. Each type of logical channel is defined by the type of information it carries. Logical channels provided by

MAC are divided into two distinct sets, namely *Control Channels* and *Traffic Channels*. Control channels provide means to transfer control information. Broadcast Control Channel (BCCH) and Paging Control Channel (PCCH) are examples of control channels. Traffic channels provide means to transfer user plane information. The Dedicated Traffic Channel is an example of traffic channel. Table 5.3 lists the logical channels provided by the MAC layer and how it maps to the transport channels provided by the physical layer. For example, the Broadcast Control Channel (BCCH) at MAC layer maps to Broadcast Channel (BCH) or Forward Access Channel (FACH) at the physical layer. This mapping applies only to downlink direction. Note that broadcast information is sent by the network to UEs. There is no flow of broadcast information from UE to network. Thus, the uplink mapping of BCCH does not apply. Similar arguments are used to understand the logical channels and their mapping with physical channels.

The MAC layer is detailed in 3GPP TS 25.321.

5.7.2.1 MAC Layer Architecture

Based on the type of logical channels provided, the MAC layer is further divided into four different logical entities, as follows:

- **MAC-b:** This controls access to the Broadcast CHannel (BCH).

- **MAC-hs:** The MAC-hs controls access to the High-Speed Downlink Shared CHannel (HS-DSCH).

- **MAC-c/sh:** It controls access to the common and shared channels (except the broadcast and high-speed downlink shared channel). This includes control of Common Packet CHannel (CPCH), Random Access CHannel (RACH), Forward Access CHannel (FACH), Downlink Shared CHannel (DSCH), Uplink Shared CHannel (USCH) and Paging CHannel (PCH).

- **MAC-d:** The MAC-d controls access to the Dedicated CHannel (DCH).

Figure 5.11 depicts the MAC layer architecture comprising MAC-b, MAC-hs, MAC-c/sh and MAC-d. As shown in the figure, the MAC maps various logical channels to the transport channels.

5.7.2.2 MAC Layer Functions

The MAC layer performs various functions of which the important ones are summarized as follows (for detailed description of MAC functions, the reader is referred to 3GPP TS 25.301):

- **Mapping between Logical Channels and Transport Channels:** As illustrated in Table 5.3, the MAC layer maps logical channels (between RLC and the MAC layer) to the transport channels (between MAC and the physical layer).

Table 5.3 Logical Channels and Mapping with Transport Channels***

Type	Name of Logical Channel	Abbr.	Description	Mapping to Transport Channel	
				Uplink	Downlink
Control Channel	Broadcast Control Channel	BCCH	A downlink channel for broadcasting system control information.	N/A	BCH/FACH
	Paging Control Channel	PCCH	A downlink channel that transfers paging information. This channel is used when the network does not know the location cell of the UE, or, the UE is in the cell-connected state (utilizing UE sleep mode procedures).	N/A	PCH
	Common Control Channel	CCCH	Bi-directional channel for transmitting control information between the network and UEs. This channel is commonly used by the UEs that have no RRC connection with the network as well as by the UEs that use common transport channels when accessing a new cell, after cell re-selection.	RACH	FACH
	Dedicated Control Channel	DCCH	A point-to-point bi-directional channel that transmits dedicated control information between a UE and the network. This channel is established through RRC connection setup procedure.	RACH/CPCH (FDD)/DCH/USCH (TDD)	FACH/DSCH/HS-DSCH/DCH
	Shared Channel Control Channel	SHCCH	Bi-directional channel that transmits control information for uplink and downlink shared channels between network and UEs. This channel is for TDD only.	RACH (TDD)/USCH (TDD)	FACH (TDD)/DSCH (TDD)
Traffic Channel	Dedicated Traffic Channel	DTCH	It is a point-to-point channel, dedicated to one UE, for the transfer of user information. A DTCH can exist in both uplink and downlink.	RACH/CPCH (FDD)/DCH/USCH (TDD)	FACH/DSCH/HS-DSCH/DCH
	Common Traffic Channel	CTCH	A point-to-multipoint unidirectional channel for transfer of dedicated user information for all or a group of specified UEs.	N/A	FACH

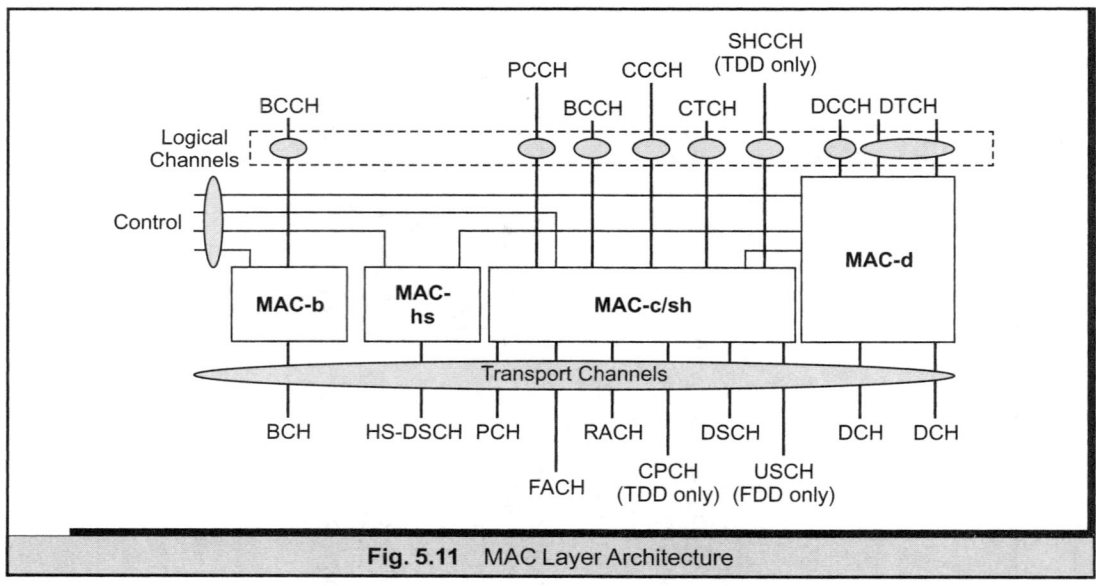

Fig. 5.11 MAC Layer Architecture

- **Selection of appropriate Transport Format:** Each transport channel is associated with either a single transport format (for transport channels with a fixed or slow changing rate) or a set of *transport formats* (for transport channels with fast changing rate). A transport format is defined as a combination of encoding, interleaving, bit rate and mapping onto physical channels. For example, a variable rate DCH has a set of transport formats, one transport format for each rate, whereas a fixed rate DCH has a single transport format. The MAC layer selects a transport format for each channel depending upon the instantaneous source rate. The control of transport formats ensures efficient use of transport channels.

- **Priority handling between data flows of one UE:** When selecting a transport format, the MAC layer can prioritize data flows to a UE. The priority is determined using the RLC buffer status and the priority of the logical channel on which the request is received. The RRC determines the priority of a logical channel at the time the radio bearer is setup/reconfigured. Through priority handling, the higher priority flows can be given higher bit rate combinations and lower priority flows can be given lower bit rate combinations (even zero rate).

- **Priority handling between UEs by means of Dynamic Scheduling:** In order to utilize the spectrum resources efficiently for bursty transfer, a dynamic scheduling function may be applied. MAC realizes priority handling on common and shared transport channels.

- **Identification of UEs on common Transport Channels:** In certain cases, like when a particular UE is addressed on a common downlink channel, or when a UE is using the RACH, the MAC layer identifies the received message, i.e. whether it is for the UE or not. This is done by using the UE identification field in the MAC PDU header placed for this purpose. If the message is indeed for this UE, the MAC forwards it to the RLC layer from where it is routed to higher layers.

- **Multiplexing/de-multiplexing of upper layer PDUs into/from Transport Blocks delivered to/from the Physical Layer on Common Transport Channels:** For common transport channels, MAC provides this service, as it is not provided by the physical layer.

- **Multiplexing/de-multiplexing of upper layer PDUs into/from Transport Block Sets delivered to/from the physical layer on Dedicated Transport Channels:** The MAC also provides the means to multiplex/de-multiplex dedicated transport channels (i.e. DTCH) on the same transport channel (i.e. DCH). This is done to enhance the efficiency of information transfer.

- **Traffic Volume Measurement:** MAC monitors the traffic volume on logical channels and reports this to RRC layer. Based on the required volume, the RRC can decrease or increase the allocated capacity.

- **Transport Channel type switching:** This refers to the switching between the common and dedicated transport channels, based on a switching decision derived by RRC.

- **Ciphering:** Ciphering refers to encrypting the data to prevent unauthorized users from accessing this data. Ciphering is done by MAC only when RLC uses the transparent mode. In other RLC modes (like acknowledge and unacknowledged mode), the ciphering function is performed by the RLC layer.

5.7.3 Radio Link Control (RLC)

Though Medium Access Control (MAC) layer provides some of the link layer functions, it does not provide all services needed from a link layer. For example, it does not provide reliable and acknowledged service. Thus, the Radio Link Control (RLC) is required at the link layer. The RLC provides various services including reliable data transfer, segmentation and reassembly, flow control, error control, and sequence numbering.

The protocol layers over the RLC layer differ for the user plane and the control plane. In the user plane, RLC provides services to Packet Data Convergence Protocol (PDCP) and Broadcast/Multicast Control Protocol (BMC). RLC also allows higher layer data to be carried directly over it.

In the control plane, RLC provides services to the Radio Resource Control (RRC) protocol.

The RLC layer is detailed in 3GPP TS 25.322.

5.7.3.1 RLC Layer Architecture

In order to provide only those functions as required by the upper layer, RLC provides three different modes of operation:

- RLC Transparent Mode (TM)
- RLC Unacknowledged Mode (UM)
- RLC Acknowledged Mode (AM)

5.7.3.1.1 RLC Transparent Mode (TM)

The Transparent Mode (TM) is the most basic mode of operation. As RLC performs a very primitive function in this mode, it is said to operate in transparent mode. The term 'transparent' is also used because the RLC mode does not add any header or trailer.

In TM mode, the transmitting side has a transmission buffer and the receiving side has a reception buffer. In addition, if segmentation is supported, there is a segmentation and reassembly function at the transmitting and receiving side respectively. The RLC operation in Transparent Mode (TM) is depicted in Figure 5.12. Note that in this mode the ciphering function is not performed at the RLC layer. Instead, the MAC layer performs ciphering for RLC in TM mode.

The RLC TM operates over BCCH, DCCH, PCCH, CCCH, SHCCH or a DTCH logical channel.

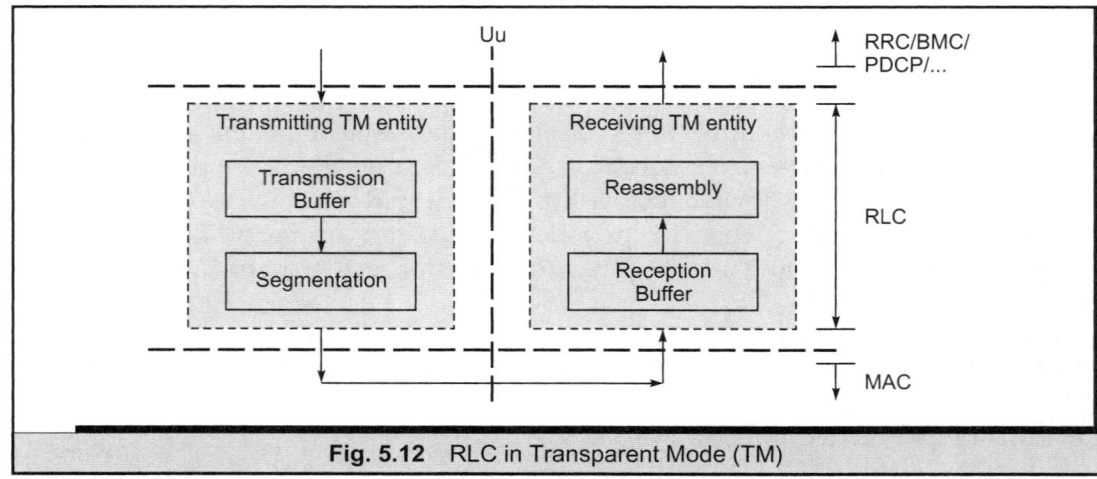

Fig. 5.12 RLC in Transparent Mode (TM)

5.7.3.1.2 RLC Unacknowledged Mode (UM)

The Unacknowledged Mode (UM) goes a step further from Transparent Mode. In the unacknowledged mode, an RLC header is added to the RLC PDU. This PDU header has a sequence number and a length indicator. The sequence number is used for sequence numbering and to guarantee the integrity of reassembled PDU. The length indicator is used to indicate the last octet of each RLC SDU ending within an RLC PDU. Note that in a RLC PDU, there can be more than one concatenated RLC SDU. The length indicator is used to identify the boundaries of the RLC SDU.

Figure 5.13 shows the RLC operation in Unacknowledged Mode (UM). The enhancement over transparent mode is the addition and removal of RLC header and the ciphering/de-ciphering function.

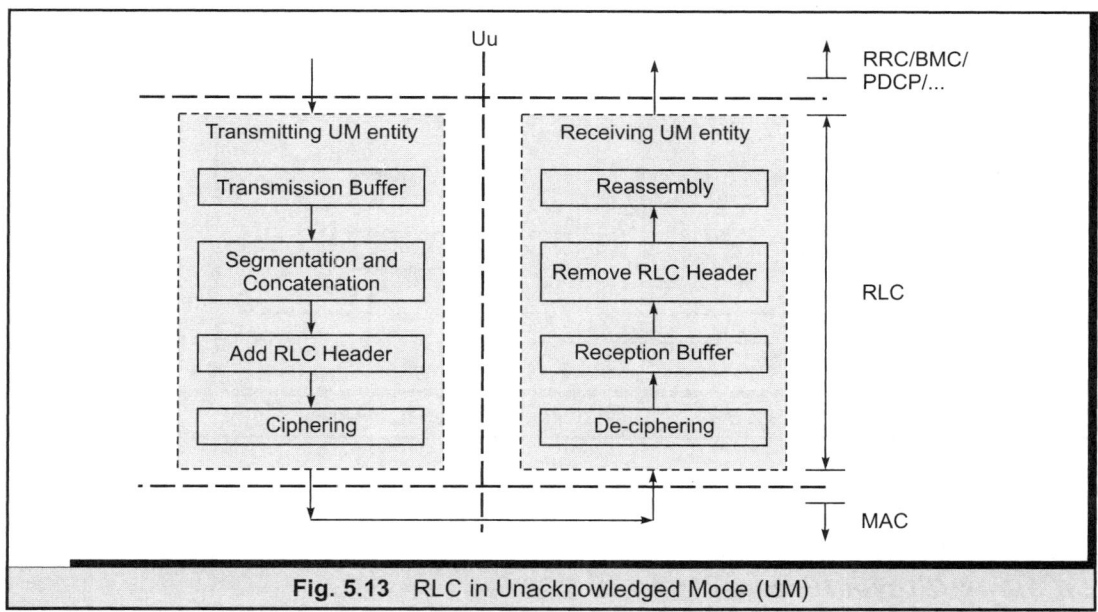

Fig. 5.13 RLC in Unacknowledged Mode (UM)

The RLC UM operates over CCCH, SHCCH, DCCH, CTCH or a DTCH logical channel.

5.7.3.1.3 RLC Acknowledged Mode (AM)

The functionality of RLC in Acknowledged Mode (AM) is most comprehensive. It includes segmentation and reassembly, concatenation, error correction, flow control, and ciphering. The explanation of each of these functions is provided in the next section. The RLC operation in Acknowledged Mode (AM) is depicted in Figure 5.14.

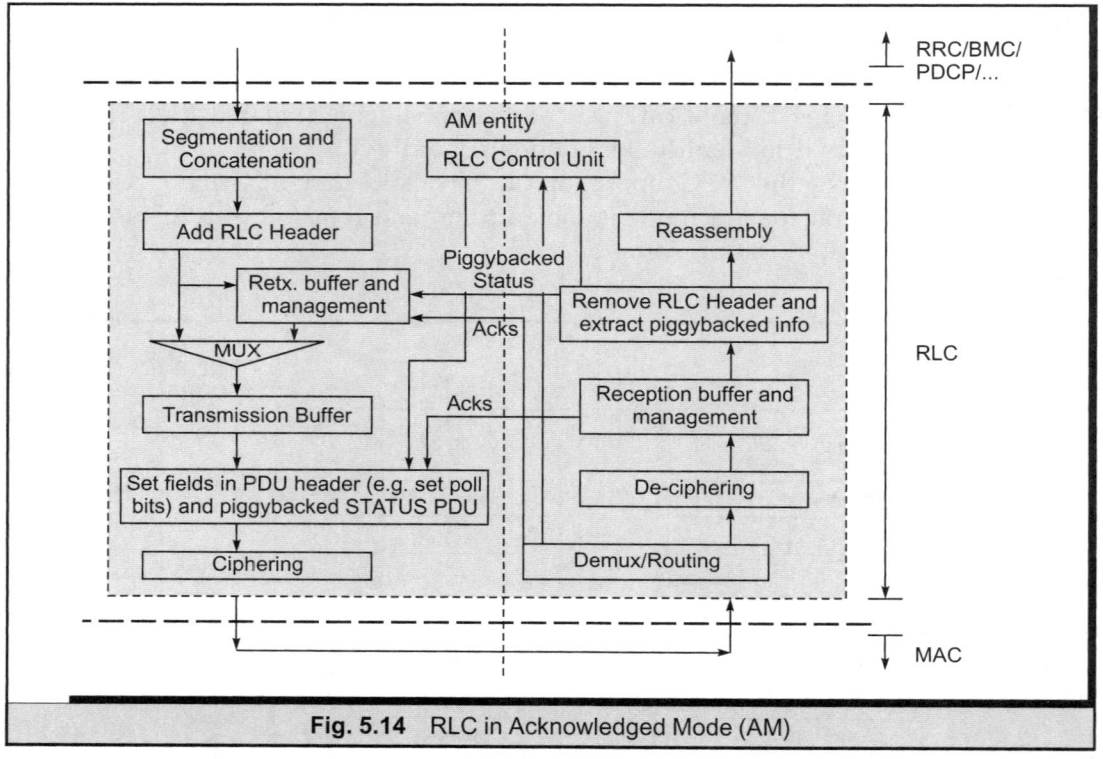

Fig. 5.14 RLC in Acknowledged Mode (AM)

The RLC AM operates over DCCH or DTCH logical channel.

5.7.3.2 *RLC Layer Functions*

The RLC layer performs various functions. Its exact function depends upon the mode in which it operates (i.e. TM, UM or AM). The RLC layer functions include typical link layer functions like segmentation and reassembly, flow control, error control and sequence numbering. A generic listing of functions of RLC is as follows:

- **Transfer of user data:** This is the most important function of RLC layer. RLC provides three modes of transfer, namely Transparent Mode (TM), Unacknowledged Mode (UM) and Acknowledged Mode (AM). During data transfer, the data is segmented and reassembled, and, if required, concatenated.

- **Segmentation and reassembly:** This refers to segmenting a RLC SDU into smaller RLC PDUs at the transmitting side and reassembling them at the receiving side. To ensure data integrity, sequence numbers are added to the RLC PDU header (for UM and AM). The RLC PDU size is adjustable to the actual set of transport formats.

- **Concatenation:** RLC layer provides the facility to concatenate more than one RLC SDU in a RLC PDU. The length indicators included in the RLC header are used to identify the RLC SDU boundaries.

- **Padding:** In case the concatenation option is not applicable and the received SDU cannot fill the entire RLC PDU, then padding is used to fill the remaining payload.

- **Error correction:** In acknowledged mode, there is a provision to retransmit erroneous PDUs using one of the available techniques (e.g. Selective Repeat, Go Back N, or a Stop-and-Wait ARQ).

- **In-sequence delivery of upper layer PDUs:** RLC ensures that the order in which SDUs are delivered to the upper layer is same as that in which they are received from the upper layer of the transmitting side.

- **Duplicate Detection:** Using sequence numbers, the RLC layer ensures that duplicate PDUs are not delivered to the RLC users.

- **Flow control:** This refers to the ability of RLC receiver to control the rate at which the RLC transmitting entity may send information.

- **Sequence number check:** In the unacknowledged mode where the retransmission option is not available, sequence numbering is used to guarantee the integrity of reassembled PDUs.

- **Ciphering:** In unacknowledged and acknowledged mode, RLC provides the option to prevent unauthorized read of user data by encrypting/ciphering the data. If this option is used, only the data portion is ciphered; the header portion is sent as plaintext. Note that for the transparent mode, the ciphering function is performed by MAC.

5.7.4 Packet Data Convergence Protocol (PDCP)

The Packet Data Convergence Protocol (PDCP) layer is used to carry user plane information for the PS domain. PDCP does not exist in the control plane. PDCP carries data protocols like IP and PPP. Typically, network layer protocols like IP reside directly over the link layer protocol (RLC in this case). However, a wireless network requires that the bandwidth in the radio interface be used optimally. Thus, compression of redundant header information forms one of the most important functions of PDCP protocol. Apart from this, PDCP provides means to transfer PDCP SDU received from non-access stratum using the RLC layer. Consequently, the functions of PDCP are similar to those of Sub-Network Dependent Convergence Protocol (SNDCP) used in GPRS.

The PDCP layer is detailed in 3GPP TS 25.323.

5.7.4.1 PDCP Layer Architecture

The PDCP architecture is not very complicated. It provides means to transfer user plane information using one of the modes of RLC layer (TM, UM or AM). The RRC layer controls the behaviour of the PDCP layer. Figure 5.15 depicts the PDCP layer architecture.

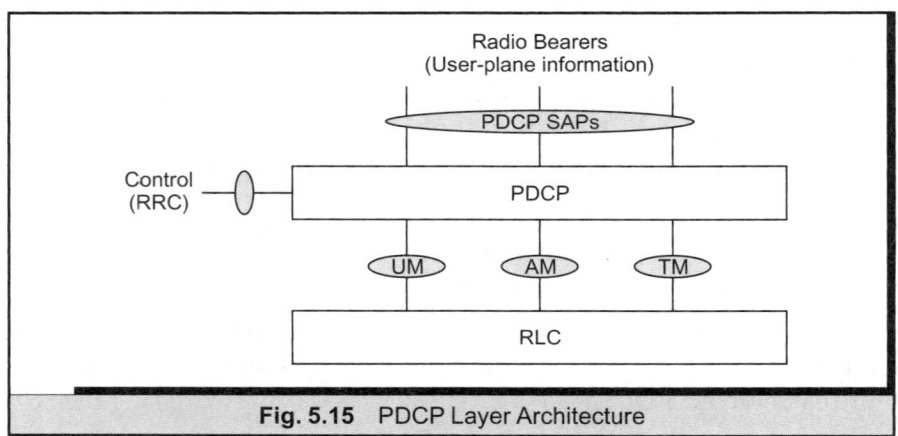

Fig. 5.15 PDCP Layer Architecture

5.7.4.2 PDCP Layer Functions

The PDCP layer performs three important functions. These are summarized as follows:

- **Header Compression and Decompression:** This is the most important function performed by the PDCP layer. The objective is to improve the efficiency of data transfer (i.e. ratio of the amount of useful data transferred versus the total data transferred). This is because bandwidth over the radio interface is a scarce resource and must be used efficiently. The PDCP protocol defines various optimization methods, each identified by a packet identifier or a PID, which forms a part of the PDCP PDU header. The PID numbers range from 0 to 31 out of which numbers 15 onwards are unassigned. The assigned numbers are primarily based on two protocols "IP Header Compression" specified in RFC2507 and 'RObust Header Compression (ROHC)' specified in RFC3095.

- **Transfer of User Data:** The next important function of PDCP is to transfer user plane information for the PS domain. For this, PDCP uses one of the modes of RLC layer (TM, UM or AM).

- **Maintenance of PDCP sequence numbers:** This procedure is used to support 'lossless SRNS relocation procedure'. The PDCP sequence numbers need to be maintained to ensure in-order delivery of PDCP PDUs for the acknowledged mode.

5.7.5 Broadcast/Multicast Control (BMC)

The Broadcast/Multicast Control (BMC) layer is used to carry user plane information in the downlink direction. The information is broadcast or multicast in nature and is sent in the unacknowledged mode.

The BMC layer is detailed in 3GPP TS 25.324.

5.7.5.1 BMC Layer Architecture

Like the PDCP architecture, the BMC architecture is simple. It provides means to transfer user plane information using the UM mode of RLC layer. Note that unlike the PDCP layer, it does not use AM. This is intuitive because BMC provides broadcast/multicast functionality, which does not require acknowledged delivery. The operations of BMC are controlled by RRC using control SAPs provided by the BMC layer. Figure 5.16 depicts the BMC layer architecture.

5.7.5.2 BMC Layer Functions

The BMC layer performs various functions. These are summarized as follows:

- **Storage of Cell Broadcast Message:** When BMC receives a Cell Broadcast (CB) message for transmission, it does not send it immediately. Rather, it stores the message for some time so that the same can be sent as per a schedule.

- **Scheduling of BMC messages:** As mentioned in the previous point, the BMC messages are sent as per a schedule. For this, the BMC sends a special message called the Schedule Message. At the UE side, the BMC sends these schedule messages to the RRC, which then knows when to receive the CB messages. When the UE is not expecting a CB message, it can enter the power saving mode.

- **Traffic Volume Monitoring:** While BMC is responsible for transfer of CB messages, RRC is responsible for controlling the radio resources. Thus, BMC periodically predicts the expected amount of CB traffic volume (in kbps) and indicates this to RRC. Based on this traffic measurement, RRC allocates radio resources for CB traffic.

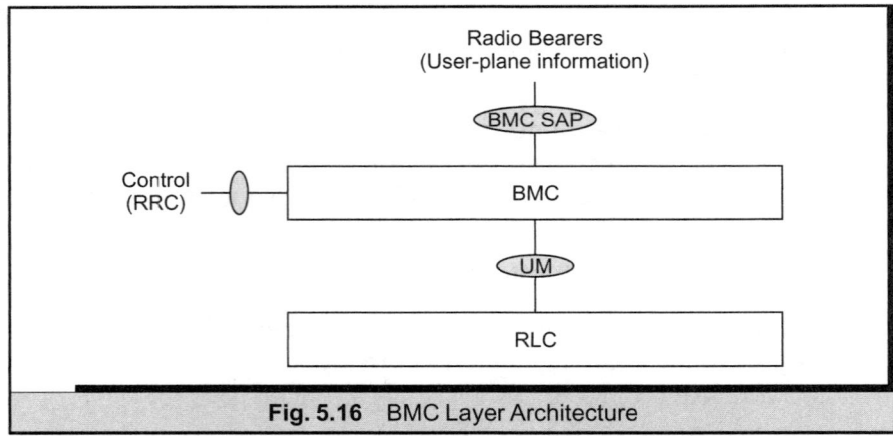

Fig. 5.16   BMC Layer Architecture

- **Transmission of BMC messages to UE:** The BMC sends the BMC messages (both the CB and Schedule) to UE.

- **Delivery of Cell Broadcast Messages to Upper Layer (NAS):** This is about the delivery of CB messages to the upper layer at the UE. Note that this function applies only to the UE. In the UTRAN, the CB messages are only transmitted; they are not received.

5.7.6 Radio Resource Control (RRC)

The Radio Resource Control (RRC) is the most important of the radio interface protocols. It is used for signalling between UTRAN and the UE. It controls the lower layers including PDCP, BMC, RLC, MAC and the physical layer. RRC is used in the control plane; it is not used in the user plane.

The RRC layer is detailed in 3GPP TS 25.331.

5.7.6.1 RRC Layer Architecture

The RRC layer exists in the control plane. It provides services to the Non-Access Stratum (NAS) using three different Service Access Points (SAPs). These SAPs (Figure 5.17) are detailed as follows:

- **General Control (GC):** This SAP is used by NAS to broadcast information to all UEs in a certain geographical area. The service over GC SAP is characterized by use of the Unacknowledged Mode (UM) of RLC. Since the broadcast information is sent periodically, the lack of reliable data transfer for the GC sap is not seen as an impediment to quality service.

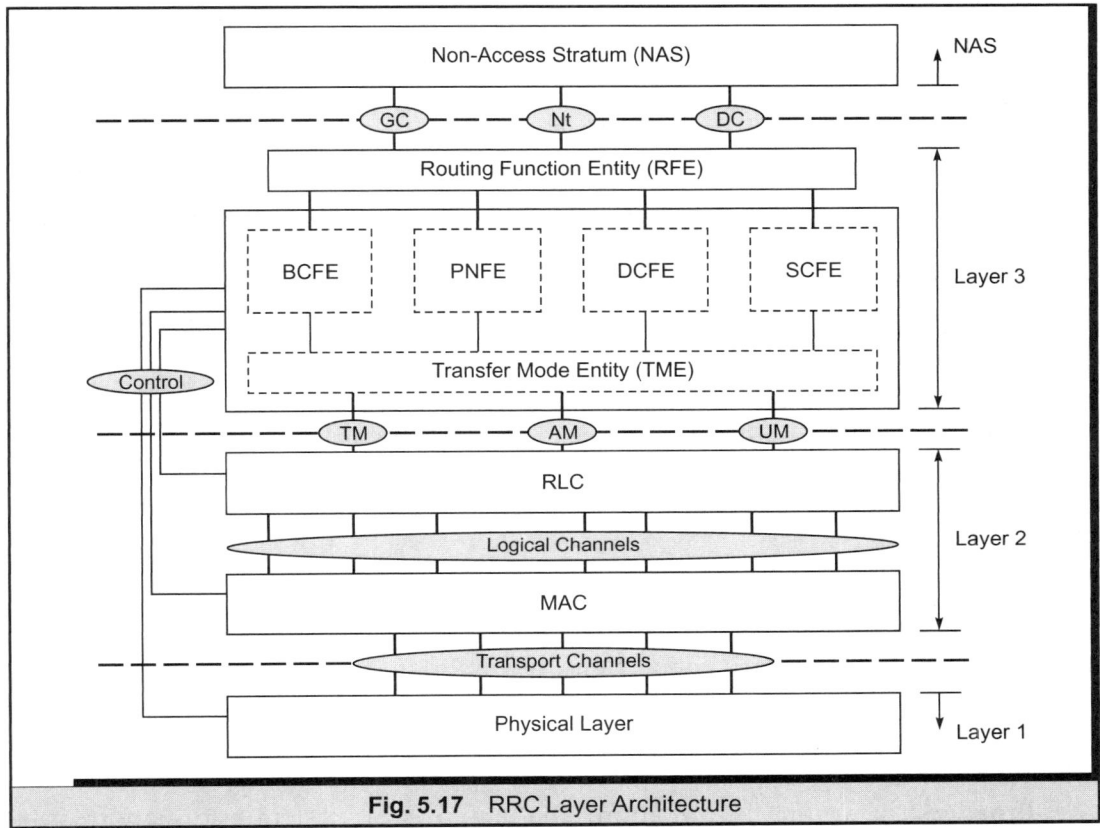

Fig. 5.17 RRC Layer Architecture

- **Notification (Nt):** This is used by NAS to send paging information in a certain geographical area to one or more UEs. Like GC, the service over Nt SAP is characterized by use of the Unacknowledged Mode (UM) of RLC.

- **Dedicated Control (DC):** NAS uses this SAP to establish and release a connection and to transfer messages using this connection. Each connection is associated with Quality of Service (QoS) requirements. The service over DC is characterized by in-sequence transfer, priority handling and duplication avoidance. Depending upon the scope of functionality required, one of the RLC modes, i.e. TM, UM or AM is used for DC SAP.

To provide services over various SAPs as discussed above, the RRC layer is logically divided into four different entities. These entities are as follows:

- **Broadcast Control Function Entity (BCFE):** This entity performs the functions related to broadcast of system information from RNC to UE. The broadcast is carried out using the TM or UM mode of the RLC layer. In the NAS side, this entity provides services corresponding to the GC SAP.

- **Paging and Notification Control Function Entity (PNFE):** This entity performs functions related to paging, from RNC to UE, using the TM or UM modes of RLC layer. In the NAS side, this entity provides services corresponding to the Nt SAP.

- **Dedicated Control Function Entity (DCFE):** This entity performs functions specific to one UE. It uses any of the RLC modes — TM, UM or AM. In the NAS side, this entity provides services corresponding to the DC SAP.

- **Shared Control Function Entity (SCFE):** It is used by the DCFE in the TDD mode. SCFE uses TM or UM modes of RLC layer.

The Transfer Mode Entity (TME) maps the requirements of the higher layer entities to one of the RLC modes. In the NAS side, another entity called the Routing Function Entity (RFE) is used to map the services provided by RRC to one of the three SAPs used by NAS. Though RFE is shown outside the RRC module, it is logically a part of the RRC.

5.7.6.2 RRC Layer Functions

The RRC layer performs various functions. Its important functions are summarized as follows (for detailed description of RRC functions, the reader is referred to 3GPP TS 25.301 and to Chapter 7, which provides details of various procedures involving the RRC):

- **Broadcast of system information:** The broadcast of system information is the means by which, information about the system as well as the serving cell is periodically broadcasted to all UEs within a certain cell. Since this information is common to all UEs within a cell, it is possible to transmit it in a broadcast fashion. The system information is broadcast in blocks, called the System Information Blocks (SIBs). A SIB groups together system information of similar nature. In the UTRAN, the RRC layer broadcasts the SIBs from the network to all UEs. The information is broadcasted periodically by RRC to ensure that the information used by UE is up to date. The broadcast takes place as per a schedule maintained by RRC. The broadcasting function applies to information provided by both the non-access stratum (i.e. Core Network) as well as the access-stratum (i.e. Access Network). The difference between the two types of information is that the former generally contains network information (e.g. PLMN identifier), while the latter contains cell-specific information (e.g. parameters for cell selection and re-selection).

- **Establishment, re-establishment, maintenance and release of an RRC connection:** In UMTS, the concept of RRC connection (or radio connection) is de-linked from the concept of a radio bearer. An RRC connection in UMTS is a static concept, which is established only once and exists until it is released. The radio bearer, on the other hand, defines the properties of the radio connection, and can be reconfigured during the lifetime of the RRC connection. Further, while only a maximum of one RRC connection is allowed to exist per UE, there can be multiple radio bearers per RRC connection, each with its own data transfer characteristics.

The RRC layer is responsible for establishment, re-establishment, maintenance and release of an RRC connection. The establishment of an RRC connection is initiated by a request from higher layer (non-access stratum) at the UE side. In contrast, the RRC connection is always released by the network. The re-establishment of the RRC connection is initiated by the UE in case of connection loss.

- **Establishment, reconfiguration and release of Radio Bearers:** As discussed in the previous point, the notion of RRC connection is different from radio bearers that are used to carry actual information. It is the responsibility of RRC to act upon the request of the higher layer (non-access stratum) and to establish, reconfigure and release the radio bearers in the user plane. More than one radio bearer can be established by a UE, at the same time.

- **Assignment, reconfiguration and release of radio resources for the RRC connection:** For a given RRC connection, RRC assigns the radio resources (e.g. codes and channels). The assignment is done both for user plane as well as for the control plane. RRC also coordinates the mapping of multiple radio bearers to a given RRC connection.

- **RRC connection mobility functions:** The mobility function refers to tracking the location of a UE while it has an established RRC connection. This function includes procedures like Cell update, URA update, UTRAN Mobility Information, Active Set update, Soft handover, Hard handover and Inter-system Handover. These procedures are detailed in Chapter 7.

- **Paging/notification:** Paging is the procedure used by RRC to transmit paging information to selected UEs. Paging is used for various reasons. First, to establish a signaling connection between UE and CN. In this case, the paging procedure is used by the RRC to indicate to the UE that there is an incoming call waiting. Second, it is used to initiate a Cell Update Procedure. In this case, the paging procedure is used by the UTRAN to indicate that it has some downlink data to be sent to the UE. Third, paging is used to initiate reading of the Master Information Block. It is also used to establish a new signalling connection, in cases where the originating CN is other than the current serving CN. The paging procedures are detailed in Chapter 7.

- **Routing of higher layer PDUs:** The RRC layer routes a PDU received for a higher layer to the entity for which it is destined (e.g. the RANAP entity at RNC).

- **Control of requested QoS:** This refers to the responsibility of RRC whereby it ensures that the QoS requested for the radio bearers can be met. This is done by allocation of sufficient radio resources.

- **UE measurement reporting and control:** The network can request the UE to perform measurements, which can aid the network in providing certain services. Such measurements could include measurement of the traffic volume that a UE is generating/receiving. This can be used by the network to decide whether to allocate or remove resources for the UE. Similarly, core network services like location services require the UE to perform certain measurements to aid in identifying the location of the UE. RRC at RNC controls the measurements performed by the UE. This control includes: What to measure, When to measure, and How to report. The RRC at UE reports these measurements to the network.

- **Security Management:** Of the two important security management functions of the RRC, the first relates to control of ciphering (i.e. whether ciphering is on or off). In case ciphering is required, the RLC performs ciphering in acknowledged/ unacknowledged mode while MAC performs it in the transparent mode. The second function relates to integrity protection. This is done by inserting a Message Authentication Code (MAC-I) in those RRC messages that are considered sensitive.

- **Initial cell selection and re-selection in idle mode:** The RRC facilitates the selection of most suitable cell to camp on. Even when a cell is selected, a UE on the move must ascertain that the camped-on cell is the most suitable one. This is referred to as re-selection. The re-selection procedure is based on the system information received from the network.

- **Configuration and Resource Allocation for CBS:** While the Cell Broadcast (CB) messages are transferred to UE using BMC, the RRC controls the configuration for the CBS. The RRC also allocates radio resources for CB messages based on traffic volume requirements indicated by BMC. The radio resource allocation set by RRC is indicated to BMC, which in turn uses this information to generate schedule messages. The CB messages are then sent as per schedule. Another function related to CBS is the configuring of the lower layers of UE so as to enable it to save power and receive the CB messages in a discontinuous manner.

5.7.6.3 RRC Messages

The RRC protocol is used to exchange signaling messages between RNC and UE. Table 5.4 shows important RRC signaling messages (Refer to Chapter 7 and 8 for use of RRC messages in various procedures).

| **Table 5.4** | Important RRC Messages*** |

Type	*Messages*	*Description*
Mobility	Active Set Update/Active Set Update Complete	Used by the UTRAN to update the Active Set of the UE. The Node Bs to which the UE is simultaneously connected is called the UE's "Active Set".
	Cell Update/Cell Update Confirm	Used by the UE to update the UTRAN of the current cell that the UE is camping on after cell re-selection, or, as a supervision mechanism by means of periodic Cell Updates.
	URA Update/URA Update Confirm	Used by the UE to retrieve a new URA Identity after cell re-selection, when the new selected cell does not belong to the URA where the UE was initially residing, or, as a supervision mechanism by means of periodic URA Updates.
	UTRAN Mobility Information/ UTRAN Mobility Information Confirm	Used by the UTRAN to send new mobility information to the UE. The information consists of any one or a combination of a new C-RNTI, U-RNTI or any other mobility information.
	Handover from UTRAN Command	Used to transfer a connection between the UE and UTRAN to another radio access technology (e.g. GSM).
	Handover to UTRAN Command/ Handover to UTRAN Complete	Used to transfer a connection between a UE and another radio access technology (e.g. GSM) to UTRAN.
Paging	Paging Type 1	Used by the UTRAN for paging a UE in RRC Idle Mode. Also used in the RRC connected mode for CELL_PCH and URA_PCH states.
	Paging Type 2	Used by the UTRAN for paging a UE in CELL_DCH and CELL_FACH states of RRC Connected Mode.
Connection Management	RRC Connection Request	Used by the UE to initiate the process of RRC connection establishment.
	RRC Connection Setup/ RRC Connection Setup Complete	Used by the UTRAN to setup an RRC connection on request from the UE.
	RRC Connection Release/ RRC Connection Release Complete	Used by the UTRAN to release an RRC connection to a UE.
RB Management	RB Setup/RB Setup Complete	Used by the UTRAN to establish a new radio bearer for a UE.
	RB Release/ RB Release Complete	Used by the UTRAN to release one or more radio bearers for a UE.

Contd.

| Table 5.4 | Contd. |

Type	*Messages*	*Description*
	RB Reconfiguration/ RB Reconfiguration Complete	Used by the UTRAN to reconfigure the radio bearer for a UE. This may result in a reconfiguration in the lower layer parameters.
Data transfer	Initial Direct Transfer	Used by the UE to transmit a NAS layer message to the UTRAN when no signalling connection exists.
	Uplink Direct Transfer	Used by the UE to transmit a NAS layer message to the UTRAN when a signalling connection already exists.
	Downlink Direct Transfer	Used by the UTRAN to transmit a NAS layer message to the UE when a signalling connection already exists.
Measurement control	Measurement Control	Used by the UTRAN to inform the UE about the measurements that it wishes the UE to perform.
	Measurement Report	Used by the UE to report the measurements collected by it to the UTRAN.
Others	System Information	Used by the UTRAN to broadcast system information to UEs within a cell.
	Physical Channel Reconfiguration/ Physical Channel Reconfiguration Complete	Used by the UTRAN to establish, modify or release physical channels for a UE. The decision of the UTRAN is based on the current state of the network resources.
	Transport Channel Reconfiguration/ Transport Channel Reconfiguration Complete	Used by the UTRAN to reconfigure the transport channel parameters by modifying the Transport Format Set.
	UE Capability Information/ UE Capability Information Confirm	Used by the UE to convey UE specific capability information to the UTRAN.

5.8   ATM-BASED TRANSPORT NETWORK PROTOCOLS

In Section 5.5, the UTRAN protocol architecture was discussed, and the transport network layer protocols based on ATM were briefly mentioned. This section elaborates upon these protocols; in particular, the following:

- Asynchronous Transfer Mode (ATM)
- ATM Adaptation Layer 2 (AAL2)
- ATM Adaptation Layer 5 (AAL5)
- Service Specific Connection Oriented Protocol (SSCOP)
- Service Specific Co-ordination Function for NNI (SSCF-NNI)

- Service Specific Co-ordination Function for UNI (SSCF-UNI)
- Message Transfer Part 3 for Broadband (MTP3b)
- Signaling Transport Converter
- AAL2 Signaling

Each of these protocols is discussed in the following sub-sections. Note that some protocols discussed in Section 5.5 are not covered in this list, which include SCCP and GTP-U. These are discussed in the next chapter. Further, the SIGTRAN protocols (like M3UA/SCTP/IP) are also not covered in this chapter. They are discussed in Chapter 15.

5.8.1 Asynchronous Transfer Mode (ATM)

ATM or 'Asynchronous Transfer Mode' is defined as 'a transfer mode in which the information is organized into cells; it is asynchronous in the sense that the recurrence of cells containing information is not periodic.' This definition encompasses three basic terms, as follows:

- **Transfer Mode:** This refers to the techniques used to transmit, switch and multiplex information. In other words, transfer mode is the means of packaging, sending and receiving information on the network. Circuit switching (as embodied by telephony networks) and packet switching (as embodied by frame relay and IP networks) best describe the two extremities of transfer modes. This is because in circuit switching, it is sent as bit streams, while in packet switching, information is sent as large frames. ATM fits in between these two extremes because it uses a very small sized frame (53bytes). By using this small size frame (precisely, a cell!), ATM retains the speed of circuit switching while still offering the flexibility of packet switching. Hence, ATM is also referred to as fast packet-switching technology.

- **Asynchronous Nature:** ATM is asynchronous in the sense that the recurrence of cells containing information is not periodic. Usually, the terms synchronous and asynchronous refer to the way the data is transmitted. In the synchronous mode, the transmitter and receiver clocks are synchronized and frames are sent/ received periodically. In asynchronous mode, timing information is derived from the data itself, and the transmitter is not compelled to send data periodically. This definition applies to ATM. Note that the term 'asynchronous' is used for the ATM layer and not for the physical layer; that is, the multiplexing of cells on to the physical medium is asynchronous, not the transmission of cells. Unlike Time Division Multiplexing, no slot is reserved for a logical channel in ATM and cells are transmitted as and when they arrive. This makes the transfer of cells for a particular channel non-periodic, which is why ATM is an asynchronous transfer mode.

- **Cell-based transfer:** Information in ATM is 'organized into cells', which means that lowest unit of information in ATM is a cell. A cell is a fixed size frame of 53 bytes, with 5 bytes of header and 48 bytes of payload. The header carries information required to switch cells, while payload contains the actual information to be exchanged. Figure 5.18 illustrates the concept of cell-based transfer. Information from various sources is multiplexed and segmented, resulting in a stream of ATM cells. Each cell is transmitted and received independent of other cells. A cell is identified by the labels carried in the header. Here, label refers to the Virtual Channel Identifier (VCI) and Virtual Path Identifier (VPI) fields. These fields

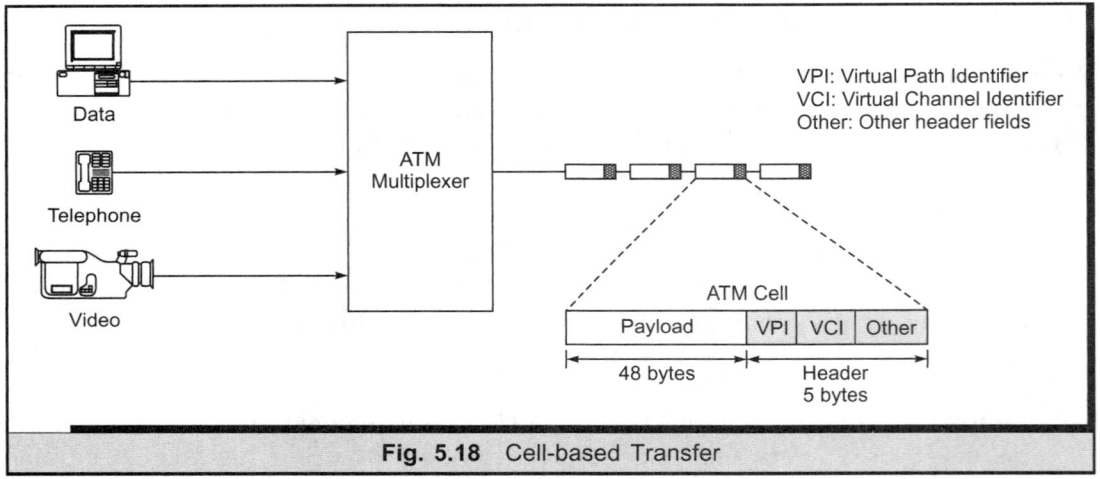

Fig. 5.18 Cell-based Transfer

identify the virtual circuit to which a cell belongs. In essence, ATM is a cell-based technology, which employs virtual-circuit concepts to forward information streams. The ATM virtual connection is characterized as follows:

- ATM uses virtual connections to switch cells from one node to another. The virtual circuits are identified by VPI and VCI. The VPI/VCI fields are carried in the ATM cell headers and are called connection identifiers.
- Cells belonging to a virtual connection follow the same path. Thus, cell sequence is maintained implicitly.
- The bandwidth allocated to a virtual connection is assigned at the time of connection set-up and is based on the requirements of the source and the available capacity.
- Each virtual connection is provided with a Quality of Service (QoS). The connections are bi-directional in nature. The bandwidth allocated in forward and backward directions may or may not be the same (symmetric/

asymmetric connection). The bandwidth in the backward direction can even be zero (rendering the connection unidirectional).

– Same VPI/VCI value is used across a link in both the directions.

5.8.1.1 *ATM Layer Architecture*

The ATM layer architecture is a three-dimensional model (Figure 5.19). This model has three planes, each corresponding to a set of functionality required of the protocol stack. The three planes are as follows:

- **User plane:** The user plane (or U-plane) is concerned with the transfer of user information. At the transmitting side, this plane is responsible for packing user information into cells and transmitting the cells using the underlying physical medium. At the receiving side, it performs reverse operation and derives the higher layer information in exactly the same format as it was received from the user at the transmitting side.

- **Control plane:** The control plane (or C-plane) is responsible for establishing and releasing connections between a given source and destination. When a new connection is established, the control plane establishes a mapping at the intermediate switches between incoming VPI/VCI and outgoing VPI/VCI. The mapping so derived is used to switch cells. When the same connection is released, the control plane removes the mapping stored within the intermediate nodes.

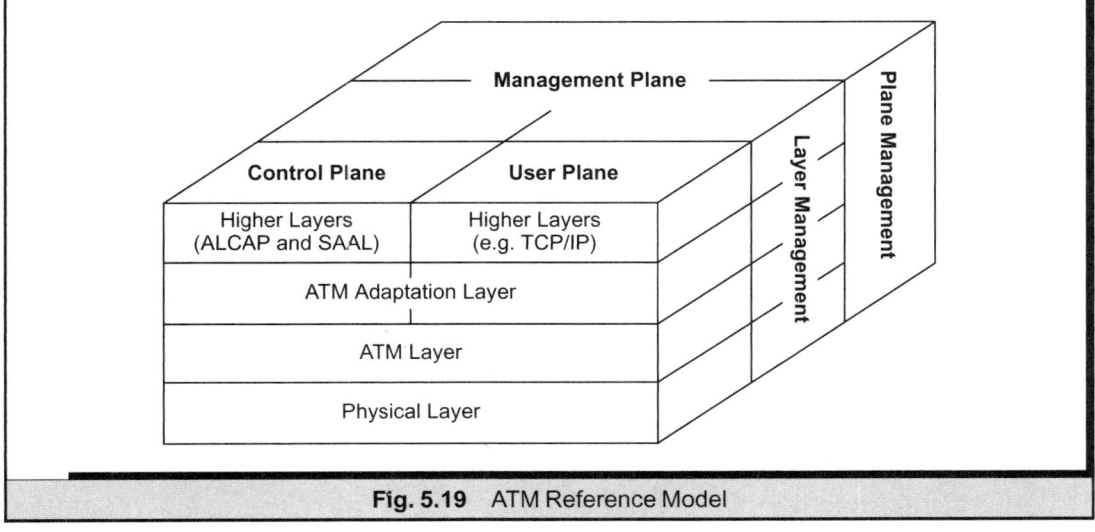

Fig. 5.19 ATM Reference Model

- **Management plane:** This plane is responsible for managing the individual layers in the protocol stack and providing coordination between the layers. The management plane is divided into layer management and plane management. Layer management is responsible for managing each of the layers, including their administration, maintenance and configuration. Plane management, which cuts through all the layers, is responsible for coordination among different planes.

Irrespective of the planes, the three lower layers of ATM layer architecture are the ATM physical layer, ATM layer and ATM Adaptation Layer (AAL). The lowest layer is the physical Layer, which is concerned with carrying cells over the physical media across the network. This layer is not discussed any further in this book.

The layer above the physical layer is the ATM layer. This is where the core function of ATM resides. It is the ATM layer that deals with cells. The ATM layer receives 48 byte packets from AAL, attaches a 5-byte header and sends it to the lower layer. Apart from generation and multiplexing of cells, a host of other functionality is implemented at this layer. These functions are explained in the following sub-section.

The ATM Adaptation Layer (AAL) resides over the ATM layer. The AAL is responsible for handling different types of data and mapping the requirements of the applications to the services provided by the lower layer. In order to support a variety of applications, four AALs are defined. These are AAL1, AAL2, AAL 3/4 and AAL5, each serving the requirements of a separate class of applications. The AAL2 and AAL5, which are used in UTRAN, are explained in the following sections.

5.8.1.2 ATM Layer Functions

The main function of ATM layer is to switch ATM cells using the connection identifiers (i.e. VPI and VCI). Apart from this, the ATM layer performs functions like cell construction, congestion control, traffic monitoring, and providing QoS. Some of the important ATM layer functions are:

- **Cell Relaying and Forwarding using VPI/VCI:** The basic function of the ATM layer is to replace the VPI/VCI value of an incoming cell with a new VPI/VCI value. Depending upon the type of connection (virtual path connection or virtual channel connection), either only the VPI value or both VPI/VCI values are swapped. The mapping between incoming and outgoing VPI/VCI values is maintained within an ATM switch.

- **Cell Construction:** The construction of the cell is done at the ATM layer. The ATM layer receives from the upper layer a 48-byte payload to which it appends a 5-byte header, fills in the necessary values in the header and sends it to the physical layer. The VPI/VCI values are determined using the per-connection information stored in the translation table at the ATM layer.

- **Cell Multiplexing and De-multiplexing:** The ATM layer multiplexes cells from various logical connections using the VPI/VCI field on the underlying physical medium. The allocation of bandwidth depends upon the connection parameters fixed at the time of connection establishment. Any free bandwidth is distributed to various contending connections using some fair algorithm. At the receiving end, the ATM layer de-multiplexes cells and hands the same over to appropriate higher layer protocols.

- **Support for Multiple QoS Classes:** ATM provides multiple QoS classes to the users. This allows applications with varied requirements to be supported. The QoS required by the application is agreed upon at the time of connection establishment. The intermediate nodes from the source to the destination maintain, besides other information, a mapping from the connection identifiers to the QoS parameters. The cells arriving on a link from different connections get a differential treatment depending upon the QoS parameters associated with the connection to which the cells belong.

- **Usage Parameter Control (UPC):** The ATM layer monitors each connection for the amount of cells injected in the network. This amount must fall below a certain threshold value depending upon the parameters of the connection. If the end-system injects more cells than the value agreed upon, the ATM network takes appropriate measures. These measures include dropping the cells sent in excess to the agreed value.

5.8.2 ATM Adaptation Layer 2 (AAL2)

The AAL2 resides over the ATM layer. It provides the means for bandwidth-efficient transmission of low-rate, short and variable length packets in delay sensitive applications. AAL2 was introduced in 1997 to overcome the limitations of AAL1 (e.g. inefficient use of bandwidth for variable and small length packets) some four years after other AALs were standardized. The delay in standardization was due to the inherent complexity involved in defining a standard that could support transmission of compressed time-sensitive data.

AAL2 is detailed in ITU-T I.363.2.

5.8.2.1 AAL2 Layer Architecture

The AAL2 layer is divided into two parts, namely Common Part Sublayer (CPS) and Service Specific Convergence Sublayer (SSCS). The structure of AAL2 is shown in Figure 5.20.

The CPS sublayer provides the basic functionality of AAL2, which includes packaging the variable payload into cells and providing error correction.

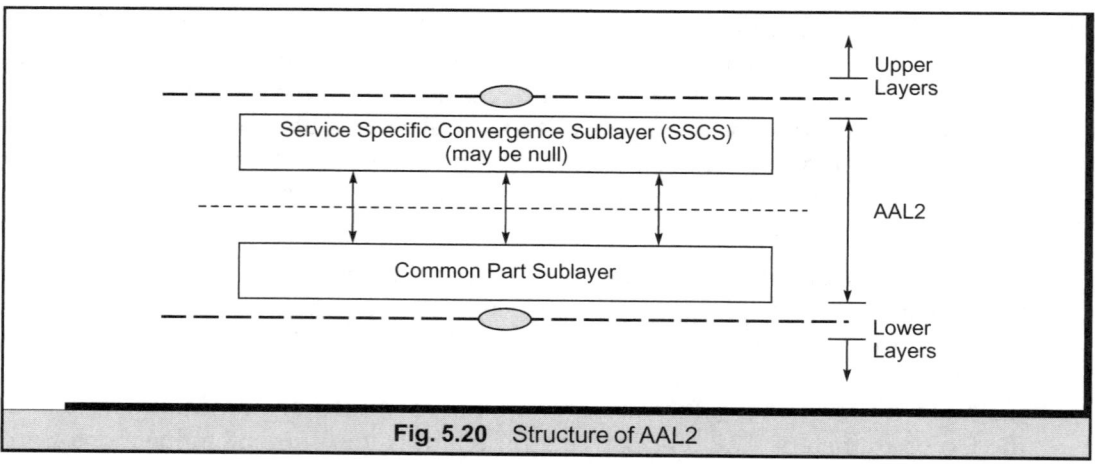

Fig. 5.20 Structure of AAL2

The SSCS sublayer directly interacts with the AAL user. This layer may be used to enhance the services provided by CPS or to tailor the services provided by CPS to support specific AAL2 user services. This sublayer may even be null and only provide a mapping between AAL user primitives and the CPS service primitives. One of the SSCS layers standardized by ITU-T is defined in I.366.1. This specification defines the Segmentation and Reassembly functions of Service Specific Convergence Sublayer (SAR SSCS). The SAR SSCS is divided into three parts as follows:

- **Service Specific Segmentation and Reassembly Sublayer (SSSAR):** This is the basic function of SAR SSCS. The functionality includes data transfer of SSSAR-SDUs of up to 65568 octets.

- **Service Specific Transmission Error Detection Sublayer (SSTED):** The SSSAR does not provide any error detection function. To provide this functionality, the SSTED function may optionally be used. The role of SSTED is to detect corrupted SSTED-SDUs. These corrupted SSTED-SDUs are not delivered to the SSTED user. However, the corrupted or lost SSTED-SDUs are not re-transmitted.

- **Service Specific Assured Data Transfer Sublayer (SSADT):** To provide support for re-transmission, the SSADT function may be used over and above the SSSAR and SSTED function.

In the UTRAN, only the SSSAR function of SAR SSCS is used.

5.8.2.2 AAL2 Layer Functions

The functions and characteristic features of AAL2 are listed as follows:

- AAL2 provides the means for bandwidth-efficient transmission of low-rate, short, and variable length packets in delay sensitive applications.

- It provides for the transfer of time-sensitive constant bit rate traffic as well as variable bit rate traffic.
- It provides the means for compression and silence suppression techniques.
- AAL2 enables multiple user channels on a single ATM virtual circuit and allows for varying traffic load on each virtual circuit as well as each user channel on the virtual circuit.

5.8.3 ATM Adaptation Layer 5 (AAL5)

The AAL5 provides an extremely simple mode of data transfer with minimal overheads. Unlike AAL2, the AAL5 does not provide the means for transfer of delay-sensitive data. Its use is limited to data applications that do not have stringent delay requirements.

To carry large size PDUs, a single AAL5 SDU is concatenated with a trailer and then segmented into multiple SAR PDUs that are then carried in the payload part of different ATM cells. To identify the beginning and end of SAR SDUs in a stream of ATM cells, a one bit identifier is used, which is placed in the cell header. Thus, SAR-PDU in AAL5 contains 48 bytes of data with no byte for overhead (except for the last SAR-PDU which also contains the trailer). Given its extremely simple implementation, AAL5 is also sometimes called the Simple and Efficient Adaptation Layer (SEAL).

AAL5 is detailed in ITU-T I.363.5.

5.8.3.1 AAL5 Layer Architecture

The AAL5 layer is divided into two sublayers, namely the Convergence Sublayer (CS) and the Segmentation and Reassembly sublayer (SAR). The CS itself is divided into two parts: the Common Part Convergence Sublayer (CPCS) and Service Specific Convergence Sublayer (SSCS). The structure of AAL5 is depicted in Figure 5.21. The SSCS sublayer may be used to support certain added functionality required by some applications. It may also be null in which case it just provides a mapping to the primitives of the underlying CPCS sublayer. Irrespective of the type of SSCS, the CPCS and the SAR sublayers are always the same. Therefore, they are together termed as Common Part AAL. The SAR and the CPCS sublayers are explained as follows:

- **Segmentation and Reassembly (SAR):** The segmentation and reassembly sublayer in AAL5 is very simple. This layer does not add any header or trailer to the SAR-SDU. It just breaks down the SAR-SDU into 48 bytes SAR-PDUs, which in turn form the payload of the cells. The beginning and end of the SAR-SDU is indicated through one of the bits in ATM cell header (called the ATM-User-to-ATM-User indication bit). A value of 1 of this bit indicates the end of a SAR-SDU. On the other hand, a value of 0 indicates the beginning or continuation of an SAR-PDU. This value is passed on to the ATM layer along with the SAR-PDU.

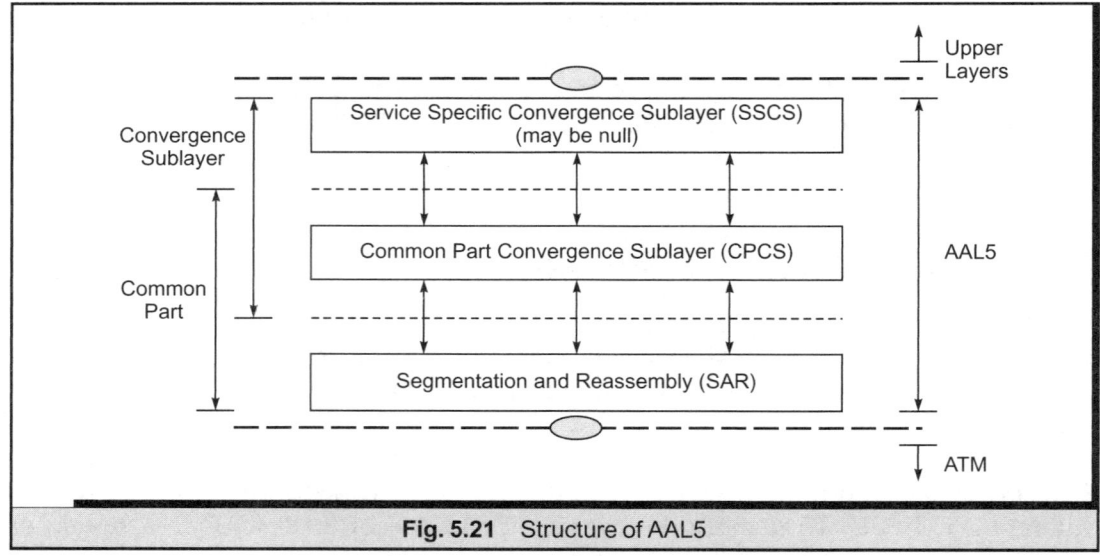

Fig. 5.21   Structure of AAL5

- **Common Part Convergence Sublayer (CPCS):** The CPCS of AAL5 provides two modes of data transfer, namely the *message mode* and the *streaming mode*. In the message mode of data transfer, an entire CPCS-SDU is received from the upper layer and only then transferred to the SAR sublayer. On the other hand, the streaming mode allows the CPCS to start transferring data before it has received the complete CPCS-SDU from the upper layer. In such a case, the CPCS transfers the partially submitted CPCS-SDU to SAR sublayer for transfer. The CPCS adds an 8-byte trailer. This is used for various functions, including error checking and length determination.

5.8.3.2   AAL5 Layer Functions

The functions and characteristic features of AAL5 are as follows:

- **Simple Implementation for Data Transfer:** AAL5 is the simplest of AALs and provides efficient means to transfer data. It is presumed that in the long run, most of the applications (barring certain delay sensitive applications) will use AAL5. The AAL5 is also used to carry messages of signaling protocols.

- **Trailer-based:** The header format in AAL5 is extremely simple. The SAR layer does not add any header or trailer, thereby removing the overheads associated with it. The CPCS-PDU adds a trailer, which is of 8 bytes.

- **Minimal Overhead:** Overhead in AAL5 is the least among all AALs. For a CPCS-PDU of maximum size (65,536 bytes), the overhead is .01 percent. Even for a modest CPCS-PDU of 1500 bytes, the overhead is .5 percent, which is quite nominal.

5.8.4 Service Specific Connection Oriented Protocol (SSCOP)

As discussed in previous sections, ATM has various adaptation layers (or AALs). Prominent among these are AAL2 and AAL5. AAL2 is the layer used to carry voice while AAL5 carries data. Apart from these, there is another AAL that is used to carry signaling information. It is called the Signaling ATM Adaptation Layer (SAAL). The peculiar characteristic that distinguishes it from AAL5 is that it provides reliable delivery, which is not a feature of AAL5.

As AAL5 does not provide reliable delivery, a service specific layer that provides assured delivery is required. For this purpose, the Service Specific Connection Oriented Protocol (SSCOP) layer is used above AAL5. The functions of SSCOP are similar to those of the acknowledged-mode of RLC layer (e.g. providing sequence numbering, flow control and error control).

SSCOP is detailed in ITU-T Q.2110.

5.8.4.1 SSCOP Layer Architecture

The SSCOP forms a part of SAAL. The latter consists of three sub-layers, namely the Service Specific Co-ordination Function (SSCF), the Service Specific Connection Oriented Protocol (SSCOP) and the AAL5. The SSCF in turn has two variants, one for User-Network Interface (UNI) and another for Network-Node Interface (NNI). In the UTRAN, both the SSCFs are used. The SSCF for NNI is used in Iu_CS, Iu_PS and the Iur interface. The SSCF for UNI is used for the Iub interface.

Depending upon whether SSCOP is used in SSCF-NNI or SSCF-UNI, the layer architecture of SSCOP differs. Sections 5.8.5.1 and 5.8.6.1 discuss the position of SSCOP with respect to SSCF-NNI and SSCF-UNI respectively.

5.8.4.2 SSCOP Layer Functions

The SSCOP has various functions, of which the important ones are summarized as follows:

- **Transfer of user data:** This refers to the transfer of user data between SSCOP users. The SSCOP supports both assured and unassured data transfer.

- **Sequence Integrity:** SSCOP ensures that the order of SSCOP delivered to its user at the receiving side is the same as that delivered by the SSCOP user at the transmitting side.

- **Error correction by selective re-transmission:** By delivering packets with sequence numbers, missing PDUs are detected and thereafter are selectively re-transmitted.

- **Flow control:** This refers to the ability of the SSCOP receiver to control the rate at which the SSCOP transmitting entity may send information.

- **Keep alive:** In case there are prolonged periods in which no data is exchanged between SSCOP peers, there is a keep-alive mechanism to ensure that the connection is kept alive.

- **Local data retrieval:** This refers to the mechanism whereby the local SSCOP user is allowed to retrieve the in-sequence SDUs that have not yet been released by the SSCOP entity.

- **Connection control:** This refers to establishment, release and re-synchronization of SSCOP connection.

5.8.5 Service-Specific Co-ordination Function for NNI (SSCF-NNI)

The SSCF-NNI is a lightweight protocol. Its primary function is to map the particular requirements of the Layer 3 protocol (in this case MTP3b) to the services provided by SSCOP.

SSCF for NNI is used as defined in ITU-T standard Q.2140.

5.8.5.1 SSCF-NNI Layer Architecture

Figure 5.22 shows the layer architecture of SSCF-NNI. Note that the SSCF-NNI interfaces with SSCOP using *signals* (and not through *primitives*). The difference between signals and primitives is that the former are exchanges between two sub-layers of a layer, while primitives are exchanged between two layers. SSCF-NNI in turn provides services to MTP3b.

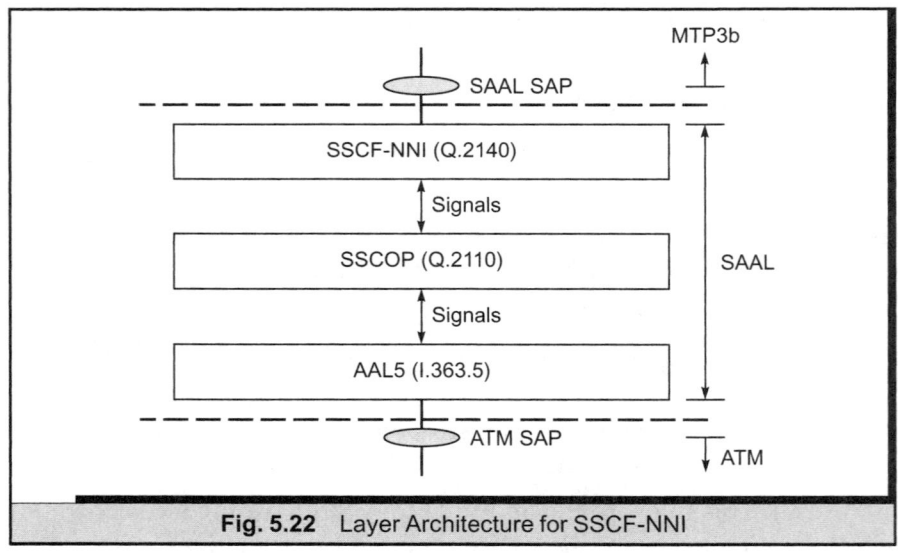

Fig. 5.22   Layer Architecture for SSCF-NNI

5.8.5.2 *SSCF-NNI Layer Functions*

The SSCF provides SAAL primitives to the SAAL user (i.e. MTP3b). The user sees the SAAL as a single entity while internally, the SAAL is divided into three sub-layers (i.e. SSCF, SSCOP and AAL5). The functions of SAAL are represented by the functions of SSCOP that have already been discussed. Suffice it to say that the function of SSCF-NNI is to map the services provided by SSCOP to MTP3b.

5.8.6 Service-Specific Co-ordination Function for UNI (SSCF-UNI)

Like SSCF-NNI, the primary function of SSCF-UNI is to map the particular requirements of the UNI Layer 3 protocol (in this case NBAP) to the services provided by SSCOP. Note that the SSCF-UNI protocol is used between Node B and RNC. This is unlike SSCF-NNI, which is used between RNC and CN (over Iu interface) or between two RNCs (over Iur interface). The reason for this is that the Iub interface between Node B and RNC can be seen as an equivalent of a User Network Interface (UNI). Thus, in this case, the SSCF-NNI is not applicable.

SSCF for UNI is used as defined in ITU-T standard Q.2130.

5.8.6.1 *SSCF-UNI Layer Architecture*

Figure 5.23 shows the layer architecture of SSCF-UNI. Note that the exchange of SSCF-UNI with SSCOP takes place using *signals* (and not *primitives*). In turn, SSCF-UNI provides services to the NBAP layer.

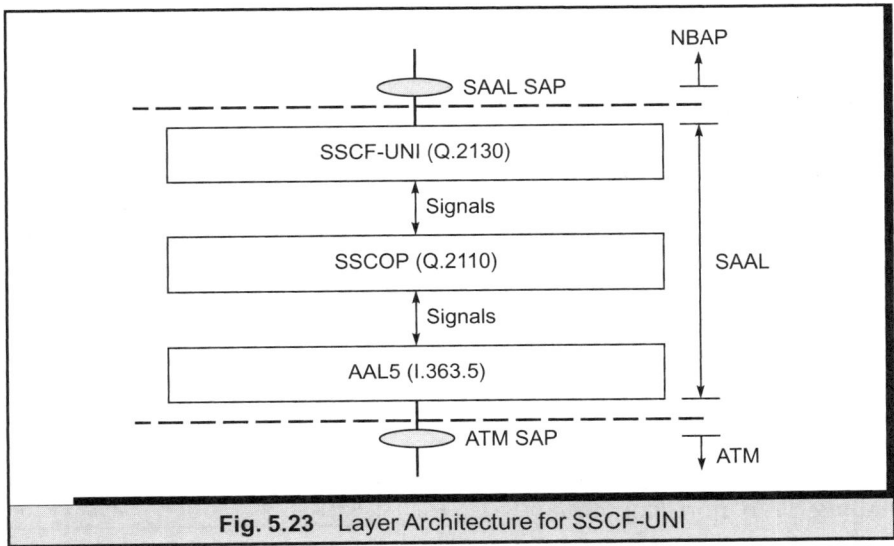

Fig. 5.23 Layer Architecture for SSCF-UNI

5.8.6.2 SSCF-UNI Layer Functions

The SSCF provides the SAAL primitives to SAAL user (i.e. NBAP). The following functions are provided by SSCF at the UNI:

- **Transfer of Data:** Two modes of data transfer are supported by SSCF, namely the *Unacknowledged transfer* and *Assured transfer* of data. The data is assumed to be octet aligned, with a maximum of 4096 octets. Note that the assured data transfer service of SSCF-UNI is provided in conjunction with SSCOP because SSCF-UNI uses the services of SSCOP.

- **Establishment and Release of SAAL Connections:** For assured transfer, SSCF provides the means to establish and release SAAL connections. SAAL connections are used to exchange assured data between SAAL peers.

- **Transparency of Transferred Information:** The SSCF provides transparent transfer of information in the sense that there is no restriction on its content, format or coding. Further, the information is transparent to SAAL as SAAL cannot interpret the transferred information.

5.8.7 Message Transfer Part 3 for Broadband (MTP3b)

The Message Transfer Part 3 (MTP3) is a Layer 3 protocol used in SS7 networks to connect various Core Network entities. In the access network, however, the ATM is used below SAAL, instead of SS7. In particular, ATM is used for the Iu interface (i.e. Iu_CS and Iu_PS interfaces) and the Iur interface. Since ATM is also referred to as a broadband technology, the equivalent for MTP3 that runs over SAAL/ATM is referred to as Message Transfer Part 3 for Broadband (MTP3b).

The MTP3b is defined in ITU-T specification Q.2210. Essentially, this protocol defines how the MTP3 protocol (as defined in ITU-T specification Q.704) can be adapted for signaling links that provide services as per the Q.2140 recommendation.

5.8.7.1 MTP3b Layer Architecture

As stated in the previous section, MTP3b is designed for control of signaling links that provide the services of Recommendation Q.2140. Thus, the MTP3b resides over SSCF-NNI (Figure 5.24).

In turn, MTP3b provides services to its service users. In the UTRAN, the service users are SCCP (for Iu_CS/Iu_PS/Iur interface) and Signalling Transport Converter (STC) for

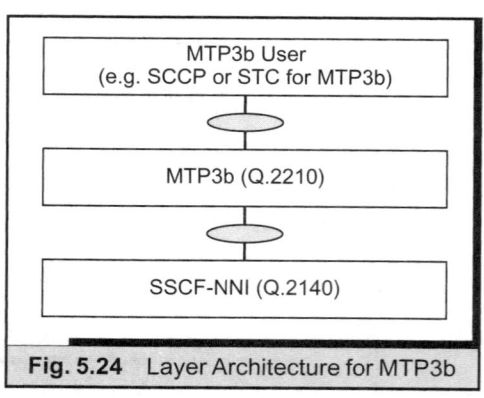

Fig. 5.24 Layer Architecture for MTP3b

MTP3b (applicable for Iu_CS and Iur) interface. The Signalling Transport Converter (STC) for MTP3b is explained in Section 5.8.8.

5.8.7.2 MTP3b Layer Functions

The functions of MTP3b are similar to those provided by MTP3. These include indications of whether the link is in-service or out-of-service and whether there is congestion or not. Apart from this, MTP3b provides data transfer facility. For further details on this, the reader is referred to MTP3 section of Chapter 6.

5.8.8 Signaling Transport Converter (STC)

On some of the UTRAN interfaces where AAL2 is used as bearer, there is the need for a signaling protocol for establishment, maintenance and release of these AAL2 connections. These connections are established over an existing ATM virtual connection and are managed using the AAL2 signaling protocol. In the context of UTRAN, the AAL2 signaling protocol is also referred to as Access Link Control Application Part (ALCAP).

Now, the ALCAP can reside over SSCF-UNI (for Iub interface) and MTP3b (for Iu_CS/Iur interface). Given this, one option is to adapt AAL2 signaling for every possible signaling transport. Alternatively, the services used by AAL2 signaling can be kept fixed; further, for each signaling transport, a converter function can be used that maps the services of the underlying transport to the fixed primitives expected by AAL2 signaling. The second option leads to the notion of the Signaling Transport Converter (STC). The ITU-T standards define the two STCs as follows:

- **STC for MTP3b:** This STC is used for the Iu_CS and Iur interface. In this case, the STC resides over MTP3b. The STC for MTP3b is defined in ITU-T standard Q.2150.1.

- **STC for SSCF-UNI:** This STC is used for the Iub interface. In this case, the STC resides over SSCF-UNI. The STC for SSCF-UNI is defined in ITU-T standard Q.2150.2.

5.8.8.1 STC Layer Architecture

Figure 5.25 shows the layer architecture of STC for SSCF-UNI and for MTP3b. The AAL2 signaling entity expects the following from any STC:

1. IN-SERVICE.indication
2. OUT-OF-SERVICE.indication
3. CONGESTION.indication
4. TRANSFER.request
5. TRANSFER.indication

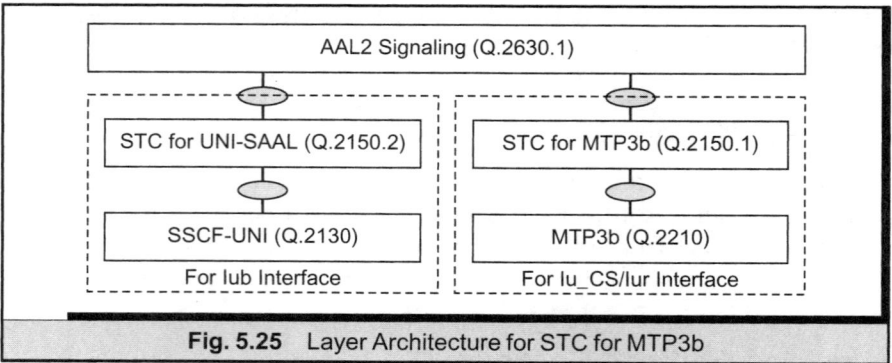

Fig. 5.25 Layer Architecture for STC for MTP3b

To provide the above services, a particular STC maps the services provided by the lower layer to one of the aforementioned services. In particular, the STC for SSCF-UNI uses the service primitives of SSCF-UNI (e.g. AAL-ESTABLISH, AAL-RELEASE, AAL-DATA and AAL-UNITDATA) to provide a service to ALCAP. Similarly, the STC for MTP3b uses the MTP3b service primitives (like MTP-TRANSFER, MTP-PAUSE, MTP-RESUME and MTP-STATUS) to provide service to ALCAP.

5.8.8.2 STC Layer Functions

STC functions are as follows:

- **Data transfer service availability:** This involves reporting to the AAL2 signaling entity the availability/non-availability of the underlying transport network for transfer of AAL2 signaling messages.

- **Data transfer function:** It involves carrying actual signaling messages.

- **Congestion reporting:** Reporting to the AAL2 signaling entity the level of congestion in the underlying transport network.

- **Maximum length indication:** This refers to reporting to the AAL2 signaling entity the maximum length of PDU that can be transferred by it (using the underlying transport network).

5.8.9 AAL2 Signaling

Generally, signaling is used between the user and the network, or between two network elements to establish, monitor, and release connections by exchanging control information. In the context of AAL2, the AAL2 signaling protocol encompasses the control plane functions to establish, clear and maintain AAL2 connections. An AAL2 connection can be viewed as the logical concatenation of one or more AAL2 links between two AAL2 service end-points. An AAL2 link is a communication path between two AAL2 nodes that are uniquely identified by a Channel IDentifier (CID).

Contrary to popular perception, AAL2 signaling is totally independent of ATM signaling. Out of all the AALs, only AAL2 has its separate signaling protocol. For other AALs (including AAL5), the paths are established either through configuration (i.e. Permanent Virtual Circuits (PVCs)) or through ATM signaling procedures (i.e. Switched Virtual Circuits (SVCs)). For AAL2, however, it was felt that extensions to ATM signaling would not be sufficient. This was primarily because, first, if ATM signaling protocol were to be extended, then the AAL2 network would get coupled to the way the ATM connection is established (i.e. the network has to be SVC and not PVC). Note that in RAN, there is predominant use of PVC, so the merging of AAL2 signaling with ATM signaling was not a viable solution.

Secondly, since the ATM network has many signaling protocols like ATM Forum's UNI 4.0 and PNNI, ITU-T's Q.2931/Q.2971, and ITU-T's B-ISUP, incorporating AAL2 into these specifications requires extensions in all these protocols. This is practically very difficult and also time-consuming.

Third, there are obvious consequences of decoupling, in that it could allow the existence of multiple AAL2 overlay networks operating over a single ATM network, each with its own addressing and routing plan. The independence becomes more crucial when the operator of the ATM and AAL2 network is not the same organization, as this allows both parties to be in charge of the addressing within their own network.

Finally, AAL2 defines multiplexing of many AAL2 channels within an ATM virtual connection. To establish and release these connections, a dynamic protocol is necessary. The requirement is fulfilled by AAL2 signaling. Similar requirement does not arise in other AALs because AAL1 and AAL5 do not have multiplexing support.

To summarize, the AAL2 signaling protocol is a protocol completely independent of ATM signaling. The former assumes the existence of ATM virtual channels on which AAL2 channels are then established. How these ATM virtual channels are established is not within the purview of AAL2 signaling.

The ITU-T recommendation Q.2630.1 defines the procedures for AAL2 signaling for Capability Set 1. There is another protocol specified in ITU-T Q.2630.2 for Capability Set 2. The latter provides some optional additions over the former.

In the context of UTRAN, the AAL2 Signaling protocol is also referred to as Access Link Control Application Part (ALCAP).

5.8.9.1 *AAL2 Signaling Layer Architecture*

AAL2 Signaling defines an overlay network over the existing ATM network. The overlay network of AAL2 assumes an addressing, signaling and routing mechanism of its own. However, standards only exist for signaling. There are no standards for routing and addressing.

The architecture for AAL2 Signaling is shown in Figure 5.26. Note that the AAL2 signaling entity either resides at end-points or at intermediate switches. At the end-point, the AAL2 Signaling entity provides services (like connection establishment and release) to the served user. In the intermediate nodes (i.e. switches), the AAL2 stack provides bridging and routing support. The underlying layer of AAL2 Signaling is a Signaling Transport Converter (STC). This converter is specific to the type of transport mechanism. For example, if AAL2 signaling stack resides over UNI-SAAL the signaling transport converter is based on ITU-T Recommendation Q.2150.1. Similarly, for MTP3b/NNI-SAAL the signaling transport converter is based on ITU-T Recommendation Q.2150.1. The different STCs defined for AAL2 signaling have already been discussed in this chapter.

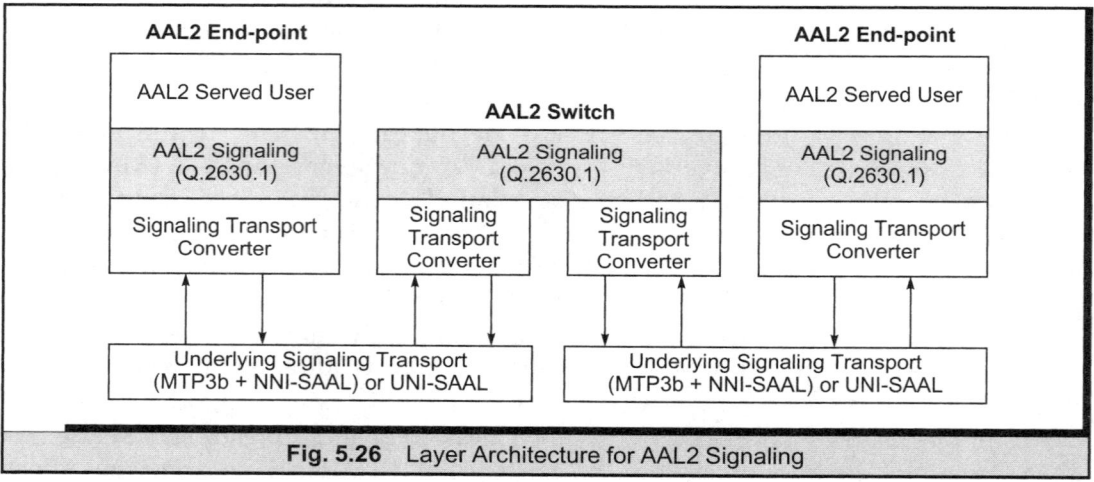

Fig. 5.26   Layer Architecture for AAL2 Signaling

5.8.9.2 *AAL2 Signaling Layer Functions*

AAL2 signaling is used to provide various functions. Of these, the important ones are summarized as follows:

- **Connection Management:** The primary function of AAL2 signaling entity is to provide the means to establish and release AAL2 connections.

- **Error Handling:** AAL2 signaling also provides mechanisms for detecting and reporting signaling procedural errors or other failures—detected by the AAL2 signaling end-point—to AAL2 management.

- **Reset:** To handle abnormal scenarios, AAL2 signaling provides a reset facility. The three types of resets are:

    (1) reset all AAL2 paths associated with an end-point

(2) reset a single AAL2 path and its channels

(3) reset a particular channel in an AAL2 path in an interface.

- **Block/Unblock:** AAL2 signaling also provides the means to block and unblock AAL2 paths between adjacent nodes. A blocked path is not used for carrying new connections, other than test connections.

5.8.9.3 AAL2 Signaling Messages

Most message exchanges in AAL2 signaling involve a two-way handshake. The entity initiating the procedure sends a *request* message. The receiving entity responds with a *confirm* message. Table 5.5 shows important AAL2 signaling messages. The Establish Request message is used to initiate connection establishment. The peer entity responds by Establish Confirm. To release an established connection, the Release Request/ Confirm messages are used. Other messages shown in the Table 5.5 are used for layer management functions.

Table 5.5 Important AAL2 Signaling Messages

Type	Messages	Description
Signaling	Establish Request/Confirm	Used to establish an AAL2 connection between AAL2 end-points. The request message contains the destination end-point address (e.g. E.164 address), link characteristics and other service specific information (e.g. audio profiles).
	Release Request/Confirm	Used to release an existing AAL2 connection between AAL2 end-points.
Management	Block Request/Confirm	Used to block AAL2 paths from carrying new connections other than test connections.
	Unblock Request/Confirm	Used to unblock a blocked AAL2 path so that it is available again for carrying new connections.
	Reset Request/Confirm	Used to release all the affected channels and associated resources specified in the reset request.

5.9 APPLICATION LAYER PROTOCOLS

In Section 5.5, the UTRAN protocol architecture was discussed. In that section, the radio network layer protocols used in the UTRAN were briefly mentioned. This section elaborates upon these protocols; in particular, the following:

- Radio Access Network Application Part (RANAP)
- Radio Network Sub-system Application Part (RNSAP)

- Node B Application Part (NBAP)
- Iu User Plane Protocol
- Iu Framing Protocol

5.9.1 Radio Access Network Application Part (RANAP)

The Iu interface connects the Core Network and Access network. Since the Core Network is further divided into the CS and PS domain, the Iu interface with these domains is called the Iu_CS and the Iu_PS interface respectively.

While the user plane differs for both these domains (Iu_CS uses AAL2 and Iu_PS uses AAL5), the protocol developers endeavoured to keep the signaling protocol independent of the underlying transport. The result is Radio Access Network Application Part (RANAP) protocol that is used for Iu_CS and Iu_PS. The RANAP protocol is used for signaling exchanges between RNC and MSC/VLR over the Iu_CS and between RNC and SGSN over the Iu_PS.

The 3GPP TS 25.413 defines the RANAP protocol.

5.9.1.1 RANAP Layer Architecture

The RANAP protocol is used for signaling over the Iu interface. Figure 5.27 depicts the context in which the RANAP protocol operates. RANAP resides over Iu transport for signaling bearers, which can be purely ATM (SCCP/MTP3b/SSCF-NNI/SSCOP/AAL5) or IP (SCCP/M3UA/SCTP/IP) over any link layer (e.g. ATM or Frame Relay).

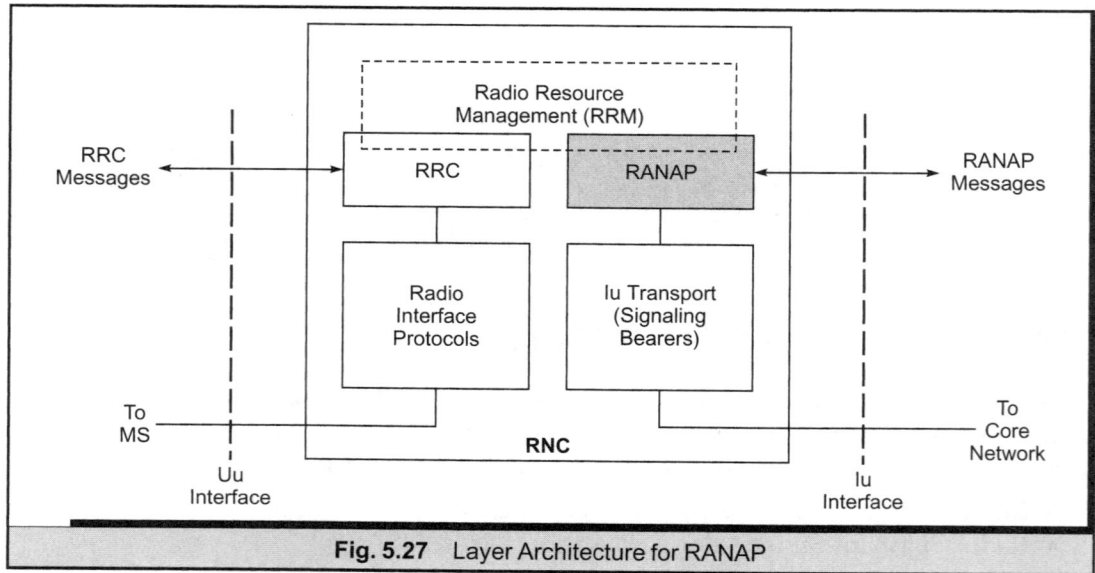

Fig. 5.27 Layer Architecture for RANAP

The RANAP interacts with a Radio Resource Management (RRM) function. The RRM contains the logic for allocating radio resources, for which the RRM interacts with the RRC layer. The RRC, in turn, controls other radio protocols like RLC and MAC.

5.9.1.2 RANAP Layer Functions

RANAP signaling has various functions of which the important ones are summarized as follows:

- **Radio Access Bearer (RAB) Management:** A RAB is the means to provide a bearer (with a specified QoS) between the UE and Core Network. The RANAP protocol manages RABs (i.e. it sets up, modifies, and releases RABs). The operations are initiated by CN. However, the UTRAN can request the CN to release a RAB. Further, if the request cannot be served immediately, it is possible to queue it. The option of queuing the setup of RABs is provided by RANAP. A RAB is associated with a priority and a pre-emption capability. This information is provided by CN to the UTRAN. The UTRAN in turn uses it to queue or pre-empt RABs accordingly.

- **Iu Management:** The RANAP is also used to manage the Iu link. This involves establishment and release of resources for Iu connection and managing the overload.

- **Radio Resource Management:** As discussed in the previous section, the RANAP interacts with the Radio Resource Management (RRM) function to analyze the current usage of radio resources and accept/reject new requests for RAB establishment/modification.

- **Transport of NAS signaling information:** RANAP provides means to transport NAS signaling information between UE and CN. For this, it uses the 'Initial UE' message when there is no Iu connection with the CN; it uses the 'Direct Transfer' message over an existing Iu signaling connection.

- **Mobility Management:** RANAP provides various mobility management functions, which include paging, SRNS relocation and handovers. These functions are elaborated in Chapter 8.

- **Iu User Plane Management:** As discussed in Section 5.9.5.1, the Iu User Plane (UP) protocol operates in two modes: the Transparent Mode and the Support Mode. The CN uses the RANAP to set the mode for Iu UP protocol. Apart from this, the Iu UP also requires an initialization function, which is also provided by RANAP.

- **Security Mode Control:** RANAP provides the means for the CN to send the security keys (Cipher Key (CK) and Integrity Key (IK)) to the UTRAN. It also provides the means to control the security mode of the UTRAN.

- **Other Functions:** Apart from the functions mentioned above, RANAP supports certain other functions. These include tracing, which allows the tracing of various events related to the MS, and the activities of the MS. Apart from tracing, RANAP is also used by CN to collect reports for the volume of unacknowledged data. This report is sent to CN for accounting purposes. In addition, RANAP also provides the means to the report geographical position of the MS to CN. The CN controls the mode in which UTRAN reports the location of the MS.

5.9.1.3 RANAP Messages

RANAP protocol is used to exchange signaling messages between RAN and CN over the Iu interface. Table 5.6 shows important RANAP signaling messages.

5.9.2 Radio Network Sub-system Application Part (RNSAP)

The RNSAP protocol is used in the UTRAN between two RNCs over the Iur interface. Here, one RNC takes the role of Serving RNC (SRNC) and the other takes the role of Drift RNC (DRNC). The two RNCs work in the master-slave mode, with SRNC acting as the master and DRNC as the slave. The RNSAP protocol is used primarily for inter-RNC soft handover.

This protocol is detailed in the 3GPP TS 25.423 specification.

5.9.2.1 RNSAP Layer Architecture

RNSAP protocol defines the Iur interface signaling procedures. These procedures are essentially divided into four distinct sets:

- **RNSAP Basic Mobility Procedures:** This set of procedures is used to handle mobility within the UTRAN. This is the most important of the RNSAP procedures. The procedures belonging to this set include SRNC relocation, inter-RNC cell update and UTRAN registration area update.

- **RNSAP DCH Procedures:** This set of procedures is used to handle dedicated channel traffic (it includes DCH, DSCH and TDD USCH) between two RNCs. Unlike the basic mobility procedures which is used only for signaling, this set of procedures provides support for data transfer over the Iur interface. The data transfer takes place using a frame protocol (refer Section 5.9.6). The procedures belonging to this set include establishment, modification and release of dedicated channel in the DRNC due to hard and soft handover, set-up/release of dedicated transport connections over Iur interface and data transfer for dedicated channels.

- **RNSAP Common Transport Channel Procedures:** This set of procedures is used to handle common and shared channel traffic (it excludes DCH, DSCH and TDD

| Table 5.6 | Important RANAP Messages |

Type	*Messages*	*Description*
Paging	Paging	Used by the CN to request UTRAN to contact the UE. The paging procedure may be initiated for the UE when there is an incoming call for it.
Data Transfer	Initial UE Message	Used to establish the Iu signalling connection between the CN and the RNC, and to transfer the initial NAS-PDU received from the UE to the CN domain.
	Direct Transfer	Used to carry the UE-CN signaling messages over the Iu interface after the Iu signaling connection is established. The UE-CN signaling messages are not interpreted by UTRAN.
Iu Connection	Iu Release Request	Used by UTRAN to request the CN to release the Iu connection for a particular UE due to some UTRAN generated reason.
	Iu Release Command / Iu Release Complete	Used by CN to release the Iu connection and all UTRAN resources related only to the connection to be released.
RAB Management	RAB Assignment Request/ RAB Assignment Response	To establish new RABs and/or to enable modifications to RABs and/or to release already established RABs.
	RAB Release Request	To enable UTRAN to request the CN to release one or several RABs
Relocation	Relocation Required / Relocation Command	Sent from the SRNC to the CN, to prepare for relocation of the SRNS. The Relocation Command is the response from the CN on successful handling of Relocation Required message from UTRAN.
	Relocation Request/ Relocation Request Acknowledge	To request the target RNS to allocate resources for relocation of SRNS
	Relocation Detect	Used by the Target RNC to indicate to the CN that the relocation execution trigger has been received. The target RNC, after sending this message, starts the SRNC operation.
	Relocation Complete	Used by the Target RNC to indicate to the CN that the new SRNC-Id has been indicated to the UE, and that the relocation procedure is now complete.
Others	Security Mode Command/ Security Mode Complete	Used by CN to send security information to the RNC.
	CN Invoke Trace	Used by CN to inform the RNC that it should begin a trace record for a UE.

USCH) between two RNCs. In particular, this set of procedures facilitates the set-up and release of common channel transport connections over the Iur interface.

- **RNSAP Global Procedures:** Unlike the sets of procedures mentioned above, this set does not apply to a UE. Implementation of this is considered optional. The procedures belonging to this set include transfer of cell measurements and Node B timing information between two RNCs.

5.9.2.2   RNSAP Layer Functions

RNSAP protocol is used to provide various functions, of which the important ones are summarized as follows:

- **Radio Link Management and Supervision:** The Radio Link Management function is used by the SRNC to manage the radio links using dedicated resources in the Drift Radio Network Sub-system (DRNS). The Radio Link Supervision mechanism allows the DRNC to report any failures or restorations of a radio link to the SRNC.

- **Physical Channel Reconfiguration:** This function is used by the DRNC to request the SRNC for reconfiguration of the physical channel resources corresponding to a radio link.

- **Measurements on Dedicated Resources:** This procedure enables the SRNC to initiate measurements in DRNS for dedicated resources, and DRNC to report those measurements.

- **Downlink Power Drifting Correction:** This allows the SRNC to adjust the downlink power level of one or more radio links to avoid power drifting between the radio links.

- **Rate Control:** This procedure is used by the DRNC to control the uplink and downlink transfer rate for each DCH configured for the radio links of the UE.

- **Common Control Channel Signaling Transfer:** This function is used by the DRNC to forward a Uu message, received on the Common Control Channel (CCCH), to the SRNC. It is also used by the SRNC to request the DRNC to transfer a Uu message over a CCCH controlled by the DRNC.

- **Paging:** This function allows the SRNC to page a UE in a URA, or a cell within the Drift RNS.

- **Common Transport Channel Resource Management:** This procedure is used by the SRNC to utilize common transport channel resources within a Drift RNS.

- **Relocation:** This allows the SRNC to execute a relocation that has been previously prepared.

- **Reporting of General Error Situations:** This allows the reporting of general error conditions between the SRNC and the DRNC, and is used for functions which do not have in-built error reporting mechanism.

- **Measurements on Common Resources:** This procedure is used between the DRNC and the SRNC to initiate measurements on common resources, and for the reporting of these measurements.

- **Information Exchange:** This is used between the SRNC and the DRNC to request for specific information as well as to report this information.

- **Iur Interface Reset:** This procedure is used to reset the Iur interface, either partly or completely. The reset procedure is used to align the resources between the RNCs in the event of an abnormal failure.

5.9.2.3 RNSAP Messages

RNSAP protocol is used to exchange signaling messages between two RNCs over the Iur interface. Table 5.7 shows the important RNSAP signaling messages.

5.9.3 Node B Application Part (NBAP)

The Node B Application Part (NBAP) protocol is used in the radio network control plane between the RNC and the Node B, over the Iub interface. This protocol is used by the RNC to control the resources of Node B, and also by Node B to send measurement reports to RNC. Besides this, NBAP protocol is used for fault management purposes.

The NBAP protocol is detailed in the 3GPP TS 25.433 specification.

5.9.3.1 NBAP Layer Architecture

The NBAP protocol defines the Iub interface signaling procedures. These procedures are essentially divided into two distinct sets:

- **NBAP Common Procedures:** These are not specific to any UE (save one exception). They carry out the general operation and management functions for Node B. These procedures include cell configuration, configuration of the Node B common resources, initialization and reporting of the cell or Node B specific measurements. Besides this, fault management procedures are also part of the common procedures. However, the exception is the establishment of the first radio link with any UE. Though this procedure is specific to a particular UE, it is handled as part of the common procedures.

- **NBAP Dedicated Procedures:** These procedures handle the UE specific procedures (besides the first RL set up, which is a part of the common procedures). They

Table 5.7 Important RNSAP Messages

Type	Messages	Description
Mobility	Uplink Signaling Transfer Indication	Used by DRNC to forward a Uu message from the UE to the SRNC.
	Downlink Signaling Transfer Request	Used by SRNC to request the DRNC to transfer a Uu message in a cell.
	Paging Request	Used by SRNC to request the CRNC to page the UE in a cell or URA under control of the CRNC.
	Relocation Commit	Used by SRNC to execute the relocation procedure. Sent to the Target RNC.
Radio Link Procedures	Radio Link Set-up Request/ Response	Used by SRNC to request the DRNC to reserve necessary resources for one or more radio links.
	Radio Link Addition Request	Used by SRNC to request the DRNC to reserve necessary resources for one or more additional radio links for the UE.
	Radio Link Deletion Request/ Response	Used by SRNC to release resources for one or more radio links in the DRNS.
	Radio Link Restore Indication	Used by DRNC to notify SRNC of the establishment and re-establishment of UL synchronization on the Uu interface
	Radio Link Reconfiguration Prepare/Ready	Used by SRNC during the Synchronized Radio Link Reconfiguration Preparation procedure to prepare a new configuration of radio links, related to one UE-UTRAN connection, within a DRNS.
	Radio Link Reconfiguration Commit	Used by SRNC to order DRNS to switch to the new configuration for the radio links, prepared previously by the Synchronized Radio Link Re-configuration Preparation procedure.
	Radio Link Reconfiguration Request/ Response	Used by SRNC during the Unsynchronized Radio Link Reconfiguration procedure to reconfigure the radio links, related to one UE-UTRAN connection, within the DRNS.
Common Transport Channel	Common Transport Channel Resources Request/Response	Used by SRNC to initialize the Common Transport Channel User Plane towards the DRNC, and/ or, to initialize the Common Transport Channel resources in the DRNC to be used by a UE.
	Common Transport Channel Resources Release Request	Used by SRNC to request DRNC to release the Common Transport Channel resources for a given UE.
Power Control	DL Power Control Request	Used by SRNC to initiate the Downlink Power Control procedure, which is used to balance the downlink transmission powers of the radio links for one UE.

include addition, release and reconfiguration of the radio links for a UE. They also include initialization and reporting of radio link specific measurements. Handling of dedicated and shared channels is done as part of NBAP dedicated procedures.

Figure 5.28 depicts the layer architecture for NBAP, which consists of *Common Procedures* and *Dedicated Procedures*.

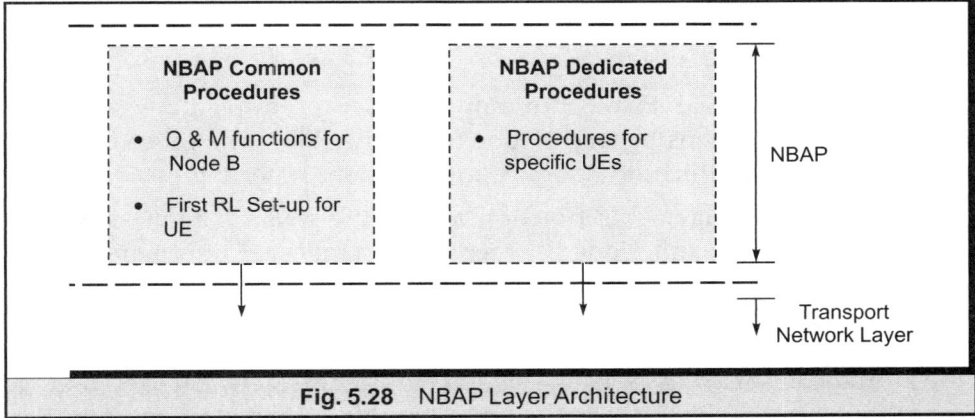

Fig. 5.28   NBAP Layer Architecture

5.9.3.2   *NBAP Layer Functions*

The functions of the NBAP Layer, which consist of functions within the common procedures and dedicated procedures, are as follows:

- **Cell Configuration Management:** This procedure is used by RNC to manage the cell configuration information in a Node B.

- **System Information Management:** This is used by RNC to manage the scheduling of system information to be broadcast within a cell.

- **Common Transport Channel Management:** The RNC uses this procedure to manage the configuration of the common transport channels in a Node B.

- **Resource Event Management:** This function is used by Node B to inform the RNC about the status of the Node B resources.

- **Configuration Alignment:** This procedure enables the RNC and Node B to verify the radio resource configuration information available with each node.

- **Measurements on Common and Dedicated Resources:** This procedure enables the RNC to initiate measurements in Node B, and Node B to report those measurements.

- **Radio Link Management and Supervision:** This function is used by the RNC to manage the radio links in the Node B, and to report failures and restoration of radio links.

- **Downlink Power Drifting Correction:** This procedure allows the RNC to adjust the downlink power level of one or more radio links to avoid power drifting between the radio links.

- **Bearer Re-arrangement:** This function allows Node B to indicate the need for bearer re-arrangement for a particular UE. It also allows the RNC to re-arrange the bearers for the UE.

- **Reporting of General Error Situations:** This procedure allows reporting of general error conditions between Node B and the RNC. The procedures here are used for functions which do not have in-built error reporting mechanism.

- **Information Exchange:** This function allows the RNC to request Node B for specific information, and Node B to report the requested information.

5.9.3.3 NBAP Messages

The NBAP protocol is used to exchange signaling messages between RNC and Node B over the Iub interface. Table 5.8 lists the important NBAP signaling messages.

Table 5.8 Important NBAP Messages

Type	Messages	Description
Cell Management	Cell Setup Request/ Cell Setup Response	Used by CRNC to setup a cell in the Node B, and to indicate to Node B to reserve the required resources for the cell.
	Cell Reconfiguration Request/Response	Used by CRNC to reconfigure a cell in the Node B.
	Cell Deletion Request/ Response	Used by CRNC to delete a cell in the Node B. The Node B would also delete any remaining dedicated and common channels within the cell.
Common Transfer Channel	Common Transport Channel Setup Request/Response	Used by CRNC to request Node B to establish the necessary resources for Common Transport Channels.
	Common Transport Channel Reconfiguration Request/Response	Used by CRNC to request Node B to reconfigure Common Transport Channels and/or Common Physical Channels, while they still might be in operation.
	Common Transport Channel Deletion Request/Response	Used by CRNC to request Node B to delete of Common Transport Channels and Common Physical Channels.

Contd.

| Table 5.8 | Contd. |
| | |

Type	*Messages*	*Description*
Radio Link Protocol	Radio Link Setup Request/ Response	Used by CRNC to request Node B to reserve necessary resources for one or more radio links, and configure these new radio links.
	Radio Link Addition Request/Response	Used by CRNC to request Node B to reserve necessary resources for one or more additional radio links towards the UE.
	Radio Link Deletion Request/Response	Used by CRNC to release resources in Node B for one or more established radio links towards the UE.
	Radio Link Restore Indication	Used by Node B to notify CRNC of the achievement and re-achievement of UL synchronization on one or more radio link sets on the Uu interface
	Radio Link Reconfiguration Prepare/Ready	Used by CRNC during the Synchronized Radio Link Reconfiguration Preparation procedure to prepare a new configuration of radio links, related to one UE-UTRAN connection, within a Node B.
	Radio Link Reconfiguration Commit	Used by CRNC to order Node B to switch to the new configuration for radio links, prepared previously by the Synchronized Radio Link Reconfiguration Preparation procedure.
	Radio Link Reconfiguration Request/Response	Used by CRNC during the Unsynchronized Radio Link Reconfiguration procedure to reconfigure the radio links, related to one UE-UTRAN connection, within the Node B.
	Radio Link Pre-emption Required Indication	Used by Node B to indicate to the CRNC that resources corresponding to a radio link need to be freed, as a result of pre-emption. The CRNC should then delete the radio links that are to be pre-empted.
Others	System Information Update Request/Response	Used by CRNC to indicate certain necessary operations to Node B, in order for the Node B to apply the correct scheduling of system information, and to include the appropriate content in the system information segments broadcast on the broadcast channel.
	DL Power Control Request	Used by CRNC to initiate the Downlink Power Control procedure, which is used to balance the downlink transmission powers of the radio links for one UE.

5.9.4 Service Area Broadcast Protocol (SABP)

The SABP protocol is used between the Cell Broadcast Center (CBC) and RNC over the Iu_BC interface. It is used to control the information broadcasted by an RNC in a Service Area.

The SABP protocol is detailed in the 3GPP TS 25.419 specification.

5.9.4.1 SABP Layer Architecture

The SABP layer architecture is simple. As highlighted in Figure 5.9, the SAB resides over TCP/IP protocol. IP in turn uses AAL5 over ATM.

5.9.4.2 SABP Layer Functions

SABP is a relatively simple protocol. Its important functions are summarized as follows:

- **Message Handling:** This function handles the broadcast of new messages, amendment of existing broadcasted messages and stopping of the broadcast of specific messages. This is done using the Write-Replace message.

- **Load Handling:** This refers to determining the load of the broadcast channels at any particular point in time. This is done using the Load-Query message.

- **Reset:** This function permits the CBC to end broadcasting in one or more Service Areas. This is done by using the Reset message.

5.9.4.3 SABP Messages

The SABP protocol is used to exchange messages between Cell Broadcast Center and RNC over the Iu_BC interface. Table 5.9 shows important SABP messages.

Table 5.9   Important SABP Messages

Type	Messages	Description
SAB	Write-replace Write-replace Complete	Used to broadcast new information or replace a message already broadcast to a chosen Service Area(s).
	Load-Query Load-Query Complete	Used to obtain the current permissible bandwidth available for broadcast within particular Service Area(s).
	Reset Reset Complete	The purpose of the Reset procedure is to end broadcasting in one or more Service Areas in the RNC.

5.9.5 Iu User Plane (UP) Protocol

The Iu UP protocol is used in the radio network user plane for the transport of user data between UTRAN and the CN over the Iu interface. One instance of the Iu UP protocol is associated with one Radio Access Bearer (RAB) and carries the data for that RAB. It is used at both the Iu_CS and the Iu_PS interfaces. Nevertheless, the protocol is designed in such a way that it is independent of the CN domain to which it is interfacing.

The Iu UP protocol is detailed in the 3GPP TS 25.415 specification.

5.9.5.1   Iu UP Layer Architecture

The Iu UP layer can operate in two modes. The mode to be used for a particular RAB is selected by the CN. The following are the Iu UP modes of operation:

- Transparent Mode (TM)
- Support Mode (SM) for Pre-defined SDU Sizes.

5.9.5.1.1   Transparent Mode (TM)

The Iu UP Transparent Mode (TM) of operation does not perform any framing or control operation. It simply transfers the user data across the Iu interface, providing a truly transparent operation. Figure 5.29 shows the Iu UP operation in the Transparent Mode. The data is transparently passed on to the Transport Network Layer (TNL) without any framing at the Iu UP.

5.9.5.1.2   Support Mode (SM)

The Iu UP Support Mode (SM) contains a Frame Handler function that performs framing of the user data into segments of pre-defined size. It also provides functions

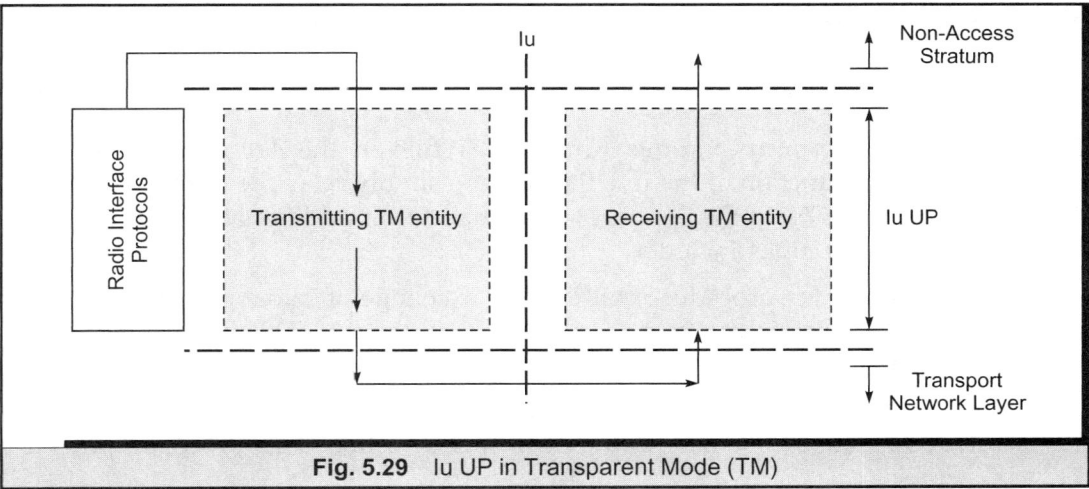

Fig. 5.29   Iu UP in Transparent Mode (TM)

for procedure control, which includes procedures for initialization of Iu UP instance, rate control, and time alignment. Initialization procedures for Iu UP Support Mode are used to negotiate the version of the Iu UP mode to be used, besides initialization of other parameters used for the operation of the Iu UP instance. The rate control procedure controls the maximum rate at which traffic can flow in the downlink direction, while the time alignment procedure controls the timing of the downlink data to the RNC over the Iu interface. Besides these control procedures, the Support Mode also provides NAS Data Stream Specific Functions, which include a function to classify the frame quality. Classification of the frame quality can be based on, for example, the CRC of the radio frame received. Figure 5.30 shows the Iu UP operation in the Support Mode (SM).

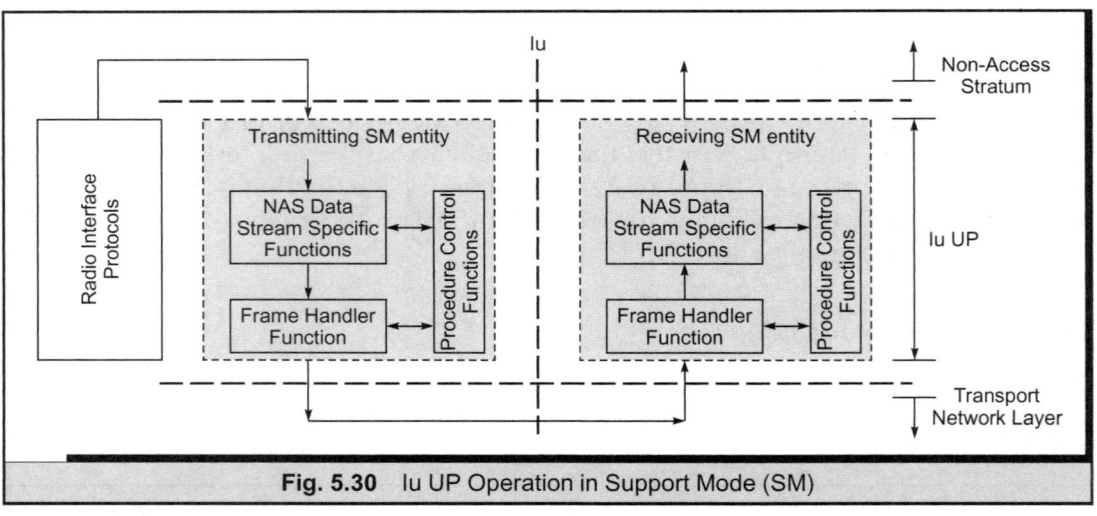

Fig. 5.30   Iu UP Operation in Support Mode (SM)

5.9.5.2   *Iu UP Layer Functions*

The Iu UP layer performs various functions. While in the Transparent Mode of operation, the only function of the Iu UP layer is to simply relay the user data between the UTRAN and the CN; in the Support Mode, it performs additional functions besides data relay. These functions include:

- **Initialization:** This procedure controls the exchange of initialization information between the Iu UP instance at the UTRAN and the CN. This initialization information is required for the operation of the Iu UP instance in support mode.

- **Rate Control:** The rate control procedure controls the maximum rate at which user data can be sent in the downlink direction. This procedure interacts with functions outside the Iu UP protocol layer to achieve the required functionality.

- **Time Alignment:** This procedure controls the timing of the downlink data to the RNC over the Iu interface. The time alignment procedure also requires interacting with functions outside the Iu UP protocol layer to achieve the required functionality.

- **Frame Quality Classification:** This function classifies the quality of the received frames. The frame quality classification is based on radio frame classification, and on a delivery parameter, which determines whether delivery of erroneous frames is allowed or not.

5.9.6 Framing Protocols for Iub and Iur Interface

Framing protocols define frame structures for carrying user data and control information across the Iur and the Iub interface. Transport blocks (transferred over transport channels) carrying user data and control information, received at Node B over the Uu interface, are transparently relayed over the Iub interface using the frame structures defined by the framing protocols. Similarly, in the downlink direction, transport blocks received as part of the framing protocols over the Iub interface, are transparently relayed by Node B over the physical channels through the Uu interface. Framing protocols are also used over the Iur interface, between the DRNC and the SRNC, to carry user data and control information within transport blocks.

The Iu FP protocol is detailed in the 3GPP TS 25.425, GPP TS 25.427 and 3GPP TS 25.435 specifications.

5.9.6.1 *Framing Protocol Layer Architecture*

Framing Protocols are essentially divided into two distinct sets:

- **Framing Protocols for Common Transport Channels (FP-CCH):** These protocols provide transfer of transport blocks for common transport channels, which include RACH, FACH, CPCH, USCH, DSCH, HS-DSCH and PCH. Besides data transfer, FP- CCH also offers control mechanisms for node synchronization and transport channel synchronization between the RNC and Node B. FP-CCH is covered in 3GPP TS 25.425 (for Iur interface) and TS 25.435 (for Iub interface).

- **Framing Protocols for Dedicated Transport Channels (FP-DCH):** These channels provide transfer of transport blocks for the Dedicated Transport Channel (DCH). Like in the case of FP-CCH, FP-DCH provides control mechanisms for node synchronization and transport channel synchronization. Besides this, FP-DCH also provides mechanisms for transport of other radio interface parameters between the SRNC and Node B. FP-DCH is covered in 3GPP TS 25.427 (for both Iur and Iub interfaces).

Figure 5.31 depicts the layer architecture for framing protocols. The two components shown in the diagram collectively provide the framing mechanisms for common and dedicated transport channels.

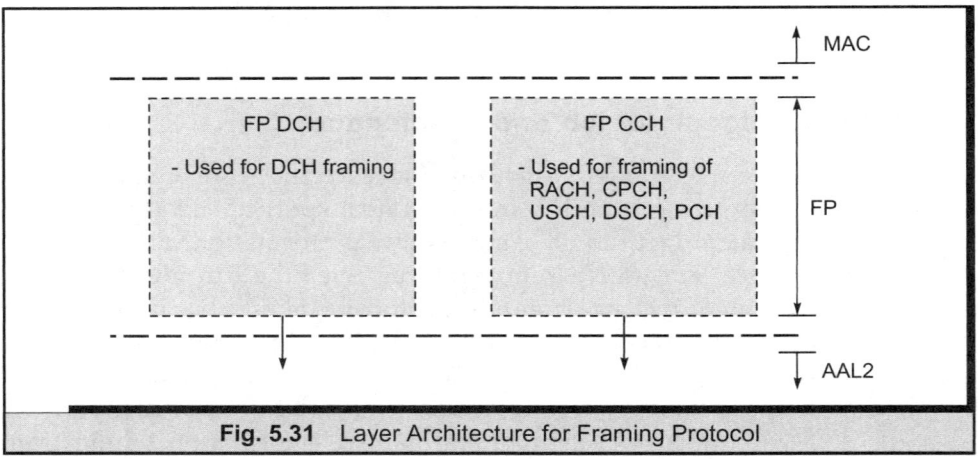

Fig. 5.31   Layer Architecture for Framing Protocol

5.9.6.2 *Framing Protocol Functions*

The following are the functions of the framing protocols:

- **Data Transfer:** This procedure is used for the transfer of transport blocks between the RNC and Node B.

- **Node Synchronization Mechanism:** Framing protocols provide the means for node synchronization between the RNC and Node B.

- **Transport Channel Synchronization Mechanism:** Framing protocols provide the means for synchronization of transport channels between the RNC and Node B.

- **Transfer of Power Control Information:** The FP-DCH provides the means for transfer of outer loop power control information between the SRNC and Node B.

SUMMARY

This chapter provides an overview of the Access Network entities (like RNC and Node B) and Access Network protocols (like RRC, RANAP, RNSAP, and NBAP). The RRC along with the lower layer protocols like RLC and MAC form the radio interface protocols that are used for signaling between UE and UTRAN. In a UTRAN, the NBAP protocol is used between RNC and Node B whereas the RNSAP protocol is used between the two RNCs. The RANAP protocol is used for signaling between the RNC and Core Network over the Iu interface. Part 3 of the book in built upon the topics introduced in this chapter explaining the various Access Network procedures like RRC procedures (Chapter 7) and UTRAN procedures (Chapter 8).

CORE NETWORK

6.1 INTRODUCTION

In Chapter 3, an overview of the UMTS Network architecture was presented, the focus being on its three basic components, i.e. the User Equipment, Access Network and Core Network. Chapters 4 and 5 elaborated upon the User Equipment and Access Network, respectively. This chapter discusses the last of the three basic elements of the UMTS Network architecture, namely the Core Network.

To explain the various aspects of Core Network, the information in this chapter is organized as per the outline provided in Figure 6.1.

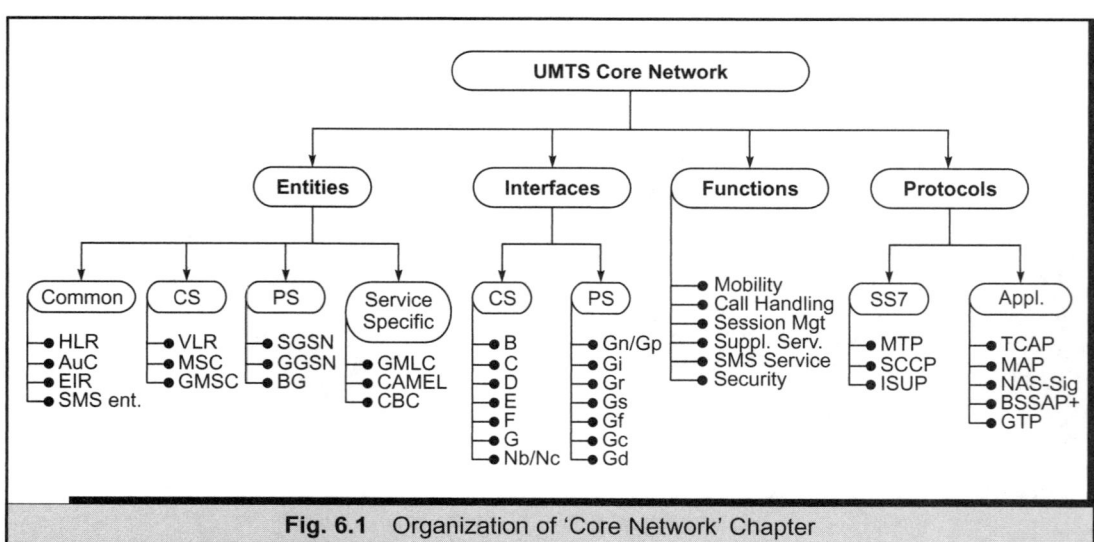

Fig. 6.1 Organization of 'Core Network' Chapter

6.2 ENTITIES COMMON TO CS AND PS DOMAIN

The Core Network is logically divided into two distinct domains: CS domain and PS domain. There are some entities, that which are specific to these domains, and some that are used by both the CS and PS domain. This section details the entities that are common to both the domains.

The common entities are used to manage and provide subscription information, authentication information and equipment identity information. Apart from this, they provide specialized services like Short Message Service (SMS). These common entities are as follows:

- Home Location Register (HLR)
- Authentication Center (AuC)
- Equipment Identity Register (EIR)
- Short Message Service (SMS) entities. These include:
    - Short Message Service Center (SM-SC)
    - SMS Gateway MSC (SMS-GMSC)
    - SMS Interworking MSC (SMS-IWMSC)

6.2.1 Home Location Register (HLR)

The Home Location Register (HLR) is the master database for a subscriber (i.e. it holds the subscriber data). The HLR maintains and provides subscriber data to other network entities on demand (i.e. the data is *pulled* by network entities). In certain cases, the subscriber data is sent to the network entities (for example, when the data is modified by the operator, the HLR *pushes* it to network entities). Subscriber data is maintained both for the CS and PS domain.

Typically, the subscriber data is provisioned by the network operator. Such data, is referred to as Permanent Data. HLR also maintains Temporary Data (or dynamic data), which is obtained by HLR dynamically from other network entities. The temporary data is used for dynamic procedures (for example, the VLR address is a dynamic data maintained by HLR and is used for routing a mobile-terminated call).

The following are some of the important information elements maintained by HLR:

- International Mobile Subscriber Identity (IMSI)
- One or more Mobile Subscriber ISDN (MSISDN) associated with an IMSI
- Network Access Mode (identifies the type of subscription, which can be for the CS domain, PS domain, or for both the domains)
- Roaming Restrictions
- VLR and SGSN address
- Barring Status for various services

- Bearer Service and Tele Service Information
- Supplementary Service Information
- CAMEL Subscription Information.

Section 6.11 provides the details of various elements of subscriber data and the elements stored at HLR. Based on the stored information, the HLR performs the following functions:

- **Mobility Management:** This includes maintaining the location at which a user is registered. It also includes deleting registration information of an MS from VLR/SGSN if the MS has moved out of the area controlled by VLR/SGSN.

- **Call and/or Session Establishment support:** The HLR supports the call and/or session establishment procedures in CS and PS domains. For example, during a mobile-terminating session, HLR provides the address of SGSN where a MS is currently registered.

- **Access Authorization:** This refers to allowing access to only authorized users. For example, it is checked whether the user is allowed to roam in a visited network or not.

- **Facilitates a Host of Services:** HLR facilitates a host of services including Short Message Service (SMS), Supplementary Service and CAMEL Service. For example, HLR communicates with the gsmSCF to support the CAMEL Services related to the CS and PS domains.

6.2.2 Authentication Center (AuC)

The Authentication Center (AuC) holds authentication information. This information is used for authentication and other security-related functions. The AuC is often depicted as a part of HLR. Thus, the term AuC/HLR is used to represent the entity that performs the functions of HLR and AuC. Note that the interface between HLR and AuC is called the H-Interface. This is a non-standardized proprietary interface. Given this, the AuC is merged with HLR and may also be referred to as HLR/AuC or simply HLR.

At the center of all AuC functions is a secret key that is shared between the AuC and the USIM of the UE. This key is associated with an IMSI and is used for authentication. It is shared between the USIM and AuC, and facilitates roaming for the subscriber because even in a visited network, the VLR SGSN can authenticate the user by obtaining security information from AuC.

The AuC itself does not take part in the actual authentication. This function is performed by the VLR for CS domain and SGSN for the PS domain. The role of the AuC is to provide VLR/SGSN with the necessary information, which can be used by the latter for authentication.

6.2.3 Equipment Identity Register (EIR)

In order to monitor the legitimacy of a UE in the UMTS network, an Equipment Identity Register (EIR) is used. This EIR holds the list of International Mobile Equipment Identity (IMEI) used in the UMTS network. The IMEI is used for identifying a user equipment.

A UE can be classified under one of the following three categories:

- **White list:** This includes all the number series of IMEI that are permitted for use.

- **Black list:** Includes all the number series of IMEI that are barred from use.

- **Grey list:** Apart from the white and black lists, there is a grey list that is not barred. However, this is tracked by the network for various reasons (e.g. for evaluation).

The aforementioned categories determine whether a UE can be used in the network or not.

6.2.4 Short Message Service (SMS) entities

In order to provide SMS, the Core Network has three distinct entities. The first of these is the Short Message Service Center (SM-SC). The SM-SC is responsible for the relaying and store-and-forwarding of a short message between an MS and a Short Message Entity (SME). The SME is any entity that can send or receive SMS messages. The SM-SC is considered to lie outside the UMTS PLMN. Thus, the functionality of SM-SC is not formally standardized by the 3GPP specifications.

Apart from SM-SC, there are two gateway entities that relay messages to and from SM-SC. These entities are SMS Gateway MSC (SMS-GMSC) and SMS Interworking MSC (SMS-IWMSC).

The function of SMS Gateway MSC (SMS-GMSC) is to submit short messages to MS from the SM-SC.

While SMS-GMSC enables the delivery of a short message from SM-SC to MS, the SMS Interworking MSC (SMS-IWMSC) enables such a message to be delivered from MS to SM-SC. Collectively, SMS-GMSC and SMS-IWMSC provide the facility to deliver short messages from SM-SC to MS and from MS to SM-SC respectively.

6.3 ENTITIES SPECIFIC TO THE CS DOMAIN

Apart from the common entities, the CS domain has certain entities that are specific to it. They perform important functions like mobility management and call establishment. These CS-specific entities are as follows:

- Visitor Location Register (VLR)

- Mobile Switching Center (MSC)
- Gateway Mobile Switching Center (GMSC)
- Interworking Function (IWF)

6.3.1   Visitor Location Register (VLR)

Like HLR, Visitor Location Register (VLR) is a repository of information, which is obtained by the VLR from the HLR. This information is used by the VLR to handle mobile-originated (MO) and mobile-terminated (MT) calls. The information stored at the VLR is temporary and is deleted when the subscriber leaves the VLR area. This is done through mobility management procedures.

The following are some of the important information elements maintained by VLR:

- International Mobile Subscriber Identity (IMSI)
- One or more MS International ISDN (MSISDN) associated with an IMSI
- Mobile Station Roaming Number (MSRN)
- Temporary Mobile Subscriber Identity (TMSI)
- Local Mobile Station Identity (LMSI) (this is applicable only when VLR uses a LMSI).
- Location Area where the MS is registered.
- SGSN address where the MS is registered (this is applicable only when there is a Gs interface between MSC/VLR and SGSN).
- Supplementary Service Information.

The most important function of VLR is to maintain the registration status of MS, perform location updates, and obtain subscription information from HLR. Based on this information, VLR performs mobile-originated (MO) and mobile-terminated (MT) call handling.

A VLR controls one or more MSC area(s).

6.3.2   Mobile Switching Center (MSC)

The MSC is the node that interfaces between the Access Network and the Core Network. It performs all functions necessary to handle the circuit-switched services to and from the MS: in particular, the switching and signaling functions for the MS located in the MSC area.

The MSC can interface with multiple access networks. Thus, it has an interface with one or more Radio Network Sub-system (RNS). To provide backward compatibility with GSM systems, an MSC has interface with one or more Base Station Sub-system (BSS).

It is customary to represent VLR and MSC as a single unit and term this unit as MSC/VLR. The motivation here is that an MSC needs to interface with a VLR where the subscriber data is maintained. This interface is internal to MSC and VLR, and is non-standardized. Hence, this non-standardized interface is collapsed and the two units are collectively referred to as MSC/VLR. The MAP specification 3GPP TS 29.002 refers to the MSC-VLR interface as MAP B interface and defines a few MAP messages between MSC and VLR (e.g. MAP_CHECK_IMEI). However, the MAP specification also states that the MAP B interface is not fully specified. Hence, it is strongly recommended that the B-interface should not be implemented as an external interface. In summary, even though the MAP protocol is used between MSC and VLR, the interface between these two entities is non-standardized.

6.3.3 Gateway Mobile Switching Center (GMSC)

Apart from MSC/VLR, the CS domain has the Gateway Mobile Switching Center (GMSC). The GMSC provides connectivity to external CS networks (including the CS domain of other UMTS networks and PSTN networks). It also provides the means for an incoming call to be delivered to the MSC where the MS is registered. To route the call to the actual location of the MS, the GMSC interrogates the HLR. The HLR in turn interrogates VLR to obtain the Mobile Station Roaming Number (MSRN), after which it returns the MSRN to GMSC. Using the MSRN, the GMSC routes the call to the destined MSC.

In order to allow interworking between a PLMN and the fixed networks (i.e. ISDN, PSTN and PDNs), an Interworking Function (IWF) is defined. The IWF is a logical entity that resides at MSC, and is required to convert the protocols used in the PLMN to those used in the appropriate fixed network. The IWF may have no role to play when service implementation in the PLMN is directly compatible with the service at the fixed network. The interworking functions are described in 3GPP TS 29.007.

6.4 ENTITIES SPECIFIC TO THE PS DOMAIN

Apart from the common entities described in Section 6.2, the PS domain has certain entities that are specific to it. These entities, which perform important functions like mobility management and session management, are as follows:

- Serving GPRS Support Node (SGSN)
- Gateway GPRS Support Node (GGSN)
- Border Gateway (BG)

6.4.1 Serving GPRS Support Node (SGSN)

The function of SGSN in the PS domain is quite similar to that of MSC/VLR in CS domain. Thus, an SGSN maintains subscriber information obtained from the HLR.

The following are some of the important information elements maintained by the SGSN:

- International Mobile Subscriber Identity (IMSI)
- One or more Packet-Temporary Mobile Subscriber Identity (P-TMSI)
- Zero or more Packet Data Protocol (PDP) Addresses
- Routing Area where the MS is registered
- VLR number where the MS is registered (this is applicable only when there is a Gs interface between MSC/VLR and SGSN)
- GGSN address of each GGSN for which an active PDP context exists

The most important function of SGSN is to maintain the registration status of MS, perform routing updates, obtain subscription information from HLR and activation/deactivation of PDP context. SGSN also delivers packets to and from MS.

An SGSN controls one or more routing area(s).

6.4.2 Gateway GPRS Support Node (GGSN)

The GGSN provides an interface with Packet Data Networks (PDN). It converts the IP packets received from SGSN into the appropriate format of the external network (typically IP networks). In the reverse path, the GGSN converts the incoming packet to the IP packets and delivers it to the destined MS using the PDP context stored by it. The GGSN connects with the SGSN through an IP-backbone over which the packets are tunneled, using the GPRS Tunneling Protocol (GTP).

6.4.3 Border Gateway (BG)

A BG is required to connect two PLMNs providing PS domain services. The BG is a gateway connecting a PLMN to an inter-PLMN backbone network (Figure 6.4). Its function is to provide the appropriate level of security for incoming and outgoing packets.

6.5 SERVICE-SPECIFIC ENTITIES OF THE CORE NETWORK

Apart from the various entities discussed in the preceding sections, there are other entities in the Core Network that provide specific services. These service-specific entities of the Core Network are as follows:

- Gateway Mobile Location Center (GMLC)
- CAMEL entities
- Cell Broadcast Center (CBC)

6.5.1 Gateway Mobile Location Center (GMLC)

One of the possible killer applications in 3G is the location-based service. In order to provide this service, the Core Network has a specific entity called the Gateway Mobile Location Center (GMLC). The GMLC provides the interface for the outside world. Its main function is to receive positioning requests from Location Service (LCS) clients, authenticate the client and authorize that the client has the permission to make this request. Another important function of GMLC is to convert the positioning results into the desired format.

The LCS and the functionality of GMLC is explained in detail in Chapter 13.

6.5.2 CAMEL Entities

Customized Application for Mobile network Enhanced Logic (CAMEL) is used to provide specialized services (e.g. pre-paid service). CAMEL entities include:

- GSM Service Control Function (gsmSCF)
- GSM Service Switching Function (gsmSSF)
- GSM Specialized Resource Function (gsmSRF)
- GPRS Service Switching Function (gprsSSF)

The CAMEL service is briefly explained in Chapter 13.

6.5.3 Cell Broadcast Center (CBC)

The Cell Broadcast Service (CBS) is a Teleservice that enables an Information Provider to submit short messages for broadcasting to a specified area within the PLMN. To provide CBS, the Core Network has an entity called the Cell Broadcast Center (CBC). The CBC manages the CBS messages and delivers the same to RNC from where they are broadcast to the specified area of the PLMN. A CBC may be connected to more than one BSC/RNC.

The CBS and the functionality of CBC are explained in detail in Chapter 13.

6.6 NETWORK INTERFACES OF CS DOMAIN

The network architecture and interfaces for the CS domain are shown in Figure 6.2. As depicted in the figure, the core components of CS domain are the MSC, VLR and GMSC. The MSC/VLR interfaces with BSS over the A interface and with the RNS (of UTRAN) over the Iu_CS interface.

The VLR and GMSC communicate with HLR to obtain subscriber information. VLR obtains subscriber data from HLR over the D interface; the GMSC obtains the MSRN and other information from HLR using the C interface.

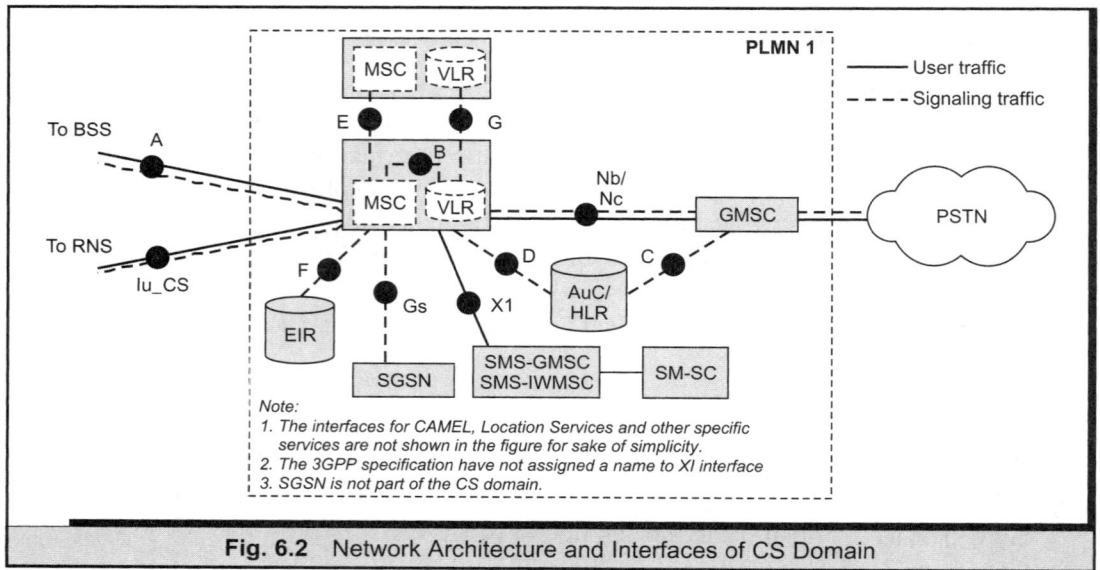

Fig. 6.2   Network Architecture and Interfaces of CS Domain

Two VLRs communicate with each other over the G interface. Two MSCs communicate with each other over the E interface. MSC and VLR communicate with each other over the B interface.

To provide SMS services, the MSC/VLR interfaces with the SMS-GMSC for delivering SMS from SM-SC to MS. For transferring SMS from MS to SM-SC, the MSC/VLR interfaces with SMS-IWMSC. There is no name assigned to these interfaces by the 3GPP specifications and hence they are collectively referred to as X1 interface in this book.

To verify equipment identity, the MSC interfaces with EIR over the F interface.

All the CS-domain interfaces are listed in Table 6.1. These interfaces are further elaborated upon in the following sub-sections. Note that certain interfaces like A/Iu interface are not listed in this table as they have already been covered in the previous chapter. Some other interfaces, like the Gs interface, are covered along with the interfaces of PS domain.

6.6.1   B Interface between MSC and VLR

The VLR is the information store for an MSC. It maintains the subscriber information for all the subscribers that are controlled by the associated MSC. Whenever the MSC requires information related to a subscriber controlled by it, it interrogates the VLR. This interrogation is done over the B interface that exists between the MSC and VLR. Typically, during a location update, the VLR obtains subscriber information from the

| Table 6.1 | | CS-domain Interfaces |

Interface	Between	Description	Protocol	Specification
B	MSC – VLR	Used by the MSC to interrogate VLR for subscriber information.	Non-standardized (Note 1)	Non-standardized (Note 1)
C	GMSC – HLR	Used by the GMSC to retrieve location information from HLR.	MAP	29.002
D	VLR – HLR	Used by the VLR to obtain sub-scriber information from HLR.	MAP	29.002
E	MSC – MSC	Used by two MSCs to exchange information during handover.	MAP	29.002
F	MSC – EIR	Used by MSC to enable EIR to verify the IMEI retrieved from MS.	MAP	29.002
G	VLR – VLR	Used by a VLR to obtain IMSI and authentication parameter from old VLR.	MAP	29.002
Nb	MSC – GMSC	Used for user data transport between MSC and GMSC.	AAL2 RTP/UDP	I.363.2 RFC1889
Nc	MSC – GMSC	Used for signaling transport between MSC and GMSC.	ISUP	Q.76x
X1	VLR – SMS-IWMSC VLR – SMS-GMSC	Used to deliver short message from MS to SM Service Center, and vice versa.	MAP	29.002

Note 1: Even though the B interface uses the MAP protocol to exchange certain messages, it is an internal interface (see Sections 6.3.2 and 6.6.1).

HLR. There-after, this information is maintained at VLR till the MS moves out of the VLR area.

The B interface is non-standardized and hence is not detailed in any 3GPP specification. Note that the B interface is also referred to in MAP specification (3GPP TS 29.002) as MAP B interface. Thus, there are messages exchanged between MSC and VLR using the MAP protocol. Despite this, it is recommended that the B interface be implemented as an internal interface. Hence, it is also said to be a non-standardized interface.

6.6.2 C Interface between GMSC and HLR

To route a mobile-terminated call, the GMSC needs to know the MSRN. For this, it uses the C interface to interrogate the HLR, which in turn requests the VLR to provide the MSRN. The response is relayed to the GMSC via HLR.

The MAP protocol, specified in 3GPP TS 29.002, is used for the C interface.

6.6.3 D Interface between VLR and HLR

The D interface is used by VLR to maintain data related to the location of an MS at the HLR. This interface is also used by VLR to obtain subscriber data from HLR.

When a MS roams into a new MSC area managed by a different VLR, the new VLR updates the HLR on this movement. The HLR then sends to the new VLR the subscriber data required to provide various CS-domain services. It also asks the old VLR to delete any record maintained for the said subscriber. In case only the MSC area changes but the VLR remains the same, the HLR is still informed about this change. However, as the VLR is not changed, there is no need to provide subscriber information or to delete information from the old VLR.

Apart from the crucial function of mobility management, the VLR also uses this interface to obtain authentication information from the HLR. This information is then used to authenticate the subscriber.

The D interface is also used by VLR to purge a MS. Once the MS is marked 'Purged' by HLR, the subscriber is unreachable for mobile-terminated (MT) requests received by HLR.

The MAP protocol, specified in 3GPP TS 29.002, is used for the D interface.

6.6.4 E Interface between MSC and MSC

During a call, when a mobile station moves from one MSC area to another, a handover procedure is performed. For performing this procedure, two MSCs exchange information over the E interface.

The MAP protocol, specified in 3GPP TS 29.002, is used for the E interface.

6.6.5 F Interface between MSC and EIR

The EIR is the central database for maintaining the list of International Mobile Equipment Identity (IMEI) used in the UMTS network. In order to ascertain the legitimacy of a UE, MSC may use the F interface, requesting EIR to check the IMEI retrieved from the MS.

The MAP protocol, specified in 3GPP TS 29.002, is used for the F interface.

6.6.6 G Interface between VLR and VLR

When a MS moves from one VLR area to another, the new VLR uses the G interface to obtain IMSI and authentication information from the old VLR. This is done to prevent the MS from having to send permanent identity IMSI to the new VLR over the air interface again.

The MAP protocol, specified in 3GPP TS 29.002, is used for the G interface.

6.6.7 Nb/Nc Interface between MSC and GMSC

The interface between the MSC and GMSC is divided into two parts; one for the signaling plane and another for the user plane. The interface for the signaling plane is referred to as the Nc interface, for which different options are possible—the important one being ISUP.

The Nb interface is for bearer control and user transport. Here too, multiple options are possible, including RTP/UDP/IP and AAL2.

6.6.8 Interface between VLR and SMS-MSC

To provide short message service, the VLR interfaces with the SMS-MSC (SMS-GMSC and SMS-IWMSC). Since no name is assigned for this interface by the 3GPP specifications, it is referred to as X1 in this book. The X1 interface is based on the MAP protocol specified in 3GPP TS 29.002.

6.7 INTERFACES OF PS DOMAIN

The PS domain provides the means to transfer high-speed and low-speed data in an efficient manner. The PS domain interfaces with various types of Access Networks like BSS, GERAN and UTRAN. In this section, the interfaces with BSS and UTRAN are discussed; the GERAN-CN interface is not discussed.

The network architecture and interfaces for the PS domain are depicted in Figure 6.3. As shown in the figure, the core components of the PS domain are the GPRS Support Nodes (GSN). The two types of GSN are: the Serving GPRS Support Nodes (SGSN) and the Gateway GPRS Support Nodes (GGSN). The SGSN interfaces with BSS over the Gb interface and with the RNS (of UTRAN) over the Iu_PS interface.

Another GSN, the Gateway GPRS Support Nodes (GGSN) provides the interface with Packet Data Networks (PDN). The SGSN and GGSN connect with each other using an IP-backbone over which the packets are tunneled using the GPRS Tunneling Protocol (GTP). There are two types of interfaces between various GSNs. Within a PLMN, the interface between the GSNs is called the Gn interface. Between two PLMNs, the GSNs communicate using the Gp interface.

The SGSN and GGSN communicate with HLR to obtain information. SGSN obtains subscriber data from HLR over the Gr interface; the GGSN obtains the SGSN address and other information from HLR using the Gc interface. The Gc interface is optional. In

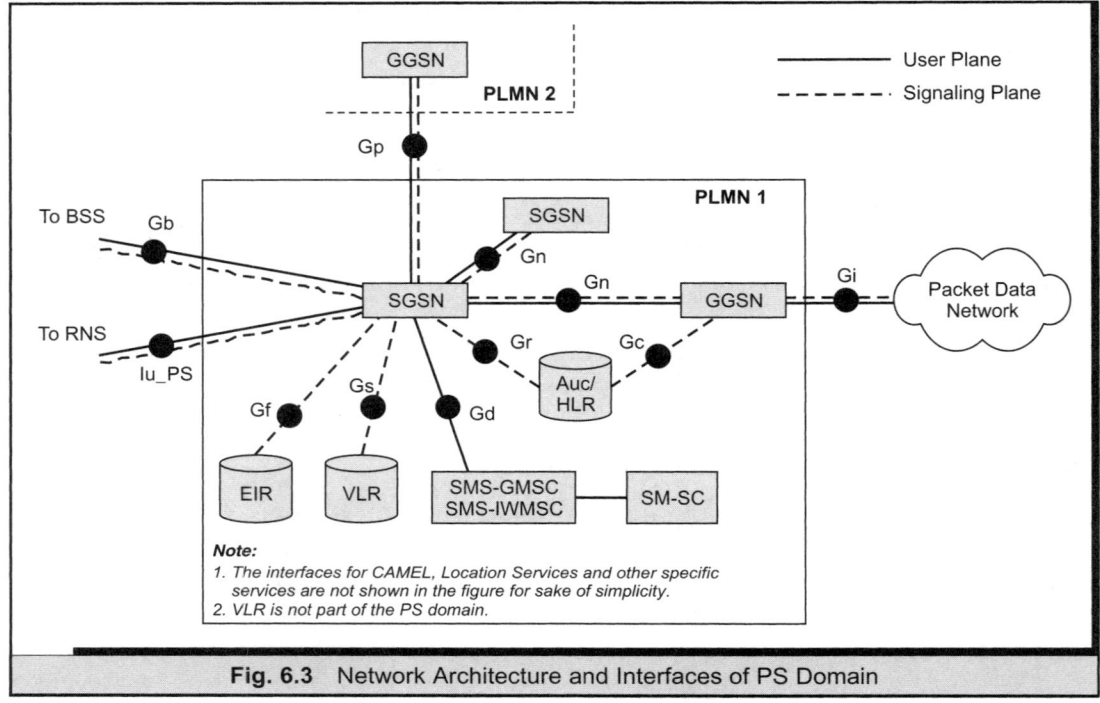

Fig. 6.3 Network Architecture and Interfaces of PS Domain

case the GGSN does not support the Gc interface, it can use GTP protocol to communicate with any GSN in the same PLMN that has an SS7 interface. This GSN acts as a 'GTP to MAP protocol' converter and provides connectivity with HLR (see Section 6.9.2.4).

To provide SMS services, the SGSN interfaces with the SMS-GMSC for delivering SMS from SM–SC to MS. For transferring SMS from MS to SM–SC, the SGSN interfaces with SMS-IWMSC. The interfaces between SGSN and SMS-GMSC and those between SGSN and SMS-IWMSC are collectively referred to as Gd interface.

To verify equipment identity, the SGSN interfaces with EIR over the Gf interface.

All the PS-domain interfaces are listed in Table 6.2. These interfaces are further elaborated upon in the following sub-sections. Note that certain interfaces like the Gb interface are not listed in this table as they have already been covered in the previous chapter.

6.7.1 Gn/Gp Interface between two GSNs

Two GPRS Support Nodes (GSN) are connected to each other using a backbone. Depending upon whether the backbone is within a PLMN or spans across a PLMN, the interface between GSNs is called the Gn interface and Gp interface respectively. As

Table 6.2	PS-domain Interfaces

Interface	Between	Description	Protocol	Specification
Gn	Two GSNs	Used to support mobility. Applicable when GSNs are located in the same PLMN.	GTP	29.060
Gp	Two GSNs	Used to support mobility. Applicable when GSNs are located in different PLMNs.	GTP	29.060
Gr	SGSN–HLR	Used by the SGSN to obtain subscriber information from HLR.	MAP	29.002
Gs	SGSN–VLR	Used for coordinating the functions of SGSN and VLR when an MS has both CS and PS domain services. This interface is optional.	BSSAP+	29.018/ 29.016
Gf	SGSN–EIR	Used by SGSN to enable EIR to verify the IMEI retrieved from MS.	MAP	29.002
Gc	GGSN–HLR	Used by the GGSN to retrieve information about the location and supported services for the MS, to be able to activate a packet data network address. The interface is optional.	MAP (Note 1)	29.002
Gi	GGSN–PDN	Used to exchange data with external packet data network.	IP	29.061
Gd	SGSN–SMS-IWMSC SGSN–SMS-GMSC	Used to deliver short message from MS to SM Service Center, and vice versa.	MAP	29.002

Note 1: If there is no SS7 interface in the GGSN, any GSN in the same PLMN that has an SS7 interface can be used as a 'GTP to MAP protocol' converter, thereby providing a signaling path between the GGSN and the HLR (see Section 6.9.2.4).

shown in Figure 6.4, the SGSN and the GGSN within a PLMN connect using an Intra-PLMN backbone over the Gn interface. For connecting GSNs of different PLMNs, an Inter-PLMN backbone is used. This provides connectivity between two GSNs over the Gp interface. Irrespective of the nature of the interface, two GSNs use the GPRS Tunneling Protocol (GTP) to communicate with each other.

The GTP protocol is defined in 3GPP TS 29.060.

It is also customary to use a Border Gateway (BG) to connect with an Inter-PLMN backbone. The BG is a Gateway connecting a PLMN with another PLMN, using an Inter-PLMN, backbone. The function of the BG is to provide the appropriate level of security for incoming and outgoing packets.

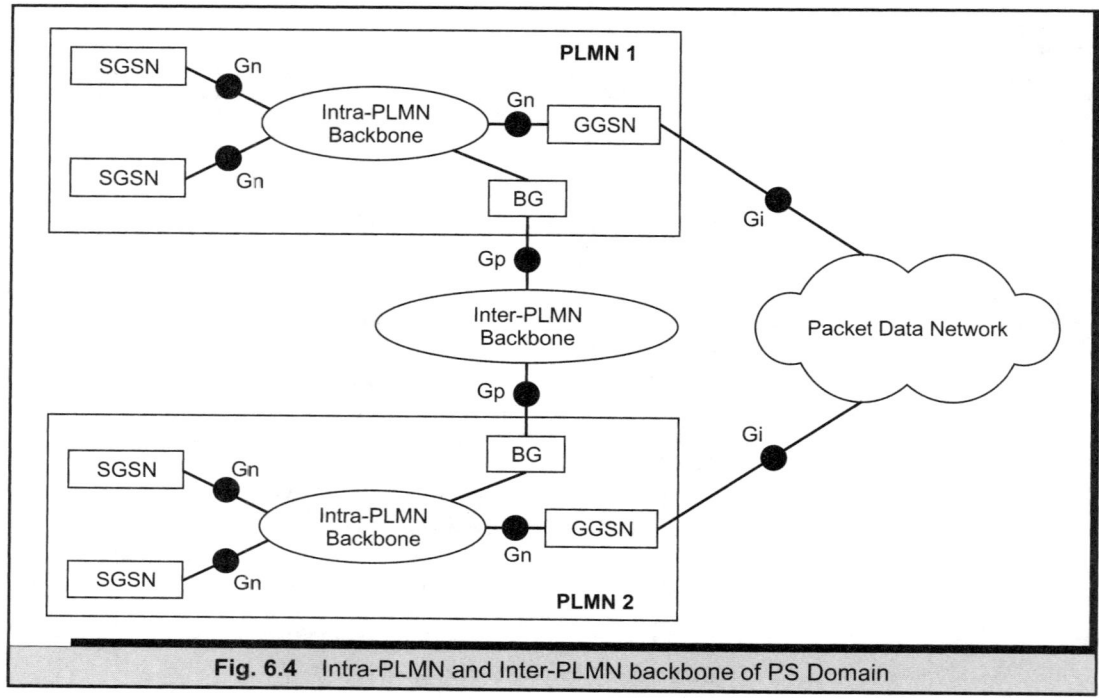

Fig. 6.4 Intra-PLMN and Inter-PLMN backbone of PS Domain

6.7.2 Gi Interface between GGSN and PDN

The GGSN is the Gateway of the PLMN. It interfaces with the PDN over the Gi interface. For external network, the GGSN is a router which routes incoming packets. The important difference between a normal IP router and the routing functionality of the GGSN is that the IP packets are tunneled to the MS and not routed as normal IP datagrams. This is required because the ease of fixed-line routing is not possible in the mobility supported by the PS domain.

Figure 6.4 shows the connectivity of a GGSN with external Packet Data Network (PDN). Typically, the PDN could be the Internet, so that the PS-domain subscriber could access it using the PS-domain services. In the 3GPP specifications, support is provided for both IPv4 and IPv6 based services.

3GPP TS 29.061 defines various scenarios wherein packet services could be accessed using the IPv4, IPv6 and PPP over the Gi interface.

6.7.3 Gr Interface between SGSN and HLR

The Gr interface between SGSN and HLR is equivalent to the D interface between VLR and HLR. The Gr interface is used by SGSN to update data related to the location of a

MS at the HLR. The interface is also used by SGSN to obtain subscriber data from HLR.

When an MS roams into a new SGSN area managed by a different SGSN, the new SGSN updates the HLR about this movement. The HLR then sends to the new SGSN subscriber data required to provide various PS-domain services. The HLR also asks the old SGSN to delete any record maintained for the said subscriber. In case only the routing area changes but the SGSN remains the same, the HLR is still informed about this change. However, as the SGSN is not changed, there is no need to provide subscriber information or to delete information from the old SGSN.

Apart from the crucial function of mobility management, SGSN also uses this interface to obtain authentication information from the HLR. This information is then used to authenticate the subscriber.

The Gr interface is also used by SGSN to purge an MS. Once an MS is marked 'Purged' by the HLR, the subscriber is unreachable for MS-terminated requests received by HLR (e.g. network requested PDP-context activation).

The MAP protocol, specified in 3GPP TS 29.002, is used for the Gr interface.

6.7.4 Gs Interface between SGSN and MSC/VLR

The functions of MSC/VLR in the CS domain and SGSN in the PS domain are somewhat similar. For example, both MSC/VLR and SGSN handle IMSI attach/detach functions, albeit for different domains. This is just one example, as there are many other commonalities as well. Given this scenario, if the functions of the two entities could somehow be combined, there could be vital savings in terms of radio resources.

In order to do this, the optional Gs interface is defined between SGSN and MSC/VLR. The Gs interface allows various procedures like IMSI attach/detach via SGSN, paging for CS-connection via SGSN, and co-ordination of Routing-Area/Location-Area (RA/LA) Update. In Chapter 9, the use of Gs interface is elaborated upon by explaining the Combined RA/LA Update.

The SGSN and MSC/VLR communicating over the Gs interface maintain an association with each other. This association is created/updated during various message exchanges between the two. The exchanges take place using the Base Station Subsystem Application Part + (BSSAP+) protocol.

The BSSAP+ protocol is specified in 3GPP TS 29.018. Another specification, 3GPP TS 29.016, provides the lower layer requirements for the Gs interface.

6.7.5 Gf Interface between SGSN and EIR

The EIR is the central database for maintaining the list of International Mobile Equipment Identity (IMEI) used in the UMTS network. In order to ascertain the legitimacy of a UE, SGSN may use the Gf interface requesting EIR to check the IMEI retrieved from the MS.

The MAP protocol, defined in 3GPP TS 29.002, is used for the Gf interface.

6.7.6 Gc Interface between GGSN and HLR

To carry out the network requested PDP-context activation procedure, the GGSN needs to know the SGSN with which an MS is registered. For this, the GGSN uses the Gc interface to request the HLR to provide the SGSN address. Further, if the network requested PDP-context activation fails, GGSN sends a failure report to HLR over the Gc interface.

The MAP protocol, defined in 3GPP TS 29.002, is used for the Gc interface. The implementation of this interface is optional. In case there is no SS7 interface in the GGSN, any GSN in the same PLMN that has this interface can be used as a 'GTP to MAP protocol' converter, thereby providing a signaling path between the GGSN and the HLR. Section 6.9.2.4 provides more details on this scenario.

6.7.7 Gd Interface between SGSN and SMS-MSC

To provide short message service, the SGSN interfaces with the SMS-MSC (SMS-GMSC and SMS-IWMSC), using, the Gd interface.

The Gd interface is based on MAP protocol, which is specified in 3GPP TS 29.002.

6.8 CS DOMAIN PROTOCOL ARCHITECTURE

The CS domain protocol architecture can be divided into two distinct categories, one for user plane and another for control plane. This is explained in greater detail in the following sub-sections.

Note that the protocol architecture used for communication between CS (or PS) domain and the Access Network depends upon the type of Access Network (BSS or UTRAN). In case the Access Network is BSS, the user plane is defined as 'User plane for A/Gb mode'. In case it is UTRAN, the user plane is defined as 'User plane for Iu mode'. In this chapter, only the Iu mode is discussed. For the A/Gb mode, the reader is referred to the appropriate 3GPP specifications.

6.8.1 User Plane

Figure 6.5 shows the protocol architecture of user plane for the Iu interface (CS domain). In the access side, the MS communicates with RNS (in particular the Node B)

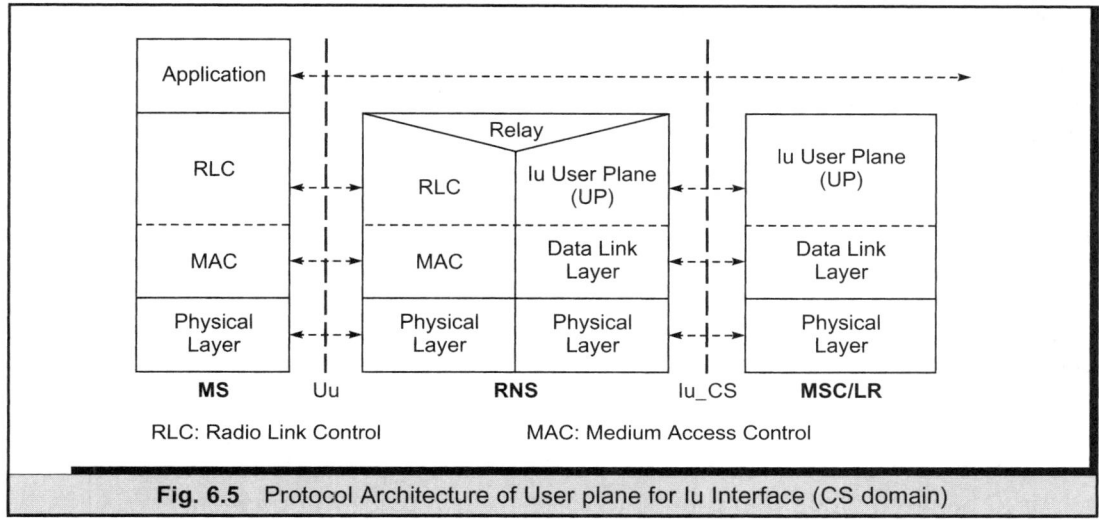

Fig. 6.5 Protocol Architecture of User plane for Iu Interface (CS domain)

over the WCDMA air interface. The WCDMA based physical layer is the lower-most layer at the MS. The Medium Access Control (MAC) resides over the physical layer. The MAC controls the access signaling for the radio channel. The Radio Link Control (RLC), which provides reliable link layer functionality resides over the MAC layer. The RLC/MAC are defined in 3GPP TS 25.322 and 3GPP TS 25.321 respectively.

The RLC/MAC protocols were explained in Chapter 5. User applications reside above the RLC/MAC layer.

On the access side, the air interface terminates at the Node B, while the RLC and MAC protocols terminate at the RNC. The framing protocols and AAL2 protocols are used between Node B and RNC to carry user plane information.

Between RNC and MSC/VLR, the Iu UP protocol is used. This protocol is carried over any data link layer protocol (e.g. AAL2 and ATM).

6.8.2 Control Plane

In the Control Plane for CS domain, different protocol architectures are applicable, depending upon the type of interface. This section discusses the protocol architectures for the following interfaces:

- Control Plane for Iu Interface
- Control Plane for MAP-based Interfaces
- Control Plane for Nc Interface

6.8.2.1   *Control Plane for Iu Interface*

Figure 6.6 shows the protocol architecture of control plane for Iu interface (CS domain). At the MS, the lower three layers of the control plane comprise of RLC, MAC and the Physical Layer (WCDMA air interface). The Radio Resource Control (RRC) layer, which resides over the RLC, is the core signaling protocol between MS and RNC. The RRC protocol, defined in 3GPP TS 25.331, is an Access Stratum protocol. There is a Non-Access Stratum (NAS) signaling protocol over the RRC protocol, which is referred to as NAS-Sig protocol in this book. The NAS-Sig protocol comprises of Layer 3 signaling between MS and the Core Network (i.e. MSC/VLR). NAS-Sig provides functions like mobility management, connection management, call control and session management functionality. The NAS-Sig protocol is defined in 3GPP TS 24.008.

On the access side, the air interface terminates at the Node B, while the RLC, MAC and RRC protocols terminate at the RNC. The framing protocols and AAL2 protocols are used between Node B and RNC to carry control plane information.

Between RNC and MSC/VLR, the Radio Access Network Application Part (RANAP) signaling protocol is used. This protocol handles the signaling between the MSC/VLR and the UTRAN. The RANAP protocol, defined in 3GPP TS 25.413, runs

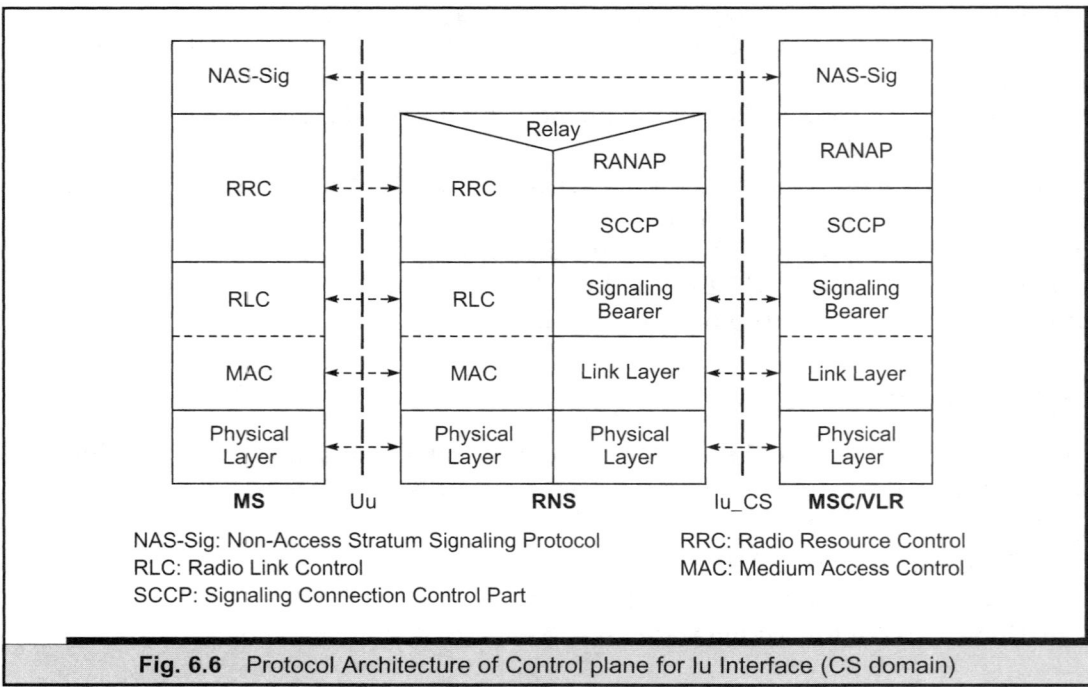

NAS-Sig: Non-Access Stratum Signaling Protocol   RRC: Radio Resource Control
RLC: Radio Link Control   MAC: Medium Access Control
SCCP: Signaling Connection Control Part

Fig. 6.6   Protocol Architecture of Control plane for Iu Interface (CS domain)

over the Signaling Connection Control Part (SCCP) protocol. The SCCP resides over the signaling bearer (e.g. MTP protocols).

6.8.2.2 Control Plane for MAP-based Interfaces

Figure 6.6 depicted the interfaces between the MS, the Access network and the MSC/VLR for the control plane. There are other control plane interfaces, most of which are based on Mobile Application Part (MAP) protocol. The MAP is one of the most important Core Network protocols. It is used on the C, D, E, F, G and X1 interfaces. Figure 6.7 depicts the MAP-based Interfaces for VLR (i.e. D interface between VLR–HLR and G interface between VLR–VLR). Other MAP-based interfaces follow the same protocol architecture.

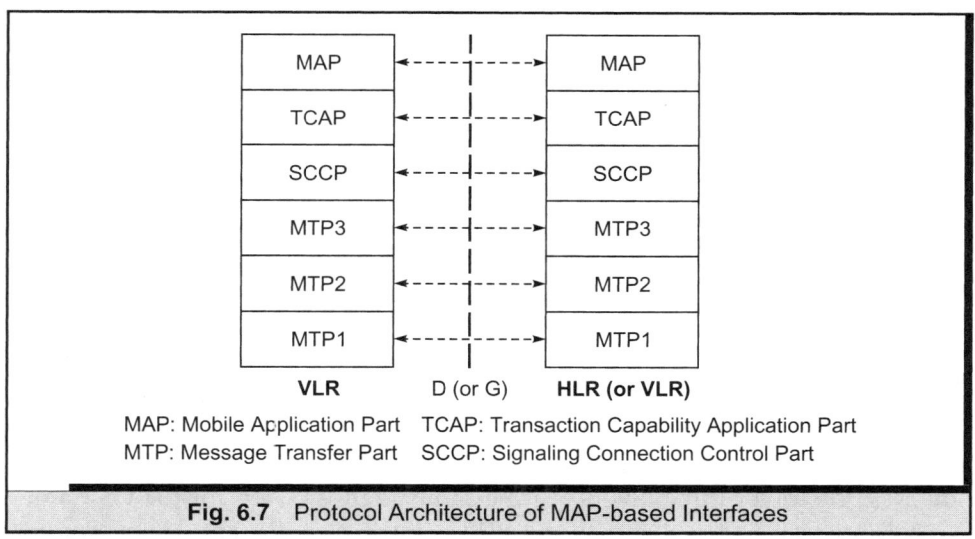

Fig. 6.7 Protocol Architecture of MAP-based Interfaces

The MAP protocol is defined in 3GPP TS 29.002. MAP resides over Transaction Capability Application Part (TCAP) protocol. MAP/TCAP use the SS7-based network with MTP1/2/3 as bearer for information exchange.

6.8.2.3 Control Plane for Nc Interface

The Nc interface between MSC and GMSC is based on the ISDN User Part (ISUP) protocol (Figure 6.8). This interface is primarily used for setup, maintenance and tear-down of transport bearers between MSC and GMSC. The ISUP protocol is defined in ITU-T Q.76*x* specifications.

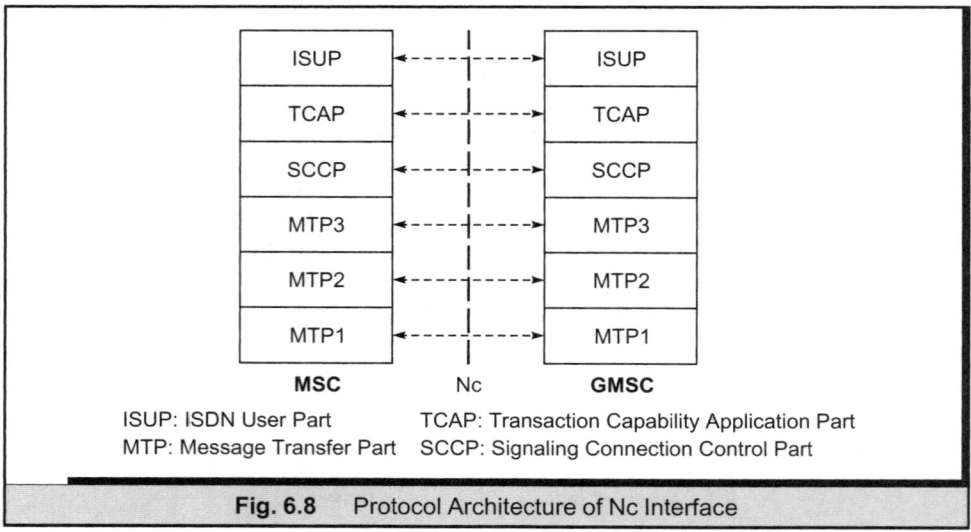

Fig. 6.8   Protocol Architecture of Nc Interface

6.9   PS DOMAIN PROTOCOL ARCHITECTURE

Like the CS domain, the PS domain has distinct protocol architectures for user plane and control plane. The protocol architecture for both these categories is explained in the following sub-sections.

6.9.1   User Plane

Figure 6.9 shows the protocol architecture of user plane for Iu mode (PS domain). This is quite different from the user plane of CS domain (Figure 6.5).

At the MS, the lower three layers are the same as those used for the CS domain (i.e. RLC/MAC/Physical layer). Above the RLC layer, the Packet Data Convergence Protocol (PDCP) is used. This protocol, as the name suggests, is specific to the PS domain. The PDCP is used to carry data protocols like IP and PPP. Typically, network layer protocols like IP reside directly over the link layer protocol (RLC in this case). However, a wireless network requires that the bandwidth in the radio interface be used optimally. Thus, one of the most important functions of PDCP protocol is the compression of redundant header information. PDCP is defined in 3GPP TS 25.323. A network layer protocol (e.g. IP) resides over the PDCP; user applications reside over the network layer protocol.

On the access side, the air interface terminates at the Node B, while the RLC, MAC and PDCP protocols terminate at the RNC. The framing protocols and AAL2 protocols are used between Node B and RNC to carry user plane information.

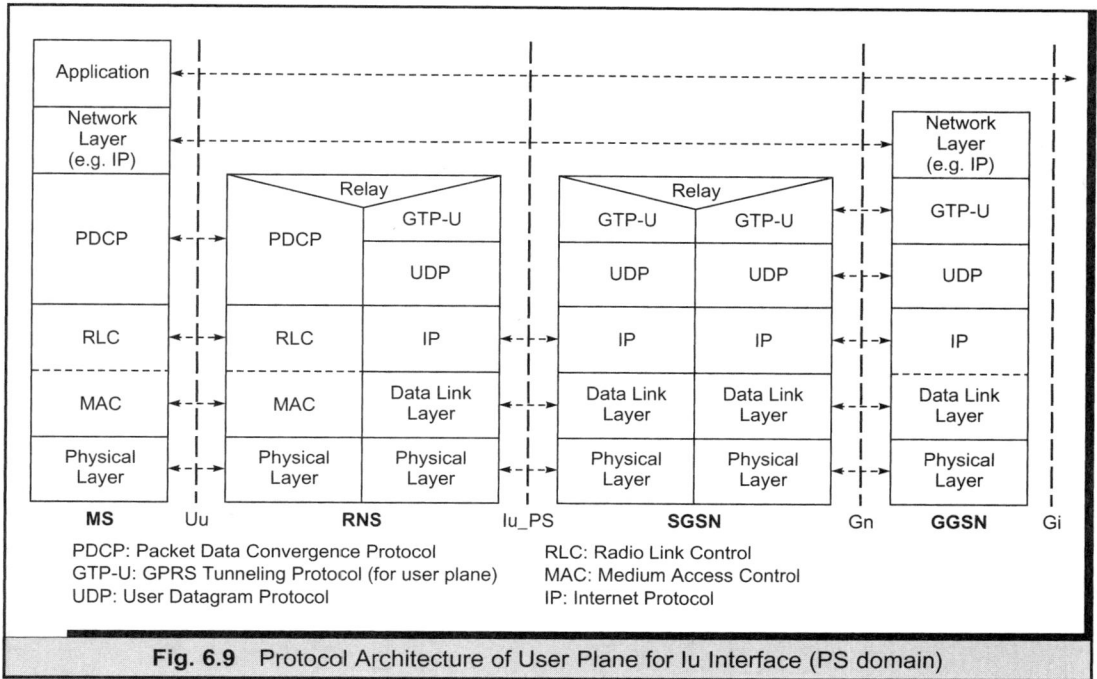

Fig. 6.9 Protocol Architecture of User Plane for Iu Interface (PS domain)

Between RNC and SGSN, the GPRS Tunneling Protocol—User Plane (GTP-U)—is used. The GTP protocol, which includes messages for both the user plane (GTP-U) and the control plane (GTP-C), is defined in 3GPP TS 29.060. The GTP-U is carried over UDP/IP and any lower layer protocol (e.g. ATM). The GTP-U protocol is also used between SGSN and GGSN. The GTP-C, which is not shown in the figure, is used for Signaling between two GSNs.

6.9.2 Control Plane

In the control plane for PS domain, different protocol architectures are applicable, depending upon the type of interface. This section discusses the protocol architectures for the following interfaces:

- Control Plane for Iu Interface
- Control Plane for Gn/Gp Interface
- Control Plane for Gr/Gf Interface
- Control Plane for GTP-MAP Protocol Converter
- Control Plane for Gs Interface

6.9.2.1 Control Plane for Iu Interface

Figure 6.10 shows the protocol architecture of control plane for Iu interface (PS domain). This protocol architecture is similar to that depicted in Figure 6.6 for CS domain. The only difference arises in NAS-Sig protocol where the procedures for CS and PS domain are different.

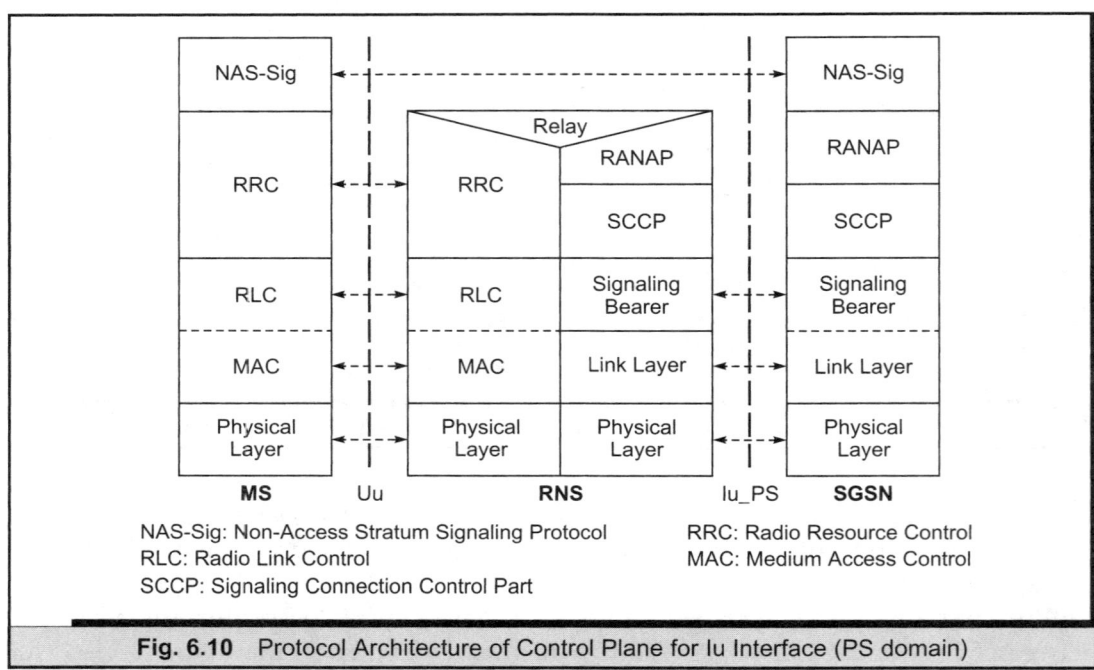

Fig. 6.10 Protocol Architecture of Control Plane for Iu Interface (PS domain)

6.9.2.2 Control Plane for Gn/Gp Interface

The Gn/Gp interface between GSNs is based on the GPRS Tunneling Protocol-Control Plane (GTP-C) protocol (Figure 6.11). This interface is primarily used for creation, modification and deletion of PDP context. It is also used to obtain SGSN context information maintained at an SGSN during inter-SGSN routing area update. The GTP-C along with GTP-U is defined in 3GPP TS 29.060.

6.9.2.3 Control Plane for Gr/Gf Interface

The Gr interface between SGSN and HLR is based on the Mobile Application Part (MAP) protocol (Figure 6.12). MAP is also used between SGSN and EIR over the Gf

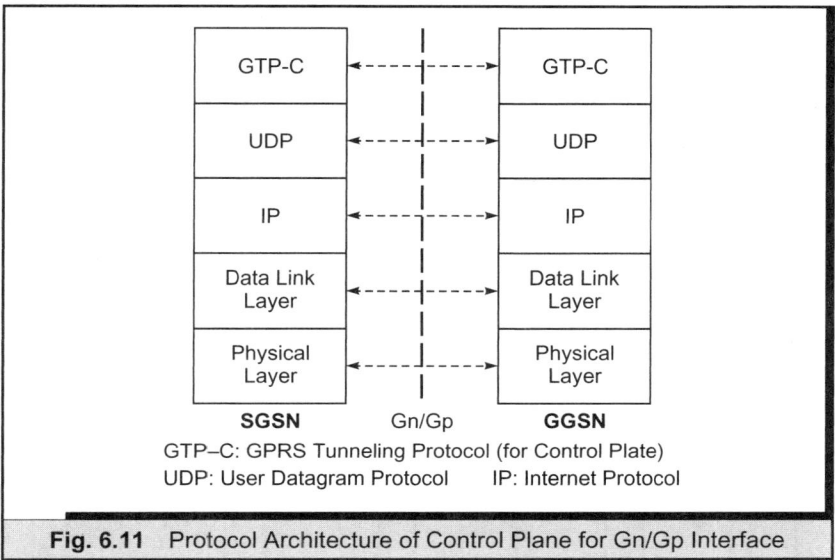

Fig. 6.11   Protocol Architecture of Control Plane for Gn/Gp Interface

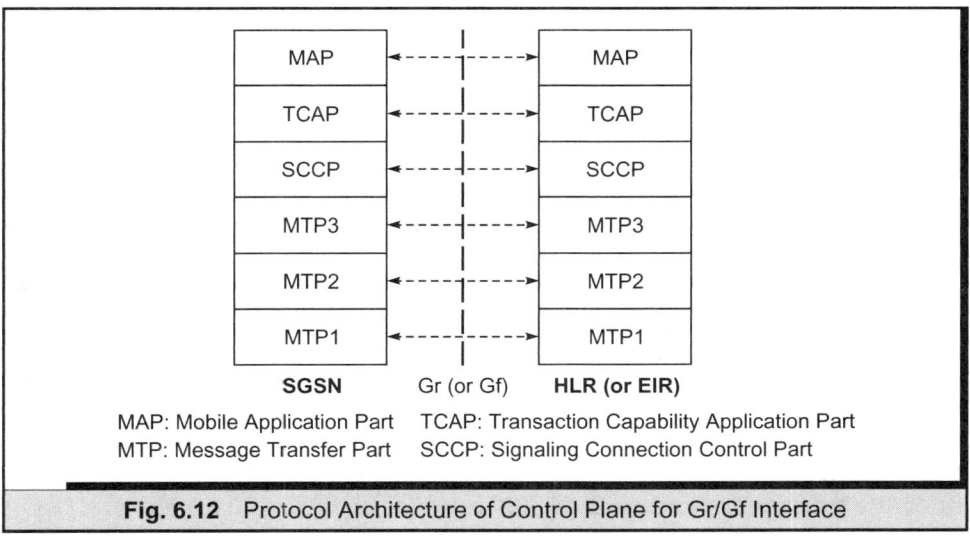

Fig. 6.12   Protocol Architecture of Control Plane for Gr/Gf Interface

interface and between GGSN and HLR over the Gc interface. The MAP protocol is defined in 3GPP TS 29.002. MAP resides over Transaction Capability Application Part (TCAP) protocol. MAP/TCAP uses SS7-based network with MTP1/2/3 as bearer for information exchange. The MAP/TCAP protocol is also used between SGSN and SMS-GMSC and between SGSN and SMS-IWMSC for providing SMS service.

6.9.2.4   *Control Plane for GTP-MAP Protocol Converter*

As stated earlier, the Gc interface between GGSN and HLR is based on MAP. MAP is a computationally complex protocol and increases the software and hardware cost of a GGSN. Thus, to simplify the implementation of GGSN and to remove the SS7 interface from it, 3GPP specifications make the implementation of Gc interface for a GGSN optional. However, this option does not imply that the GGSN does not need to communicate with HLR; this is unavoidable. What it means is that GGSN must have alternate means to obtain information from HLR without actually implementing the SS7-based Gc interface. To provide such means, the GTP-C protocol has additional messages that are used by GGSN to communicate with HLR via a GTP-MAP protocol converter (Figure 6.13).

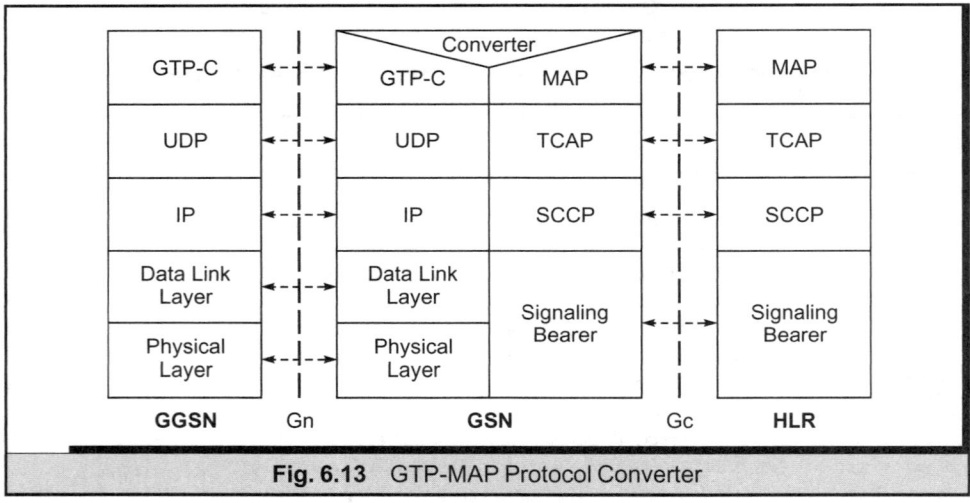

Fig. 6.13   GTP-MAP Protocol Converter

The benefit of such a GTP-MAP protocol converter is that it enables many GGSNs to communicate with a single GSN that supports such protocol conversion. In such a scenario, the GGSNs only need to support an IP interface with the GSN. The GSN can then do protocol conversion for all GGSNs.

A question that may arise is: why is such an interface not specified for SGSN, which too interfaces with HLR. The answer is as follows: SGSN uses MAP to communicate with HLR, EIR and SMS-GMSC/SMS-IWMSC. Now, if SGSN has to do away with the MAP interface, it would need a protocol converter for all these interfaces. This strategy leads to standardization complexity and thus has not been adopted by the 3GPP specifications.

6.9.2.5 *Control Plane for Gs Interface*

As mentioned in Section 6.7.4, there is an optional interface between SGSN and MSC/VLR. The Gs interface allows various procedures like IMSI attach/detach via SGSN, paging for CS-connection via SGSN and co-ordination of Routing-Area/Location-Area (RA/LA) Update.

The Gs interface uses the Base Station Sub-system Application Part + (BSSAP+) protocol defined in 3GPP TS 29.018. Another specification, 3GPP TS 29.016, provides the lower layer requirements for the Gs interface.

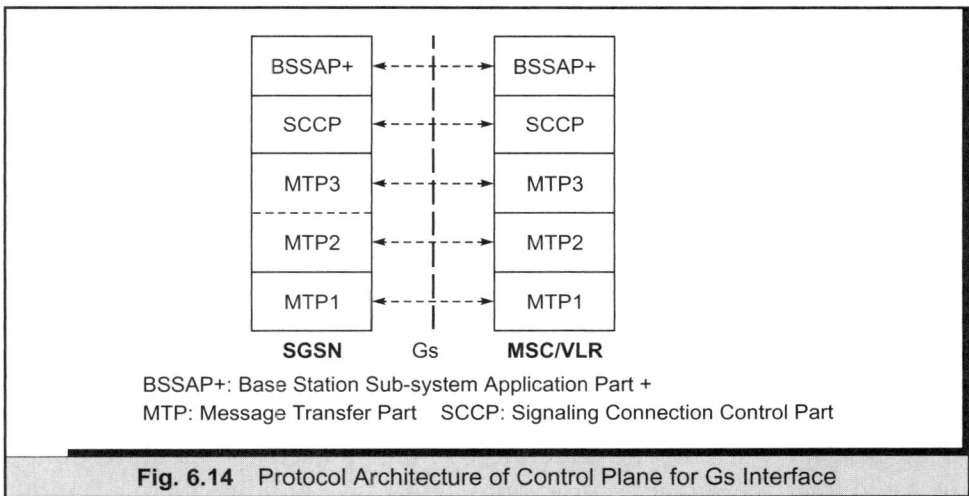

Fig. 6.14 Protocol Architecture of Control Plane for Gs Interface

6.10 CORE NETWORK FUNCTIONS

So far, the chapter has covered in detail the various CN entities, the interfaces between these entities and the protocols residing thereon. This section briefly touches upon the functions of the Core Network. These functions have been explained in greater detail in the following chapters. The idea here is to present a summary of the functions of the Core Network.

6.10.1 Mobility Management

The objective of mobility management is to support the mobility of user terminals, such as informing the network of their present location. Mobility management also ensures that the identity of the user is kept confidential and that only authenticated users can avail of network services.

The mobility management functions in the Core Network are divided into two categories:

- For the CS domain, they are categorized under the heading Mobility Management (MM) procedures.
- For the PS domain, these functions come under the GPRS Mobility Management (GMM) procedures.

The MM/GMM procedures are defined in 3GPP TS 24.008 and 3GPP TS 23.012. A brief summary of these procedures is provided in the following sub-sections.

6.10.1.1 MM Procedures

The MM procedures fall under three categories, depending upon when they can be initiated,

- MM common procedures
- MM specific procedures, and
- MM connection management procedures.

6.10.1.1.1 MM Common Procedures

These procedures are always initiated when there is a radio resource connection. The MM common procedures includes, but are not limited to the following sub-procedures:

- **TMSI re-allocation procedure:** In order to hide the permanent identity (IMSI) of a subscriber, a temporary identity (TMSI) is used for identification within the radio-interface signaling procedures. A TMSI can be re-allocated at different times, but the re-allocation is necessarily done during a change of location area.

- **Identification procedure:** As mentioned in the TMSI re-allocation procedure, a subscriber is identified by a temporary identity (TMSI). However, there may be scenarios when the IMSI is not known to the network. Such scenarios require that the network request an MS to provide its IMSI or IMEI. This request is termed as identification procedure and is used only when absolutely necessary.

- **Authentication procedure:** The authentication procedures are explained in Section 6.10.6.

- **IMSI detach procedure:** This procedure can be invoked by the MS if the MS is de-activated or if the SIM is detached from it. Once IMSI is detached, the MS becomes inactive in the network.

6.10.1.1.2 MM Specific Procedures

The MM specific procedures can only be started if no other MM specific procedure is running or no MM connection exists between the network and the MS. A MM connection exists between two peer MM entities (one at MS and the other at MSC or SGSN). Several MM connections may be active at the same time. A MM connection always requires an underlying RR connection. In case there are multiple MM connections, all of them use the same RR connection. In order to establish the MM connection, the RR connection is first established (provided it is not already established). When MM connections using the same RR connection are released, the associated RR connection is also released. The MM connection is detailed in Chapter 9.

A new MM specific procedure can be started only when a running MM specific procedure ends or when all MM connections are released. However, a MM common procedure can be initiated during a MM specific procedure.

The MM specific procedures includes, but are not limited to the following sub-procedures:

- **Normal Location Update:** This procedure is used to update the location area of the MS in the network. The various situations in which this procedure is invoked are described in Chapter 9. One such situation is when the MS changes its location area.

- **Periodic Location Update:** This is used to periodically notify the presence of the MS in the network (for the CS domain).

- **IMSI attach:** The IMSI attach procedure is used to indicate that the MS is active in the network. It is mandatory for an MS to support IMSI attach/detach; however it is optional for the network to support this. The network indicates the support, or lack of it, through broadcast information.

These procedures are explained in detail in Chapter 9.

6.10.1.1.3 MM Connection Management Procedures

These procedures are used to establish, maintain and release a MM connection between the MS and the network. The Connection Management layer exchanges information over the MM connection. More than one MM connection may exist at the same time. However, all MM connection(s) use the same RR connection. The MM connection management procedure is detailed in Chapter 9.

6.10.1.2 GMM Procedures

GMM procedures are the counterpart of MM procedures for the PS domain. Depending upon how and when they can be initiated, the GMM procedures fall under two categories:

- GMM common procedures
- GMM specific procedures

6.10.1.2.1 GMM Common Procedures

These procedures are always initiated when a PS-signaling connection exists (i.e. a connection exists between MS and SGSN of the PS domain). The GMM common procedures include, but are not limited to the following sub-procedures:

- **P-TMSI re-allocation procedure:** The P-TMSI is the PS-domain equivalent of TMSI in CS domain. Thus, the relevance and applicability of P-TMSI is similar to that of TMSI. The only difference is that TMSI applies to a location area, while P-TMSI applies to a routing area.

- **GPRS Identification procedure:** This procedure is similar to the MM identification procedure. It is used to obtain the IMSI or IMEI from the MS when the network does not have it.

- **GPRS Authentication procedure:** The GPRS authentication procedures are explained in Section 6.10.6.

6.10.1.2.2 GMM Specific Procedures

The GMM specific procedures are similar to the MM specific procedures. They include, but are not limited to the following sub-procedures:

- **Normal Routing Area Update:** This procedure is used to update the routing area of an MS in the network.

- **Periodic Routing Area Update:** This is used to periodically notify the presence of the MS in the network (for the PS domain).

- **GPRS attach:** The GPRS attach procedure is used to indicate that the MS is active in the network.

- **GPRS detach:** This procedure can be invoked by the MS if the MS is de-activated or if the SIM is detached from it. Once GPRS detached the MS becomes inactive in the network.

Apart from the above, a few GMM procedures are defined that optimize the operations when both CS and PS domain services are used. These procedures are as follows:

- **Combined Routing Area Update:** This procedure is used to collectively perform the location area and routing area update. In this procedure, the SGSN conveys the received information to the VLR so that the MS does not have to separately communicate with VLR.

- **Combined GPRS attach/detach:** Like combined routing area update, the combined GPRS attach/detach enables an MS to indicate its activity in both CS and PS domain.

The above procedures, including the simple as well as optimized GMM-specific procedures, are explained in detail in Chapter 9.

6.10.2 Call Handling

The mobility management procedures are essentially those that are used to track the location and activity of a mobile in the UMTS network. After registering its location information successfully, the MS is ready to initiate and receive calls, which are of two types:

- **Mobile-Originated (MO) Calls:** These are calls initiated by the MS. An MO-Call is handled by the MSC/VLR of the visited network.

- **Mobile-Terminated (MT) Calls:** These are calls received by the MS. An MT-Call is handled by GMSC, HLR, and MSC/VLR.

The procedures involved in handling the MO and MT calls of the MS are clubbed under the 'Call Handling' functionality. A basic mobile-to-mobile (end-to-end) call can be treated as a combination of MO and MT call. The details on MO and MT call procedures are provided in Chapter 10.

6.10.3 Session Management

In the PS domain, there is no concept of calls. Thus, the call handling procedure mentioned in the previous section is not applicable to this domain. However, an analogous concept, termed as *sessions*, is applicable in this case. A session can be viewed as a PDP context maintained by the MS and the GSN for information exchange in the PS domain. It may be recalled (Chapter 3) that a PDP context contains the PDP type, PDP address and Access Point Name, among others. This PDP context is used to communicate with external Packet Data Networks.

The Session Management (SM) function of Core Network enables the activation, de-activation and modification of a PDP context. The SM procedures can be performed only when a GMM context exists between the MS and the network. If a context does not exist, it must first be created.

The SM procedures include the following:

- **PDP Context Activation:** This procedure is used to create a PDP context between the MS and the network. The creation of PDP context is initiated by the MS. The network may also ask the MS to initiate the activation of a PDP context (this may happen when a PDU is received by GGSN for MS and no PDP context exists in the GGSN database).

- **PDP Context Modification:** This procedure is used by the network, or by the MS, to modify the attributes associated with a PDP context. The modifications may be for the QoS or for the priority, or for other parameters negotiated during the activation of PDP context.

- **PDP Context De-activation:** This procedure is used by the network or by the MS to de-activate an existing PDP context.

The Session Management procedures are explained in detail in Chapter 11.

6.10.4 Supplementary Services

As mentioned in Chapter 3, services offered in a UMTS mobile network are divided into two broad categories: *Basic Services* and *Supplementary Services*. Basic services are the primary services that a subscriber requires from a mobile network. These include Voice Telephony, Fax, SMS, Circuit-Switched Data Service and GPRS-based Packet services. Supplementary services, on the other hand, modify or supplement the basic services, and independently do not hold any meaning. Examples of supplementary services are Call Forwarding, Call Barring, etc. A network operator may or may not offer supplementary services to the subscribers.

A number of operations can be performed on a supplementary service. These operations are as follows:

- **Provisioning:** Provisioning is the mechanism by which a supplementary service is subscribed to by a subscriber.

- **Withdrawal:** Withdrawal is the opposite of provisioning, and involves making the supplementary service unavailable to the subscriber.

- **Registration:** Some supplementary services require that certain information related to the service be registered with the network, before the subscriber can use the service. As an example, the 'Forwarded-to-number (FTN)' must be registered with the service provider, before the Call Forwarding supplementary service can be used by the subscriber.

- **Erasure:** The process of removal of the information registered for a supplementary service is called Erasure. Erasure plays a reverse role to that of the registration operation.

- **Activation:** A supplementary service provisioned for a subscriber, and for which registration (if required) has already taken place, can be activated and made operational. This is called the Activation of a supplementary service.

- **Deactivation:** The reverse of the Activation process is the process of deactivation.

- **Interrogation:** Interrogation is the process by which a subscriber is allowed to query the state of a particular supplementary service.

- **Invocation:** This is the process of invoking a supplementary service, either through a mobile subscriber request, or automatically as a result of fulfillment of certain conditions.

The control or management of the supplementary services is called the Supplementary Services Management Procedures, which are explained in detail in Chapter 12.

6.10.5 Short Message Service

Short message service is the means by which a short text (up to 160 characters) can be exchanged between the UE and a Short Message Service Center (SMSC). Like incoming and outgoing calls, the short message service includes both mobile-terminated (SMS-MT) and mobile-originated (SMS-MO) SMS. The SMS can be sent and received using both the CS and PS domains. Short Message Service is explained in detail in Chapter 13.

6.10.6 Security Functions

The Core Network provides various security functions, which are categorized as follows:

- **User identity confidentiality:** One of the important security requirements is to ensure that the user identity (IMSI) is kept confidential and that it cannot be eavesdropped on the radio access link. To serve this requirement, the TMSI re-allocation procedure and identification procedure are used. Both these procedures have already been explained.

- **User Authentication:** Apart form user identity confidentiality, the most important security function of the Core Network is User Authentication. The user authentication procedure corroborates the identity of the user.

- **Network Authentication:** The Core Network also provides the user with the means to authenticate the network. Thus, user authentication and network authentication together lead to form *mutual authentication*.

- **Key Generation:** Apart from mutual authentication, the Core Network also facilitates the generation of the ciphering key and integrity key. These keys are used to provide confidentiality and data integrity. Note that these functions are not within the purview of the Core Network functions. The Core Network only provides the means to generate these keys. The confidentiality and data integrity actually takes place between the MS and Access network.

The above-mentioned procedures, along with other security procedures applicable in UMTS network, are explained in detail in Chapter 14.

6.11   SUBSCRIBER DATA

To provide the various services to a subscriber, both for CS and PS domain, a subscriber is associated with various data elements, referred to as Subscriber Data. Subscriber data is required for various purposes, including subscriber identification, authentication, routing, call handling, charging, operation and maintenance purposes. Elements of subscriber data fall under two categories: Permanent Subscriber Data and Temporary Subscriber Data.

Permanent Subscriber Data is provisioned by the network operator and cannot be changed dynamically (e.g. the subscriber identity IMSI is a permanent data). Temporary Subscriber Data, on the other hand, is data that may change dynamically. The SGSN address is an example of a temporary data as it may change with the change in SGSN area.

In the Core Network, different network entities maintain different subsets of subscriber data required for their functioning. Table 6.3 lists the important elements of subscriber data and the applicability of this data at HLR, VLR, SGSN, and GGSN. For details of these elements, reader is referred to 3GPP TS23.008.

Table 6.3   Subscriber Data maintained in the Core Network

Type	Element	Relevant at	Description
Address	IMSI	HLR, VLR, SGSN, GGSN	IMSI uniquely identifies a subscriber.
	MSISDN	HLR, VLR and SGSN	MSISDN uniquely identifies a service.
	Network Access Mode (NAM)	HLR	Specifies whether the subscriber is allowed to access the CS domain, the PS domain or both.
	TMSI/P-TMSI	VLR/SGSN	Used to hide the permanent identity IMSI for CS/PS domain.
	IMEI	SGSN	Used to identify User Equipment.
Authentication	K	AuC (HLR)	Long-term secret key; used to generate Authentication Vectors.
	SQN	AuC (HLR)	Sequence number; used for synchronization.
	Authentication Vector	VLR/SGSN	Consists of (1) Random Challenge (RAND), (2) Expected Response (XRES), (3) Cipher Key (CK), (4) Integrity Key (IK) and (5) Authentication Token (AUTN). These are used for authentication, confidentiality and data integrity.

Contd.

Table 6.3 Contd.

Type	Element	Relevant at	Description
	Key Set Identifier (KSI)	VLR/SGSN	Used to ensure consistency of UMTS authentication information (CK and IK) between the MS and the VLR/SGSN.
Routing	MSRN	VLR	MSRN is used to route calls to the MS.
	LAI/RAI	VLR, SGSN	Used to identify Location/Routing area.
	MSC number	VLR, HLR	Identifies the MSC with which a subscriber is registered.
	VLR/SGSN number	HLR	Identifies the VLR/SGSN with which a subscriber is registered.
	GGSN number	HLR	Identifies the GGSNs that are to be contacted when activity from the MS is detected and the flag MNRG is set.
	HLR number	VLR, SGSN	Used to identify the HLR associated with a subscriber.
	Subscriber Restriction	HLR	Used to determine whether or not certain restrictions apply to the subscription (e.g. this field defines whether service is available in all GSM PLMNs, or in one national and all foreign GSM PLMNs or is regionally restricted, etc.).
	RSZI	HLR, VLR, SGSN	Identifies the regional zones where the subscriber is allowed to roam within a PLMN. The regional zones are identified at HLR using a list of Regional Subscription Zone Identity (RSZI) maintained per (CC, NDC) pair. The VLR/SGSN store a part of the zone list according to their specific (CC, NDC) information.
	Cell Global ID or Service Area ID	VLR/SGSN	Indicates the Cell Global Identity (CGI) of the cell in GSM or the Service Area Identification of the service area in UMTS in which the MS is currently in radio contact or was last in radio contact.
Basic Service	Bearer service	HLR, VLR	Identifies whether a bearer service is provisioned for the MS or not.
	Tele-service	HLR, VLR, SGSN	Identifies whether a tele-service is provisioned for the MS or not.

Contd.

Table 6.3 Contd.

Type	Element	Relevant at	Description
	Transfer of SM option	HLR	Indicates which path should be used for transfer of Terminating Short Message when GPRS is not supported by the GMSC. Two options are available, one via MSC and the other via SGSN.
MS Status	IMSI detached flag	VLR	Indicates that the MS is in the detached state (i.e. no longer reachable).
	MS Not Reachable for GPRS (MNRG)	HLR, SGSN, GGSN	In HLR, it indicates whether the MS is marked as GPRS detached or GPRS not reachable in the SGSN and possibly in the GGSN.
			In SGSN, it indicates whether activity from the MS shall be reported to the HLR.
			In GGSN, it indicates that the MS is marked as GPRS detached in the SGSN.
	MS purged for GPRS/non-GPRS flag	HLR	MS purged for GPRS/non-GPRS flag is set in the HLR per IMSI in order to indicate that the subscriber data for the MS concerned was purged in the SGSN VLR.
	MS Not Reachable Reason (MNRR)	HLR	Stores the reason for an MS being absent when an attempt to deliver a short message to it fails at the MSC/SGSN with a cause of Absent Subscriber.
Operator Determined Barring (ODB)	Subscriber status	HLR, VLR, SGSN	Indicates whether the subscriber is subject to ODB.
	General Data for ODB	HLR, VLR, SGSN	Set of data that indicates which categories of ODB applies to the subscriber (e.g. "Barring of all outgoing international calls").
	Notification to CSE flag	HLR	Indicates whether the change of ODB data shall trigger Notification on Change of Subscriber Data or not.
	gsmSCF address list	HLR	Contains the list of gsmSCF addresses to which Notification on Change of Subscriber Data is to be sent.
Short Message	Messages Waiting Data (MWD)	HLR	Consists of an address list of the SM-SC which have messages waiting to be delivered to the MS.

Contd.

Table 6.3	Contd.

Type	Element	Relevant at	Description
	Mobile Station Not Reachable Flag (MNRF)	HLR, VLR	Boolean parameter, indicates if the address list of MWD contains one or more entries because an attempt to deliver a short message to an MS has failed with a cause of Absent Subscriber.
	Memory Capacity Exceeded Flag (MCEF)	HLR	Boolean parameter, indicates if the address list of MWD contains one or more entries because an attempt to deliver a short message to an MS has failed with a cause of MS Memory Capacity Exceeded.
GPRS NAM	PDP Type	HLR, SGSN, GGSN	Indicates the type of protocol used by the MS for a certain service (e.g. IP and X.25).
	PDP Address	HLR, SGSN, GGSN	Holds the address of the MS for a certain service (e.g. an IPv4 address). If dynamic addressing is allowed, PDP Address is empty in the HLR, and, before the PDP context is activated, empty in the SGSN.
	New SGSN Address	SGSN	IP-address of the new SGSN to which PDUs should be forwarded from the old SGSN after an inter-SGSN routing update.
	APN	HLR, SGSN, GGSN	Used to Access a GPRS service.
	GGSN Address in Use	SGSN	Indicates the IP address of the GGSN currently used by a certain PDP Address of the MS.
	Dynamic Address	GGSN	Indicates whether the address of the MS is dynamic.
	SGSN Address	HLR, SGSN	Indicates the IP Address of the SGSN currently serving the MS.
	GGSN-list	HLR	Defines the GGSNs to be contacted when activity from the MS is detected and MNRG is set. It contains the GGSN number and optionally the GGSN IP address.
	QoS	HLR, SGSN, GGSN	Defines the Quality of Service for a PDP context. Includes Subscribed QoS, Requested QoS and Negotiated QoS.

Contd.

Table 6.3   Contd.

Type	Element	Relevant at	Description
Super-Charger	Age Indicator	HLR	Indicates the age of the subscription data provided by the HLR, e.g. the date and time at which the subscriber data was last modified in the HLR.
Others	Camel Subscription Information (CSI)	HLR, VLR GMSC, SGSN	Refers to various types of CSI (e.g. O-CSI, T-CSI, SS-CSI, etc.) that are used for different CAMEL procedures.
	Trace Information	HLR, VLR, SGSN	Refers to informations used for subscriber tracing

6.12   SS7 PROTOCOLS

In Section 6.8 and 6.9, the protocol architecture of the CS and PS domain were discussed, and the various protocols used in the architecture were briefly mentioned. In this section, the focus is on these protocols, particularly the protocols used in 3G UMTS. The GSM protocols are not covered in this chapter; for details on these, the reader is referred to the appropriate specifications.

The 3G Core Network protocols are covered under two sections: SS7 protocols and application protocols. This section elaborates upon the following SS7 protocols:

- Message Transfer Part
- Signaling Connection Control Part (SCCP)
- ISDN User Part (ISUP)

6.12.1   Message Transfer Part (MTP)

The MTP provides Layer 1 to Layer 3 services as per the OSI reference model. The MTP acts as a transport system and provides reliable transfer of signaling messages between peer entities. It provides services to many users including Signaling Connection Control Part (SCCP), ISDN User Part (ISUP) and the Telephone User Part (TUP), among others. MTP is defined in ITU-T Q.70x series of protocols.

6.12.1.1   MTP Layer Architecture

The MTP Layer Architecture is depicted in Figure 6.15. As shown in the figure, the MTP consists of three layers: MTP1, MTP2 and MTP3. These protocols are also called the Level 1, Level 2 and Level 3 protocols respectively. Various Level 4 protocols reside over MTP3. The Level 4 protocols are also called *user part* protocols. Important

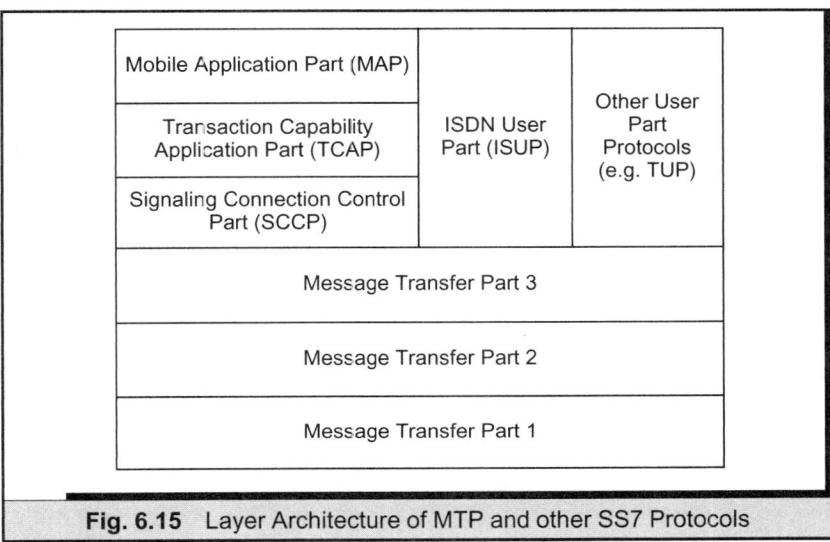

Fig. 6.15 Layer Architecture of MTP and other SS7 Protocols

user part protocols are the Signaling Connection Control Part (SCCP) and ISDN User Part (ISUP).

6.12.1.2 MTP Layer Functions

The MTP layer functions are as follows:

- **MTP1:** This provides the physical layer functionality for SS7-based networks. The MTP1 is based on European hierarchy (56Kbps and multiples thereof) or North American digital hierarchy (64Kbps and multiples thereof).

- **MTP2:** The MTP2 provides the link layer functionality for SS7-based networks. The link layer ensures reliable exchange of signaling messages including error checking, flow control, and sequence checking.

- **MTP3:** The MTP3 provides the network layer functionality, i.e. addressing, routing and congestion control, to SS7-based networks. The MTP3 layer ensures that messages can be delivered between signaling points regardless of whether they are directly connected or not. In case of failures, the protocol includes the necessary procedures to inform the remote entity of the consequences. Such failures lead to the reconfiguration of the network so that messages can be delivered reliably. Each SS7 node using MTP3 is uniquely identified within a network using an SS7 address called a *Point Code*. European networks, use 14 bit point codes while in North American networks, 24 bit point codes are used.

6.12.2   Signaling Connection Control Part (SCCP)

The SCCP provides certain enhanced functions that are lacking in MTP3. One of the most important functions of SCCP is to provide the means whereby applications residing on a node using SCCP can be uniquely identified. For this, the SCCP uses the notion of Sub-System Numbers (SSN). Apart from this, it provides connectionless and connection-oriented network services to transfer signaling information.

SCCP is defined in ITU-T Q.71x series of specifications.

6.12.2.1   SCCP Layer Architecture

The SCCP layer architecture is depicted in Figure 6.15. As shown in the figure, the SCCP layer uses the services of MTP3. In turn, it provides services to various users (e.g. TCAP). Apart from TCAP, the SCCP also provides services to other user protocols like Operations Maintenance and Administration Part (OMAP) protocol.

6.12.2.2   SCCP Functions

The SCCP provides connectionless and connection-oriented network services to transfer signaling information. Apart from this, it provides two important functions: *Sub-system Numbering* and *GT Translation*. These functions are explained in the following sub-sections.

6.12.2.2.1   Sub-system Numbering

Though MTP allows two MTP3 nodes to exchange messages, this is inadequate when more than one application resides on a SS7 node. For this, a means is required to identify different applications on an SS7 node. This functionality is provided by Sub-System Numbers (SSN) that are used to identify applications within network entities that use SCCP signaling. Table 6.4 lists the Global and National SSN defined for UMTS (also refer to 3GPP TS 23.003).

6.12.2.2.2   Global Title Translation (GTT)

Apart from sub-system numbering, SCCP also provides an advanced addressing capability using Global Title Translation (GTT). The *GT* is an address (such as dialled digits) that does not explicitly contain information that would allow routing in the signaling network. In order to route the signaling messages, a translation mechanism is required. The GTT is used for such a translation. The use of GTT frees the originator of a message from the burden of knowing the Destination Point Code (DPC) and the SSN. This reduces the requirements for nodes to know the point codes of distant

| Table 6.4 | Sub-System Numbers (SSN) |

Type	Binary Value	Decimal Value	Protocol
Globally Standardized SSN	0000 0110	6	HLR (MAP)
	0000 0111	7	VLR (MAP)
	0000 1000	8	MSC (MAP)
	0000 1001	9	EIR (MAP)
	0000 1010	10	Allocated for evolution (possible Authentication Center)
National SSN	1111 1001	249	PCAP
	1111 1010	250	BSC (BSSAP-LE)
	1111 1011	251	MSC (BSSAP-LE)
	1111 1100	252	SMLC (BSSAP-LE)
	1111 1101	253	BSS O&M (A interface)
	1111 1110	254	BSSAP (A interface)
	1000 1110	142	RANAP
	1000 1111	143	RNSAP
	1001 0001	145	GMLC (MAP)
	1001 0010	146	CAP
	1001 0011	147	gsmSCF (MAP) or IM-SSF (MAP)
	1001 0100	148	SIWF (MAP)
	1001 0101	149	SGSN (MAP)
	1001 0110	150	GGSN (MAP)

nodes. In such cases, the origination may request to route the message using Global Titles. Depending on network topology, Global Titles are translated either at a Signal Transfer Point (STP) or at a gateway exchange where a network has an inter-working function with an adjacent network.

Thus, the addressing information delivered to SCCP for message routing contains one or more of the three elements: Destination Point Code, SSN and GT. For successful message transmission, the minimum requirement is for a Destination Point Code in order for the message to leave the SCCP node. If none is present, the called address information is submitted for GT Translation. This produces as a minimum a Destination Point Code and optionally a SSN or a new GT. This new information is then used for further routing. Section 6.13.2.5.2 provides more information on routing using Global Titles.

6.12.3 ISDN User Part (ISUP)

ISDN User Part (ISUP) is used to setup, manage and release circuits that carry voice and data over circuit-switched networks. In other words, the ISUP protocol provides the signaling functions required to support basic bearer services and supplementary services for voice and non-voice applications in an Integrated Services Digital Network. ISUP is used for both ISDN and non-ISDN calls.

ISUP is defined in a set of standards. ITU-T Q.761 provides a functional description of ISUP. ITU-T Q.762 provides information on general functions of messages and signals in ISUP. The ITU-T Q.763 provides information on message formats and message coding. The signaling procedures are detailed in ITU-T Q.764.

6.12.3.1 ISUP Layer Architecture

ISUP protocol operates between GMSC and MSC. Figure 6.8 depicts the context in which ISUP operates. ISUP uses the services of SCCP, which in turn uses the SS7 transport protocols (MTP3/2/1).

6.12.3.2 ISUP Layer Functions

The ISUP layer performs various functions. While a detailed description of its functions is beyond the scope of this chapter, suffice it to say that it is used for call setup, maintenance and release. The use of ISUP for signaling in UMTS is explained in Chapter 10.

For a detailed description of ISUP functions, the reader is referred to ITU-T Q.761 to Q.764.

6.12.3.3 ISUP Messages

The ISUP protocol is used to exchange signaling messages between GMSC and MSC. Table 6.5 shows important ISUP messages.

6.13   APPLICATION PROTOCOLS

This section disscusses application protocols and also details the NAS Signaling protocol between UE and the Core Network. The main focus is on the following protocols:

- Transaction Capabilities (TCAP)
- Mobile Application Part (MAP)
- NAS Signaling
- GPRS Tunneling Protocol (GTP)
- Base Station Sub-system Application Part + (BSSAP+)

6.13.1   Transaction Capabilities (TCAP)

This refers to a set of communication capabilities that provide a generic interface between applications and a network service layer. TCAP provides a mechanism for

Table 6.5	Basic ISUP Messages***

Message	Description
Initial Address Message (IAM)	This is sent in the forward direction to initiate reservation of an outgoing circuit and to transmit number and other information relating to the routing and handling of a call. The IAM includes the Originating Point Code (OPC), Destination Point Code (DPC), Circuit Identification Code, dialed digits and optionally, the calling party number and name.
Subsequent Address Message (SAM)	This may be sent in the forward direction following an IAM to convey additional information.
Address Complete Message (ACM)	This is sent in the backward direction to indicate that all the address signals required for routing the call to the called party have been received.
Answer Message (ANM)	This is sent in the backward direction to indicate that the call has been answered.
Release Message (REL)	This is sent in either direction to indicate that the circuit is being released due to the reason supplied and is ready to be put into the idle state on receipt of the release complete message.
Release Complete Message (RLC)	This is sent in either direction in response to the receipt of a release message when the circuit concerned has been brought into the idle condition.

applications residing on a node in a SS7 network, to exchange information with peer applications on remote nodes. The applications need not be concerned with the underlying network layer service, since TCAP provides a single generic interface to them. TCAP can be used by a wide range of applications spread across multiple elements of the telecommunication network. One such category of applications is the group of 'mobile service applications' that include the Mobile Application Part (MAP), and the MAP applications (MAP-Users). In turn, TCAP uses the services of SCCP and MTP3 for availing the network layer service. While SCCP/MTP3 are the only protocols defined to provide network layer service to TCAP, new network layer services may be by ITU-T or other standardization bodies. TCAP is defined in ITU-T Q.77*x* series of specifications.

6.13.1.1 *TCAP Layer Architecture*

The TCAP layer is decomposed into two sub-layers: the *Transaction sub-layer*, and the *Component sub-layer*. Figure 6.16 depicts the decomposition of TCAP into these two layers, which are explained in the following sub-sections.

6.13.1.1.1 *Component Sub-layer*

The component sub-layer deals with the concept of 'Components'. Components are Application Protocol Data Units (APDUs). They are the means by which TCAP conveys a request (received from a TCAP-User) for an operation, or a response to the operation. This operation is required to be performed by a remote end, and in most cases, the response to this operation is expected from the remote end. On the local node, these components are exchanged between the TCAP-User and the component sub-layer of TCAP. A TCAP-User may send multiple components to the component sub-layer, before these are encoded and sent to the

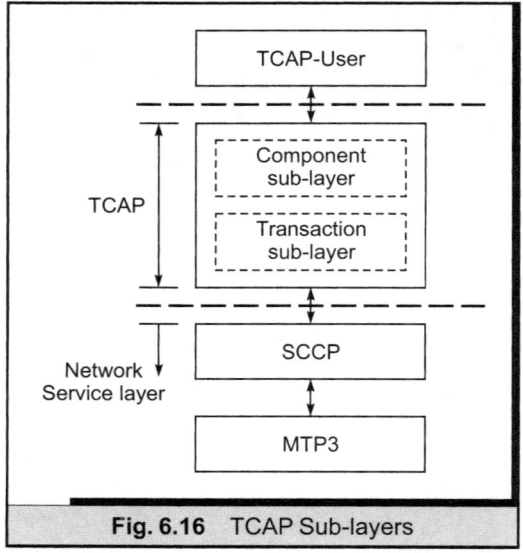

Fig. 6.16   TCAP Sub-layers

remote end in a single TCAP message. However, TCAP preserves the order of the components within a single TCAP message, which means that at the destination end, the components are delivered to the TCAP-User in the same order in which they were received at the sending node.

Generally, multiple TCAP components are required to be exchanged between peer application entities (which are TCAP-Users) to complete a particular scenario. As an example of such a scenario, consider the case of a mobile station migrating to a new location. A sequence of steps is required to be followed for the mobile station to inform HPLMN of its new location. In other words, a group of components is required to be exchanged between the mobile station and the HPLMN for the scenario to be complete. These components, which are part of a single scenario, are grouped together into 'Dialogs'. Figure 6.17 depicts the relationship between Dialogs, TCAP messages and components.

As shown in the figure, multiple components of the same dialog may be bundled into one TCAP message that is sent to the remote node. A complete dialog may consist

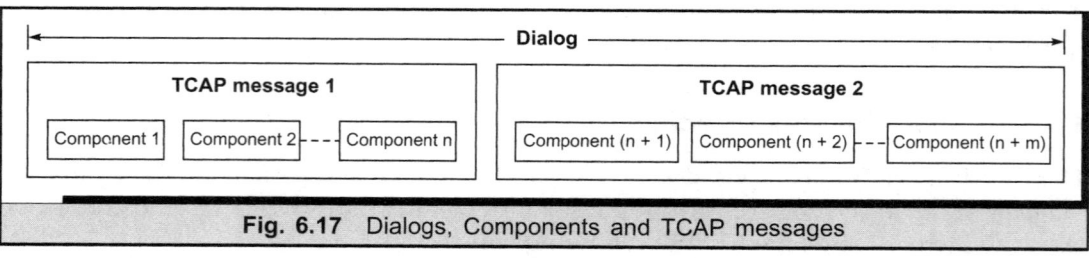

Fig. 6.17   Dialogs, Components and TCAP messages

of multiple TCAP messages, each possibly containing multiple components. TCAP messages exchanged between peer nodes may optionally contain portions of the dialog handling protocol (not shown in figure), besides the components (recall that components are APDUs). For a complete Dialog, multiple TCAP messages may be exchanged with the remote node, each with one or more TCAP components, and optionally, portions of the dialog-handling protocol. The exact breakup of the components of a dialog into TCAP messages is dependent on the TCAP-user, and is hence application-specific. The component sub-layer of TCAP and the TCAP-user identify each TCAP dialog by a dialog-Id. The dialog-Id is used between the TCAP-user and the TCAP component sub-layer to uniquely identify a dialogue. Further, each component within a dialog is identified using an Invoke-Id. Here, invoke stands for invocation; that is, each component is treated as an invocation of an operation carried out in that component.

6.13.1.1.2 *Transaction Sub-layer*

The transaction sub-layer of TCAP provides the capability for the exchange of TCAP components with the remote nodes. Thus, while the component sub-layer of TCAP is involved in the formation of the components, the transaction sub-layer is involved in the transport of these components to the remote nodes. The transaction sub-layer of TCAP does not look into the contents of these components. In fact, it deals with the TCAP messages as a unit of information transfer, and does not look into the contents of the messages. Thus, the transaction sub-layer is involved in the transport of TCAP messages for a TCAP dialog. It uses a Transaction-Id to identify a TCAP dialog. There exists a one-to-one mapping between the dialog-Id at the component sub-layer, and the transaction-Id at the transaction sub-layer. Table 6.6 depicts the relationship between Dialog-Id, Invoke-Id and Transaction-Id.

6.13.1.2 *TCAP Interfaces and Service Primitives*

Figure 6.18 depicts the interfaces between the TCAP-user, TCAP and the network service layer. The internal interface of TCAP, that is between the component and transaction sub-layer of TCAP, is also depicted in the figure.

As shown in the figure, the TCAP-user interacts with the component sub-layer of TCAP using the 'TC Service Primitives'. These primitives include the Dialog-Id and the Invoke-Id for each component exchanged on the interface. Since the component sub-layer of TCAP deals with both the dialogs, as well as the components within the dialog, the TC service primitives include the dialog handling primitives (summarized in Table 6.7) and the component handling primitives (Table 6.8). Note that the tables only provide a list of the important service primitives.

Table 6.6 Dialogs, Components, Transactions and Invocations

Component Sub-layer		Transaction Sub-layer
Dialogs *(Identified using Dialog-Id)*	*Components* *(Identified using Invoke-Id)*	*Transactions* *(Identified using Transaction-Id)*
Dialog-Id 1	Invoke-Id 1 Invoke-Id 2	Transaction-Id 4
Dialog-Id 2	Invoke-Id 1 Invoke-Id 2 Invoke-Id 3	Transaction-Id 5

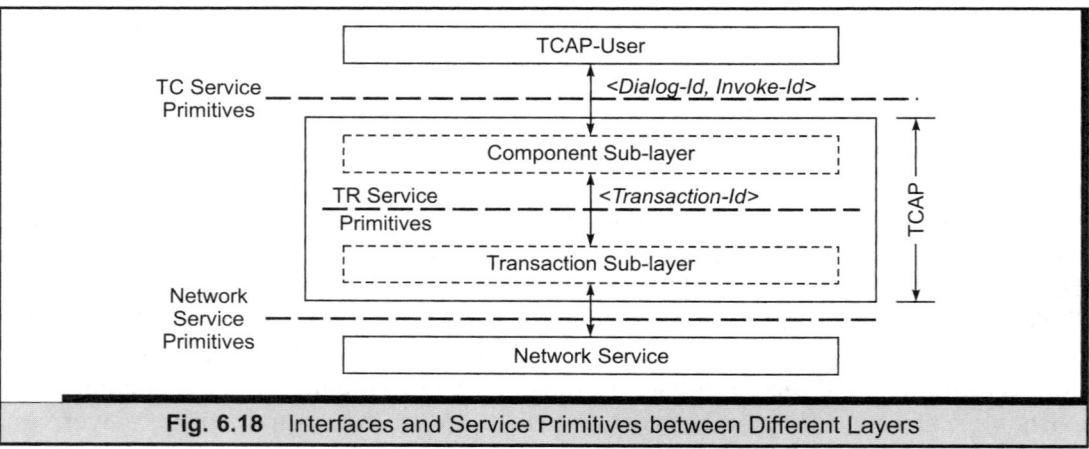

Fig. 6.18 Interfaces and Service Primitives between Different Layers

Table 6.7 Dialog Handling Primitives of Component Sub-layer

Name	Type	Description
TC-BEGIN	Request, Indication	Begins a dialog
TC-CONTINUE	Request, Indication	Continues a dialog
TC-END	Request, Indication	Ends a dialog

The component sub-layer of TCAP interacts with the transaction sub-layer of TCAP using the 'TR Service Primitives'. These service primitives include the Transaction-Id for each TCAP message exchanged on the interface. Since the transaction sub-layer of TCAP deals only with dialogs (or transactions), the TR service primitives include only the dialog-handling primitives (summarized in Table 6.9). Again, the tables only provide a list of the important primitives.

Table 6.8 Component Handling Primitives of Component Sub-layer

Name	Type	Description
TC-INVOKE	Request, Indication	Invocation of an operation
TC-RESULT-L	Request, Indication	Either only the result or the last part of the segmented result of a successfully executed operation.
TC-RESULT-NL	Request, Indication	Non-final part of the segmented result of a successfully executed operation.
TC-U-ERROR	Request, Indication	Reply to a previously invoked operation, indicating that the operation execution has failed.

Table 6.9 Dialog Handling Primitives of Transaction Sub-layer

Name	Type	Description
TR-BEGIN	Request, Indication	Starts a transaction.
TR-CONTINUE	Request, Indication	Continues a transaction.
TR-END	Request, Indication	Ends a transaction.

6.13.2 Mobile Application Part (MAP)

The MAP protocol is the heart of the Core Network. It was developed specifically to cater to the requirements of a mobile network environment. MAP allows mobile service applications on nodes in a mobile network to communicate with each other using the SS7-based protocols (MTP1, MTP2, MTP3 and SCCP) and TCAP. MAP can also reside over IP-based SIG TRAN, details of which is provided in Chapter 15.

The MAP protocol consists of a collection of messages and procedures, which are required by mobile service applications to fulfill their signaling needs. These procedures are grouped into specific categories, which include Mobility Management, Call Handling, Supplementary Service Management, and SMS Management, among others. Each of these categories of procedures could be used by multiple applications providing specialized services.

The MAP protocol is defined in 3GPP TS 29.002.

6.13.2.1 MAP Layer Architecture

Figure 6.19 depicts the MAP layer architecture. The MAP layer acts as a TCAP-user, and uses the services of the component sub-layer of TCAP. It acts as a service provider to the Multiple MAP user applications. Each of the MAP user applications performs a specialized function that requires exchange of one or more MAP messages between nodes in the core network. In Figure 6.19, only three MAP applications, namely Mobility Management, Call Handling and Supplementary Service Management are depicted. Apart from these, there are many other MAP applications.

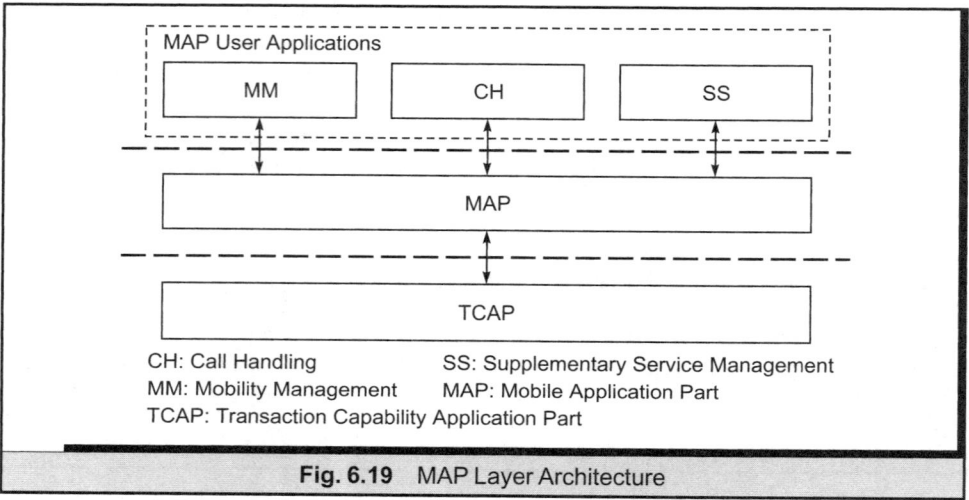

CH: Call Handling   SS: Supplementary Service Management
MM: Mobility Management   MAP: Mobile Application Part
TCAP: Transaction Capability Application Part

Fig. 6.19 MAP Layer Architecture

To understand MAP applications, take for instance a scenario where a mobile station moves to a new location and registers its location with the HPLMN (scenario was briefly discussed in Section 6.13.1.1.1). In this example, the required message handling and subsequent processing is catered to by the 'Mobility Management (MM)' application of MAP. The complete scenario requires exchange of multiple MAP messages. In particular, MAP_UPDATE_LOCATION and MAP_INSERT_SUBSCRIBER_DATA messages are exchanged between VLR and HLR in this scenario. This scenario is discussed in Section 6.13.2.4. The different MAP messages are explained in Section 6.13.2.3.

A particular exchange of set of MAP messages is identified between the MAP and the MM application, as well as between MAP and TCAP, using a unique Dialog-Id. Each MAP message within the dialog is identified using an Invoke-Id. Thus, it may be said that each MAP message within a dialog is actually one component for that dialog at the TCAP component sub-layer.

6.13.2.2 MAP Layer Functions

As stated earlier, there are a number of MAP users residing over MAP. Thus, MAP can be viewed as a stack that provides a set of procedures, each catering to the needs of a particular user application. These procedures or MAP functions are summarized as follows (for a more comprehensive list of MAP messages and associated descriptions, refer to Table 6.11):

- **Mobility Management procedures:** This includes procedures for updating the location information stored in HLR, deleting subscriber record from VLR/SGSN, and other mobility related procedures.

- **Authentication procedures:** This includes procedures to retrieve authentication information from the HLR and to inform the HLR of an authentication failure.

- **Call Handling procedures:** This includes procedures whereby GMSC obtains the MSRN from VLR via HLR, and also procedures related to mobile-originated (MO) call handling and Immediate Service Termination (IST).

- **IMEI check:** This includes procedures to verify the IMEI at the EIR.

- **Subscriber Tracing procedures:** This includes procedures to activate/deactivate the subscriber tracing at the VLR/SGSN.

- **Subscriber Data Management procedures:** This includes procedure to update/delete subscriber data maintained at VLR/SGSN.

- **Fault Recovery procedures:** This includes procedures to recover after a restart so that data maintained at various network entities (e.g. HLR, VLR and SGSN) is consistent.

- **CAMEL procedures:** This includes procedures to obtain and modify CAMEL related information.

- **SMS procedures:** This includes procedures to obtain routing information for SMS, indicating the success status of delivery and other related procedures.

- **Supplementary Service procedures:** This includes various procedures required to provide Supplementary Service. Some of the procedures are: register for SS, erase data for SS, activate/deactivate SS, interrogate SS and invoke SS, among others.

- **PDP Context Activation procedures:** This includes procedures necessary for network-initiated PDP context activation.

Note that the above list of procedures is by no means comprehensive. It only lists the important procedures. For a complete list of procedures, the reader is referred to 3GPP TS 29.002.

Besides thess service-specific procedures, MAP also provides common procedures, which are used by all service applications. These procedures are related to handling of dialogs, which includes opening a new dialog, continuing a dialog, and terminating a dialog. These dialog-handling procedures of MAP find a one-to-one mapping with the dialog-handling primitives of the component sub-layer of TCAP. Table 6.10 gives a list of the important common procedures of MAP. Note that this is not the complete list of procedures; there are other procedures dealing with abnormal conditions that are not covered here.

Table 6.10   Common Procedures of MAP

Name	*Type*	*Description*	*Mapping to TCAP Primitive*
MAP-OPEN	Request, Indication, Response, Confirm	Opens a new dialog.	TC-BEGIN
MAP-DELIMITER	Request, Indication	Transmits all MAP messages received from the user application. These MAP messages are transmitted as TCAP components in one TCAP message.	TC-CONTINUE
MAP-CLOSE	Request, Indication	Ends a dialog.	TC-END

6.13.2.3 MAP Messages

As mentioned in the previous section, there are many users of MAP. Each user performs a very specific function. The important categories of MAP messages are Mobility Management, Authentication, Call Handling, IMEI checks, Tracing, Subscriber Data handling, Fault Recovery, CAMEL data handing, SMS management, Supplementary Service Management and network-requested PDP context activation. Table 6.11 lists the important MAP messages associated with each of these categories.

Table 6.11   Important MAP Messages

Type	*Message*	*Between*	*Description*
Mobility Management	MAP_UPDATE_LOCATION	VLR-HLR	Used by VLR to update the location information stored in the HLR.
	MAP_CANCEL_LOCATION	VLR-HLR, SGSN-HLR	Used by HLR to delete a subscriber record from the VLR or SGSN.
	MAP_SEND_IDENTIFICATION	VLR-VLR	Used between a current and previous VLR to retrieve IMSI and authentication data for a subscriber registering afresh in that VLR.
	MAP_PURGE_MS	VLR-HLR	Used between the VLR and HLR to cause the HLR to mark its data for an

Contd.

| **Table 6.11** | Contd. |

Type	*Message*	*Between*	*Description*
			MS so that any request for routing information for a MT call or a MT short message is treated as if the MS is not reachable.
	MAP_UPDATE_ GPRS_LOCATION	SGSN-HLR	Used by SGSN to update the location information stored in the HLR.
Authentication	MAP_SEND_ AUTH_INFO	VLR-HLR, SGSN-HLR	Used by VLR or SGSN to retrieve authentication information from the HLR.
	MAP_ AUTH_FAILURE_ REPORT	VLR-HLR, SGSN-HLR	Used between the VLR/SGSN and HLR to report authentication failures.
	MAP_AUTHENTI CATE	MSC-VLR	Used when the VLR receives a MAP service indication from the MSC concerning a location registration, call setup, operation on a supplementary service or a request from the MSC to initiate authentication.
Call Handling	MAP_SEND_ ROUTING_INFO	GMSC-HLR	Used by the Gateway MSC to perform the interrogation of the HLR in order to route a call towards the called MS.
	MAP_PROVIDE_ ROAMING_NUMBER	VLR-HLR	Used by HLR to request a VLR to send back a roaming number to enable it to instruct the GMSC to route an incoming call to the called MS.
	MAP_IST_ALERT	MSC-HLR	Used between the MSC and the HLR, to report that the IST timer running for a call for the subscriber has expired.
	MAP_IST_COMMAND	SGSN-HLR	Used by HLR to instruct the MSC to terminate ongoing call activities for a specific subscriber.
	MAP_RESUME_C ALL_HANDLING	MSC- GMSC	Used by the terminating VMSC to request the GMSC to resume handling the call and forward it to the specified destination.
IMEI check	MAP_CHECK_IMEI	SGSN-EIR, MSC-EIR	Used by MSC or the SGSN requesting EIR to check IMEI.
	MAP_OBTAIN_IMEI	MSC-VLR	Used between the VLR and MSC to request for IMEI. If the IMEI is not available in the MSC, it is requested from the MS.

Contd.

Table 6.11 Contd.

Type	Message	Between	Description
Subscriber Tracing	MAP-ACTIVATE-TRACE-MODE	VLR-HLR SGSN-HLR	Used by HLR to activate subscriber tracing in the VLR or SGSN.
	MAP-DEACTIVATE-TRACE-MODE	VLR-HLR SGSN-HLR	Used by HLR to de-activate subscriber tracing in the VLR or SGSN.
Subscriber Data	MAP-INSERT-SUBS-DATA	VLR-HLR, SGSN-HLR	Used by HLR to insert/update subscriber data at VLR or SGSN.
	MAP-DELETE-SUBS-DATA	VLR-HLR, SGSN-HLR	Used by HLR to delete subscriber data at VLR or SGSN.
Fault Recovery	MAP_RESET	VLR-HLR, SGSN-HLR	Used by HLR, after a restart, to indicate to a list of VLRs or SGSNs that a failure occurred.
	MAP_RESTORE_DATA	VLR-HLR	Used to update the LMSI in the HLR, if provided, and to request the HLR to send all data to the VLR that are to be stored in the subscriber's record at VLR.
CAMEL data	MAP-ANY-TIME-INTERROGATION	gsmSCF-HLR	Used by the gsmSCF to request information (e.g. subscriber state and location) from the HLR.
	MAP-PROVIDE-SUBS-INFO	VLR-HLR SGSN-HLR	Used by HLR to request information (e.g. subscriber state and location) from the VLR or SGSN at any time.
	MAP-ANY-TIME-SUBSCRIPTION-INTERROGATION	gsmSCF-HLR	Used by gsmSCF to request subscription information (e.g. call forwarding supplementary service data or CSI) from the HLR at any time.
	MAP-ANY-TIME-MODIFICATION	gsmSCF-HLR	Used by gsmSCF to modify information of the HLR at any time.
	MAP-NOTE-SUBS-DATA-MODIFIED	gsmSCF-HLR	Used by HLR to inform the gsmSCF that subscriber data has been modified.
SMS	MAP_SEND_ROUTING_INFO_FOR_SM	GMSC-HLR	Used between the Gateway MSC and the HLR to interrogate routing information from HLR in order to route a SM towards the serving MSC.
	MAP-MO-FORWARD-SHORT-MESSAGE	MSC-IWMSC, SGSN-IWMSC	Used to forward mobile originated short messages to the SMS Interworking MSC.
	MAP-MT-FORWARD-SHORT-MESSAGE	MSC-GMSC, SGSN-GMSC	Used by the Gateway MSC to forward short messages to the serving MSC or the SGSN.

Contd.

Table 6.11	Contd.

Type	Message	Between	Description
	MAP-REPORT-SM-DELIVERY-STATUS	GMSC-HLR	Used by the Gateway MSC to set the Message Waiting Data into the HLR or to inform the HLR of successful SM transfer after polling.
	MAP-SEND-INFO-FOR-MT-SMS	MSC-VLR	Used by the MSC receiving a MT SM to request subscriber related information from the VLR.
	MAP-SEND-INFO-FOR-MO-SMS	MSC-VLR	Used by the MSC which has to handle a MO SM request to get the subscriber related information from the VLR.
	MAP-READY-FOR-SM	VLR-HLR, SGSN-HLR	Used by VLR/SGSN to indicate to the HLR that the MS is now ready to receive an SM. (e.g. when a radio contact is established with the MS).
	MAP-ALERT-SERVICE-CENTRE	HLR-MSC	Used when HLR detects that a subscriber, whose MSISDN is in the Message Waiting Data file, is active or the MS has memory available.
Supplementary Service Management	MAP_REGISTER_SS	VLR-HLR	Used to register data related to a SS.
	MAP_ERASE_SS	VLR-HLR	Used to erase data related to a SS.
	MAP_ACTIVATE_SS	VLR-HLR	Used to activate a SS.
	MAP_DEACTIVATE_SS	VLR-HLR	Used to de-activate a SS.
	MAP_INTERROGATE_SS	VLR-HLR	Used to retrieve information for a SS.
	MAP_INVOKE_SS	MSC-VLR	Used between the MSC and the VLR to check the subscriber's subscription to a given SS in the VLR.
	MAP_REGISTER_PASSWORD	VLR-HLR	Used when MS requests to register a new password for a particular SS.
	MAP_GET_PASSWORD	VLR-HLR	Used by HLR to get the password from the subscriber for a particular SS.
PDP Context Act.	MAP_SEND_ROUTING_INFO_FOR_GPRS	GGSN-HLR	Used by GGSN to request GPRS routing information from the HLR.
	MAP_FAILURE_REPORT	GGSN-HLR	Used by GGSN to inform the HLR that network requested PDP-context activation has failed.
	MAP_NOTE_MS_PRESENT_FOR_GPRS	GGSN-HLR	Used by HLR to inform GGSN that the MS is present for GPRS again.

Contd.

| **Table 6.11** | Contd. | | |

Type	Message	Between	Description
Others	MAP_PROCESS_ ACCESS_REQ	MSC-VLR	Used between MSC and VLR to initiate processing of an MS access to the network, e.g. in case of mobile originated call set-up or after being paged by the network.
	MAP-FORWARD- NEW-TMSI	MSC-VLR	Used by a VLR to allocate, via MSC, a new TMSI to a subscriber during an ongoing transaction (e.g. call set-up, location updating or supplementary services operation).
	MAP_PAGE	MSC-VLR	Used between VLR and MSC to initiate paging of an MS for mobile terminated call set-up, mobile terminated short message or unstructured SS notification.

6.13.2.4   *Example of MAP Message Exchange*

For a MAP user application that uses common procedures as well as service-specific procedures, the normal flow of messages for a scenario follows the following sequence:

1. Open a new MAP dialog using the MAP-OPEN common primitive.
2. Transmit service-specific messages to MAP using the MAP service-specific procedures.
3. Push the messages transmitted so far to TCAP (and eventually to the remote node) using the MAP-DELIMITER common primitive.
4. Repeat steps 2 and 3 for all messages (or responses to messages) that need to be sent to the remote node.
5. Close the MAP dialog using the MAP-CLOSE common primitive.

Figure 6.20 depicts the information flow between two network nodes (HLR and VLR) for the update location scenario of mobility management. Messages exchanged between the MAP-user, MAP and TCAP are shown for both VLR and HLR.

6.13.2.5   *Routing of MAP Messages*

One of the important aspects related to understanding MAP protocol is how MAP messages are routed between MAP users (also known as Application Service Entities) using the services of the SCCP/MTP3 layer. This section discusses the routing of MAP messages.

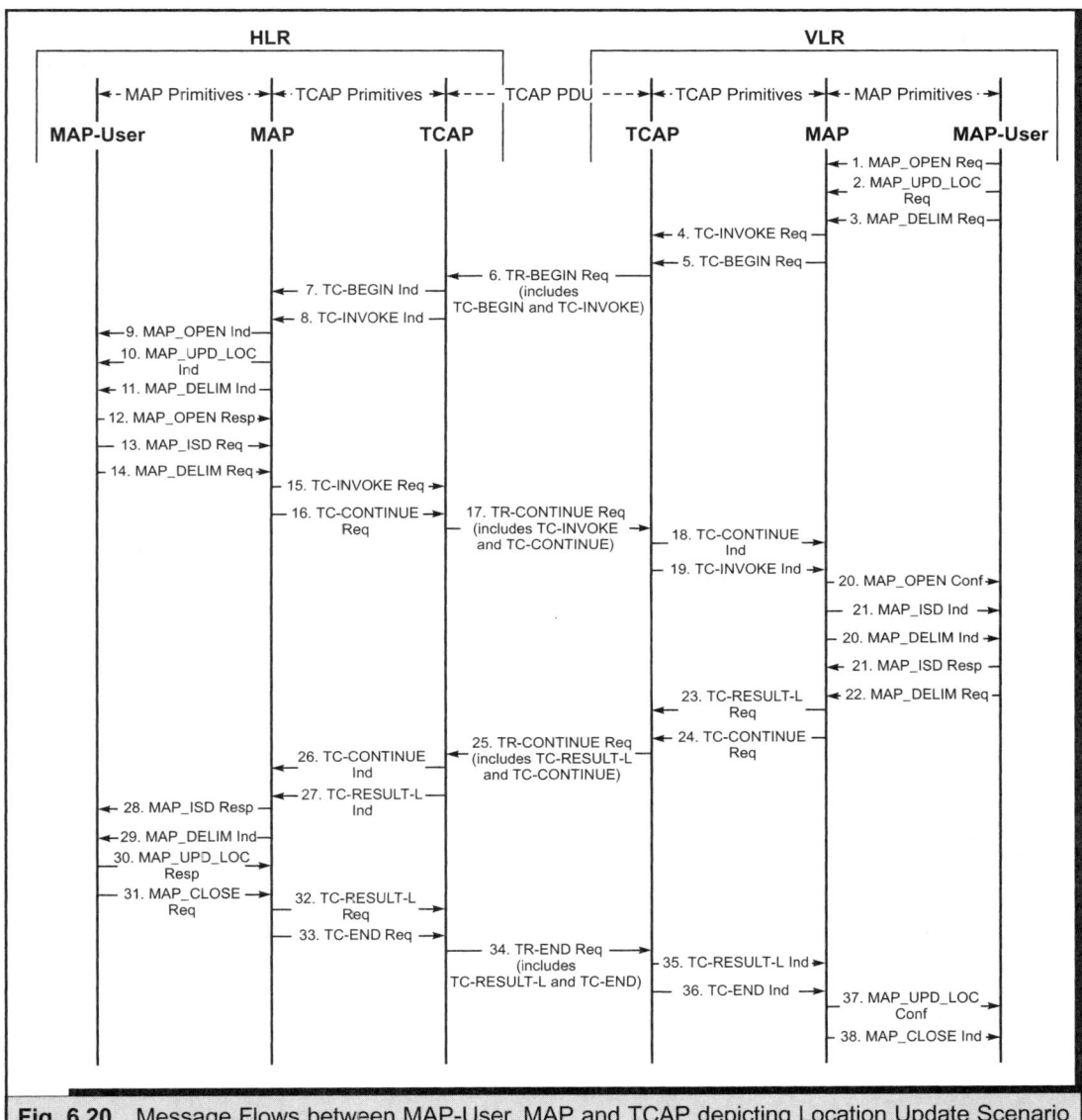

Fig. 6.20 Message Flows between MAP-User, MAP and TCAP depicting Location Update Scenario

The process of routing MAP messages can be considered to be a two-step process. The first step is to route the message from the source network node (source SS7 node) to the destination network node (destination SS7 node) across the SS7 network. The second step is to route it to the correct Application Service Entity (ASE) on the destination network node. Just as different application protocols (HTTP, FTP, etc.)

can reside over the transport layer protocols (TCP and UDP), each uniquely identified by the port numbers, similarly the ASEs can be uniquely identified by Sub-System Numbers (SSN), as discussed earlier. Thus, the second step of the process consists of using the SSN of the destination ASE to route the message to this entity, once it has reached the destination node.

The first step of the process, which requires routing the message across the SS7 network to the destination network node, is more challenging. To route the MAP message to the destination node, the services of the SCCP/MTP3 layer are used. Two different mechanisms are generally followed for routing of MAP messages between the network nodes. This depends upon whether the sending and the receiving ASEs belong to the same PLMN or to different ones. In the first case, the routing is called *Intra-PLMN routing;* in the second case it is called *Inter-PLMN routing.*

Irrespective of whether the routing is Intra-PLMN or Inter-PLMN, two general mechanisms are possible. These are explained in the following subsections.

6.13.2.5.1 *Route on Point Code (PC)*

This approach for routing is generally followed for routing of messages between ASEs within the same PLMN (Intra-PLMN). In this approach, the sending ASE provides the Point Code (PC) of the destination node to the SCCP layer along with the SSN of the destination ASE. The message is routed from source network node to the destination node by the MTP3 layer using the destination PC. Once the message reaches the destination node, it is routed within the node to the correct ASE by using the SSN.

Although this scheme can be used to route messages to any ASE in any PLMN, it is generally not possible for the sending ASE to know the PC of each destination node in all the PLMNs. At the very best, PCs for all nodes within the same PLMN can be configured at the sending ASE. Hence, this scheme can be employed for Intra-PLMN message routing.

6.13.2.5.2 *Route on Global Title (GT)*

This approach is followed for routing of messages between ASEs of different PLMNs when the PC of the destination node is not known to the sending ASE. The approach uses a Global Title (GT) to address the destination node and to route the message towards this destination node. As the name suggests, GT is a title that is known to all ASEs, whether within the same PLMN or in different PLMNs. The GT of the destination node can be in any one of the following addressing formats:

- **E.212 Numbering Plan:** This is generally used as the GT in networks that support ANSI SCCP. In this case, the subscriber IMSI, which is an E.212 number, is used as the GT to address the HLR and to route a message towards the HLR. Similarly, when a MS moves from one VLR area to another, it provides the previous Location Area Identity (LAI) to the new VLR in a registration request. The new VLR uses the E.212 numbering plan elements of this LAI to address the previous VLR and to route any message towards it.

- **E.214 Numbering Plan:** The E.214 Numbering Plan is used as the GT in networks that support ITU-T SCCP. An E.214 number can actually be considered similar to an E.212 number and is used similarly. In fact, an E.214 number is derived from an E.212 number by following a simple translation mechanism. For details on this translation function, the reader is referred to the ITU-T E.214 specifications.

- **E.164 Numbering Plan:** The E.164 Numbering Plan is used as the GT in networks that support either ANSI or ITU-T SCCP. In this case, the sending entity includes its E.164 number in the calling party address of the SCCP PDU. The called party address contains an E.164 number of the destination node. This scheme can be used to route a message to any node in the core network, since each node has an E.164 number.

Though the route on the GT approach is generally used for Inter-PLMN routing, it can also be used for Intra-PLMN routing in cases where the sending ASE is not aware of the PC of the destination node. Thus, though it is expected that the 'Route on PC' approach would be used for Intra-PLMN routing, and the 'Route on GT approach' would be used for Inter-PLMN routing, there may be no restriction on which approach is used in which scenario. Table 6.12 depicts different approaches for routing of MAP messages.

Table 6.12 Routing of MAP messages

Type of Routing	ITU-T SCCP		ANSI SCCP	
	Addressing Format	*Applicable for Nodes*	*Addressing Format*	*Applicable for Nodes*
Inter-PLMN	E.164 number of the destination node	All network nodes	E.164 number of the destination node	All network nodes
	E.164 number of MS (MSISDN)	HLR	E.164 number of MS (MSISDN)	HLR
	E.214 number (derived from E.212 IMSI/LAI)	HLR (IMSI), VLR (LAI)	E.212 Number (IMSI/LAI)	HLR (IMSI), VLR (LAI)
Intra-PLMN	Point Code of the destination node	All network nodes	Point Code of the destination node	All network nodes

6.13.3 | NAS Signaling

One way of modeling a UMTS network is by dividing it into Access Stratum (AS) and Non-Access Stratum (NAS). This division of a UMTS network into AS and NAS layers was discussed in Chapter 3. The Non-Access Stratum (NAS) protocols apply between UE and the Core Network over the Uu interface. The access stratum acts as a carrier/transport for the NAS protocols. The NAS protocols are not terminated at the UTRAN. Instead, they are terminated at the Core Network.

Protocols in the NAS layer are divided into two categories: NAS control plane protocols and NAS user plane protocols. The layer comprising of NAS control plane protocols is referred to as the NAS signaling layer. This section discusses the NAS Signaling layer architecture and functions. The NAS signaling layer is discussed in 3GPP specifications TS 24.007 and TS 24.008.

6.13.3.1 *NAS Signaling Layer Architecture*

The NAS signaling layer operates between the UE and CN. Figure 6.21 depicts the architecture of this layer. The NAS signaling layer uses the services of AS layer to transparently carry the NAS signaling layer information between the UE and the CN.

The NAS signaling layer consists of the NAS control plane protocols. These protocols include a Mobility Management (MM) sub-layer for the CS domain and a

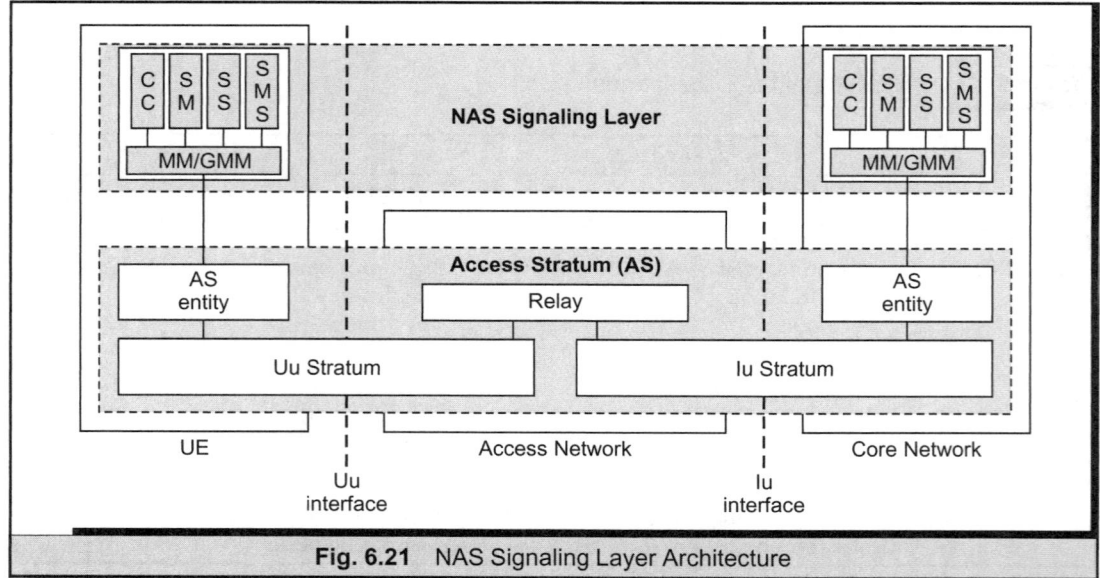

Fig. 6.21 NAS Signaling Layer Architecture

GPRS Mobility Management (GMM) sub-layer for the PS domain. A number of protocols reside over the MM/GMM sub-layer. These include the Call Control (CC), Session Management (SM), Supplementary Service (SS) and Short Message Service (SMS) protocols.

6.13.3.2 *NAS Signaling Layer Functions*

The NAS signaling layer performs various functions, of which the important ones are summarized as follows (for detailed description of NAS signaling layer functions, the reader is referred to 3GPP TS 24.008):

- **Location Management:** The MM/GMM procedures enable mobility of user terminals, such as keeping track of the present location of the subscriber. Chapter 9 provides details of the MM and GMM procedures.

- **Management of User Calls/Sessions:** The Call Control (CC) procedures involve handling of mobile-originated (MO) and mobile-terminated (MT) calls. CC procedures are also sometimes referred to as Call Handling (CH) procedures. For PS domain, the Session Management (SM) layer performs functions similar to those performed by the CC layer for CS domain. It manages the sessions in the PS domain. Chapter 10 provides details of the call handling procedures while Chapter 11 the details of session management procedures.

- **Supplementary Services (SS) Management:** The SS procedures involve management of the supplementary services subscribed to by a mobile user. Chapter 12 provides details of the SS procedures.

- **Short Message Service (SMS) Support:** Short message service is the means whereby a short text (up to 160 characters) can be exchanged between the UE and a SM-SC. The SMS procedures involve handling of mobile-originated (MO) and mobile-terminated (MT) SMS. Chapter 13 provides details of the SMS procedures.

6.13.3.3 *NAS Signaling Messages*

The NAS signaling layer is used to exchange signaling messages between the UE and the CN. Table 6.13 shows the important NAS signaling layer messages. The 'type' of messages and classification of messages in one of these types is as per the NAS signaling layer specification (3GPP TS 24.008). The first set of messages is used for MM procedures. The second set, categorized under Call Control, is used for CC and SS procedures. The next two categories define messages used for GMM procedures and SM procedures used in the PS domain.

Table 6.13	Important NAS Signaling Messages

Type	Message	Description
Mobility Management	LOCATION_UPDATING Req/Acc/Rej	Used by the MS to request update of its location information (normal or periodic) or to request IMSI attach.
	IMSI_DETACH	Used by the MS to set a de-activation indication in the network.
	AUTHENTICATION Req/ Resp/Rej/Failure	Used between the MS and the network to initiate authentication of the MS.
	IDENTITY Req / Resp	Used between the network and the MS to request the latter to submit the specified identity to the network.
	TMSI_RE-ALLOCATION Command/Complete	Used between the network and the MS to re-allo-cate or delete a TMSI.
	CM_SERVICE Req/Acc/ Rej/Prompt/Abort	Used by the MS to request a service for the connection management sub-layer entities. The network responds indicating whether the service can be accepted or not.
	CM_RE-ESTABLISHMENT Req	Used between the MS and the network to request re-establishment of a connection if the previous one has failed.
Call Control	ALERTING	Used between the MS and the network to indicate that the called user alerting has been initiated.
	CALL_CONFIRMED	This message is sent by the called MS to confirm an incoming call request.
	CALL_PROCEEDING	This message is sent by the network to the calling MS to indicate that the requested call establish-ment information has been received.
	CONNECT	Used between the network and the MS to indicate call acceptance by the called user.
	CONNECT_ACKNOWLEDGE	This message is sent by the network to the called MS to indicate that the MS has been awarded the call. It shall also be sent by the calling MS to the network to acknowledge the offered connection.
	EMERGENCY_SETUP	This message is sent from the MS to initiate emer-gency call establishment.
	SETUP	This message is sent by the network to the MS to initiate a MT call establishment. It is also sent from the MS to the network to initiate a MO call.
	MODIFY/MODIFY Complete/Rej	Used between the MS and the network to request a change in bearer capability for a call.

Contd.

Table 6.13	Contd.	

Type	*Message*	*Description*
	USER_INFORMATION	This message is sent by the MS to the network to transfer information to the remote user. It is also sent by the network to the MS to deliver information transferred from the remote user.
	HOLD/HOLD Ack/Rej	Used between the MS and the network to request the hold function for an existing call.
	RETRIEVE/RETRIEVE Ack/Rej	Used between the MS and the network to request the retrieval of a held call.
	DISCONNECT	This message is sent by the network to indicate that the end-to-end connection is cleared. This message is also sent by the MS to request the network to clear an end-to-end connection.
	RELEASE/RELEASE_ COMPLETE	This message is sent from the network to the MS to indicate that the network intends to release the transaction identifier, and that the receiving equipment shall release the transaction identifier after sending RELEASE_COMPLETE.
	NOTIFY	This message is sent either from the MS or from the network to indicate information pertaining to a call, such as User Suspended.
	STATUS_ENQUIRY	This message is sent by the MS or the network at any time to solicit a STATUS message from the peer layer 3 entity.
	STATUS	This message is sent by the MS or the network at any time during a call to report certain error conditions. It shall also be sent as a response to the STATUS_ENQUIRY message.
	FACILITY	Used between the network and the MS to request or acknowledge a supplementary service.
GPRS Mobility Management	ATTACH Req/Acc/ Comp/Rej	This message is sent by the MS to the network in order to perform a GPRS or combined GPRS attach.
	DETACH Req/Acc	Used between the MS and the network to request the release of a GMM context.
	ROUTING_AREA_UPDATE Req/Acc/Comp/Rej	This message is sent by the MS to the network either to request an update of its location information or to request an IMSI attach for CS-domain (non-GPRS) services.
	SERVICE Req/Acc/Rej	This message is sent by the MS to establish a logical association between the MS and the network.
	P_TMSI_RE-ALLOCATION Command/Complete	Used between the MS and the network to re-allocate a P-TMSI.

Contd.

Table 6.13 Contd.

Type	Message	Description
	AUTH_AND_CIPHERING Req/Resp/Rej/Failure	Used between the MS and the network to initiate authentication of the MS identity. Additionally, the ciphering mode is set, indicating whether ciphering will be performed or not.
	IDENTITY Req/Resp	This message is sent by the network to the MS to request submission of the MS identity.
Session Management	ACTIVATE_PDP_CONTEXT Req/Acc/Rej	This message is sent by the MS to the network to request activation of a PDP context.
	REQUEST_PDP_CONTEXT_ ACTIVATION	This message is sent by the network to the MS to initiate activation of a PDP context.
	DE-ACTIVATE_PDP_ CONTEXT Req/Acc	Used between the MS and the network to request de-activation of an active PDP context.
	MODIFY_PDP_CONTEXT Req/Acc/Rej	Used between the MS and the network to request modification of an active PDP context.
	ACTIVATE_SECONDARY_ PDP_CONTEXT Req/Acc/Rej	This message is sent by the MS to the network to request activation of an additional PDP context for a PDP address that is already associated with an active PDP context.

6.13.4 GPRS Tunneling Protocol (GTP)

In the PS domain, after a PDP context is activated, the packets are encapsulated and tunneled between the PLMN. For this purpose, the GPRS Support Nodes (GSNs) use the GPRS Tunneling Protocol (GTP). The GTP is used to create a GTP tunnel. It also provides the means to exchange data using the GTP tunnel. Here, a GTP tunnel is a means for GSNs to communicate with each other in a way that the PDP PDUs are encapsulated with a GTP header and exchanged using the UDP/IP protocol.

The GTP protocol is divided into two logical parts: GTP for user plane called the *GTP-U* and GTP for control plane called the *GTP-C*.

The GTP-U helps carry user data packets using encapsulation. It is used between GSNs on the Gn and Gp interface, and also over the Iu_PS interface between RNS and the SGSN.

The GTP-C protocol provides the signaling functionality in the GTP. It is used for creation, modification and deletion of GTP tunnels. GTP-C is also used for location and mobility management. These functions are further explained in the following sections. Like GTP-U, the GTP-C is used between GSNs on the Gn and Gp interface. However, GTP-C is not used over the Iu_PS interface; instead, the RANAP protocol is used. Thus, the GTP tunnels between RNC and SGSN are managed using the RANAP protocol.

GTP, which includes both GTP-U and GTP-C, is defined in 3GPP TS 29.060.

6.13.4.1 *GTP Layer Architecture*

As already mentioned, the GTP is divided into two logical components: GTP-U and GTP-C. Figure 6.9 depicts the context in which the GTP-U operates. As shown in the figure, GTP-U operates between RNC and SGSN and between SGSN and GGSN. Thus, there are two sets of tunnels: one between the Access Network (i.e. RNC) and Core Network (SGSN) and the other within the Core Network itself (i.e. between GSNs).

Figure 6.11 depicts the context in which GTP-C operates (i.e. operates between GSNs over Gn and Gp interface). As stated earlier, the GTP-C does not operate between RNC and SGSN. For this, the RANAP protocol is used. Due to presence of two different signaling protocols, a direct GTP tunnel does not exist between RNC and GGSN. Instead, there are two sets of tunnels.

6.13.4.2 *GTP Layer Functions*

The GTP layer performs various functions, of which the important ones are summarized as follows (for detailed description of GTP functions, the reader is referred to 3GPP TS 29.060):

- **Tunnel Control and Management:** This function is performed by GTP-C. Tunnel control and management allows the creation, modification and deletion of tunnels. A tunnel is identified by a Tunnel Endpoint Identifier (TEID), an IP address, and a UDP port number. The GTP tunnels are necessary for exchanging information between external packet data networks and the MS.

- **Location Management:** This function is performed by GTP-C when network-requested PDP context activation procedure is used and the GGSN does not have a SS7 interface (i.e. Gc interface). Note that in case the GGSN has a Gc interface, it uses the MAP messages to directly communicate with HLR for the network-requested PDP context activation procedure. In case the Gc interface is absent, the GTP is used to carry messages from GGSN to a GTP-MAP converter which then uses MAP to communicate with HLR (see Section 6.9.2.4).

- **Mobility Management:** The mobility management procedures are performed by GTP-C. These procedures are used between SGSNs during GPRS Attach and Inter-SGSN routing update procedure. The mobility management procedures include identification request procedure, SGSN request procedure and forward relocation procedure.

- **Path Management:** This refers to exchanging echo packets between GTP peers to find out if the peer GSN/RNC is alive. The echo packets are not sent more than one every 60 seconds.

- **Data Transfer:** This function is performed by GTP-U protocol. GTP-U is essentially an encapsulating mechanism whereby user data packets can be exchanged between GSNs and between RNC and SGSN.

The first two functions fall under 'session management' procedures, which are explained in Chapter 11. The mobility management function is explained in Chapter 9.

6.13.4.3 GTP Messages

The GTP protocol is used to exchange signaling messages between GSNs. Table 6.14 lists important GTP messages. The 'type' of messages and their classification of messages in one of the types is as per the GTP specification. The first set of messages is used for tunnel management (i.e. creation, modification and deletion of GTP tunnels). The second set, categorized under location management, is used for network-initiated PDP context activation. These messages are used by the GTP-MAP protocol converter to communicate with the HLR. The last category is used for mobility management (i.e. for keeping tracking of the MS location).

6.13.5 Base Station Sub-system Application Part + (BSSAP+)

The Core Network is divided into two domains: the CS domain and the PS domain. Certain operations for these two domains (e.g. GPRS attach and IMSI attach) are quite similar. Thus, if somehow the functions of the two entities could be combined, there could be vital savings in terms of radio resources. In order to do this, the optional Gs interface is defined between SGSN and MSC/VLR. The Gs interface allows coordination of location information for the mobile stations that are attached to both the CS and PS domain. The Gs interface also allows CS domain information to be carried via SGSN.

For this, the Base Station Sub-system Application Part + (BSSAP+) protocol is defined for the Gs interface. BSSAP+ defines the Layer 3 messages between SGSN and VLR to allow coordination between the two databases and to relay CS domain information to be carried via SGSN.

BSSAP+ is defined in 3GPP TS 29.018. Another specification, 3GPP TS 29.016, provides the lower layer requirements for the Gs interface.

6.13.5.1 BSSAP+ Layer Architecture

BSSAP+ operates between MSC/VLR and SGSN. Figure 6.14 depicts the context in which BSSAP+ operates. BSSAP+ uses the services of SCCP, which in turn uses the SS7 transport protocols (MTP3/2/1).

Table 6.14	Important GTP Messages***		

Type	*Message*	*Between*	*Description*
Tunnel Management	Create-PDP-Context Req/Resp	SGSN-GGSN	Used by SGSN to create a tunnel between SGSN and GGSN.
	Update-PDP-Context Req/Resp	SGSN-GGSN	Used by SGSN to update various parameters (e.g. QoS) of a PDP context.
	Delete-PDP-Context Req/Resp	SGSN-GGSN	Used by SGSN or GGSN to de-activate one or a set of PDP contexts associated with a PDP address assigned to a single MS.
	PDU Notification Req/Resp	SGSN-GGSN	Used by GGSN as part of the Network-Requested PDP Context Activation Procedure requesting SGSN to request the MS to activate the indicated PDP context.
Location Management	Send Routing Info for GPRS Req/Resp	GGSN-GSN (Note 1)	Used by GGSN to obtain the IP address of the SGSN where the MS is located in case no PDP context is established.
	Failure Report Req/Resp	GGSN-GSN (Note 1)	Used by GGSN requesting HLR to set the 'Mobile Not Reachable for GPRS' (MNRG) flag for IMSI.
	Note MS GPRS Present Req/Resp	GSN-GGSN (Note 1)	Used by the GSN notifying GGSN that an MS should be reachable for GPRS again.
Mobility Management	Identification Req/Resp	SGSN-SGSN	Used by an SGSN to obtain the IMSI from old SGSN.
	SGSN Context Req/Resp	SGSN-SGSN	Used by an SGSN to obtain the Mobile Management (MM) and PDP context from old SGSN.
	SGSN Context Acknowledge	SGSN-SGSN	Used by an SGSN in response to SGSN Context Response message. The receipt of this message implies that the old SGSN can forward the user data packets to new SGSN.
	Forward Relocation Req/Resp	SGSN-SGSN	Sent by old SGSN to the new SGSN to convey necessary information to perform SRNS relocation procedure between new SSGN and Target RNC.
	Forward Relocation Complete/Ack	SGSN-SGSN	Sent by new SGSN to the old SGSN to notify that the SRNS relocation procedure has been successfully completed.

Note 1: The GSN here refers to the GTP-MAP protocol converter. These messages are used in case the Gs interface is not supported.

6.13.5.2 BSSAP+ Layer Functions

The BSSAP+ layer performs various functions, of which the important ones are summarized as follows (for detailed description of BSSAP+ functions, the reader is referred to 3GPP TS 29.018.):

- **Paging for CS domain:** VLR uses the SGSN for paging for the CS domain via the BSSAP+ protocol.

- **Location Update:** SGSN carries out the normal location update or IMSI attach via the BSSAP+ protocol.

- **Alert:** BSSAP+ provides the means for VLR to detect any activity by MS in the PS domain.

- **IMSI Attach:** BSSAP+ is used by SGSN to indicate the VLR of an explicit or implicit IMSI attach for CS domain services.

- **Information exchange:** SGSN acts as a relay to transfer information from VLR to MS and vice versa.

6.13.5.3 BSSAP+ Messages

BSSAP+ protocol is used to exchange signaling messages between MSC/VLR and SGSN. Table 6.15 lists important BSSAP+ messages.

Table 6.15   Important BSSAP+ Messages

Message	Description
BSSAP+ PAGING Req Rej	Used by VLR to perform paging via SGSN. SGSN may indicate a failure in the reject message.
BSSAP+ DOWNLINK_TUNNEL Req	Used by SGSN to convey tunneling payload received from MS.
BSSAP+ UPLINK_TUNNEL Req	Used by VLR to convey tunneling payload to MS (via SGSN).
BSSAP+ LOC_UPDATE Req/Acc/Rej	Used by SGSN to request update of location or perform IMSI attach. VLR sends a positive/negative response using Ack/Rej message.
BSSAP+ TMSI_REALLOC_COMPLETE	Used by SGSN to indicate to VLR that TMSI re-allocation procedure is complete.
BSSAP+ ALERT Req/Ack/Rej	Used by VLR to request an indication when next activity from MS is detected. SGSN sends a positive/negative response using Ack/Rej message.
BSSAP+ MS_ACTIVITY Ind	Used by SGSN to indicate to VLR that an activity for an MS has been detected.

Contd.

| Table 6.15 | Important BSSAP+ Messages |

Message	Description
BSSAP+ GPRS_DETACH Ind/Ack	Used by SGSN to indicate to VLR that a GPRS detach has been performed by MS or SGSN. VLR acknowledges the indication.
BSSAP+ IMSI DETACH Ind/Ack	Used by SGSN to indicate to VLR that an IMSI detach has been performed by MS. VLR acknowledges the indication.
BSSAP+ MS INFO Req/Ack	Used by VLR to request information associated with the indicated IMSI. SGSN responds to the request.
BSSAP+ MM INFO Req	Used by VLR to provide the MS (via SGSN) with subscriber specific information.
BSSAP+ MS UNREACHABLE	Used by SGSN to indicate to VLR that MS is unreachable.

SUMMARY

This chapter provided an overview of the significant Core Network entities (like AuC/HLR, MSC/VLR, GMSC, SGSN and GGSN) and Core Network protocols (like MAP and GTP). Part 3 of the book is built upon the topics introduced in this chapter explaining the various Core Network procedures like Mobility Management (Chapter 9), Call Handing (chapter 10), Session Management, (Chapter 11), Supplementary Service Handing (Chapter 12), Value Added Services (Chapter 13) and Security (Chapter 14).

The discerning readers may note that the UMTS Core Network has evolved form the GSM and GPRS thereby providing support for the CS and PS domains. Release 5 introduces another subsystem in the Core Network called the IP Multimedia sub-system (IMS). The IMS used the services of a PS domain (or any packet-bearer like Wireless LAN for instance) to provide IP based multi-media services. In Release 5, along with the introduction of IMS, the HLR too has evolved to become. Home Subscriber Server (HSS). Chapter 16 provides the details of HSS and IMS.

Part 3

Procedures in UMTS Network

In Part 2, the UMTS network architecture and the associated protocols were discussed. This part elaborates upon the basics of the UMTS network and explains the various procedures involved. The gamut of topics covered includes RRC procedures, UTRAN procedures, Mobility Management, Call Handling, Session Management, Supplementary Service, and other 3G Value-Added Service procedures as well as security-related procedures.

To begin with, Chapter 7 explains the RRC procedures defined for the Uu interface between the UE and the RNS. The Um procedures between UE and BSS are not covered. The important RRC procedures include RRC Connection Management procedures, Radio Bearer Control procedures, RRC Connection Mobility procedures, and Measurement procedures. Each of these categories is further divided into various sub-procedures. The important sub-procedures are also covered in Chapter 7.

Moving from Air Interface to the UTRAN and towards the Core Network, the Iub, Iur and Iu procedures are explained in Chapter 8. These procedures are based on NBAP, RNSAP and RANAP protocols, respectively. The important procedures covered in this chapter include UTRAN Global Signaling procedures and UTRAN Signaling procedures for a specific UE. Again, each of these categories is further divided into various sub-procedures. The important sub-procedures are also covered in Chapter 8.

Moving on from UTRAN to the Core Network — the most important procedure relates to Mobility Management, explained in Chapter 9. The procedures for both the CS and PS domain are covered. The important procedures discussed in the chapter include IMSI Attach/Detach, GPRS Attach/Detach, Location Update, Routing Area Update and Combined Location Area-Routing Area Update.

Chapter 10 discusses the Call Handling procedures, which apply to the CS domain. The coverage includes procedures for Mobile-Originated (MO) and Mobile-Terminated (MT) calls, and also the relationship of Call Barring and Call Forwarding supplementary services with MO and MT procedures. Then, the Optimal Routing (OR) procedures are explained. These help in optimizing routes to a given destination under various situations.

In the PS domain, the notion of call (or circuit-switched call) does not exist. In place of this, the concept of 'session' is used. A session can be viewed as a logical association between different PS domain entities (i.e. MS, SGSN and GGSN). The formal term used for a session is PDP Context. Chapter 11 covers procedures related to PDP Context activation, modification and deletion.

In Chapter 12, the Supplementary Service (SS) procedures are discussed. The chapter introduces the various supplementary services defined for a UMTS network. Procedures related to Call Independent SS Management and Call Related SS Management are also discussed.

Chapter 13 covers the procedures for some of the value-added UMTS services, including a discussion on the Short Message Service, Multimedia Messaging Service and Location Services, among others.

The concluding chapter of this part, Chapter 14, details the procedures related to security management in UMTS networks. The important procedures included here are Access Security procedures that provide mutual authentication between UE and Home Environment (Authentication Center), and other security procedures related to Integrity Protection and Confidentiality. The latter take place between the UE and the UTRAN. For providing Network Domain Security, the MAPSec and IPSec protocols are used. These are also explained in Chapter 14.

❑❑

RADIO RESOURCE CONTROL PROCEDURES

7.1   INTRODUCTION

The Radio Resource Control (RRC) is the most important protocol on the radio interface between the UE and the UTRAN. It controls all the other protocol layers at the UE and the UTRAN, including RLC, MAC, PDCP and BMC protocols. RRC is used to setup, modify and release resources at Layer 1 and Layer 2 of the radio interface protocol stack. A major part of the control signaling messages between the UE and the UTRAN are carried as RRC messages. The Higher layer NAS signaling messages are also carried as part of the RRC message payload using the RRC 'Direct Transfer' messages.

   Chapter 5 provided an overview of the RRC protocol, detailing its architecture, functions and messages. This chapter elaborates upon the various aspects of the RRC protocol; in particular, the RRC protocol state model and the procedures. The important procedures covered herein include RRC Connection Management, Radio Bearer Control, RRC Connection Mobility, and RRC Measurement.

7.2   RRC PROTOCOL STATES

In order to provide various services to its users, the RRC protocol maintains a state machine. As per this machine, the RRC is in one of the two modes; the *Idle Mode* and the *Connected Mode*. The UE is considered to be in the 'Idle Mode' when there is no radio connection between the UE and the UTRAN. In the 'Idle Mode', the UE is not involved in any active session (voice call, data transfer, etc.) and the existence of the UE is not known to the UTRAN. The 'Idle Mode' tasks of the UE include cell-selection/reselection, regular monitoring of the radio environment, and reception of broadcast system information and paging messages. These procedures are also discussed in Chapter 9.

In the idle mode, when a radio connection is established between the UE and the UTRAN, there is a transition to the 'Connected Mode'. In this mode, the existence of the UE is known to the UTRAN and the latter tracks the location of the former. An RRC connection is said to exist between the UE and the UTRAN in the connected mode, and the UE can participate in a call/session.

Figure 7.1 depicts the RRC protocol states in the connected mode, and the transitions between the states in the connected mode and the idle mode. In the connected mode, the UE can be in any of the four protocol states as shown in the figure. These states are as follows:

- **The CELL_DCH State:** This is a state where a dedicated transport/physical channel is allocated for the UE in both the uplink and the downlink direction. The Dedicated Control Channel (DCCH) and, if configured, the Dedicated Traffic Channel (DTCH) (both logical channels) are available to the UE in this state. The UE transitions from the idle mode to the CELL_DCH state of the connected

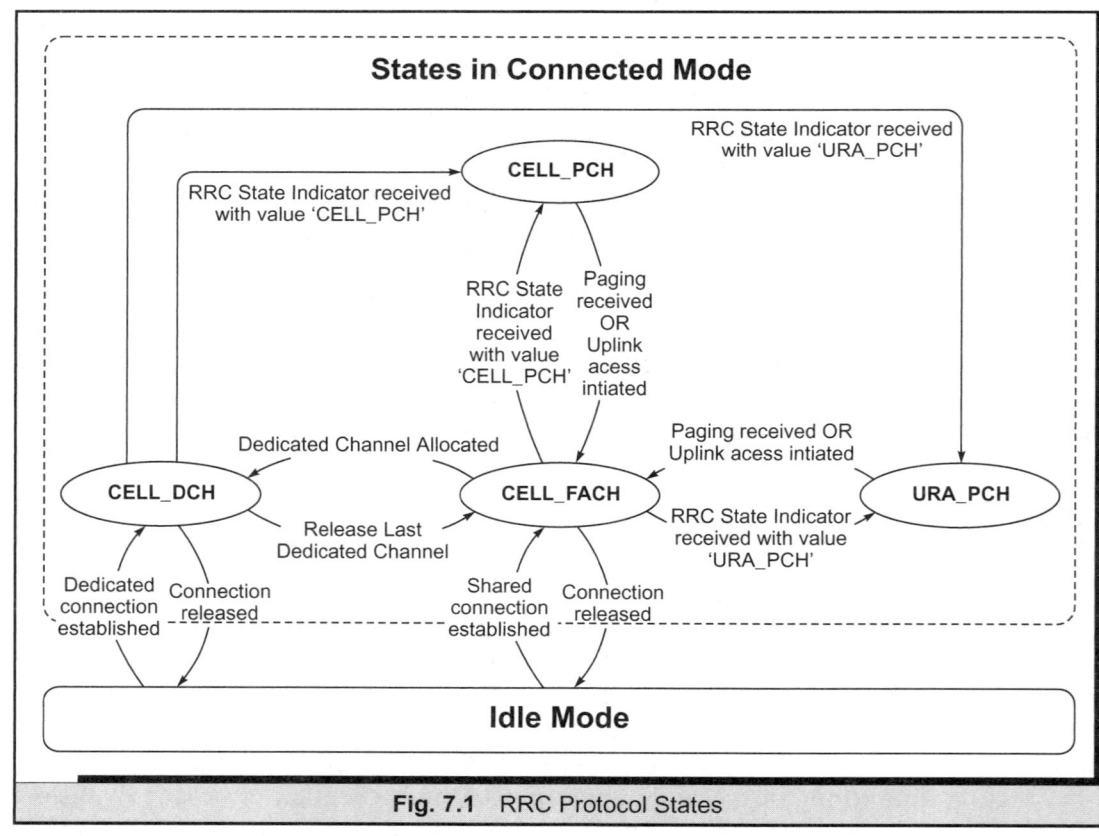

Fig. 7.1   RRC Protocol States

mode on establishment of a dedicated connection between the UE and the UTRAN. Reverse transition takes place when this connection is released.

- **CELL_FACH State:** In this state, no dedicated connection (dedicated transport/ physical channel) exists for the UE, but data can still be transferred. Data transmission is done via the common transport channels. In the uplink direction, either the RACH or the CPCH is used for this purpose. In the downlink direction, either the FACH or the DSCH (only for TDD mode) is used. The DCCH and, if configured, the DTCH are available to the UE in this state. However, both these logical channels are transported over common transport channels in the CELL_FACH state. The UE transitions from the idle mode to the CELL_FACH state of the connected mode on establishment of a shared logical connection between the UE and the UTRAN. Here too, reverse transition takes place when this logical connection is released.

  A UE in the CELL_FACH state can transition to the CELL_DCH state if the network determines that the amount of data being transferred is large enough to justify the establishment of a dedicated transport channel. Reverse transition is also possible, when the data transmission falls below a given threshold. In this case, the network may decide to withdraw the dedicated connection from the UE, and transmission of data may subsequently take place over the shared channel.

- **CELL_PCH State:** The UE can transition to this state from the CELL_DCH and CELL_FACH state, if there is no data transmission activity for a certain period in time. This transition is triggered by the UTRAN sending an RRC state indicator to the UE, signaling it to enter the CELL_PCH state in which neither the DCCH nor the DTCH are available to the UE. In the CELL_PCH state, the functions of the UE are similar to those in the idle mode (which include cell reselection, reception of system information via broadcast, monitoring of the paging channel, etc.). However, the UE is supposed to carry out periodic cell updates to allow the UTRAN to keep track of the cell where the UE is located.

  No uplink activity is possible in the CELL_PCH state. Transition from this state to the CELL_PCH state is possible on receipt of a paging message, or when some uplink activity is required. A couple of points need to be noted here. Firstly, that the transition from the CELL_PCH state to the CELL_DCH state is not allowed as per the RRC state transition model. The UE can move from the CELL_PCH state to the CELL_FACH state, and from the CELL_FACH state to the CELL_DCH state. However, direct transition from the CELL_PCH to CELL_DCH state is not allowed. Secondly, that there is no transition from the CELL_PCH state to the idle mode. This is because the connection cannot be released from the CELL_PCH state. To enter the idle mode, the UE is required to temporarily enter the

CELL_FACH state, release the established radio connection and then enter the idle mode.

- **URA_PCH State:** The state transitions to and from this state are similar to those for the CELL_PCH state. Like wise, the functionality of the UE in the URA_PCH state, and the discussion regarding transition from this state to the CELL_DCH state and the idle mode is similar to the discussion for the CELL_PCH state. The only difference between the CELL_PCH and the URA_PCH state stems from the frequency with which the UE needs to update the UTRAN with its current location. While in the CELL_PCH state, each cell update triggers an update message to the UTRAN, in the URA_PCH state, this update is triggered only on change of a UTRAN Registration Area (URA).

A trade-off exists between entering the URA_PCH state and the CELL_PCH state. Both these states are entered when there is no data transfer activity, and when the UE does not need to monitor the traffic channels. Entering the URA_PCH/CELL_PCH state helps the UE conserve its (battery) power, as in this state, it does not need to continuously monitor the traffic channels. However, using the URA_PCH state in opposition to the CELL_PCH state significantly reduces the signaling activity due to location updates. While in the CELL_PCH state, the UE needs to inform the UTRAN of its location on each cell change, in the URA_PCH state, the same is true for every URA change. On the other hand, there is also a drawback in entering the URA_PCH state, as compared to the CELL_PCH state. The drawback is when the UTRAN has some data to be sent to the UE in the URA_PCH state. In this state, the UTRAN would first have to page the UE within the URA, prior to data transfer, since the location of the UE is not accurately known at the cell level. The process of paging is discussed in Section 7.3.2.

7.3 RRC CONNECTION MANAGEMENT PROCEDURES

The RRC connection management procedures form a core part of the RRC protocol and are involved, directly or indirectly, in the transition of the UE from the idle mode to the connected mode. Some of these procedures are as follows:

- Broadcast of System Information
- Paging
- UE Dedicated Paging
- RRC Connection Establishment
- RRC Connection Release
- Signaling Connection Release Procedure
- Transmission of UE Capability Information

- Direct Transfer of NAS Messages
- Security Functions

7.3.1 Broadcast of System Information

The broadcast of system information is the means by which, information about the system as well as the serving cell is periodically broadcast to all UEs within a certain cell. Since this information is common to all UEs within the cell, it is possible to transmit it in a broadcast fashion, using 'System Information Blocks'. A system information block groups together system information of similar nature. Each block can have its own characteristics, e.g. the rate at which the information is repeated, whether the information has to be re-read by the UE, etc. Typically, the frequency of broadcast of system information depends upon how important this information is. Similarly, information that is updated more frequently requires the UE to re-read it, unlike the information that is static and can be read once.

Figure 7.2 depicts the organization of the system information. This information is maintained in the form of a tree structure, with the master information block at the root of the tree. The master information block contains scheduling information and references to the 18 different system information blocks, numbered from 1 to 18.

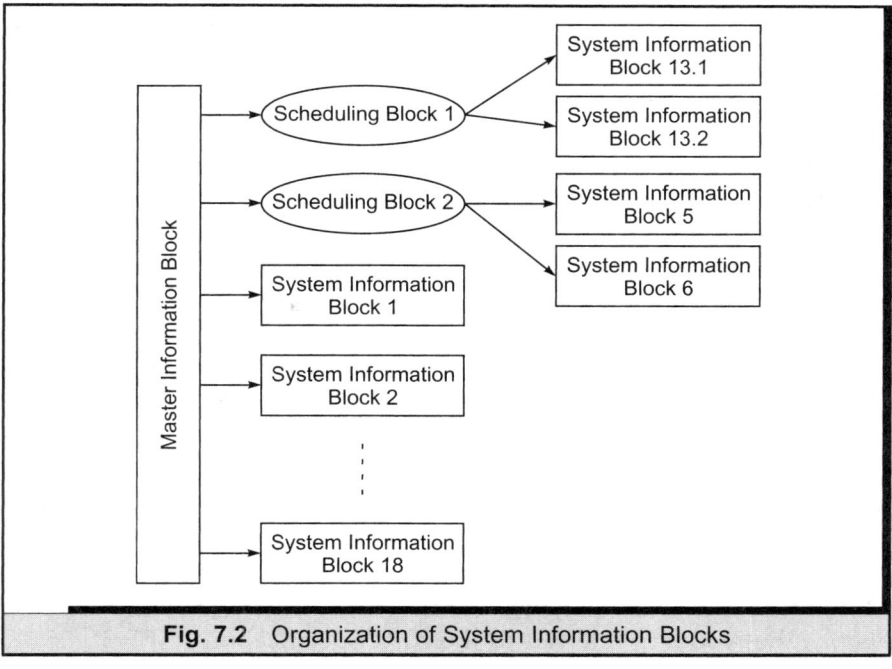

Fig. 7.2 Organization of System Information Blocks

Besides the system information blocks, the master information block can optionally contain reference to, at the most two scheduling blocks. The scheduling blocks in turn contain references and scheduling information for additional system information blocks that were not directly referred from the master information block. Thus, the scheduling information for a system information block may be obtained from either the master information block, or from one of the scheduling blocks.

The tree structure for the system information has its own advantages. As mentioned earlier, each system information block can be repeated with different rates. Hence, in any system information broadcast, it is possible to have either the complete tree, or a portion of the tree. The rate at which the information is broadcast can also be changed, based on the current load on the system. Thus, a mobile station interested only in a particular block of the system information would be required to listen to each broadcast in order to determine whether the information it requires is available in the broadcast. Maintaining the information in the form of a tree simplifies the job of the mobile station. Since the scheduling information is available in the master information block (and in some cases, the scheduling blocks), the mobile stations need to only read and decode the master information block. Once this is done, they can make out what information will be available in the broadcast, and then read what is relevant. This is particularly helpful for the mobile stations, since they do not have to read the entire information in each broadcast. They can wake up to read the master information block, and then, only the relevant information that follows it in a broadcast. However, once the mobile station has read the information, it does not even need to read and decode the master information block, unless some system information has changed since its last readings.

In such a case, the network can set the update flag for the particular system information block to indicate that the information in that block has changed. This update flag is contained in a higher block, the one that carries the scheduling information for the system information block. Thus, in most cases, the master information block contains the tag indicating the system information that has changed, and that is required to be re-read. However, this would mean that the mobile station is required to continuously read and decode the master information block. To avoid such a situation, the update flag for the Master Information block is itself sent as a tag in the paging channel. Since the mobile stations have to constantly monitor the paging channel anyway, this does not place any additional load on them.

Table 7.1 depicts the system information that is carried in the master information block and in each system information block. The table is provided for reference, and though the usage of some of the information mentioned therein is explained under the following RRC procedures, no attempt has been made to provide a comprehensive usage of the information. This has been done to avoid going into minute details. For further details, the reader is referred to the RRC Specification 3GPP TS 25.331.

Table 7.1   Organization of System Information

Information Block	*Information contained in the Block*
Master Information Block	– PLMN Identity – References to other System Information and Scheduling Blocks
System Information Block 1	– NAS System Information – Core Network Domain System Information – UE-specific Timers and Constants in 'idle mode' – UE-specific Timers and Constants in 'connected mode'
System Information Block 2	– URA Identity
System Information Block 3	– Cell Identity – Parameters for cell selection and reselection – Cell Access Restriction
System Information Block 4	– Similar to SIB 3, but used in connected mode only
System Information Block 5	– Parameters for configuration of common physical channels in the cell.
System Information Block 6	– Parameters for configuration of common and shared physical channels. To be used in con nected mode only.
System Information Block 7	– UL Interference and Dynamic Persistence Level
System Information Block 8	– Static information for Common Packet Channel (CPCH). Used for FDD mode only.
System Information Block 9	– Dynamic information for Common Packet Channel (CPCH). Used for FDD mode only.
System Information Block 10	– Information used by UEs that have DCH controlled by a 'Dynamic Resource Allocation Control' procedure. Used for FDD mode only.
System Information Block 11	– Measurement Control Information to be used in the cell
System Information Block 12	– Similar to SIB 11, but used only in connected mode
System Information Block 13	– ANSI-41 system information. Used only when the Core Network is ANSI-41.
System Information Block 14	– Parameters for common and dedicated physical channel uplink 'outer-loop power control'. Used only in TDD mode.
System Information Block 15	– Information for UE-based or UE-assisted positioning methods for location services (LCS)
System Information Block 16	– Parameters for radio bearer, transport channel and physical channel to be used by UE at the time of handover to UTRAN.
System Information Block 17	– Fast changing information for the configuration of the shared physical channels. Used only in 'connected mode' and for TDD mode only.
System Information Block 18	– PLMN Identities of neighboring cells

7.3.2 Paging

Paging is the procedure used to transmit paging information to selected UEs using the paging control channel. This information is carried by the Paging Type 1 and Paging

Type 2 messages. The use of the Paging Type 2 message is explained in Section 7.3.3. This section discusses the Paging Type 1 message, which is used when the UE is in either the 'idle mode', or in the CELL_PCH or URA_PCH state of the 'connected mode'. One paging message may contain multiple paging records, one for each UE. The paging procedure can be used for three different purposes:

- **To establish a signaling connection with the Core Network:** In this case, the paging procedure is used by the UTRAN to indicate to the mobile station that an incoming call is waiting. Since the radio connection in the UTRAN is always initiated by the UE, the paging mechanism is used to inform the UE that a connection establishment should be attempted. The paging originator in this case is the Core Network. Paging Type 1 message to establish a signaling connection is used in the 'idle mode' of the UE.

- **To initiate a Cell Update Procedure:** In this case, the paging procedure is used by the UTRAN to indicate that it has some downlink data to be sent to the UE. Here, the UE should be in the CELL_PCH or the URA_PCH state of the connected mode. On receipt of the paging message, the UE initiates a cell update procedure with cause 'Paging Response'. The paging originator in this case is the UTRAN.

- **To initiate reading of the Master Information Block:** In this case, the paging message does not contain any paging records, but only an indication that the master information block has changed (refer to Section 7.3.1). This information can be sent to the UE in either the idle mode or the CELL_PCH or the URA_PCH state of the connected mode.

7.3.3 UE Dedicated Paging

UE dedicated paging is used when the UE is in either the CELL_DCH or the CELL_FACH state of the connected mode. This means that a signaling connection is already in place, and the dedicated channel (DCCH), instead of the paging channel, can be used for the paging of the UE. Hence the name UE dedicated paging. The UE dedicated paging is used to establish a new signaling connection in cases where the originating CN is other than the current serving CN. The Paging Type 2 message is used in this case.

7.3.4 RRC Connection Establishment

In the UMTS, the concept of a RRC connection (or radio connection) has been de-linked from the concept of a radio bearer. An RRC connection in the UMTS is a static concept, which is established once and exists until it is released. On the other hand, the radio bearer defines the properties of the radio connection, and can be reconfigured during the lifetime of the RRC connection. Further, while only a maximum of one RRC connection per UE is allowed to exist, there can be multiple radio bearers for an RRC connection, each with its own data transfer characteristics. This section discusses the RRC connection establishment procedure. The radio bearer control procedures are covered in Section 7.4.

Figure 7.3 depicts the RRC connection establishment procedure. The RRC connection establishment is always initiated by the UE. In case of a UE-triggered connection establishment, the RRC connection establishment is initiated on request from non-access stratum layers on the UE, when a signaling connection is required. In this case, the process of establishing a RRC connection is initiated only if the UE is in the idle state, and hence no prior RRC connection exists. A network-triggered connection establishment occurs in case of a mobile-terminated call, where the UTRAN informs the UE of the incoming call through the paging mechanism, as discussed in Section 7.3.2. On receipt of this paging message, the UE initiates the normal RRC connection establishment procedure.

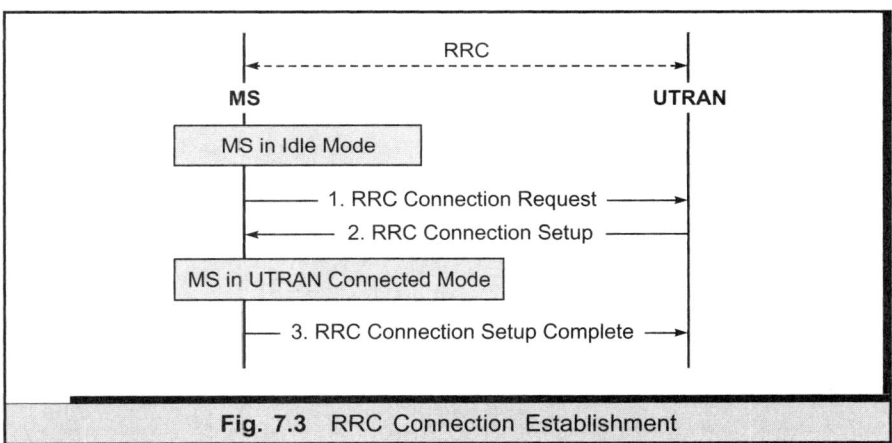

Fig. 7.3 RRC Connection Establishment

Higher (non-access stratum) layers on the UE trigger the establishment of an RRC connection when they require a signaling connection with the Core Network. A signaling connection is formed as a combination of an RRC connection in the UTRAN, and Iu connection towards the Core Network. More than one signaling connection can exist between the UE and the CN; as discussed in Section 7.3.3. However, in this case, all the signaling connections share the same RRC connection between the UE and the UTRAN.

The sequence of steps followed for the establishment of the RRC connection is as follows:

1. The RRC connection establishment is initiated by the UE by sending an RRC Connection Request message to the UTRAN. Since this message is sent on the

Random Access Channel (RACH), some form of contention resolution is required because multiple UEs may have sent the same message to the UTRAN at the same time. The contention resolution is done by including an 'Initial UE Identity' in the RRC Connection Request message. The Initial UE Identity in case of a GSM-based Core Network can be the TMSI and the LAI (TMSI + LAI), or the IMSI itself. Note, however, that the Initial UE Identity is only for the RRC connection establishment procedure, and the UTRAN can discard this information once the procedure is complete. All subsequent higher layer NAS messages include the UE identities, if required.

2. On receipt of the RRC Connection Request message, the UTRAN responds with an RRC Connection Setup message, provided the RRC connection establishment can be allowed. However, in case this is not possible, the UTRAN responds with an RRC Connection Reject message (not shown in the figure), and the procedure ends there.

   On the other hand, when the UTRAN responds with an RRC Connection Setup message, information for both the uplink and downlink transport channels is included in the message. In case the UTRAN allocates a dedicated physical channel to the UE, the latter enters the CELL_DCH state. However, if the UE is instructed to use common channels, it then enters the CELL_FACH state.

3. On successful establishment of the RRC connection, the UE responds with the RRC Connection Setup Complete message.

Both the RRC Connection Request and RRC Connection Setup messages are sent over the Common Control Channel (CCCH). However, the RRC Connection Setup Complete message is sent over the Dedicated Control Channel (DCCH), since a dedicated logical control channel has now been established for the UE. This logical channel is subsequently used for all dedicated control information exchange between the UE and the UTRAN.

The RRC connection establishment procedure also creates three, or optionally four, Signaling Radio Bearers (SRBs) over the DCCH. These SRBs are identified using Radio Bearer (RB) identities 1 to 4 (i.e. RB#1 to RB#4). They are used for RRC signaling, and later can also be reconfigured, deleted, and re-created using the radio bearer control procedures discussed in Section 7.4. The SRBs are used to carry the following RRC signaling information:

- RB#1 is used for all messages that are sent using the RLC-UM SAP.
- RB#2 is used for all messages sent using the RLC-AM SAP, with the exception of Direct Transfer messages that are sent using RB#3.
- RB#3 is for Direct Transfer messages sent using the RLC-AM SAP.
- RB#4 is optional, and if exists, is also used for Direct Transfer messages sent using the RLC-AM SAP. The reason for having two SRBs for Direct Transfer messages is to prioritize the messages. Higher priority NAS signaling messages are sent over RB#3, while those with lower priority are sent on RB#4.

Radio Bearer RB#0 is reserved for RRC messages that are sent using RLC-TM over the Common Control Channel (CCCH). These include the cell-update and URA-update messages. Note that both RB#2 and RB#3 (and RB#4, if it (exists) use the RLC-AM SAP for sending messages. The reason for using two different categories of SAPs, depending on whether the message is a 'direct transfer' message or not, is that this provides for a means of prioritize the UE-UTRAN and UE-CN messages. The message exchange between the UE and the UTRAN is given a priority over the UE-CN message exchange.

7.3.5 RRC Connection Release

The RRC connection release procedure is used to release an RRC connection, including all radio bearers (for data) and all signaling radio bearers between the UE and the UTRAN. While the RRC connection establishment procedure is always initiated by the UE, the RRC connection release procedure is always initiated by the UTRAN. As a result of the RRC connection release procedure, all signaling connections between the UE and the CN are also released, since they all share this RRC connection.

Figure 7.4 depicts the RRC connection release procedure. The following steps are taken to achieve this:

1. The RRC connection release is initiated by the UTRAN by sending a RRC Connection Release message to the UE. The downlink DCCH is used to send this message to the UE. However, if the DCCH is not available in the UTRAN and the UE is in the CELL_FACH state, then the downlink CCCH channel can be used to send this message.
2. On receipt of the RRC Connection Release message, the UE releases all its radio resources, and responds with the RRC Connection Release Complete message. It then enters the 'idle mode'.

The RRC specifications suggest that the UTRAN may transmit several RRC Connection Release messages to the UE (with the same message identification) to increase the probability of proper reception of the message. The number and periodicity of repetitions is left open as a network option. Normally, if a dedicated physical

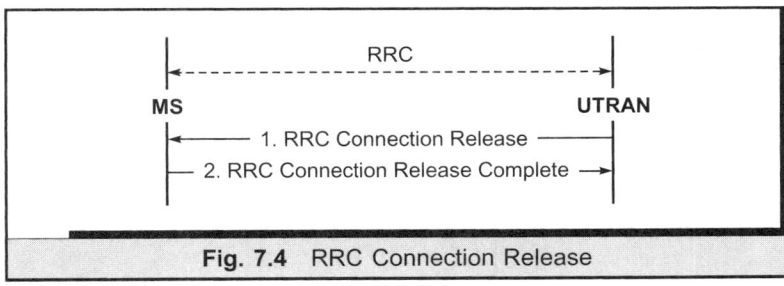

Fig. 7.4 RRC Connection Release

channel exists with the UE, the above mechanism can be employed to send multiple repetitions of the message using the unacknowledged transfer mode of RLC. However, if no dedicated physical channel exists (shared channels are used), then the message should be sent once, using the acknowledged transfer mode of RLC.

7.3.6 Signaling Connection Release Procedure

The signaling connection release message is used by the UTRAN to indicate to the UE that one of its signaling connections has been released. This procedure does not initiate the release of the RRC connection, since there may be other signaling connections sharing the same RRC connection. Figure 7.5 depicts the signaling connection release procedure.

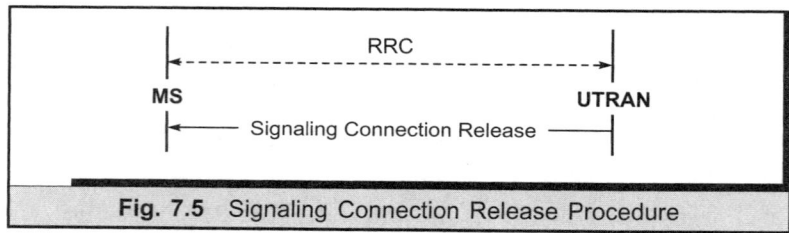

Fig. 7.5 Signaling Connection Release Procedure

Similar to the Signaling Connection Release procedure is the Signaling Connection Release Indication procedure. The latter is used by the UE to indicate to the UTRAN that one of its signaling connections has been released. This procedure may, in turn, initiate the RRC connection release procedure. The signaling connection release indication procedure is depicted in Figure 7.6.

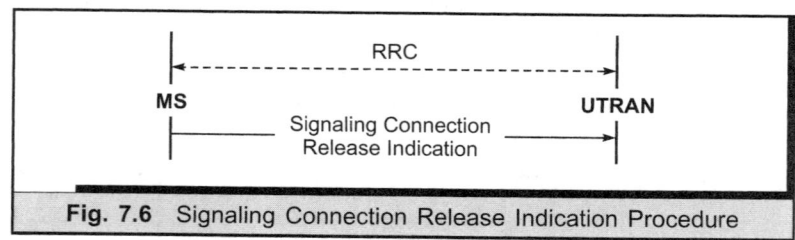

Fig. 7.6 Signaling Connection Release Indication Procedure

7.3.7 Transmission of UE Capability Information

In a 3G network, it is estimated that many different flavors of UE, with different capabilities (UE with/without multimedia support, power class of the UE, etc.) can co-exist. The UTRAN is required to know the capabilities of the UE in order to allocate resources to it. This information can be provided by the UE to the UTRAN at the time of RRC connection establishment, as a part of the RRC Connection Setup Complete message. However, the capabilities of the UE can change while in the connected mode, and these updated capabilities are required to be conveyed to the UTRAN.

Figure 7.7 depicts the procedure involved in transmission of the UE capability information to the UTRAN. The following steps are involved in the procedure:

1. In case the UE capability changes while in the connected mode, the UE sends a 'UE Capability Information' message to the UTRAN.
2. The UTRAN makes a note of the changed capabilities of the UE, and responds to this message with a 'UE Capability Information Confirm' message.

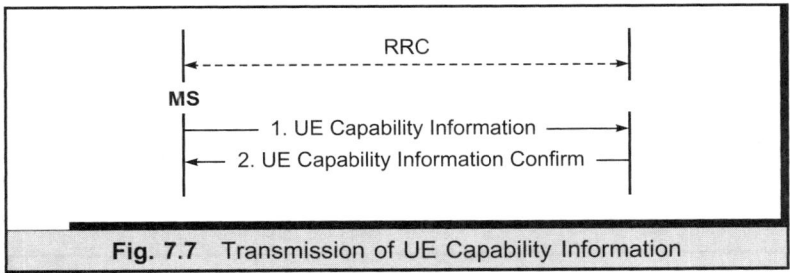

Fig. 7.7 Transmission of UE Capability Information

UE capability information can also be requested by the UTRAN from the UE. (Figure 7.8). The only difference in this case is that the procedure for the transmission of the UE capability information is triggered by the UTRAN. The UTRAN sends a 'UE Capability Enquiry' message to the UE, requesting for the UE capability information. This initiates the same steps as discussed for transfer of the UE capability information.

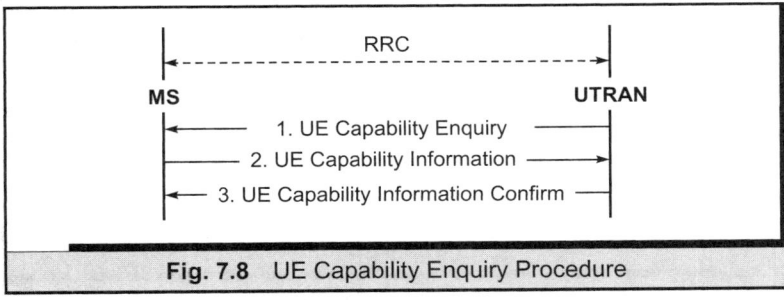

Fig. 7.8 UE Capability Enquiry Procedure

7.3.8   Direct Transfer of NAS Messages

The higher layer NAS signaling messages are sent over the air interface within the RRC Direct Transfer messages, which are of three types, as discussed in the following subsections.

7.3.8.1   Initial Direct Transfer

The Initial Direct Transfer message is used in the uplink direction when no signaling connection exists between the UE and the CN (Figure 7.9). Sending of this message from the UE to the UTRAN serves two purposes. One, it can be used to signify a request for the establishment of a signaling connection between the UE and the CN. Two, the Initial Direct Transfer message can be used to carry an initial upper layer NAS message over the radio interface, when no signaling connection exists. Consequently, the sending of the Initial Direct Transfer message is initiated when the upper layers request establishment of a signaling connection. This also includes a request for the transfer of a NAS message. The Initial Direct Transfer is used only for the first uplink NAS message, hence the name. Once a signaling connection is set up, the Downlink and Uplink Direct Transfer messages are used. The Initial Direct Transfer Message is sent on the uplink DCCH using the RLC-AM transmission mode on the signaling radio bearer RB#3.

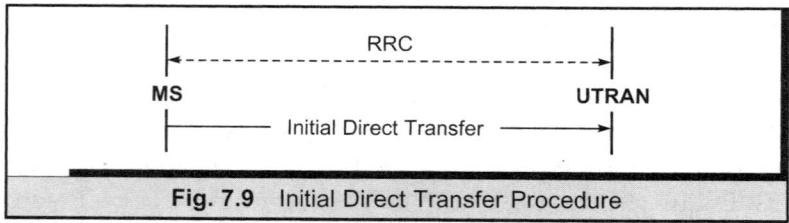

Fig. 7.9   Initial Direct Transfer Procedure

7.3.8.2   Uplink Direct Transfer

The Uplink Direct Transfer message is used in the uplink direction to carry all subsequent upper layer NAS messages (Figure 7.10). This type of Direct Transfer message is

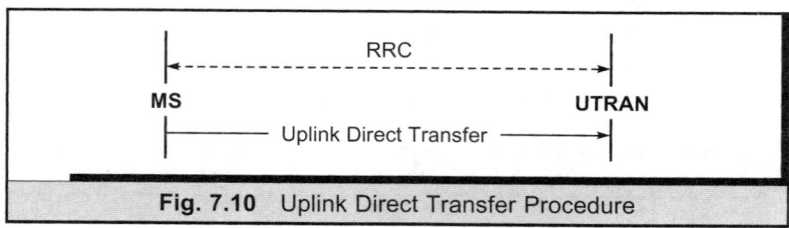

Fig. 7.10   Uplink Direct Transfer Procedure

used after the establishment of the signaling connection. The Uplink Direct Transfer message is sent on the uplink DCCH channel using the RLC-AM transmission mode. Signaling Radio Bearers RB#3 or RB#4 can be used for sending this message.

7.3.8.3 Downlink Direct Transfer

The Downlink Direct Transfer message is used in the UTRAN to UE direction to carry upper layer NAS messages over the radio interface (Figure 7.11). This message is used after the establishment of the signaling connection. It is sent on the downlink DCCH channel using the RLC-AM transmission mode. Signaling Radio Bearers RB#3 or RB#4 can be used for sending this message.

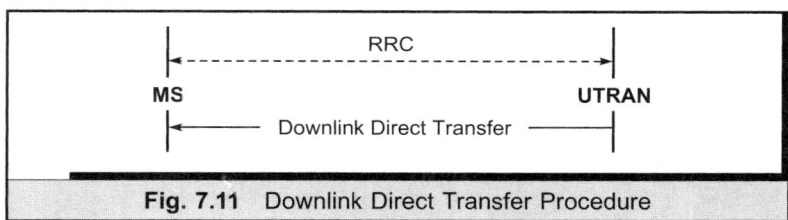

Fig. 7.11   Downlink Direct Transfer Procedure

7.3.9 Security Functions

The RRC protocol includes procedures for ensuring secure exchange of messages between the UE and the UTRAN. This includes procedures for ciphering and integrity protection. It also includes procedures to cross-check that the amount of data transmitted by the UE (or UTRAN) is the same as that received by the UTRAN (or UE). This is to ensure that no data is lost due to security lapses. This section discusses two main RRC procedures: Security Mode Control and Counter Check. These help in providing secure communication between UE and the UTRAN. For detailed procedures related to security, the reader is referred to Chapter 14.

7.3.9.1 Security Mode Control

The Security Mode Control procedure is used to trigger the start or stop of ciphering for radio bearers of a UE (for a particular CN domain), and for all signaling radio bearers. It is also used to command the restart of the ciphering mechanism with a new ciphering configuration. Besides ciphering, this procedure is used to start integrity protection for all signaling radio bearers as well as to trigger the change of integrity protection configuration during the course of the connection.

Ciphering is done by the RLC layer for services using the unacknowledged or acknowledged transfer mode of RLC. For services using the transparent mode of RLC, the ciphering is done at the MAC layer. The same ciphering algorithm is used on both the MAC and the RLC. While ciphering is done for the data bearers as well as the signaling radio bearers, integrity protection is used to guard only the signaling traffic on the air interface. Figure 7.12 depicts the security mode control procedure of the RRC. The following steps are involved in the procedure:

1. The UTRAN initiates the start/stop of ciphering, or change of the ciphering configuration by sending the Security Mode Command message to the UE. This message includes the ciphering configuration information to be used subsequently for the ciphering. The UTRAN also includes the 'activation-time' in the Security Mode Command message, which indicates the time when the UE should switch over to the new configuration.

2. The UE processes the Security Mode Command message received from the UTRAN, and applies the new configuration once the activation-time is reached. It sends a Security Mode Complete message back to the UTRAN to indicate successful handling of the message from the UTRAN.

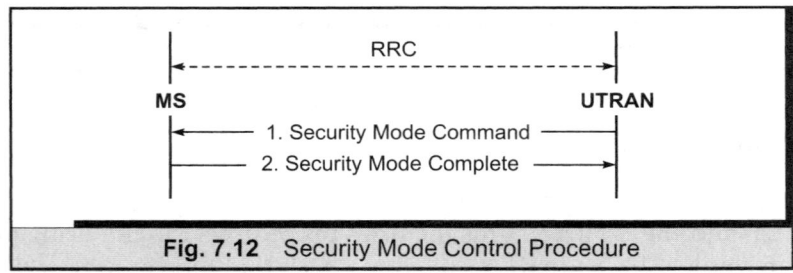

Fig. 7.12   Security Mode Control Procedure

Integrity protection is done by inserting a Message Authentication Code (MAC) into RRC PDUs. This MAC is used by the receiving RRC entity to verify the origin and integrity of the messages. Since all higher layer NAS signaling is carried in RRC Direct Transfer messages (with the exception of the Intial Direct Transfer message), they automatically become integrity protected. The trigger to start/stop the integrity protection or to change the integrity protection configuration information follows the same steps as those mentioned above for the ciphering control procedure.

7.3.9.2 *Counter Check Procedure*

The counter check procedure is used to ascertain that the amount of data transferred in both directions (uplink and downlink) over the duration of the RRC connection is identical at both the UE and the UTRAN. This ensures that there is no data loss due to a possible intruder in-between. The counter check procedure is applicable only to the radio bearers, and not to the signaling radio bearers. Thus, while integrity protection is done only in the control plane (on the signaling radio bearers), the counter check procedure is carried out only in the user plane.

Figure 7.13 depicts the counter check procedure, which involves the following steps:

1. The UTRAN maintains a count of the number of messages sent on each radio bearer configured for the UE. When any of these values reaches a critical checking value (or the threshold, defined by the UTRAN itself), the UTRAN sends a Counter Check message to the UE with the count for each radio bearer.

2. On receipt of this message, the UE compares the count for each radio bearer with the value stored within the UE for the corresponding radio bearers. It then sends a Counter Check Response message back to the UTRAN, including the number of mismatches in the response.

3. When the UTRAN receives the Counter Check Response message, it verifies the number of mismatches to figure out if anything suspicious is going on in the air interface. The UTRAN may release the RRC connection (not shown in figure) if multiple mismatches are observed in the values between the UE and the UTRAN.

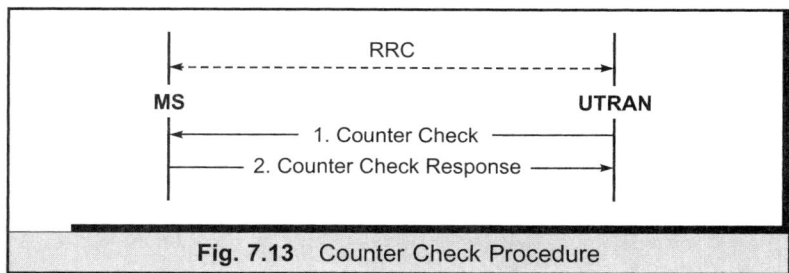

Fig. 7.13   Counter Check Procedure

7.4   RADIO BEARER CONTROL PROCEDURES

As mentioned earlier, the concept of a Radio Bearer is different from that of a Radio Connection. Multiple radio bearers with different characteristics can exist on the same radio connection. Once a radio connection is established, radio bearers can independently be established, reconfigured and released over this connection. Signaling Radio Bearers[1], that are normally setup during the RRC connection establishment, can also be established, reconfigured or released using the radio bearer control procedures. This section covers procedures specific to the establishment, reconfiguration and release of radio bearers. The transport channel and physical channel reconfiguration, covered in this section, is also carried out as part of the radio bearer procedures.

---

[1]The term 'Radio Bearer' is used to denote data bearers. For 'signaling Radio Bearers' is used.

In particular, the radio bearer control includes the following procedures:

- Radio bearer establishment
- Radio bearer reconfiguration
- Radio bearer release
- Transport channel reconfiguration
- Physical channel reconfiguration

7.4.1 Radio Bearer Establishment

The Radio Bearer Establishment procedure (Figure 7.14) is always initiated by the UTRAN. The reason being that the knowledge of network resources is available with the UTRAN. Hence, the UTRAN is the best judge of what resources can be offered to the UE for the radio bearer.

The following steps are involved in the establishment of radio bearers:

1. On receipt of a higher layer message for the establishment of a radio bearer, the RRC at the UTRAN carries out admission control, and selects Layer 2 and Layer 1 parameters for the radio bearer. It sends the Radio Bearer Setup message to the UE; this message also includes the transport channel and physical channel parameters.
2. On receipt of the Radio Bearer Setup message, the UE responds with a Radio Bearer Setup Complete message, completing the procedure for the establishment of the radio bearer.

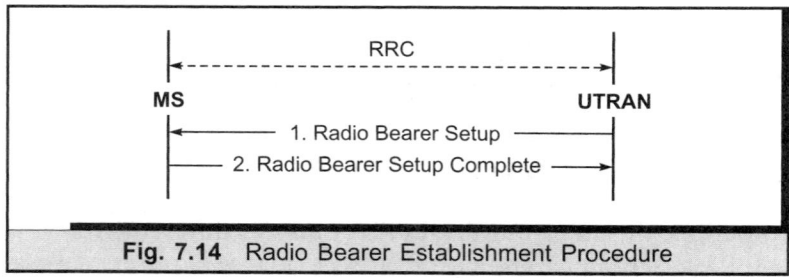

Fig. 7.14 Radio Bearer Establishment Procedure

Radio bearers can be established without a dedicated physical connection. Normally, circuit-switched and real-time services require the presence of a dedicated physical connection to provide the quality of service required. However, packet-switched services, and other non-real-time services can make use of common/shared physical channels; they do not require a dedicated physical connection. Based on these criteria, the following types of radio bearer establishment procedures are possible:

- Radio bearer setup with a dedicated physical channel allocation.

- Radio bearer setup on an existing physical channel. The physical channel is modified to accommodate the new radio bearer.
- Radio bearer setup without a dedicated physical channel allocation.

The reader is referred to the RRC specification 3GPP TS 25.331 for details on each of these procedures.

7.4.2 Radio Bearer Reconfiguration

The Radio Bearer Reconfiguration procedure is used to reconfigure the radio bearer in case the QoS parameters change, or if the measured traffic over the radio bearer increases or decreases. The UMTS provides immense flexibility in terms of the properties of a radio bearer. In UMTS, the properties of a radio bearer are allowed to change during its lifetime. This situation would occur frequently in a Third Generation mobile network, when multimedia-enabled UE participate in multi-media calls, where the traffic requirements are not fixed and may vary. The RRC radio bearer reconfiguration procedure enables the reconfiguration of the radio bearer parameters, like the quality of service, volume of traffic, etc., without the need to release and establish a new radio bearer with the required characteristics.

Figure 7.15 depicts the process of radio bearer reconfiguration. The sequence of steps involved in this reconfiguration process is as follows:

1. The Radio Bearer Reconfiguration message, initiated by the UTRAN, is sent to the UE. This message contains the new configuration parameters for the radio bearer as well as for the lower layers.
2. The UE acknowledges the receipt of this message by sending the Radio Bearer Reconfiguration Complete message to the UTRAN. This completes the reconfiguration procedure.

The radio bearer reconfiguration procedure is categorized under two types: synchronized and unsynchronized reconfiguration:

- **Synchronized reconfiguration:** In this case, the Radio Bearer Reconfiguration message from the UTRAN to the UE includes an 'activation-time' parameter,

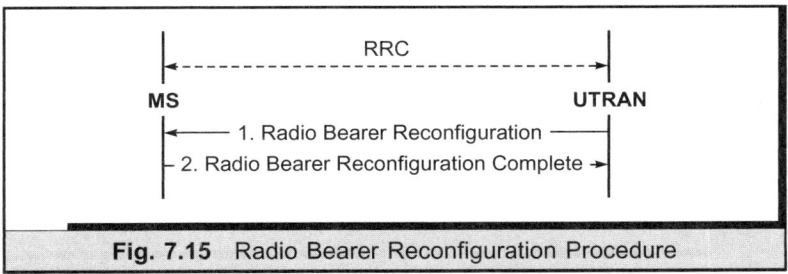

Fig. 7.15 Radio Bearer Reconfiguration Procedure

which determines the starting time for the application of the new configuration carried in the message. Thus, when the activation time is reached, both the UE and the UTRAN simultaneously switch over to the new configuration. This prevents a transient state where the UE and the UTRAN communicate with different configurations—one with the old and the other with the new configuration.

- **Unsynchronized reconfiguration:** In case of unsynchronized reconfiguration, the concept of 'activation-time' does not exist, and the configuration is changed immediately. For a transient state in-between, both the older and the newer configurations can co-exist. This means that in the transient state it is possible for the UE to be using a different configuration (the older one) from the UTRAN (which is using the newer configuration). For unsynchronized reconfiguration to be possible, both the new and the old configurations must be compatible with each other.

7.4.3 Radio Bearer Release

Radio bearer release may occur independent of the RRC connection release, which may or may not follow the process of radio bearer release. Hence, the two procedures are distinct from the UE point of view.

The radio bearer release procedure is depicted in Figure 7.16. It involves the following steps:

1. The RRC at the UTRAN sends a Radio Bearer Release message to the UE, indicating that the latter should release one or more radio bearers. Release of radio bearers may also involve modification or deactivation of a physical connection.
2. The UE responds with the Radio Bearer Release Complete message, indicating that the message from the UTRAN to the UE has been successfully handled.

Like the radio bearer reconfiguration procedure, the radio bearer release procedure too can either be synchronized or unsynchronized. Here again, unsynchronized release would mean that there does not have to be a timing synchronization at the UE and the UTRAN, and that they can independently release the radio bearer. However,

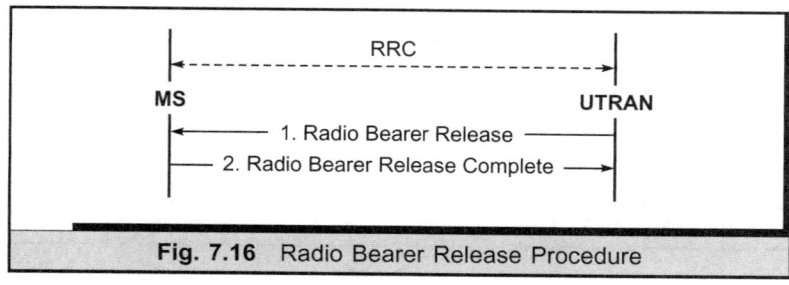

Fig. 7.16 Radio Bearer Release Procedure

in case of a synchronized release, the activation time is important, since it ensures a simultaneous release in both the UE and the UTRAN.

7.4.4 Transport Channel Reconfiguration

The properties of the transport channel define how the data is actually transmitted by the physical layer. They include the block size for transfer, Transmission Timing Interval (TTI), error protection scheme, etc. All these properties are grouped together into what is called the Transport Format. Since the properties of the transport channel can change rapidly, depending on how the higher layers send data over it, each transport channel is associated with a set of transport formats. In other words, a transport channel has a transport format set associated with it, and a particular transport format from this set, with its associated characteristics (block size, transmission rate, error protection, etc.), is used at one point in time for data transmission over the transport channel. The transport format used for this purpose can change with each TTI. This provides flexibility in the choice of the transport format, depending on the characteristics of the data received from the higher layer.

The RRC transport channel reconfiguration procedure is used to reconfigure the transport channel parameters by modifying the transport format set. This procedure is always initiated by the UTRAN, and can either be synchronized or unsynchronized, as in the case of radio bearer reconfiguration.

Figure 7.17 depicts the transport channel reconfiguration procedure, which involves the following steps:

1. The UTRAN sends a Transport Channel Reconfiguration message to the UE, with information for the modification of the transport channel parameters.
2. The UE responds with a Transport Channel Reconfiguration Complete message to indicate successful handling of the message from the UTRAN.

7.4.5 Physical Channel Reconfiguration

The Physical Channel Reconfiguration procedure is used to establish, reconfigure, and release physical channels. The decision to carry out physical channel reconfiguration is

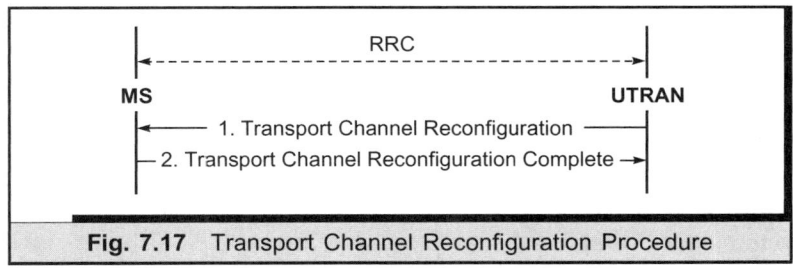

Fig. 7.17 Transport Channel Reconfiguration Procedure

made by the UTRAN, based on the state of the current network resources and/or the measurement report received from the UE.

Figure 7.18 depicts the process of reconfiguration of the physical channels. The sequence of steps involved in the procedure is as follows:

1. On the basis of the decision made at the UTRAN, the RRC layer at the UTRAN sends a Physical Channel Reconfiguration message to the UE. This message may include the activation-time as a parameter, if synchronized physical channel reconfiguration is being used.

2. On receipt of the Physical Channel Reconfiguration message, the RRC layer at the UE configures the physical layer with the modified properties. This configuration may either be done immediately, in case of an unsynchronized reconfiguration, or when the activation time has been reached, in case of synchronized reconfiguration. The configuration should, in both cases, be done on the TTI boundaries in case active communication is taking place over the channel. This is to avoid any effect on the ongoing data transmission. To acknowledge the handling of the Physical Channel Reconfiguration message from the UTRAN, the UE responds with the Physical Channel Reconfiguration Complete message. This completes the procedure.

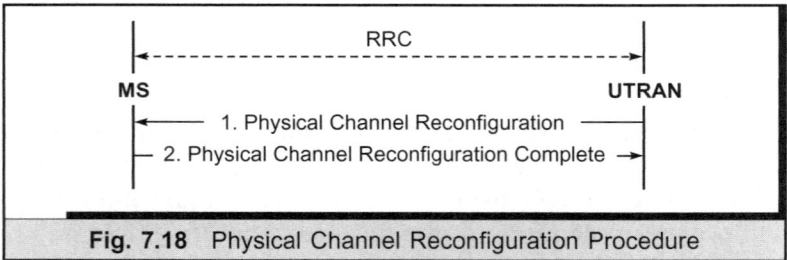

Fig. 7.18 Physical Channel Reconfiguration Procedure

7.5 RRC CONNECTION MOBILITY PROCEDURES

The RRC Connection Mobility procedures are the Mobility Management (MM) procedures of the UTRAN (refer to Chapter 9 for MM procedures). These procedures are executed for a UE in the RRC connected mode; they are not relevant to a UE in the idle mode. Some of the RRC connection mobility procedures are:

- Cell Update Procedure
- URA Update Procedure
- UTRAN Mobility Information
- Soft Handover and Active Set Update
- Hard Handover
- Inter-System Handover

7.5.1 Cell Update Procedure

The Cell Update procedure is used by the UE to inform the UTRAN of many possible conditions. It can serve several purposes, in particular, the following:

- It can act as a supervision mechanism by means of periodic cell updates. This is done in the CELL_FACH and CELL_PCH state of the RRC layer at the UE.
- It is used to update the UTRAN of the current cell that the UE is camping on after cell reselection.
- It acts on a radio link failure when in the CELL_DCH state.
- It is used to notify the UTRAN of the transition of the CELL_PCH state to the CELL_FACH state. This is done when a UTRAN initiated paging message is received, or when the higher layers request transmission of uplink data.

Figure 7.19 depicts the cell update procedure between the UE and the UTRAN. The dotted messages numbered '3' are optional. Of these optional messages, only one, or none may be sent from the UE to the UTRAN (based on the situation under which the cell update procedure is being used). The sequence of steps leading to the cell update procedure in different scenarios is as follows:

1. The cell update procedure is initiated by the UE by sending a Cell Update message to the UTRAN. The cell update could be triggered as a result of a periodic cell update, on cell re-selection, or radio link failure, or on transitioning to CELL_FACH state. The 'cause' field included in the Cell Update message indicates one of these reasons for initiating the cell update procedure.
2. The UTRAN responds to the Cell Update message received from the UE with a Cell Update Confirm message. In case none of the optional messages are sent, the procedure ends here. Else, depending on the situation, one of the optional messages is sent from the UE to the UTRAN.

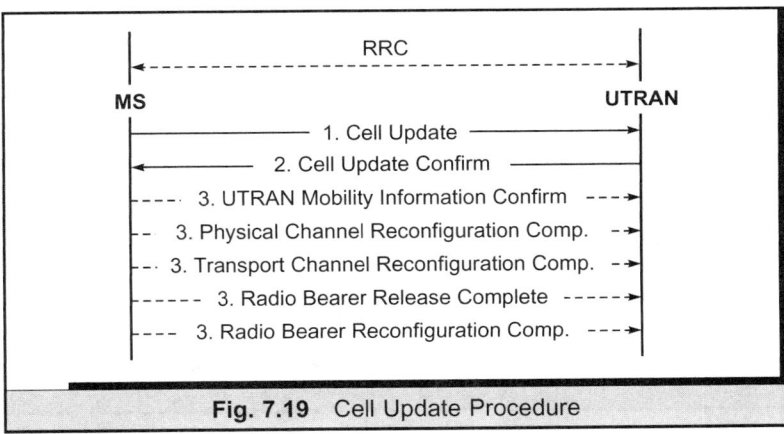

Fig. 7.19 Cell Update Procedure

3. The optional message can be either of the following:
   - In case the Cell Update Confirm message from the UTRAN to the UE includes new mobility information for the UE (new Radio Network Temporary Identifiers, U-RNTI and C-RNTI), the UE confirms the receipt and use of the new identities by sending a UTRAN Mobility Information Confirm message to the UTRAN.
   - In case the Cell Update Confirm message from the UTRAN includes a request for a physical channel reconfiguration, a transport channel reconfiguration, a radio bearer release, or a radio bearer reconfiguration, then the UE responds to these requests with a suitable message (one of the other optional messages), as indicated in the figure.

7.5.2 UTRN Registration Area (URA) Update Procedure

The UTRN Registration Area (URA) update procedure, similar to the cell update procedure, is used by the UE to convey one of the following two conditions to the UTRAN:

- It is used as a supervision mechanism, to indicate its presence by means of periodic URA updates. This is done in the CELL_FACH and URA_PCH state of the RRC layer at the UE.
- To retrieve a new URA identity after cell reselection, when the new selected cell does not belong to the URA where the UE was initially residing.

Figure 7.20 depicts the URA update procedure between the UE and the UTRAN. The sequence of steps involved in this procedure is as follows:

1. The URA update procedure is initiated by the UE by sending a URA Update message to the UTRAN. The URA update procedure could be triggered as a result of either a periodic URA update, or as a result of Cell reselection to retrieve the URA identity. The 'cause' field included in the URA Update message indicates the reason for initiating the URA update procedure.

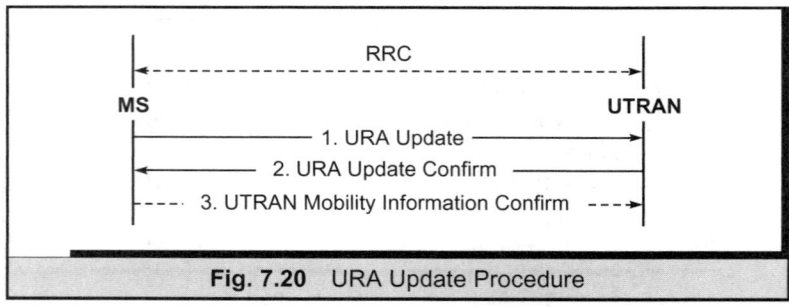

Fig. 7.20 URA Update Procedure

2. The UTRAN responds to the URA Update message received from the UE with a URA Update Confirm message.

3. This is an optional message. In case the URA Update Confirm message from the UTRAN to the UE includes new mobility information for the UE (new U-RNTI and C-RNTI), then the UE confirms the receipt and use of the new identities by sending a UTRAN Mobility Information Confirm message to the UTRAN.

7.5.3 UTRAN Mobility Information

The UTRAN Mobility Information procedure is used by the UTRAN to send new mobility information to the UE. As discussed in Sections 7.5.1 and 7.5.2, the new mobility information may also be provided to the UE as part of the Cell/URA update procedure, apart from this stand-alone mechanism. This information consists of any one or a combination of the following:

- A new C-RNTI: This is the Radio Network Temporary Identifier, allocated to the UE by the Controlling RNC.
- A new U-RNTI: This consists of the Radio Network Temporary Identifier allocated to the UE by the Source RNC (S-RNTI), plus the SRNC Identity (SRNC-Id).
- Any other mobility related information.

Figure 7.21 depicts the UTRAN mobility information exchange procedure. It consists of the UTRAN sending the information to the UE using a UTRAN Mobility Information message. The UE indicates the successful application of the new information by sending the UTRAN Mobility Information Confirm message back to the UTRAN.

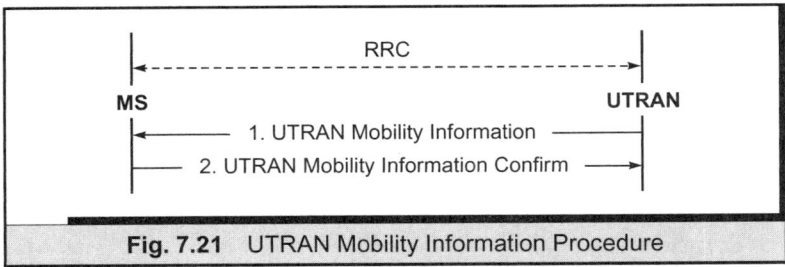

Fig. 7.21 UTRAN Mobility Information Procedure

7.5.4 Soft Handover and Active Set Update

One of the properties of a CDMA-based system is that all transceivers use the same frequency. This is quite unlike the GSM-based systems, wherein the neighbouring cells are resisted from using the same frequency to prevent interference. This CDMA property, whereby all Node Bs use the same frequency, leads to a procedure in the

UMTS called the Soft Handover (SHO) procedure. SHO is a procedure wherein a UE can maintain a radio connection with the UTRAN via two or more Node Bs simultaneously. The concept of SHO was discussed in Chapter 2.

The Node Bs to which the UE is simultaneously connected is called the UE's 'Active Set'. The RRC 'Active Set Update' procedure provides a means to update the Active Set of a UE in the CELL_DCH state.

Figure 7.22 depicts the RRC Active Set Update procedure. The sequence of steps involved in this procedure in as follows:

1. The UTRAN sends the Active Set Update message to the UE, thereby signaling it to update its active set. This message can inform the UE to perform one of the following three tasks:
   - Radio link addition
   - Radio link removal
   - Combined radio link addition and removal.

   The maximum number of radio links that a single UE is allowed is eight.

2. The UE adds and/or removes the radio links as indicated by the UTRAN in the Active Set Update message. It then sends an Active Set Update Complete message to the UTRAN indicating successful handling of the message received. In case of its failure to do so, the UE responds with an Active Set Update Failure message (not shown in figure).

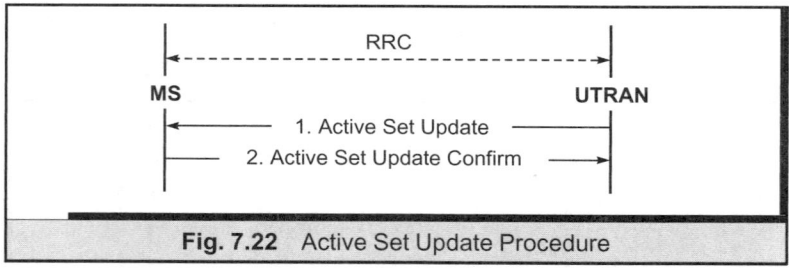

Fig. 7.22   Active Set Update Procedure

7.5.5   Hard Handover

Hard Handover (HHO) in UMTS corresponds to the normal GSM handover, which is the normal break-before-make mechanism. In this case, the original connection is broken, and a new connection, with modified parameters, is established. An audible break in voice communication is observed in this case. The HHO procedure can be used to change the radio frequency for the connection between the UE and the UTRAN. Inter-system handovers, between the UMTS and other systems can also be considered as some form of a hard handover.

Current UMTS specifications do not define any specific procedures for HHO. HHO would normally be an outcome of other procedures (defined in the UMTS) that reconfigure the air interface and that may lead to change in the radio frequency of the connection. These procedures include:

- Radio bearer establishment
- Radio bearer reconfiguration
- Radio bearer release
- Transport channel reconfiguration

7.5.6 Inter-system Handover

The inter-system handover includes the handover between the UMTS network and the other co-existing wireless networks. The most frequent handovers anticipated in the initial deployments of UMTS are expected to be between the UMTS and the GSM/GPRS networks. The UTRAN RRC specification defines procedures for inter-RAT (Radio Access Technology) handover between these wireless networks. These procedures are as follows:

- **Inter-RAT Handover to UTRAN:** This procedure is used in handover from a non-UTRAN system to the UTRAN. The network controls the process of transfer of the connection.

- **Inter-RAT Handover from UTRAN:** This effects the handover from UTRAN to another radio access technology (e.g. GSM). This procedure is used when either the UE has no established RABs, or when at least one RAB established for the UE is for the CS domain. Here again, the network controls the process of transfer of the connection.

- **Inter- RAT Cell Reselection to UTRAN:** This procedure is used when the UE, as a result of cell reselection, initiates the handover from a non-UTRAN to a UTRAN system. In this case, the UE, and to some extent the other radio access network, control the process of transfer of the connection.

- **Inter-RAT Cell Reselection from UTRAN:** Similar to the 'Inter-RAT Cell Reselection to UTRAN' procedure, except that the handover is from UTRAN to a non-UTRAN system. This process is controlled by the UE, and to some extent, by the UTRAN.

- **Inter-RAT Cell Change Order to UTRAN:** Used by a non-UTRAN based system to 'order' a UE to move to a UTRAN cell. This procedure leads to the transfer of the connection from a non-UTRAN system to UTRAN, under the control of the former.

- **Inter-RAT Cell Change Order from UTRAN:** Similar to the 'Inter RAT Cell Change Order to UTRAN' procedure, except that this results in the transfer of connection from the UTRAN to a non-UTRAN system. This procedure is controlled by the UTRAN.

Readers requiring more details on these procedures are referred to the RRC specification 3GPP TS 25.331.

7.6 MEASUREMENT PROCEDURES

In the idle mode, a UE carries out measurements that aid its cell selection/reselection process. This procedure is internal to the UE, and the UE does not report these measurements to the network. However, the network can actually request the UE to carry out some measurements, which aid the network in providing certain services. Measurements conducted in the UE could include measurement of the traffic volume that a UE is generating/receiving, which can be used by the network to decide upon the allocation or removal of resources for the UE. Similarly, services like location services require the UE to carry out certain measurements to aid in identifying the location of the UE. This section discusses the RRC procedures that are used to control the UE in terms of the measurements effected in it, and the reporting of these measurements to the network.

7.6.1 Measurement Control

The measurement control procedure is used to add, modify, or delete the measurements that the network wishes the UE to carry out. The RRC 'Measurement Control' message is used by the UTRAN to send this control information to the UE. Though the measurement control information can be provided to UEs (which are normally in Idle Mode) through the System Information Broadcast message, for UEs in the CELL_DCH state, a dedicated Measurement Control message is normally sent. The control information sent by the UTRAN to the UE in the Measurement Control message includes:

- **Measurement Identity:** This is a reference number for the measurement. It is used by the UTRAN to control, and by the UE to report the measurements.

- **Measurement Command:** This is used to add, delete, or modify a measurement.

- **Measurement Type:** This refers to the type of measurements to be carried out, from among the following:
    - Intra-frequency measurements,
    - Inter-frequency measurements,
    - Inter-system measurements,
    - Traffic volume measurements,
    - Quality measurements,

- Internal measurements,
- Measurement for location services.

- **Measurement Objects:** These are the objects that the UE would measure.

- **Measurement Reporting Criteria:** This refers to the criteria that would trigger the reporting of the measurements collected in the UE.

- **Measurement Reporting Mode:** This specifies whether the UE would use the unacknowledged or the acknowledged transfer mode of RLC for reporting the measurements collected.

On receipt of the Measurement Control message by the UE (Figure 7.23), the RRC layer configures the lower layers for the measurements. Typically, Layer 1 carries out all radio-related measurements, while the traffic volume measurement is carried out at the MAC layer. The lower layers then internally report these measurements to the RRC layer.

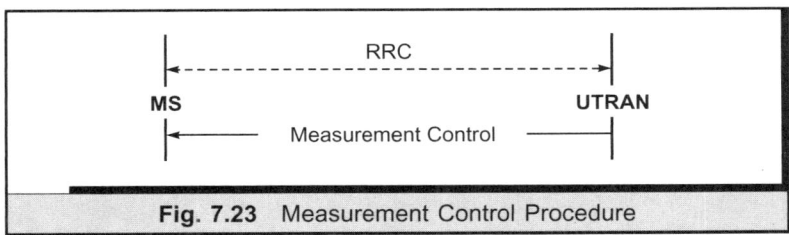

Fig. 7.23 Measurement Control Procedure

7.6.2 Measurement Report

The Measurement Report message is used by the UE to report the collected measurements to the UTRAN. This message is initiated when the measurement criteria are met. The Measurement Report message includes the measurement identity provided by the UTRAN for the measurement set, as well as the results of the measurements carried out in the UE. The Measurement Report message is sent by the UE only in the CELL_DCH or the CELL_FACH states (Figure 7.24).

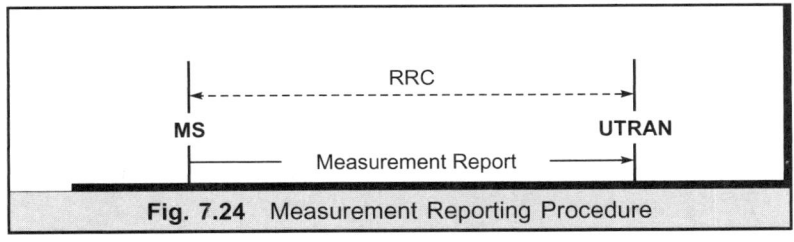

Fig. 7.24 Measurement Reporting Procedure

SUMMARY

The Radio Resource Control (RRC) is the most important protocol on the radio interface between the UE and UTRAN. It controls all other protocol layers at the UE and UTRAN, including RLC, MAC, PDCP, and BMC. RRC protocol defines a state machine, which is implemented by both UE and UTRAN. As per this state machine, the RRC can be in one of the two modes: Idle Mode and Connected Mode. In Connected Mode, RRC states includes the CELL_DCH CELL_FACH CELL_PCH and the URA_PCH states. The RRC state machine has been discussed in Section 7.2. RRC protocol defines the procedures used on the Uu interface between the UE and RNS. The important RRC procedures include RRC Connection Management Procedures, Radio Bearer Control Procedures. RRC Connection Mobility Procedures, and Measurement Procedure. Each of these categories is further divided into various sub-procedures, which have been discussed in the chapter.

UTRAN SIGNALING PROCEDURES

8.1 INTRODUCTION

While the RRC forms the main protocol on the radio interface between the UE and the UTRAN, other important protocols form the basis of all signaling that takes place within the UTRAN, and between the UTRAN and the CN. These protocols include the Radio Access Network Application Part (RANAP), Radio Network Subsystem Application Part (RNSAP), Node B Application Part (NBAP), Access Link Control Application Protocol (ALCAP) and the Service Area Broadcast Protocol (SABP). Chapter 5 provided an overview of all these protocols, including a discussion on their architecture and functions. A brief discussion on the important messages of these protocols was also provided in that chapter.

Moving on to the procedures, while the RRC procedures were covered in the last chapter, the focus of this chapter is on providing the complete picture of the signaling procedures, which involve the RANAP, RNSAP, NBAP, ALCAP and RRC signaling. The chapter carries forward from the RRC procedures, to provide an understanding of the end-to-end signaling flows: from the UE to the UTRAN to the CN, and vice-versa.

An attempt has been made to cover as many procedures and messages—discussed in Chapter 5—in the signaling examples provided through the rest of this chapter. However, the ALCAP message flow has deliberately not been shown in detail. This is to keep the focus on the messages of the UTRAN-specific protocols (RANAP, RNSAP, NBAP and RRC). Another important point needs to be noted here. While many procedures discussed in this chapter would differ depending on whether the scenario involves the Dedicated Transport Channel (DCH) or the Common Transport Channel (RACH/FACH), at most places it is assumed that it involves the dedicated transport channel. This has not been done to give any preferential treatment to the dedicated transport channel scenarios, but as a result of a conscious decision to cover as many UTRAN protocol related procedures. In most cases, as the text would suggest, the common transport channel related procedure is a subset of the dedicated transport channel procedure; hence, the decision to cover the scenarios corresponding to the dedicated transport channels.

8.2 UTRAN GLOBAL SIGNALING PROCEDURES

These consist of procedures that are not related to a specific UE. They are as follows:

- System Information Broadcasting
- Service Area Broadcast

8.2.1 System Information Broadcasting

The System Information Broadcasting procedure is used to broadcast information about the system to all UE within the broadcast area. The information that is broadcast as part of the system information broadcast was detailed in Chapter 7.

Figure 8.1 depicts the system information broadcasting procedure. The sequence of steps involved in this procedure is as follows:

1. The RNC sends the System Information Update Request message to the appropriate Node B(s) via the NBAP protocol. This message contains the master information block and system information blocks, which contain the system information to be broadcast.
2. The Node B confirms its ability to broadcast the system information by sending a System Information Update Response message to the RNC via the NBAP protocol. In case the Node B is unable to broadcast the information, it returns a System Information Update Failure message (not shown in the figure) to the RNC.
3. The Node B then broadcasts the system information over the air interface using the RRC System Information message.

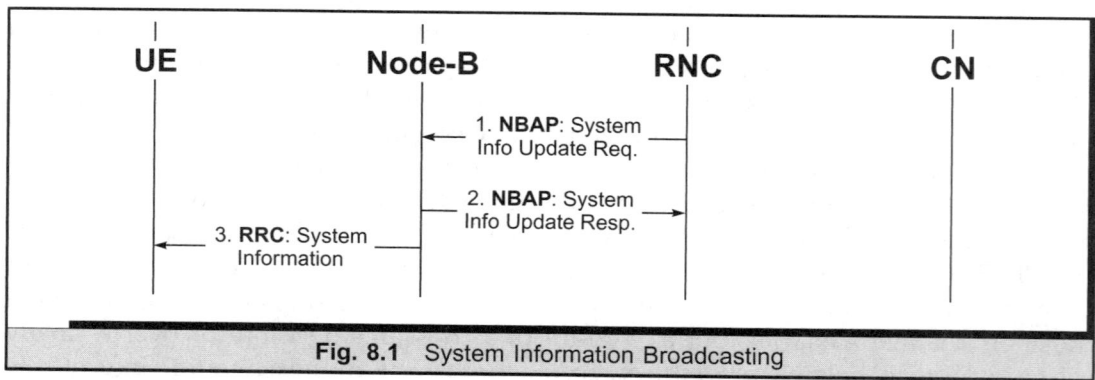

Fig. 8.1 System Information Broadcasting

8.2.2 Service Area Broadcast

The Service Area Broadcast procedure is used for the broadcast of information within a specified service area. This procedure is mainly used for the Cell Broadcast Service, discussed in Chapter 13.

Figure 8.2 depicts the steps involved in the service area broadcast procedure. These are summarized as follows:

1. The Core Network requests the RNC to broadcast information within a service area by sending a 'Write-replace' message via the SABP protocol. The Write-replace contains the broadcast message, as well as the service area list where the message is to be broadcast.

2. The RNC confirms its ability to broadcast the message by sending a 'Write-replace Complete' message to the Core Network via the SABP protocol. In case the RNC cannot broadcast the message as requested by the CN, it sends back a 'Write-replace Failure' message (not shown in the figure) to the CN.

3. The RNC then broadcasts the message over the air interface by using the 'CBS Message' provided by the BMC protocol.

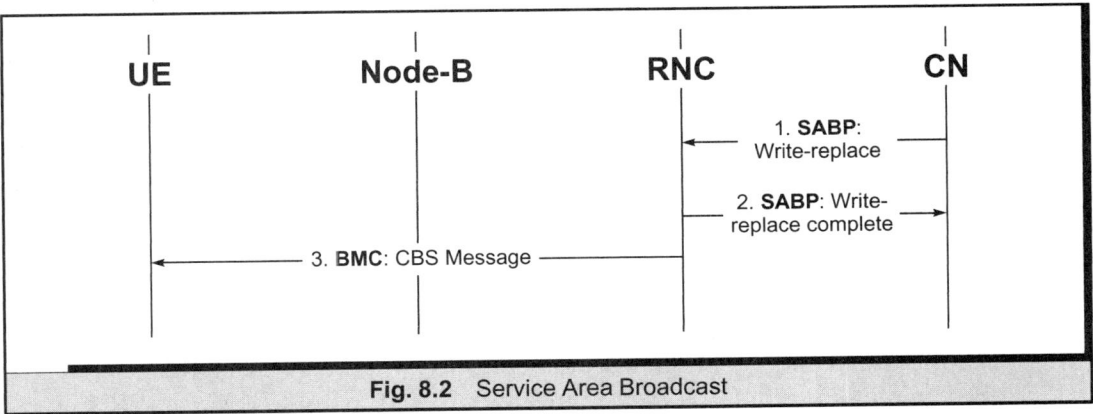

Fig. 8.2 Service Area Broadcast

8.3 UTRAN SIGNALING PROCEDURES FOR A SPECIFIC UE

Most of the UTRAN procedures covered in this chapter are specific to a particular UE. The signaling procedures specific to a UE include the following:

- Paging
- NAS Signaling Connection Establishment
- RRC Connection Establishment
- RRC Connection Release

- Radio Access Bearer Establishment
- Radio Access Bearer Release
- Physical Channel Reconfiguration
- Transport Channel Reconfiguration
- Soft Handover
- Cell Update
- URA Update
- Direct Transfer

8.3.1   Paging

The paging process is used mainly to achieve any of the following three objectives (also see Chapter 7):

- To establish a signaling connection with the UE.
- To initiate a Cell Update Procedure.
- To initiate reading of the Master Information Block by the UE.

The paging process differs depending on the state of the UE. For a UE in the RRC Idle Mode, or in the CELL_PCH and URA_PCH state of the RRC Connected Mode, the paging is carried out by broadcast of a paging message in a defined geographical area, such as a location area. This is because the exact cell position of the UE is not known. However, for a UE in the CELL_DCH or the CELL_FACH state of the RRC Connected Mode, since the exact position of the UE is known, the paging is carried out using the existing RRC Connection. Figure 8.3 depicts both these scenarios.

In case the UE is in the RRC Idle Mode, or in the CELL_PCH or URA_PCH state of the RRC Connected Mode, the sequence of steps involved in paging the UE is as follows:

1/2. The CN initiates the paging of the UE over the geographical area (LA/RA) by sending the 'Paging' message to the RNC(s) via the RANAP protocol. In this example, we assume that the geographical area consists of two RNCs: RNC1 and RNC2. Hence, in steps 1 and 2, the paging message is sent to both RNC1 and RNC2.

3/4. Both RNC1 and RNC2 broadcast the paging message to the UE via the RRC 'Paging Type 1' message, as discussed in Chapter 7.

In case the UE is in the CELL_DCH or the CELL_FACH state of the RRC Connected Mode, the dedicated UE paging procedure is used for paging the UE. This procedure uses the already established RRC connection of the UE for sending the paging message. It involves the following sequence of steps:

1'. The CN initiates the paging of the UE via the 'Paging' message of the RANAP protocol. The example assumes here that the UE is located in the area covered by RNC1, and hence, the message is sent only to RNC1, and not to RNC2.

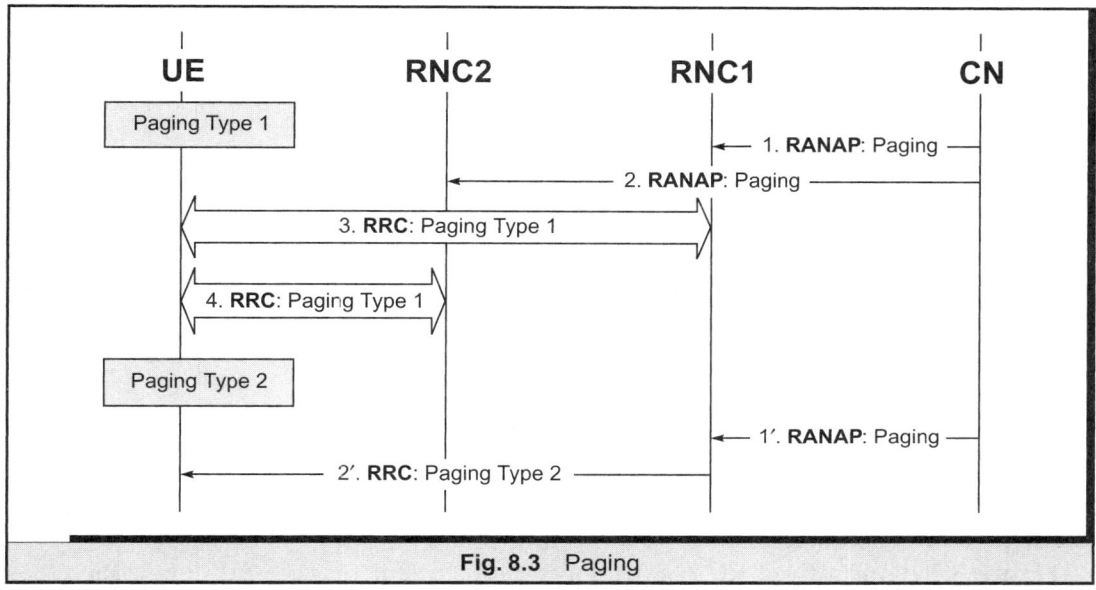

Fig. 8.3   Paging

2'. RNC1 sends the 'Paging Type 2' message to the UE via the RRC protocol. The already established RRC connection with the UE is used for sending this message.

8.3.2 NAS Signaling Connection Establishment

The NAS signaling connection establishment procedure may be initiated by the UE when it needs same service provided by the network. Alternatively, the connection establishment procedure could be triggered by a paging message received from the CN (as discussed in Section 8.3.1).

The NAS signaling connection establishment procedure is depicted in Figure 8.4. The sequence of steps followed in the procedure are:

1. The RRC connection establishment procedure is first executed to establish an RRC connection. This procedure is discussed in Section 8.3.3.
2. Next, the UE transfers the 'Initial Direct Transfer' message via the RRC procedure. This message contains the Initial NAS message for the CN (e.g. CM Service Request, Location Updating Request, etc.).
3. The serving RNC then initiates the signaling connection towards the CN, and sends the Initial UE Message to it via the RANAP protocol.

Following the establishment of the NAS signaling connection between the UE and the CN, all NAS signaling messages can be sent over this connection.

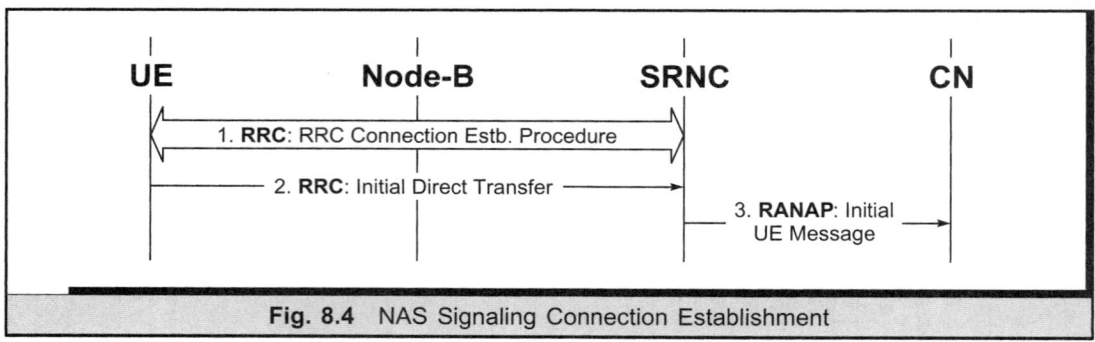

Fig. 8.4 NAS Signaling Connection Establishment

8.3.3 RRC Connection Establishment

The RRC messages exchanged for RRC Connection Establishment were discussed in Chapter 7. This section discusses the other protocols invaded in the establishment of the RRC connection state.

Figure 8.5 depicts the RRC connection establishment procedure. The sequence of steps followed is:

1. The UE initiates the establishment of the RRC connection by sending the 'RRC Connection Request' message to the serving RNC via the RRC protocol. This message includes the initial UE identity used to identify the UE, and the reason for the establishment of the connection. Since the RRC Connection Request message, and the response to this message is sent on the common control channel, the initial UE identity is used to uniquely identify the request from and response to a UE.

2. On receipt of this message from the UE, the serving RNC makes a decision on whether to use a Dedicated Transport Channel (DCH) or a Common Transport Channel (RACH/FACH) for this connection. Here, it is assumed that the SRNC decides to use the DCH for the RRC connection. The SRNC then allocates a UTRAN Radio Network Temporary Identifier (U-RNTI) for the UE, which is subsequently used to identify the UE in all further communication between the UE and the UTRAN. The SRNC reserves the necessary radio resources for this RRC connection, and sends a Radio Link Setup Request message to Node B via the NBAP protocol.

3. On receipt of the Radio Link Setup Request message from the SRNC, the Node B allocates resources for the connection and responds to the SRNC by sending a Radio Link Setup Response message via the NBAP protocol.

4. The SRNC then initiates the ALCAP procedures for establishment of an AAL2 transport bearer over the Iub interface for the DCH. The Iub data transport

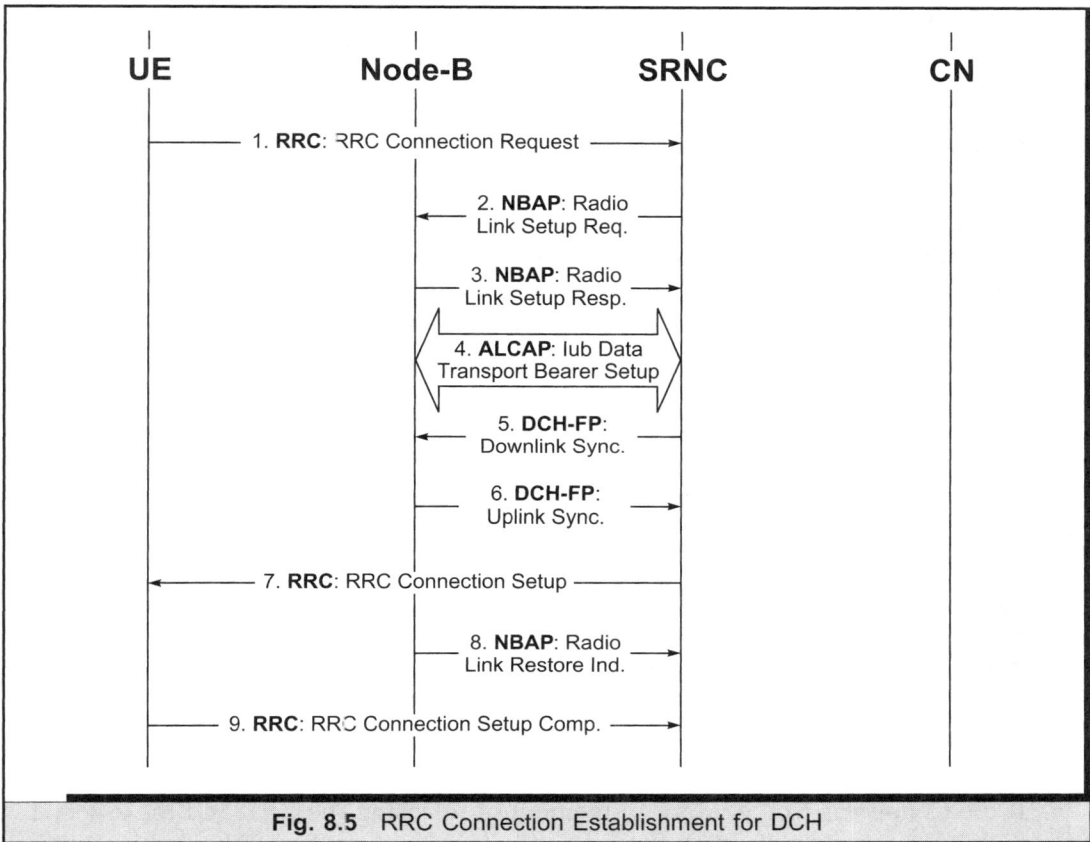

Fig. 8.5 RRC Connection Establishment for DCH

bearer thus setup is bound with the DCH used for the RRC connection. All data to be transmitted or received on this DCH would be sent or received on/from the transport bearer established in this step.

5. The DCH-Framing Protocol at the SRNC then initiates downlink synchronization of the transport bearer (with the DCH-FP at the Node B) by sending the Downlink Synchronization message.

6. The DCH-Framing Protocol at the Node B initiates uplink synchronization of the transport bearer by sending the Uplink Synchronization message. Steps 5 and 6 ensure the synchronization of the transport channel between the SRNC and the Node B. Synchronization is required for reliable transfer of information over the DCH.

7. The SRNC then responds to the UE with the 'RRC Connection Setup' message via the RRC protocol. This message includes the initial UE identity sent by the UE in

the 'RRC Connection Request' message. It also includes the U-RNTI allocated by the SRNC to be used in all subsequent communication, along with certain other connection related parameters.

8. When Node B achieves uplink synchronization with the UE over the Uu interface, it sends the 'Radio Link Restore Indication' message to the SRNC via the NBAP protocol.

9. The UE sends the 'RRC Connection Setup Complete' message to the SRNC to indicate successful handling of the 'RRC Connection Setup' message from the SRNC, and the acceptance of the U-RNTI and other connection-related parameters. This completes the procedure for RRC connection establishment.

It is quite possible that at the time of the RRC connection establishment, the required radio resources are not available. In this case, either the RRC connection establishment can be aborted, or some radio resources freed by pre-emption. Normally, if a lower priority connection exists, then the radio links associated with this connection are released to allocate resources for the new RRC connection. Pre-emption of radio links associated with a lower priority RAB for RRC connection establishment is depicted in Figure 8.6. The following steps are involved:

1-2. Same as in the case of 'RRC connection establishment without pre-emption'.

3. Node B determines that it cannot allocate resources for this RRC connection, since the required resources are not available. It sends a 'Radio Link Pre-emption Required Indication' message to the SRNC via the NBAP protocol.

4. SRNC identifies a RAB that can be released, and sends a 'RAB Release Request' message to the CN via the RANAP protocol.

5. If the CN agrees to the SRNC judgement of releasing the RAB and the associated resources, it sends a 'Iu Release Command' message to the SRNC via the RANAP protocol. This message is an indication to SRNC to release the Iu connection towards the CN for this RAB.

6. On receipt of the 'Iu Release Command' message from the CN, the SRNC initiates the release of the Iu data transport bearer using the ALCAP protocol.

7. Next, the SRNC sends the 'RRC Connection Release' message to the UE to initiate the release of the RRC connection towards the UE. The cause indicated in the release message states 'pre-emption' as the reason for this release.

8. The UE confirms the release of the RRC connection by sending the 'RRC Connection Release Complete' message to the SRNC.

9. The SRNC then initiates the release of the corresponding radio links by sending the 'Radio Link Deletion' message to Node B via the NBAP protocol.

10. The Node B confirms the release of the radio links by sending the 'Radio Link Deletion Response' message to the SRNC.

11. The Node B next initiates the release of the Iub data transport bearer towards the SRNC using the ALCAP protocol.

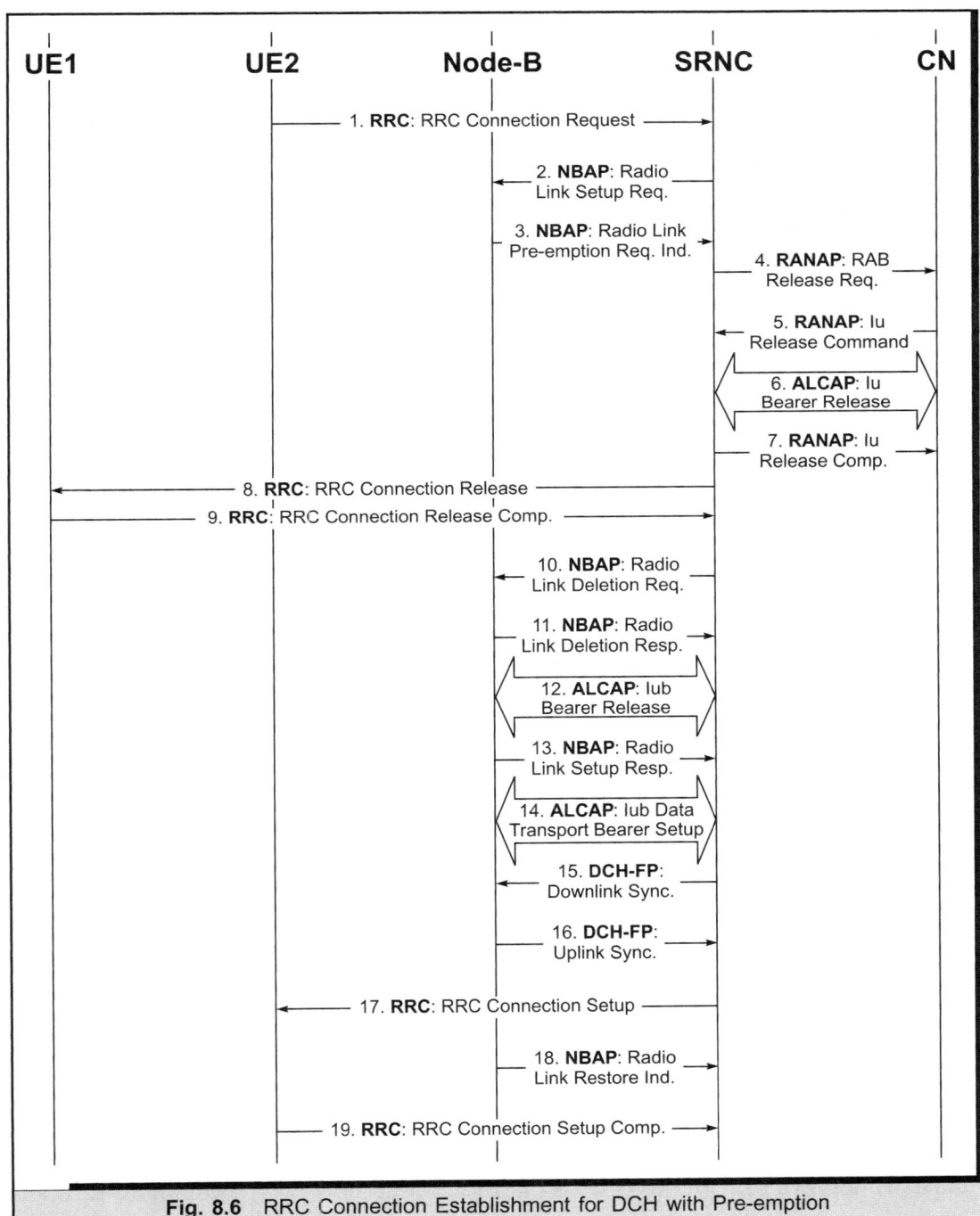

Fig. 8.6 RRC Connection Establishment for DCH with Pre-emption

12. Remaining steps are the same as steps 3 to 9 in the case of the 'RRC Connection Establishment without pre-emption'.

8.3.4 RRC Connection Release

The release of the RRC connection involves interaction between the different network elements. As a result of this procedure, all existing radio bearers for the UE are also released. Like in the case of connection establishment, the procedure for RRC connection release would differ depending on whether the scenario involves a DCH or a Common Transport Channel. The following example covers the RRC connection release procedure for a DCH. The RRC connection release procedure for a Common Transport Channel is a subset of the steps involved in the DCH case, and hence is not covered separately.

Figure 8.7 depicts the procedure for the release of the RRC connection for DCH. The sequence of steps involved in the procedure is as follows:

1. When the CN decides to release the connection, it sends an 'Iu Release Command' message to the SRNC via the RANAP protocol. This message is an indication to SRNC to release the Iu connection towards the CN and initiate the RRC connection release towards the UE. The cause for the release is indicated in the message from the CN to the SRNC.
2. The SRNC sends the 'Iu Release Complete' message to the SRNC to confirm the release of the Iu connection towards the CN.
3. Next, the SRNC initiates the release of the Iu data transport bearer using the ALCAP protocol.
4. The SRNC initiates the RRC connection release towards the UE by sending the 'RRC Connection Release' message towards the UE. This message includes the cause for the release of the RRC connection.
5. The UE confirms the release of the RRC connection by sending the 'RRC Connection Release Complete' message to the SRNC.
6. Next, the SRNC initiates the release of the radio links by sending the 'Radio Link Deletion' message to the Node B.
7. The Node B confirms the release of the radio links by sending the 'Radio Link Deletion Response' message to the SRNC.
8. The Node B then initiates the release of the Iub data transport bearer towards the SRNC using the ALCAP protocol. On completion of this step, the RRC connection release is successfully completed.

The RRC connection release for the common transport channel is simpler than it is for the DCH, since no dedicated resources exist for the UE. Steps 1 to 5 described above are followed for the release of the RRC connection for the common transport channel.

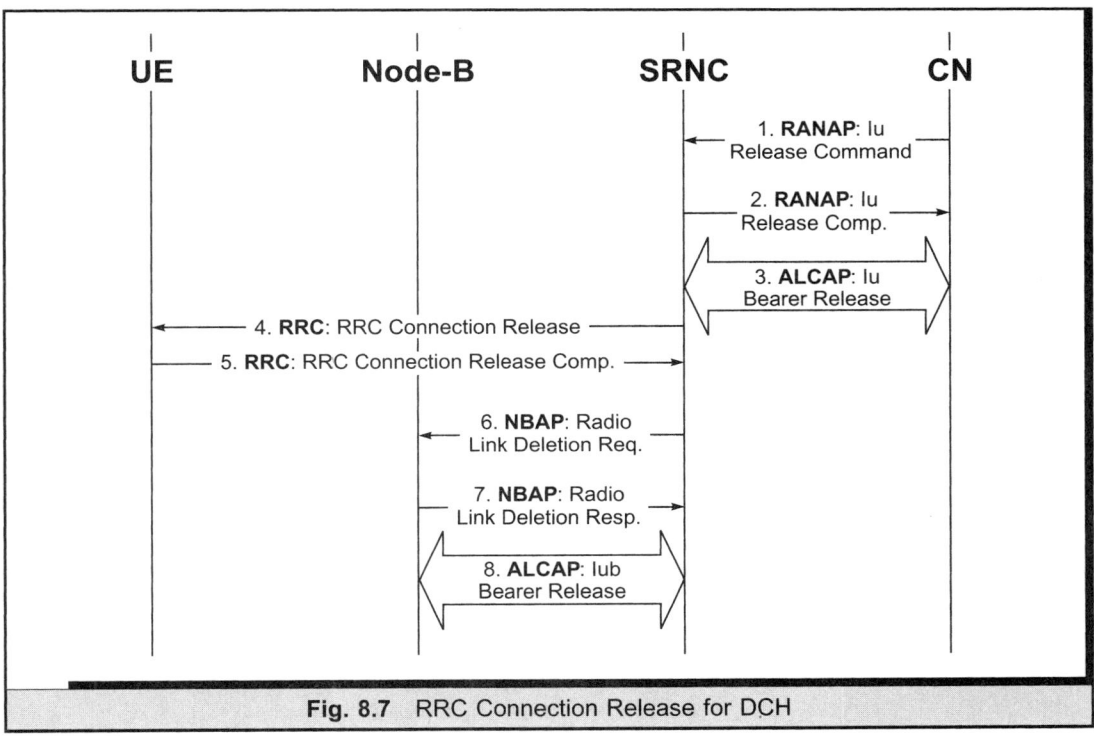

Fig. 8.7 RRC Connection Release for DCH

8.3.5 Radio Access Bearer Establishment

After the establishment of the RRC connection, radio access bearers (RABs) can be established using the radio access bearer establishment procedure. RABs can be established either on a dedicated channel (DCH) or on a common transport channel (RACH/FACH). The RAB establishment procedure can either be synchronized or unsynchronized, between the UTRAN and the UE. As discussed in Chapter 7, a synchronized procedure requires that both the UTRAN and the UE move simultaneously to the new configuration (here, the new RAB). In case of an unsynchronized procedure, a transient period can exist wherein the UE (or the UTRAN) is using the new configuration, while the UTRAN (or the UE) is still using the older configuration. This section discusses the case of RAB establishment on a DCH when in the RRC CELL_DCH state, both using the synchronized and the unsynchronized setup.

8.3.5.1 Synchronized Establishment of RAB

Figure 8.8 depicts the procedure for synchronized establishment of a RAB on a DCH, when the UE is in the CELL_DCH state. This example cites a scenario wherein the UE

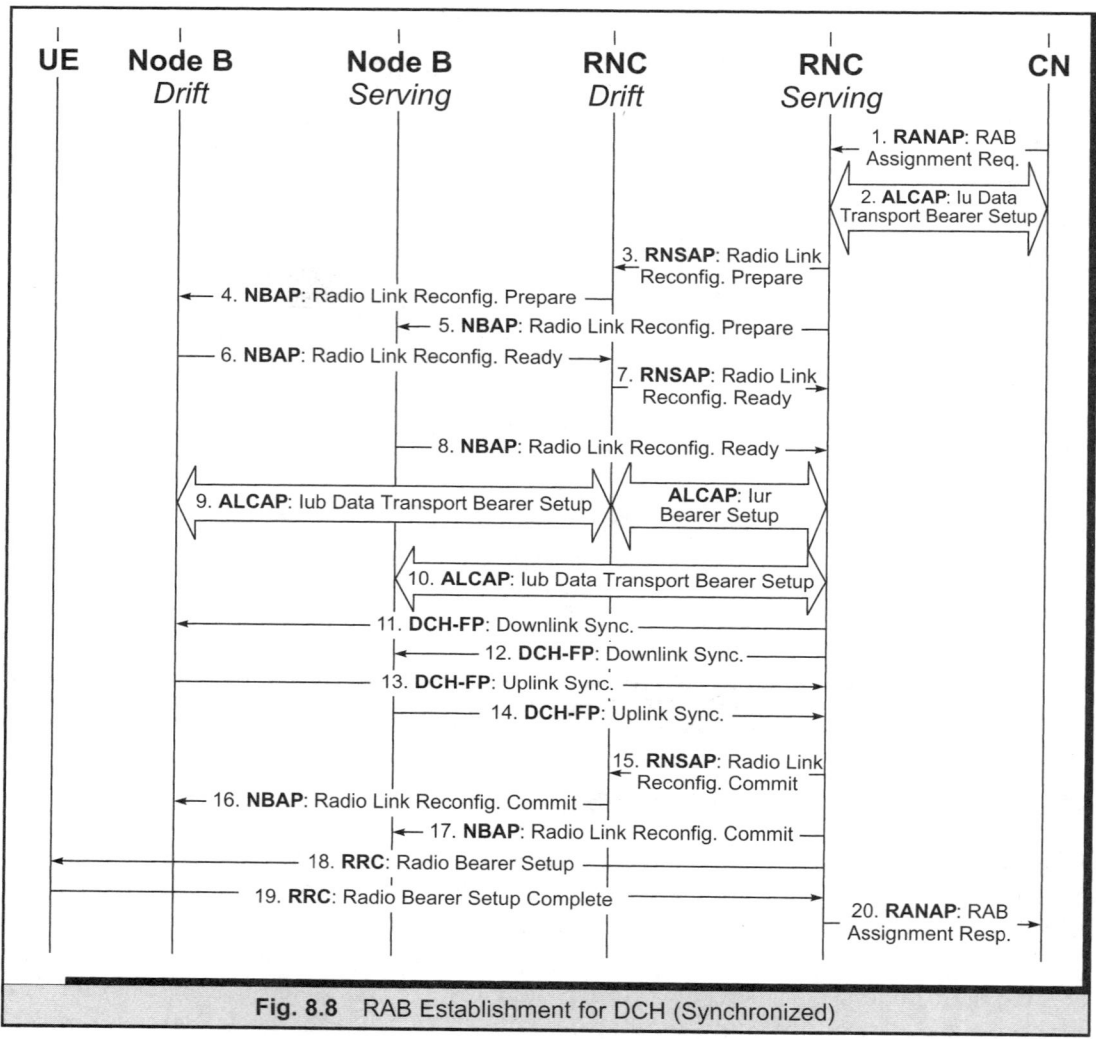

Fig. 8.8 RAB Establishment for DCH (Synchronized)

is in the region of two Node Bs, the Serving Node B and the Drift Node B, and is receiving transmission from both. This is a Soft Handover (SHO) situation. A Drift RNC controls the Drift Node B. The example assumes that the Drift Node B and the Drift RNC are distinct from the Serving Node B and Serving RNC. The reason for considering such an example is to gain insight into the RNSAP messages as well.

The sequence of steps involved in the RAB establishment procedure for the scenario cited above is as follows:

1. The RAB establishment procedure is initiated by the CN by sending the 'Radio Access Bearer Assignment Request' message via the RANAP protocol to serving

RNC. The message includes the parameters for the RAB to be established. Note that the same RANAP message is also used for the release of the connection. The contents of the message define whether the message is a request for establishment, or for release of the RAB.

2. The SRNC initiates the Iu data transport bearer setup using the ALCAP protocol.

3. Next, the SRNC sends the 'Radio Link Reconfiguration Prepare' message to the DRNC via the RNSAP protocol. This message is to request the DRNC to prepare for establishment of the DCH that would carry the RAB.

4. On receipt of this message, the DRNC forwards the 'Radio Link Reconfiguration Prepare' message to the Drift Node B via the NBAP protocol, to request the Drift Node B to prepare for establishment of the DCH.

5. Similarly, the SRNC sends the 'Radio Link Reconfiguration Prepare' message to the Serving Node B, to request it to prepare for establishment of the DCH for the RAB.

6. The Drift Node B responds to the DRNC by sending the 'Radio Link Reconfiguration Ready' message via the NBAP protocol. This indicates to the DRNC that the Drift Node B has allocated the necessary resources, and that the preparation for establishment is ready. The response from the Drift Node B includes the Transport Layer Information (AAL2 Address, etc.) that would be used to set up the Iub data transport bearer.

7. The DRNC forwards the 'Radio Link Reconfiguration Ready' message to the SRNC via the RNSAP protocol.

8. Like the Drift Node B, the Serving Node B also responds to the SRNC using the 'Radio Link Reconfiguration Ready' message via the NBAP protocol.

9. The SRNC then initiates the setup of the Iur/Iub data transport bearer (with the DRNC and the Drift Node B) using the ALCAP protocol. This transport bearer is bound to the DCH that would be carrying the RAB.

10. The SRNC also initiates the setup of the Iub data transport bearer (with the Serving Node B) using the ALCAP protocol.

11. The DCH-Framing Protocol at the SRNC then initiates the downlink synchronization of the transport bearer with the DCH-FP at the Drift Node B by sending the 'Downlink Synchronization' message.

12. Likewise, the SRNC initiates downlink synchronization of the transport bearer with the DCH-FP at the Serving Node B by sending the 'Downlink Synchronization' message.

13. The DCH-Framing Protocol at the Drift Node B initiates the uplink synchronization of the transport bearer by sending the 'Uplink Synchronization' message.

14. Similarly, the DCH-Framing Protocol at the Serving Node B initiates the uplink synchronization of the transport bearer by sending the 'Uplink Synchronization' message. After the execution of steps 11–14 the SRNC is synchronized with both the Node Bs for the transmission of data on the Iub/Iur data transport bearers.

15. The SRNC then sends the 'Radio Link Reconfiguration Commit' message to the DRNC via the RNSAP protocol to order the DRNC to switch to the new configuration of the Radio Link(s) as prepared by the 'Radio Link Reconfiguration Prepare' message.

16. The DRNC forwards the 'Radio Link Reconfiguration Commit' message to the Drift Node B via the NBAP protocol to order the Drift Node B to switch to the new configuration.

17. Similarly, the SRNC forwards the 'Radio Link Reconfiguration Commit' message to the Serving Node B to order it to switch to the new configuration.

18. Next, the SRNC sends the 'Radio Bearer Setup' message to the UE via the RRC protocol.

19. The UE responds with the 'Radio Bearer Setup Complete' message.

20. The SRNC then responds to the initial CN message initiating the RAB setup with the 'Radio Access Bearer Assignment Response' message via the RANAP protocol. This completes the procedure for the setup of the RAB.

8.3.5.2 *Unsynchronized Establishment of RAB*

Figure 8.9 depicts the procedure for unsynchronized establishment of RAB on a DCH, when the UE is in the CELL_DCH state. This example uses the same scenario as that for the synchronized RAB establishment. The sequence of steps involved in the unsynchronized RAB setup procedure is as follows:

1. The RAB establishment procedure is initiated by the CN by sending the 'Radio Access Bearer Assignment Request' message via the RANAP protocol. The message includes the parameters for the RAB to be established.

2. The SRNC initiates the Iu Data Transport Bearer Setup using the ALCAP protocol.

3. Next, the SRNC sends the 'Radio Link Reconfiguration Request' message to the DRNC via the RNSAP protocol. This message is to request the DRNC to establish a new DCH that would carry the RAB. The modification is done immediately, unlike in the case of synchronized establishment, where the DRNC waits for the 'Radio Link Reconfiguration Commit' message from the SRNC before effecting the change.

4. The DRNC sends the 'Radio Link Reconfiguration Request' message to the Drift Node B to request establishment of a new DCH.

5. Similarly, the SRNC requests the Serving Node B to set up a new DCH by sending it the 'Radio Link Reconfiguration Request' message via the NBAP protocol.

6. The Drift Node B allocates resources for this DCH and notifies the Drift RNC that the setup is complete, by sending it the 'Radio Link Reconfiguration Response' message.

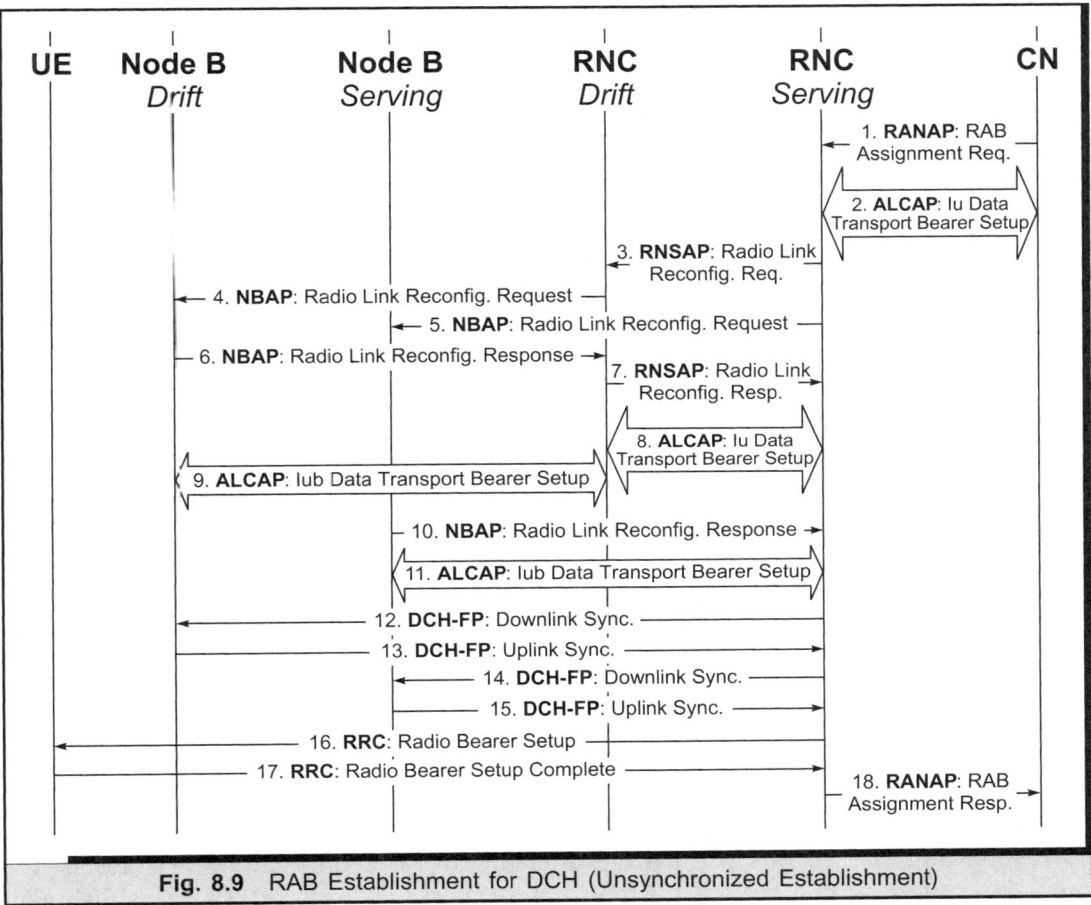

Fig. 8.9 RAB Establishment for DCH (Unsynchronized Establishment)

7. The Drift RNC then sends the 'Radio Link Reconfiguration Response' message to the SRNC to indicate successful establishment of a new DCH.

8. The SRNC initiates the setup of the Iur data transport bearer using the ALCAP protocol. This bearer is bound to the newly established DCH.

9. Next, the Iub data transport bearer is established between the Drift RNC and the Drift Node B using the ALCAP protocol. The Drift RNC performs the bridging between the Iub and the Iur data transport bearers.

10. When the Serving Node B has also allocated resources and set up the new DCH, it two informs the SRNC by sending the latter the 'Radio Link Reconfiguration Response' message.

11. The setup of the Iub data transport bearer between the Serving Node B and the SRNC is then initiated by the latter.

12. Thereafter, the DCH-Framing Protocol at the SRNC initiates the downlink synchronization of the transport bearer with the DCH-FP at the Drift Node B by sending the 'Downlink Synchronization' message.

13. The DCH-Framing Protocol at the Drift Node B initiates the uplink synchronization of the transport bearer by sending the 'Uplink Synchronization' message.

14. Similarly, the SRNC initiates the downlink synchronization of the transport bearer with the DCH-FP at the Serving Node B by sending the 'Downlink Synchronization' message to the Serving Node B.

15. The DCH-Framing Protocol at the Serving Node B initiates the uplink synchronization of the transport bearer by sending the 'Uplink Synchronization' message to the serving RNC. After the execution of steps 12 to 15 the SRNC is synchronized with both the Node Bs for the transmission of data on the Iub/Iur data transport bearers.

16. Steps 16 to 18 are similar to steps 18 to 20 in the synchronized RAB establishment case. This completes the procedure for the setup of RAB using the unsynchronized establishment mode.

8.3.6   Radio Access Bearer Release

Like the RAB establishment procedure, the RAB release procedure can also be synchronized or unsynchronized. Figures 8.10 and 8.11 depict the RAB release procedure using the synchronized and unsynchronized mechanism respectively. Both examples assume that the RAB to be released is over a DCH, and the existing RRC connection also has an associated DCH.

Note that both the procedures for RAB release use messages similar to those in the corresponding procedures for RAB establishment. The only difference lies in the message content. While the RAB establishment messages indicate RAB establishment and DCH addition as the reason for sending the messages, the RAB release messages indicate a request for RAB release and the corresponding DCH release. The message names, and the sequence of their exchange between the network entities remains largely the same. Because of the similarity between the release procedures and the establishment procedures, the detailed sequence of steps has been omitted. The reader may refer to the figures for these.

8.3.7   Physical Channel Reconfiguration

Physical channels are the lowest-layer channels that actually carry the data from one network entity to the other. While the MAC layer maps the logical channels associated with the UE RABs onto the transport channels provided by the physical layer (e.g. DTCH onto DCH), it is the responsibility of the physical layer to map the data received

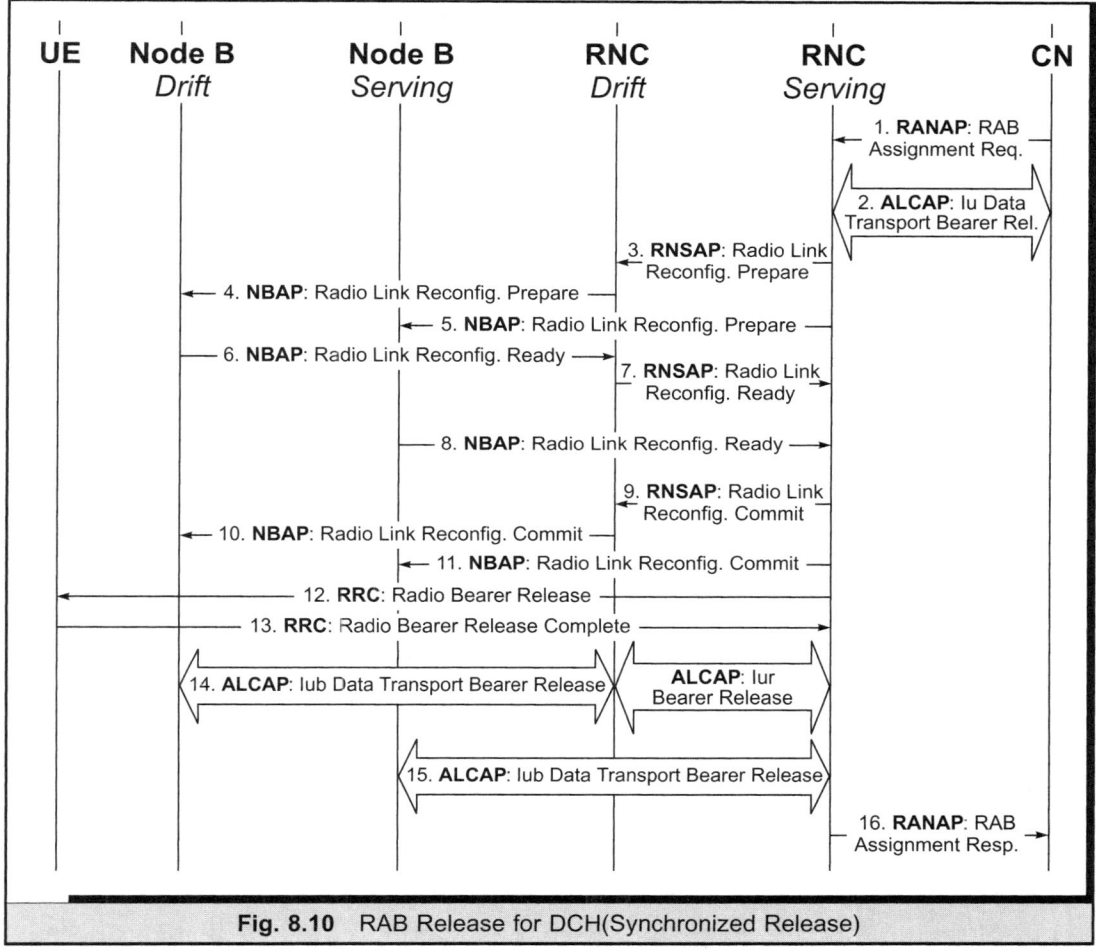

Fig. 8.10 RAB Release for DCH(Synchronized Release)

over the transport channels (DCHs) onto the physical channels. This inter-linking between the channels across layers means that the characteristics of the lower layer physical channels depend heavily on the characteristics (e.g. data transmission speed) of the transport channels configured over the physical channels. The transport channels are themselves dependent on the characteristics of the logical channels (or in some sense, the characteristics of the RABs associated with UEs). Modification in the setup of channels in the higher layers may lead to reconfiguration of the physical channels as well. The physical channel reconfiguration procedure is used to establish, reconfigure, and release the physical channels.

Figure 8.12 depicts the procedure used for reconfiguration of the physical channels.

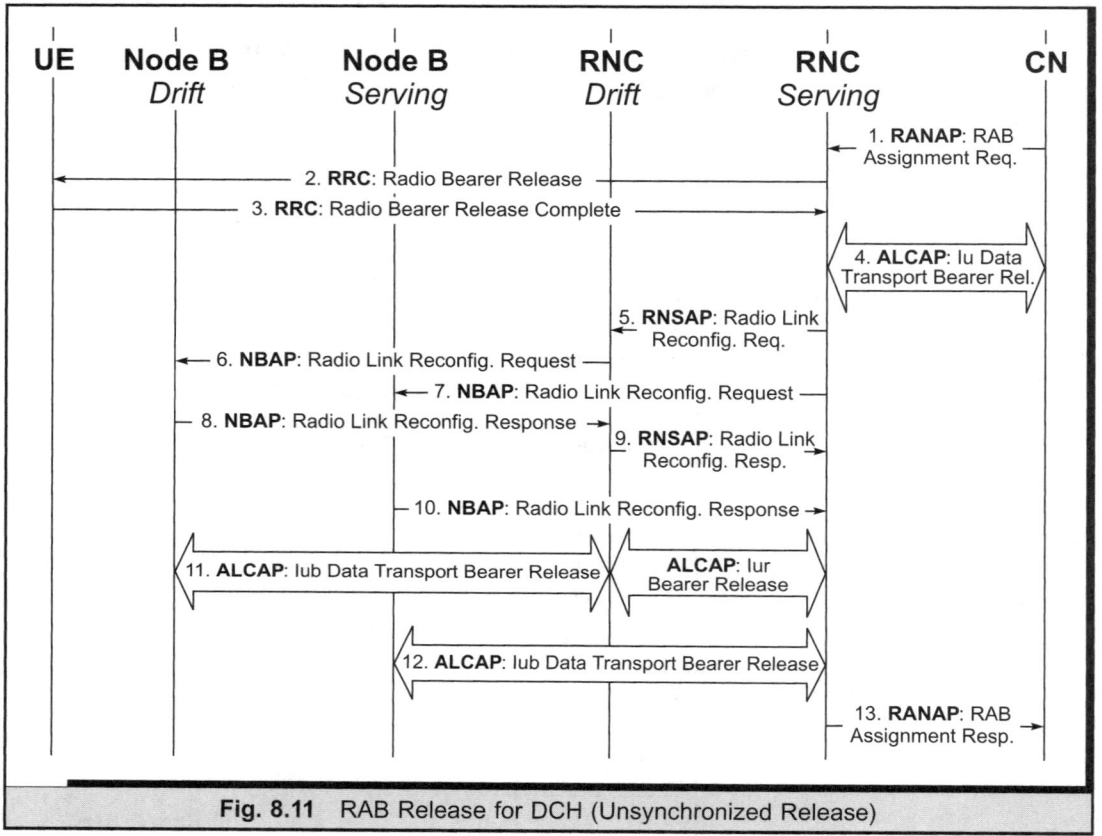

Fig. 8.11 RAB Release for DCH (Unsynchronized Release)

The sequence of steps involved is as follows:

1-9. Steps 1 to 9 use the 'Radio Link Reconfiguration Prepare', 'Radio Link Reconfiguration Ready' and the 'Radio Link Reconfiguration Commit' messages of the RNSAP and NBAP protocols to configure and switch to the new physical layer configuration. Recollect that the same set of messages, as well as the sequence of their exchange was followed for the establishment and release of RABs.

10-11. The RRC messages, 'Physical Channel Reconfiguration' and 'Physical Channel Reconfiguration Complete', are used to update the physical layer configuration at the UE, as discussed in Chapter 7.

8.3.8 Transport Channel Reconfiguration

As discussed in Section 8.3.7, change in the properties of the RAB associated with a UE may result in a requirement for reconfiguration of the physical and the transport

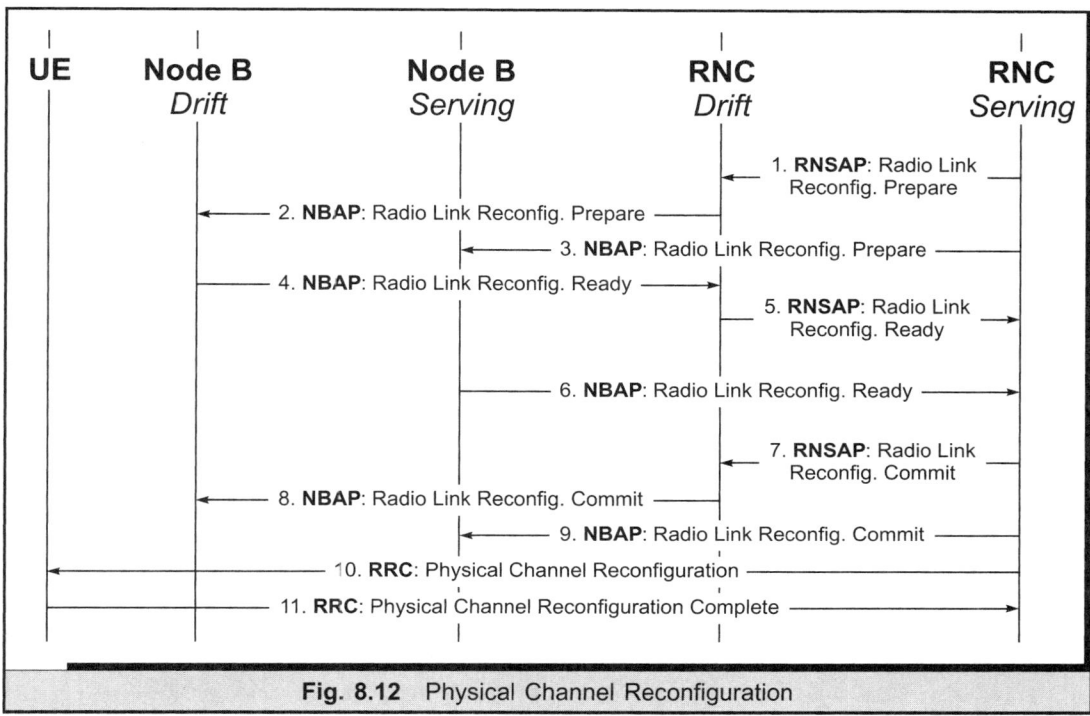

Fig. 8.12 Physical Channel Reconfiguration

channels. This section discusses the transport channel reconfiguration procedure, which like the rest of the procedures, can either be synchronized (Figure 8.13), or unsynchronized. The unsynchronized procedure is slightly different, with the three-way procedure consisting of the 'Radio Link Reconfiguration Prepare/Ready/Commit' messages beings replaced by the two-way procedure consisting of the 'Radio Link Reconfiguration Request/Response' messages.

As depicted in Figure 8.13, the transport channel reconfiguration procedure differs only slightly from the physical channel reconfiguration procedure. The difference stems from the fact that in addition to the physical channel reconfiguration related message flows, the transport channel reconfiguration procedure also requires the AAL2 transport bearer setup as well as the release procedures in order to modify the properties of the transport bearer carrying the transport channel. As shown in the figure, rather than modifying the existing AAL2 transport bearer, a new bearer with the required characteristics is first setup, mapped to the current transport channel being modified, and then the original AAL2 transport bearer is released. Alternatively, the ALCAP procedures provide an option for modification of the existing AAL2 transport bearers, in which the two steps of the setup and release of the transport bearer can be replaced by a one-step modification of the transport bearer properties.

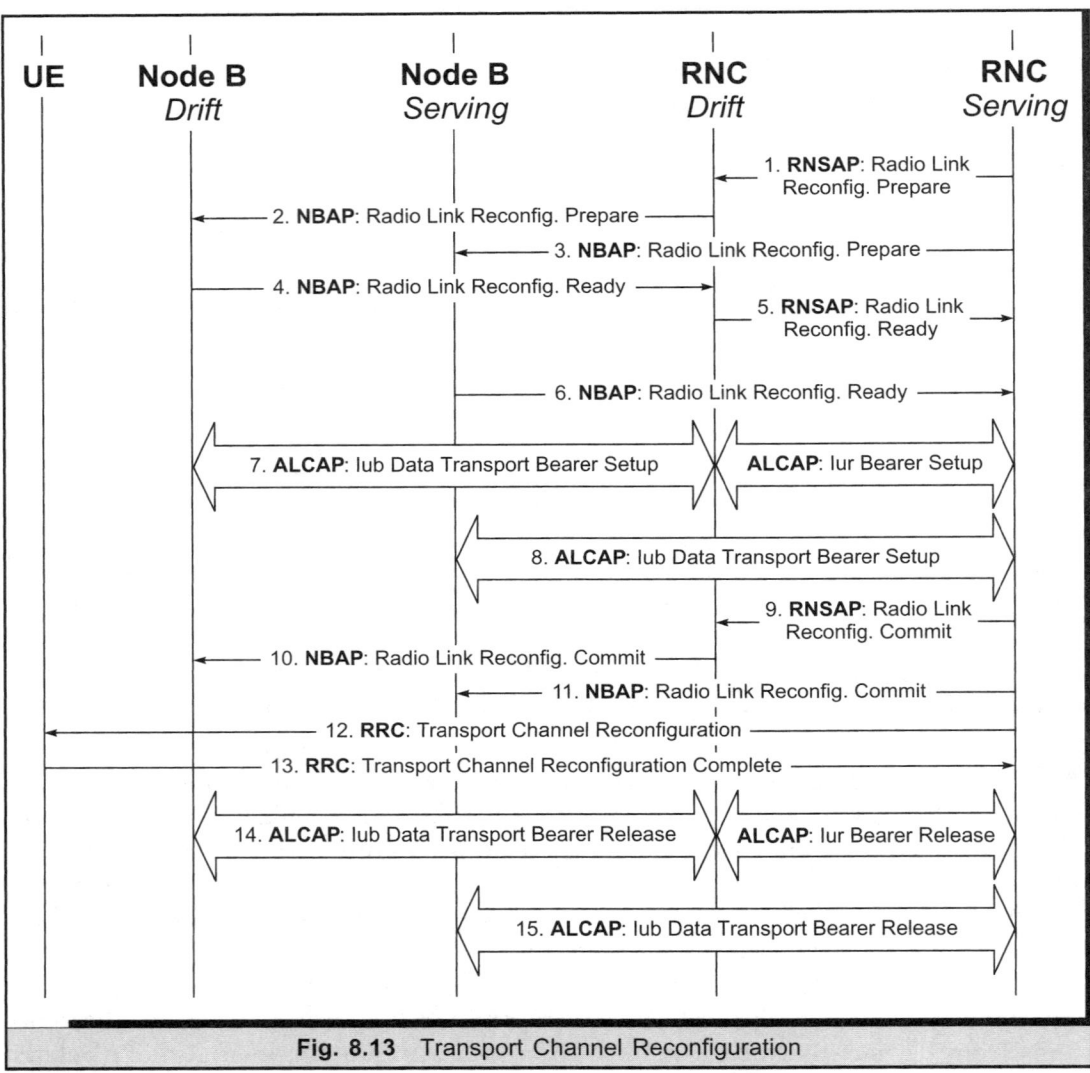

Fig. 8.13 Transport Channel Reconfiguration

The rest of the procedure for transport channel reconfiguration remains identical to the physical channel reconfiguration procedure, and hence, the step-by-step description of the message flows is omitted.

8.3.9 Soft Handover

As discussed in Chapter 7, the Soft Handover (SHO) condition occurs when the UE is in a region that is simultaneously serviced by two or more different Node Bs. In the

SHO state, the UE would have two (or more) radio links simultaneously, one with each Node B. The Node Bs to which the UE is thus connected are called the UE's 'Active Set'. Figures 8.14 and 8.15 depict the mechanism for the addition and deletion of radio links, used in the SHO condition. Both the procedures modify the UEs Active Set by adding or removing Node Bs.

8.3.9.1 Radio Link Addition

The radio link addition procedure (Figure 8.14) involves the following sequence of steps:

1. When the SRNC decides to setup a new radio link towards the UE via a different RNC, it requests the Drift RNC to reserve resources for this link by sending the 'Radio Link Setup Request' message via the RNSAP protocol.
2. If the Drift RNC determines that the necessary resources for a radio link towards the UE are available, it sends a 'Radio Link Setup Request' message to the Drift Node B via the NBAP protocol.

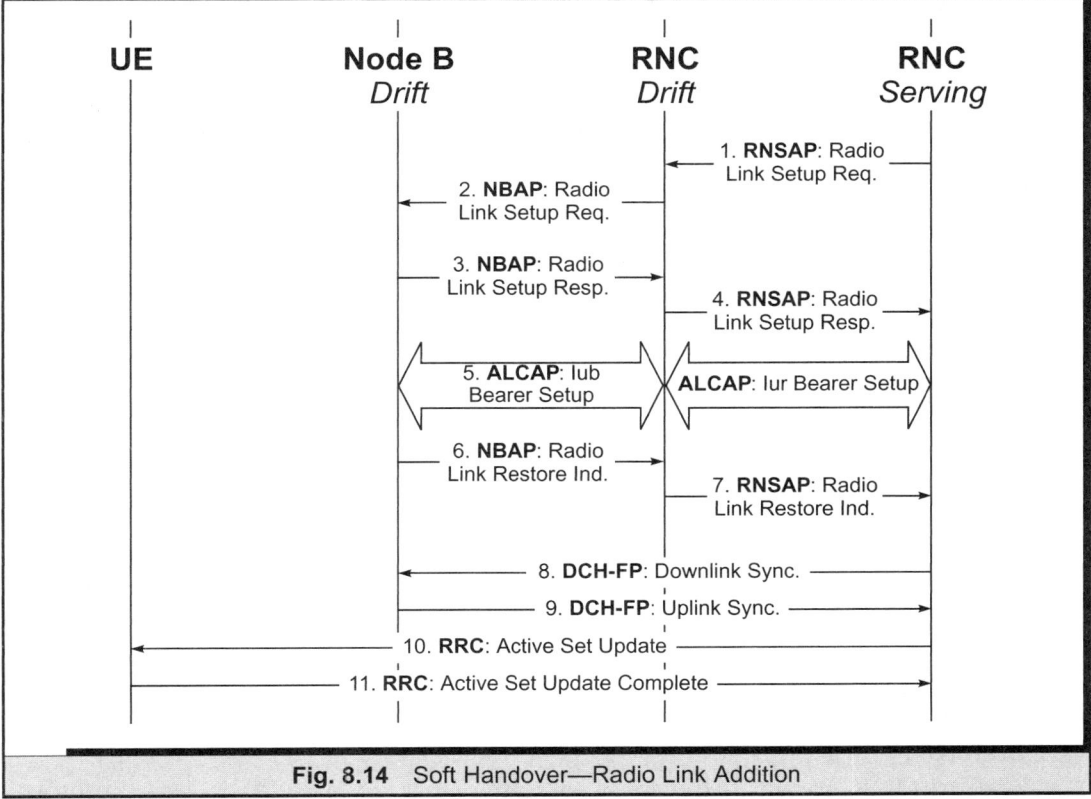

Fig. 8.14   Soft Handover—Radio Link Addition

3. On receipt of this message, the Drift Node B allocates the necessary resources and responds to the Drift RNC by sending a 'Radio Link Setup Response' message. The response includes an AAL2 transport layer address, used to create the transport bearer.

4. The Drift RNC then sends the 'Radio Link Setup Response' message to the SRNC, and includes an AAL2 transport layer address to use in the setup of the transport bearer.

5. Next, the SRNC initiates the setup of the Iur/Iub data transport bearer using the ALCAP protocol.

6. When the Drift Node B achieves uplink synchronization with the UE over the Uu interface, the Drift Node B sends the 'Radio Link Restore Indication' message to the Drift RNC via the NBAP protocol.

7. The Drift RNC informs the SRNC about the same by sending it the 'Radio Link Restore Indication' message.

8. The DCH-Framing Protocol at the SRNC then initiates the downlink synchronization of the transport bearer (with the DCH-FP at the Drift Node B) by sending the 'Downlink Synchronization' message to the Drift Node B.

9. The DCH-Framing Protocol at the Drift Node B initiates the uplink synchronization of the transport bearer by sending the 'Uplink Synchronization' message, to the serving RNC".

10. The SRNC then sends an RRC 'Active Set Update' message to the UE to modify the Active Set of the latter.

11. The UE responds by sending the 'Active Set Update Complete' message to the SRNC. This completes the procedure for the addition of a radio link in the SHO condition.

8.3.9.2 Radio Link Deletion

The radio link deletion procedure (Figure 8.15) involves the following sequence of steps:

1. When the SRNC decides to remove a radio link via a cell controlled by another RNC, it sends the RRC 'Active Set Update' message informing the UE to update its Active Set.

2. The UE in turn deactivates the downlink reception of data via the corresponding radio link, and acknowledges the SRNC message by sending it an 'Active Set Update Complete' message.

3. The SRNC then requests the Drift RNC to release the resources for the radio link by sending it the 'Radio Link Deletion Request' message via the RNSAP protocol.

4. The Drift RNC forwards this message to the Drift Node B via the NBAP protocol.

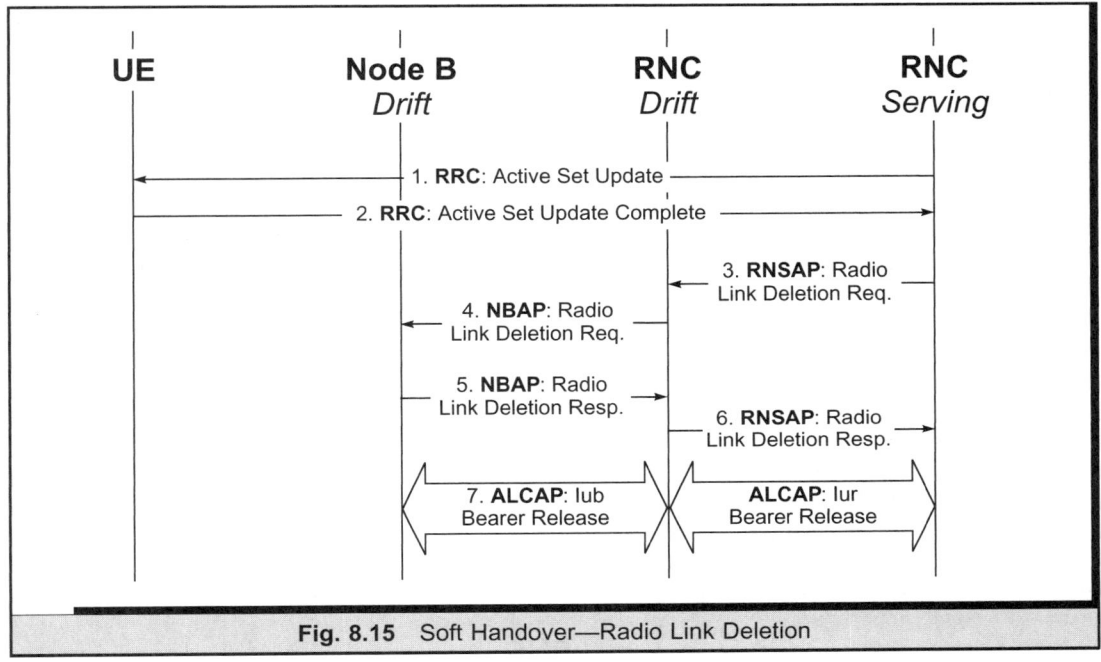

Fig. 8.15 Soft Handover—Radio Link Deletion

5. After successfully releasing the radio resources, the Drift Node B responds to the Drift RNC by sending a 'Radio Link Deletion Response' message.
6. The Drift RNC then responds to the SRNC by sending it the 'Radio Link Deletion Response' message.
7. The SRNC at this stage initiates the release of the Iur/Iub data transport bearer using the ALCAP protocol. This completes the procedure for the deletion of a radio link for a UE.

8.3.10 SRNC Relocation

As discussed in Chapter 2, SRNC Relocation is a process by which the role of the serving RNC for a UE is moved from one RNC to another. As a result, the Iu connection between the CN and the RNC is moved from the old SRNC to the new SRNC. SRNC relocation normally occurs when the UE moves out of the area serviced by the old SRNC to a region serviced by another RNC (the Drift RNC or Target RNC). The decision to effect SRNC relocation is taken solely by the existing SRNC for the UE, and is based on many factors, which include the need to reduce the network resource requirement by removing the usage of the Iur interface between the DRNC and the current SRNC. A complete list of factors to be considered in making the SRNC relocation decision is beyond the scope of this section. However, as a general guideline, the relocation decision may be taken when the UE has completely moved out of the service area of the existing SRNC, and is totally, or partly in the region of the target RNC.

The SRNC relocation procedure (Figure 8.16) involves the following sequence of steps:

1. When the SRNC decides to effect SRNC relocation, it sends a 'Relocation Required' message to the CN via the RANAP protocol. This message includes the identifier for the Target RNC, along with transparent data that the CN is required to pass on to the latter. This transparent data includes the identifier of the UE for which the SRNC relocation procedure is executed.

2. On receipt of the 'Relocation Required' message from the SRNC, the CN prepares itself for the relocation. This may involve the suspension of user data traffic and/ or signaling exchange between the CN and the UE, till the relocation procedure is completed. Once the CN has prepared itself for the relocation, it sends a 'Relocation Request' message to the Target RNC, including the transparent information received from the SRNC in the message.

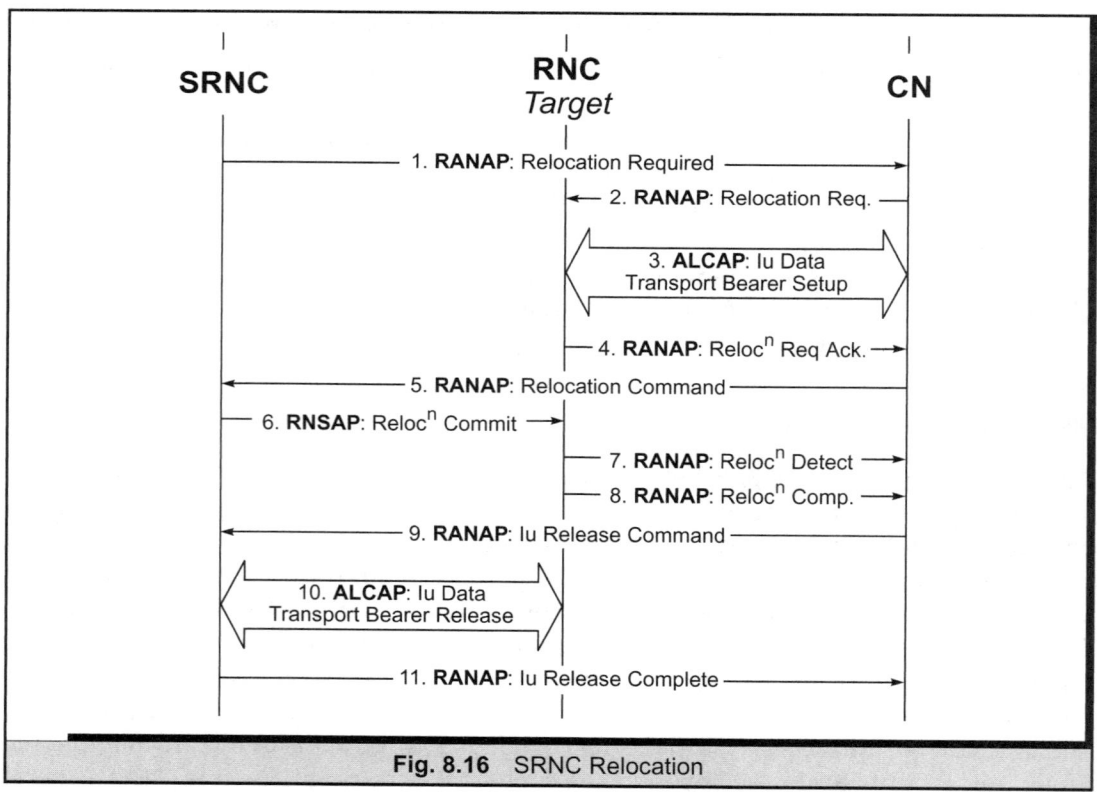

Fig. 8.16 SRNC Relocation

3. For each RAB between the UE and the CN, new Iu data transport bearers are set-up between the Target RNC and the CN using the ALCAP protocol.

4. The target RNC acknowledges the request from the CN by sending a 'Relocation Request Acknowledgement' message to the CN via the RANAP protocol. This is an indication to the CN that the target RNC is prepared to become the new SRNC for the UE.

5. The CN node then sends the 'Relocation Command' message to the SRNC to carry out SRNC relocation.

6. Next, the SRNC sends the 'Relocation Commit' message to the Target RNC via the RNSAP protocol. This is an indication for the Target RNC to proceed with the relocation.

7. The Target RNC then sends the 'Relocation Detect' message to the CN. This is an indication to the CN that the Target RNC is proceeding to switch the uplink and downlink bearers. After bearer switch, all uplink data received from the UE from the Node B (within the Target RNC area) is sent over the new Iu bearer setup between the Target RNC and the CN. All downlink data received over the new Iu bearer is sent directly to the Node B. Hence, the Iur interface is no longer used.

8. After the Target RNC has successfully carried out the bearer switch, it sends the 'Relocation Complete' message to the CN. The SRNC relocation is now complete, and the Target RNC has become the new SRNC for the UE.

9. Once the SRNC relocation is complete, the CN sends the 'Iu Release Command' to the old SRNC, informing it to release the Iu bearers between the old SRNC and the CN.

10. The ALCAP protocol is used to release the Iu bearers between the CN and old SRNC.

11. The old SRNC confirms the release of the Iu bearers between the CN and the old SRNC by sending a 'Iu Release Complete' message via the RANAP protocol.

8.3.11 Cell Update

As discussed in Chapter 7, the cell update procedure is used to serve multiple purposes; this includes the requirement to update the UTRAN of the new cell following cell reselection. Though many examples can be considered to depict the UTRAN procedure for a cell update by the UE, the one that is used in this section is considered to be most comprehensive. Other scenarios can be devised using a subset of this example. The scenario used for the cell update procedure is as follows: The UE is considered to be roaming in a region controlled by a Drift RNC (DRNC), which is distinct from the SRNC. Hence, the cell update procedure is carried out over the Iur interface. Further, it is assumed that after cell reselection, the UE has moved into a cell that is controlled by a Target DRNC, which is distinct from the Serving DRNC of the UE, where the UE was originally located before the cell reselection process.

Figure 8.17 depicts the cell update procedure for the above scenario. The steps involved in this procedure are:

1. The UE sends the RRC 'Cell Update' message to the Target DRNC after cell reselection.

2. Based on the U-RNTI received in the message from the UE, the Target DRNC identifies the SRNC for the UE. It also determines that the UE is not registered with it, before forwarding the message received from the UE to the SRNC, as part of the 'Uplink Signaling Transfer Indication' message, via the RNSAP protocol. The message also includes the Cell-Id from which the UE had sent the message, as well as the DRNC-Id of the Target DRNC. The new RNTI that the Target DRNC has allocated for the UE is also included in the message.

3. Upon receipt of the message from the Target DRNC, the SRNC initializes the UE specific information in the Target DRNC by sending it the 'Common Transport Channel Resources Initialization Request' message.

4. The Target DRNC then responds with the 'Common Transport Channel Resources Initialization Response' message to the SRNC. This message includes the transport

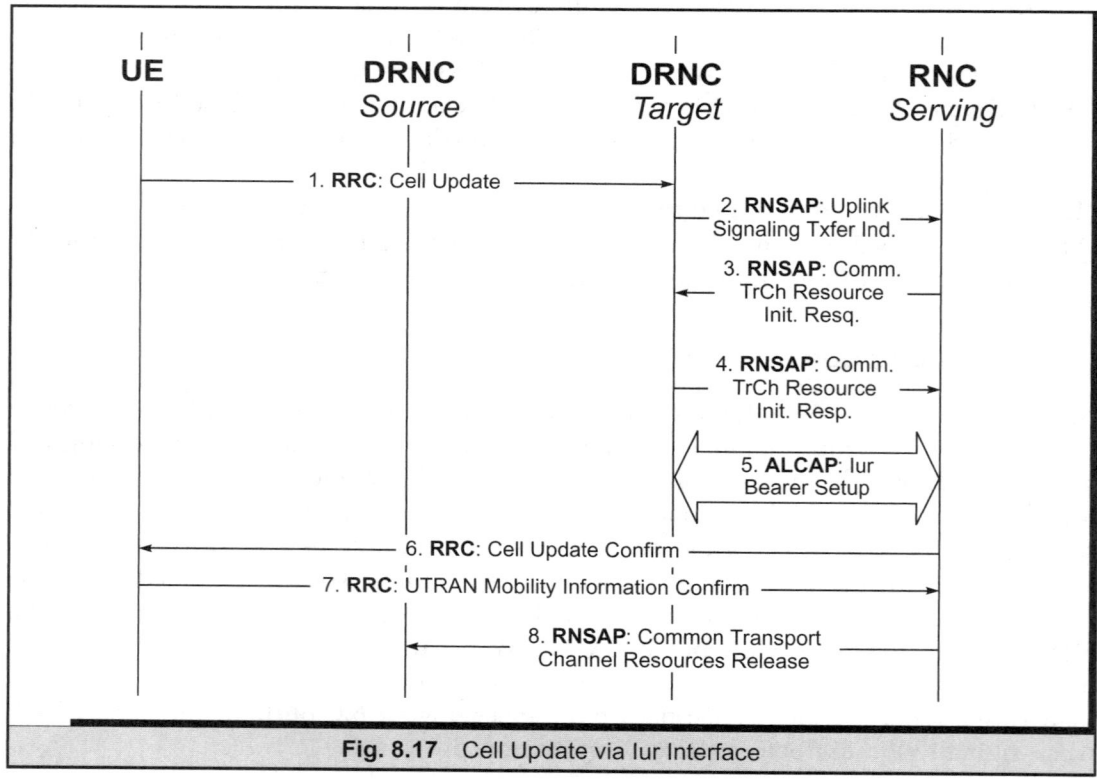

Fig. 8.17   Cell Update via Iur Interface

layer address for the setup of the transport bearer, if no appropriate Iur transport bearer, to be used for the UE, already exists.

5. Next, the SRNC initiates the setup of the Iur transport bearer using the ALCAP protocol.

6. The SRNC then sends the 'Cell Update Confirm' message to the UE via the RRC protocol. The message is sent via the Iur interface.

7. The UE confirms the receipt of the new RNTI by sending the RRC 'UTRAN Mobility Information Confirm' message to the SRNC.

8. The SRNC in turn sends a 'Common Transport Channel Resources Release' message to the Source DRNC where the UE was originally located, prior to the cell reselection process. The message is a request to the Source DRNC to release the UE related resources. This completes the cell update procedure.

8.3.12 URA Update

This section discusses the URA update procedure following cell reselection. The same scenario, as detailed in the cell update section, is used to discuss the URA update procedure.

Figure 8.18 depicts the URA Update procedure. The sequence of steps involved in the procedure is as follows:

1. The UE sends a 'URA Update' message to the UTRAN (Target DRNC) via the RRC protocol, after cell reselection. The example here assumes that as a result of cell reselection, the URA, where the UE is located, has changed.

2. As in the cell update case, the Target DRNC ascertains that the UE is not registered with it, and decodes the SRNC-Id from the U-RNTI received from the UE. It then

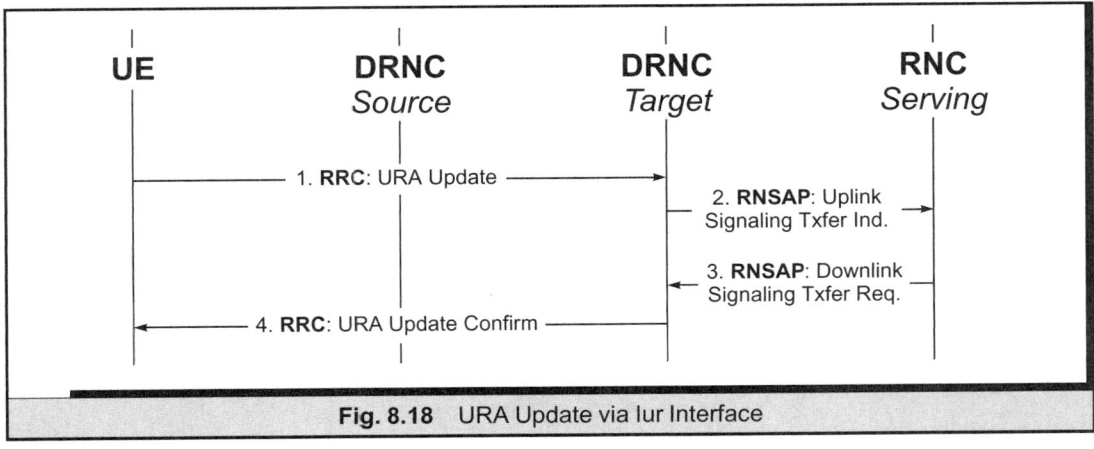

Fig. 8.18 URA Update via Iur Interface

forwards the UE message to the SRNC using the 'Uplink Signaling Transfer Indication' message via the RNSAP protocol.

3. The SRNC updates its information, and the Target DRNC becomes the new controlling RNC for the UE. The SRNC then sends the relevant information to the Target DRNC, along with the 'URA Update Confirm' message as a part of the 'Downlink Signaling Transfer Request' message.

4. The Target DRNC then forwards the 'URA Update Confirm' message to the UE. This completes the URA update procedure.

8.3.13 Direct Transfer

The concept of Direct Transfer of messages was discussed in Chapter 7, where it was explained how higher layer NAS messages are transported from the UE to the UTRAN as a part of the RRC 'Uplink/Downlink Direct Transfer' messages. This section completes the discussion; it describes how the UTRAN forwards the NAS signaling message received via the RRC protocol to the Core Network.

Figures 8.19 and 8.20 depict the procedure involved in the transporting of the higher layer NAS message in the uplink and the downlink direction, respectively.

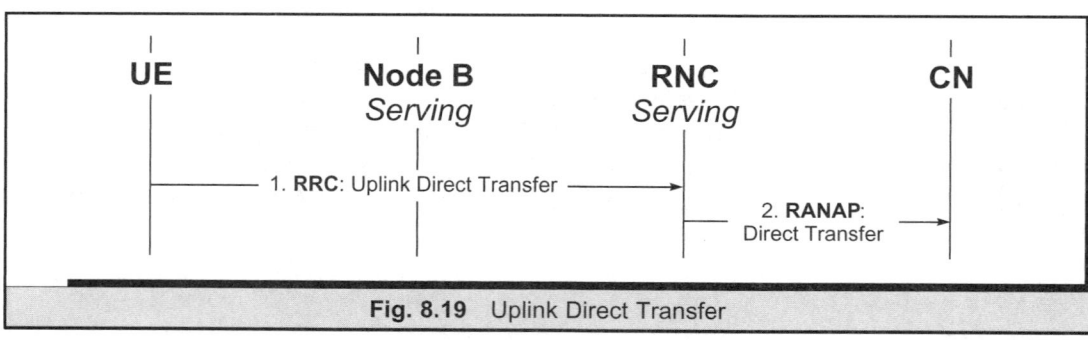

Fig. 8.19   Uplink Direct Transfer

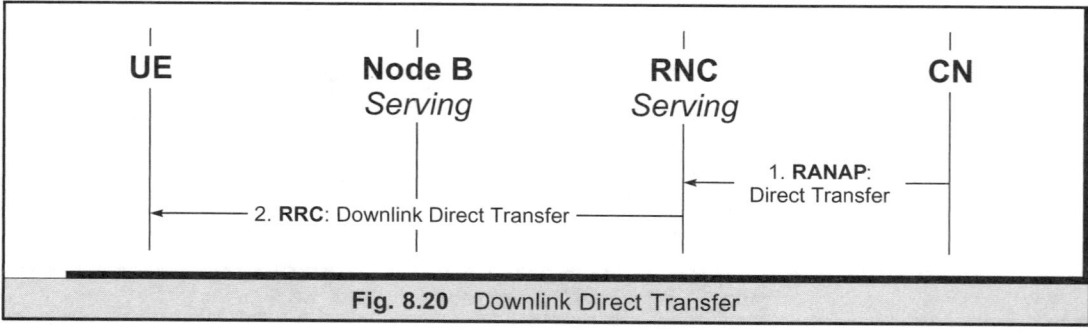

Fig. 8.20   Downlink Direct Transfer

In both the directions, the RANAP 'Direct Transfer' message is used to transfer the higher layer NAS message. While in the uplink direction the NAS message is received from the UE as a part of the RRC 'Uplink Direct Transfer' message, in the downlink direction it is sent towards the UE as a part of the RRC 'Downlink Direct Transfer' message.

SUMMARY

While the RRC is the main protocol on the radio interface between the UE and UTRAN, other important protocols form the basis of all signaling that takes place within the UTRAN, and between the UTRAN and CN. These protocols include RANAP, RNSAP, NBAP, ALCAP and SABP. This chapter discussed the Iub, Iur and Iu procedures, which are based on NBAP, RNSAP and RANAP protocols, respectively. The important procedures discussed in this chapter include UTRAN Global Signaling Procedures and UTRAN Signaling Procedures for a specific UE. Each of these categories is further divided into various sub-procedures, which have been discussed in detail in the chapter.

MOBILITY MANAGEMENT

9.1 INTRODUCTION

The very essence of a mobile network is the ability of a Mobile Station (MS) to roam within and across multiple PLMNs, and still be reachable. This is very unlike a PSTN network, where the end terminals (telephones, etc.) are considered fixed entities, and reachability to these terminals is through pre-configured fixed routes. However, this cannot be the case in a mobile network. To reach a mobile subscriber in a mobile network (e.g. in case of an incoming call for the subscriber), it becomes essential for the network to maintain the existing location of the MS. Thus, maintaining and managing location information forms the basis of the Mobility Management (MM) procedures. Besides managing location information, mobility management procedures also allow subscribers to roam into networks operated by different service providers. These concepts of mobility management are covered in detail in this chapter.

However, before proceeding, it becomes essential to distinguish between the two concepts: *location information* and *position information* of the MS. These two terms may appear to be the same, but in a UMTS network, they involve different concepts. The term 'location' implies the location of the end user (specifically, the MS) within the hierarchical structure of the network. This hierarchical structure, which is used by the network to locate the end user, was discussed in Chapter 3. The term 'position', on the other hand, implies the geographical position of the end user in terms of standardized co-ordinates. The position information is not directly used by the network; it is used by application service providers to provide specialized position services like Emergency Services/Public Safety Answering Point (PSAP). The mobility management procedures are concerned with managing the 'location' of the MS rather than its 'position'. Hence, the term location management is also sometimes used in place of mobility management. Position-based services are covered in Chapter 13, along with location services. Thus, location services deal with the 'position' of the MS, and are not to be confused with location management.

This chapter first discusses the three-state Mobility Management (MM) state model, which presents a finite state machine maintained in both the MS and the CN nodes (VLR and SGSN) for mobility management activities. This is followed by a discussion on the location information stored in various network elements and the hierarchical nature of this information. Next, the chapter covers the MM procedures, which were briefly introduced in Chapter 6. Procedures that are specific to location management vis-a-vis the MS, the Access Network and the Core Network are discussed in detail. The Core Network procedures are separately covered for the CS and the PS domain of the CN.

9.2 STATE MODEL FOR MOBILITY MANAGEMENT

To understand the functions of Mobility Management (MM), it is important to first understand the state model for MM. Figure 9.1 depicts a simple Finite State Machine (FSM) maintained in both the MS and the CN nodes (VLR and SGSN). In reality, however, the state machines are much more complicated (refer to 3GPP TS 24.008 for details on FSM). Further, the state machines defined for the CS and PS domains are not identical; they have subtle differences. Nevertheless, a simplified, easy to understand picture of the state machine is presented in Figure 9.1.

The following states are depicted in the state machine:

- **Detached State:** In this state, the location of the MS is not known to the network. In other words, the MS is not reachable by the network. This may be the case when, for example, the MS is in a 'power-off' state, or when the USIM is removed. The MS can move out from the 'Detached State' to one of the other two states by attaching itself to the network (e.g. on 'power-on'), or by registering its location with the network through the Location Registration (LR) process.

- **Idle State:** In this State, the location of the MS is known to the network (the Core Network, to be precise, to which the MS is attached at the time). However, there is no active session for the MS (i.e. no voice call, or data session, etc., in progress). Hence, the MS is assumed to be 'idle' in this state. From the 'Idle State', the MS can move to the 'Connected State' when there is an active session in which it has to participate. Participation in a session is generally characterized by the establishment of a 'connection'; hence the name. From the idle state, the MS can also move back to the detached state, for example, due to 'powering-off', or removal of USIM from the equipment, or because of failure of the network to register the MS.

- **Connected State:** Here, the location of the MS is known to the network (both the Access Network, and the Core Network). Further, there is an ongoing session active for the MS (e.g. a voice call or a data session). Thus, in this state, the MS is not considered to be idle, since it is involved in a session. From the connected state, the MS can move to the idle state on completion of the session. It can also move back to the detached state, for example, as a result of powering-off, or removal of USIM from the equipment.

Another important point needs mention here. Normally, when an MS first attaches itself to the CS domain, it transitions from the detached state to the idle state. This is because there might not be any active call/session for the MS immediately after attaching to the network. On the other hand, when an MS registers/attaches itself to the PS domain, it moves directly from the detached state to the connected state (Figure 9.1). This direct transition to the Connected State is made on the assumption that in most cases, the MS would attach to the PS domain only when it needs to participate in an active (data) session.

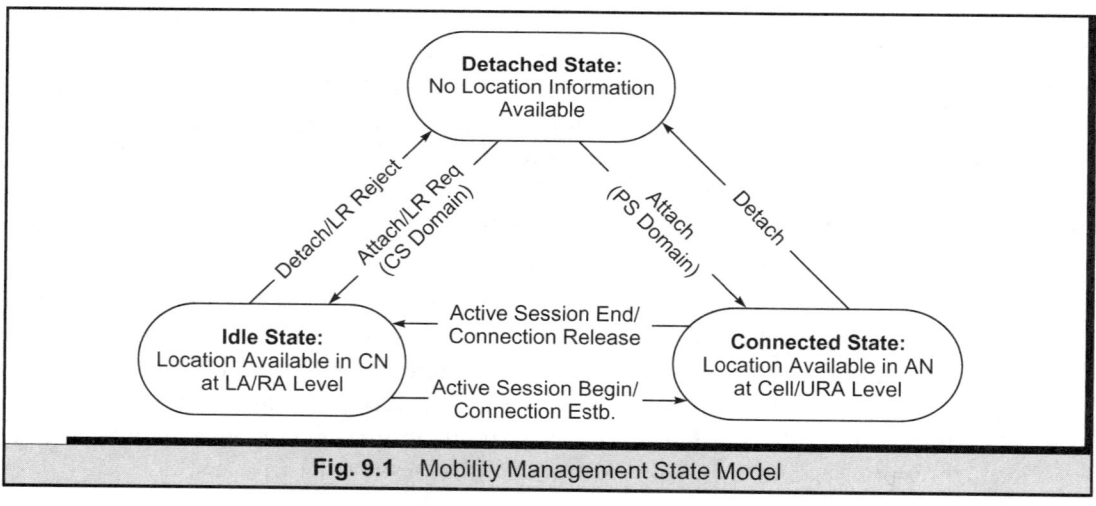

Fig. 9.1   Mobility Management State Model

Besides the state transitions, the figure also depicts the granularity of the information available to the network as regards the location of the MS (details in Section 9.3). In the detached state, no location information is available to the network, since it does not know the MS in this state. Location information is available to the network in both the connected and the idle state. However, the granularity of the location information available differs in both these states.

In the idle state, the location information of the MS is maintained by the Core Network (and not by the Access Network) only at the granularity of Location Area (LA)/Routing Area (RA). However, in the connected state, the Access Network also maintains this information, which is available at the Cell/URA level. The logic behind this design is that the exact location of the MS is only required when there is an active session for it, and not otherwise. The next section elaborates upon this aspect of hierarchical location information maintenance across network nodes. Section 9.4 introduces the concept of paging, which is used when the exact (cell-level) location of the MS is required while it is in the idle state.

9.3 HIERARCHICAL MANAGEMENT OF LOCATION INFORMATION

As discussed in Chapter 3, the UMTS network architecture is divided into hierarchical boundaries to facilitate efficient location management. At the lowest layer are the cells, which form the basic building blocks of a mobile network. A cell is uniquely and globally identified using the Cell Global Identifier (CGI). A collection of cells forms a Location Area (LA) for a CS domain, or a Routing Area (RA) for a PS domain. A routing area is normally a subset of a location area. In other words, a location area may consist of one or more routing areas. However, one routing area may only belong to one location area. The location and routing areas are identified by Location Area Identity (LAI) and Routing Area Identity (RAI) respectively. The format of these identities is explained in Chapter 3.

In addition to this, the cells in an RA are further partitioned and grouped into units called UTRAN Registration Areas (or URAs). Thus, a URA consists of one or more cells, and an RA may contain one or more URAs.

At any point in time, the exact location of an MS is known by the CGI of the cell where the MS is present. The current location of the MS is maintained and tracked by the network elements (both CN and AN). Considering the fact that the network elements are themselves hierarchically organized, the location information can also be hierarchically maintained across the network elements in a distributed fashion. The information maintained in each network element is as depicted in Table 9.1.

Since the HLR of a subscriber is the permanent store for all information about the subscriber, the location information is also maintained in the HLR. This includes

Table 9.1   Maintenance of Location Information

Location Unit	Network Element		
	MSC/VLR	SGSN	RNC
Cell	No	No	Yes*
URA	No	No	Yes*
RA	No	Yes	No
LA	Yes	No	No

* Only in 'Connected State' (refer to MM state model)

information about both the existing MSC and the VLR in which the CS domain subscriber is registered. Similarly, for a PS domain subscriber, the HLR maintains information on the SGSN in which the subscriber is registered. Thus, while the HLR stores information on the existing MSC/VLR and SGSN of the subscriber, the MSC/VLR and SGSN store information of the current LA and RA respectively. Further, for an MS in the connected state, the UTRAN (RNC, to be precise) maintains information on the exact URA and the cell in which the subscriber is available. For an MS in the idle mode, the UTRAN does not maintain any location information, and hence, the exact cell-level information is not available with the network in this state. This leads us to the concept of 'Paging', discussed in the following section.

9.4   PAGING

When a mobile station is in the 'connected state', its location information is available on a cell-level granularity. However, in the 'idle state', when the mobile station does not have any active session with the network, its location is tracked only till the level of LA/RA. In case there is an incoming call or incoming packets for the MS while in the idle state, the network is required to determine the exact location (CGI) of the mobile station. This is where the process of paging is used.

Paging is a mechanism whereby a request for location information is broadcast within the LA/RA where the mobile was last known to exist. Since the core network maintains the LA/RA level information of the MS while in the idle state, it broadcasts a message (paging message) in this LA/RA. On receipt of this message, the mobile station responds, giving its current cell location. Chapter 7 (RRC Procedures) discussed the use of the 'Paging Type 1' message for paging. Chapter 8 (UTRAN Signaling Procedures) described the message-flow within the UTRAN for the paging process.

Having understood the MM state model, and the maintenance of location information across different network elements on the basis of the state of the MS, we can now proceed to detail the procedures involved in managing this information.

9.5   MM/GMM PROCEDURES OVERVIEW

The main function of Mobility Management is to support the mobility of mobile stations, which includes functions such as informing the network of the present location of terminals. During mobility management, it is also ensured that the identity of the user is kept confidential and that only authenticated users can avail of the network service. The MM/GMM functions are performed by the MM/GMM sub-layer, which resides on top of the RRC sub-layer in the control plane (refer to CS and PS domain control plane architectures discussed in Chapter 6). Another important function of the MM sub-layer is to provide connection management services to the different entities of the upper Connection Management (CM) sub-layer.

A brief overview of MM/GMM procedures was provided in Chapter 3. Three broad categories of procedures were discussed in that chapter, namely:

- MM Connection Management Procedures
- MM/GMM Common Procedures
- MM/GMM Specific Procedures

Both the common as well as the specific procedures of MM/GMM include functions required to support the mobility of user terminals, and are used to update the location information of the MS stored in the CN. They also include procedures for authentication of users, before users can avail of the network service. These procedures form the core of this chapter, and are discussed in detail throughout, from Section 9.6 onwards. The MM connection management procedures, on the other hand, provide services to the CM entities of the upper sub-layer, and include procedures for the set up, maintenance, and release of an MM connection. A brief description of the MM con-nection management procedures is provided in this section. For detailed information, the reader is referred to 3GPP TS 24.007 and 3GPP TS 24.008 specifications.

The MM connection management procedures are used to establish, maintain and release an MM connection between the MS and the network. Figure 9.2 depicts the concept of an MM connection, which exists between two peer MM entities. MM

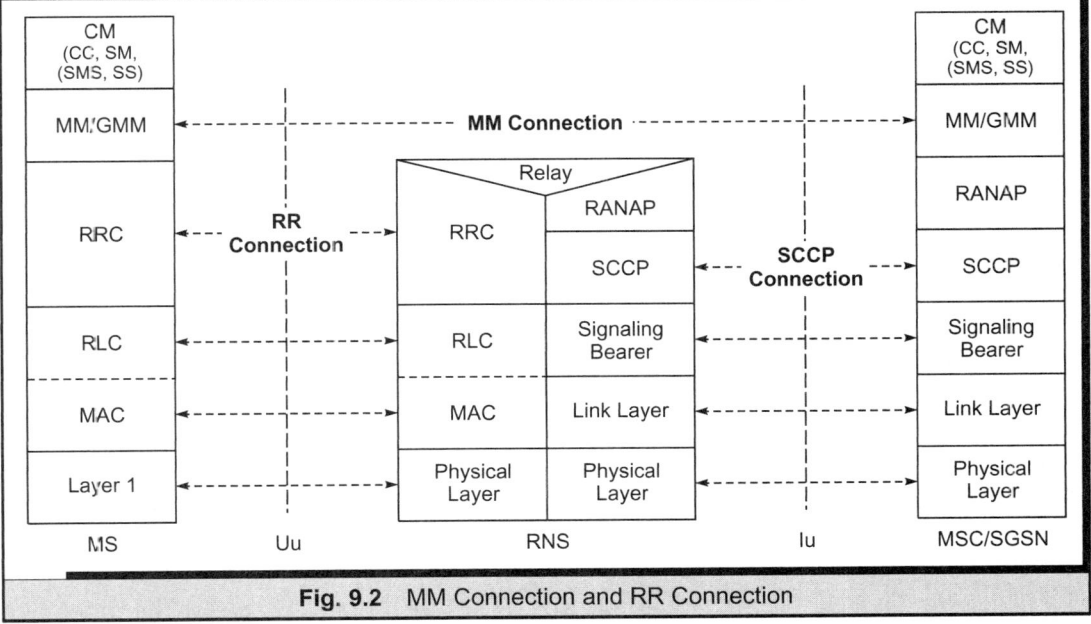

Fig. 9.2   MM Connection and RR Connection

connections are used for the exchange of Connection Management (CM) sub-layer messages between peers. Several MM connections may be active at the same time, since different CM entities communicate with their peer entities using different MM connections.

The CM sub-layer entities include Call Control (discussed in Chapter 10), Session Management (Chapter 11), Supplementary Services (Chapter 12), Short Message Service and Location Service (Chapter 13). All these entities rely on the lower MM sub-layer to establish an MM connection, on which they can exchange messages with their peer entities.

The MM sub-layer uses the 'CM Service Request' message to request establishment of the MM connection with its peer sub-layer. This message, when exchanged between peer MM sub-layers, indicates which entity of the CM sub-layer will use the MM connection. Subsequently, messages received on this MM connection can be routed by the MM sub-layer to the correct CM entity. During the establishment of an MM connection, the authentication of the MS to the network (and vice versa) is carried out through an authentication process, before message exchange can begin.

MM connections help achieve two important objectives. Firstly, they provide a mechanism for the higher layer CM entities to exchange messages with their peers, after the latter have been duly authenticated. The MM connections may also aid the MM sub-layer to route incoming messages (which it receives from the RRC sub-layer) to the correct upper sub-layer CM entity, since each CM entity is associated with a unique MM connection. Secondly, the presence of even one MM connection is an indication to the MM sub-layer that at least one CM entity is engaged in communication with the CN at that particular point in time. This prevents any MM specific function (e.g. location update) from being executed, till the time that all MM connections are released.

As depicted in Figure 9.2, an MM connection always requires an underlying RR connection. In case of multiple MM connections, all simultaneous MM connections use the same RR connection. Thus, in order to establish an MM connection, the MM sub-layer requests the RRC sub-layer to establish an RR connection, if no RR connection is already established. On release of all MM connections using the same RR connection, the associated RR connection is also released.

9.6   MM PROCEDURES IN THE MOBILE STATION

The MM procedures in the mobile station are categorized as 'Idle Mode' and 'Connected Mode' procedures. Recollect that in Section 9.2 a mobile station was said to be in one of the three broadly classified states: 'Detached State', 'Idle State' and the 'Connected State.' The idle mode procedures carried out by the MS include the following procedures:

- Procedures of the MS to transition from 'Detached State' to one of the other two states.

- Procedure of the MS when in the 'Idle State'.
- Procedures of the MS in the connected state are termed 'Connected Mode' procedures. 3GPP specifications TS 23.122 and TS 25.304 describe these procedures in detail.

9.6.1 MS 'Idle Mode' Procedures

When an MS is switched on, it tries to make contact with a PLMN. The MS selects the PLMN either automatically, or manually as a result of user selection. The MS next looks for a suitable cell of the chosen PLMN and tunes to the control channel of that cell. It then registers its presence in the registration area of the cell, if necessary, by means of a Location Registration, GPRS Attach, or IMSI Attach procedure, and moves out of the 'Detached State'.

Once in the idle state, if the MS moves out of the coverage area of the selected cell, or if it finds a more suitable cell, it reselects the more suitable cell of the selected PLMN. If the new cell is in a different registration area, a Location Registration request is made. Similarly, if the MS loses coverage of the PLMN, it selects a new PLMN, either automatically, or as a result of manual user intervention, and repeats the idle mode procedures.

The "Idle Mode" procedures of the MS consist of three steps, in the following order:

(a) PLMN Selection,
(b) Cell Selection/Reselection,
(c) Location Registration

This is as depicted in Figure 9.3. The mobile station is considered to be *registered* after a PLMN has been selected, a suitable cell identified in that PLMN, and the location registration request on that cell successfully completed.

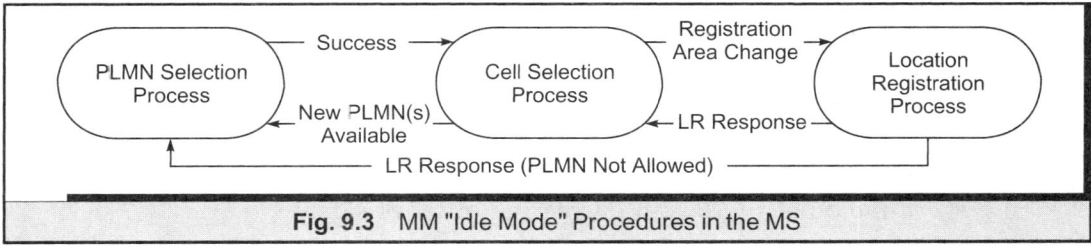

Fig. 9.3 MM "Idle Mode" Procedures in the MS

In case the location registration process for the selected cell of the selected PLMN fails, the cell or the PLMN selection process would have to be repeated. For a CS domain subscriber—when roaming within the area defined by a Location Area (LA)—the mobile station does not need to do location registration on each cell change. Location registration is required only when the MS moves from one location area to another. Similarly, for a mobile station using PS domain services, Routing Area (RA) update is required only when it moves to a different RA.

9.6.1.1   PLMN Selection

A mobile station carries out the process of PLMN selection each time it is switched on, or when it is recovering from a state of lack of coverage. This process is also executed when the mobile subscriber requests for a re-selection of the PLMN. Two modes are allowed for PLMN selection:

 (a) automatic: in which no user intervention is required
 (b) manual: in which user input is taken for selecting the PLMN

The PLMN selection mode can be configured by the subscriber on the MS. In case of automatic PLMN selection, a PLMN is selected from a list of PLMNs maintained in priority order. The highest priority PLMN that is available and 'allowable' to the mobile subscriber is selected. For an MS, an 'allowable' PLMN is the one where it is allowed to roam. In case of manual PLMN selection, the mobile station presents the mobile user with a list of available PLMNs. The mobile user makes a manual selection of the PLMN, after which the mobile station tries cell selection and registration on that PLMN.

On switch-on, or in case of recovery from loss of coverage, the mobile station first selects the last 'Registered PLMN' (if it exists). The last registered PLMN is the one that the MS had last registered on, before going into the switch-off state or the 'loss of coverage' state. If this PLMN exists, the mobile station tries the cell selection and location registration procedure in it. If this is successful, then the mobile station is said to have successfully 'camped' on the selected cell of the PLMN. If unsuccessful, the MS either follows the automatic or the manual PLMN selection procedures, based on its operating mode.

For the automatic selection mode, the sequence of steps followed by the mobile station is:

1. Select and attempt registration on other available and 'allowable' PLMNs, in the following order:
    - The Home PLMN.
    - PLMNs from the list of PLMNs stored in the USIM of the mobile station.
    - Other PLMNs with a signal level of over 85 dBm, in random order.
    - All the other PLMNs in decreasing order of signal strength.
2. On successful registration, the selected PLMN is the new 'Registered PLMN'.
3. If successful registration is not achieved because no PLMNs are available and 'allowable', then the mobile station indicates that the service is currently not available. In this case, it repeats the procedure when a new PLMN becomes available.
4. If, however, some PLMNs are available and allowable, but the Location Registration request has failed for some other reason (e.g. if the network operator has

removed the roaming facility due to non-payment of bills), then the mobile station selects the first such PLMN, and enters the 'limited service' state. In this state, the MS is only allowed to make emergency calls (Emergency calls are calls made to an emergency number, e.g. Police, Ambulance, etc.).

For manual mode operation, similar procedures as in case of automatic mode operation are followed, except that the PLMN is selected by the mobile user, based on the list provided to the latter. Hence, PLMN selection is not automatic. The PLMNs shown to the mobile user are in the same order as described in point 1 of the automatic mode operation. If, however, the user does not select any PLMN, then either the last selected PLMN (if any) is used, or any cell that is acceptable is entered; the user thus enters the limited service state.

The PLMN selection process is depicted in Figure 9.4. This process, besides the instances when the mobile station is 'switched on', is also executed whenever the user requests for reselection of PLMN. In such a case, the automatic mode operation considers the currently registered PLMN only at the end, after it has considered all other available PLMNs, and has not found any that is suitable. The manual mode operation, however, is not affected.

Whenever the mobile station is roaming in a Visited-PLMN (VPLMN) of the home country (which means that the MCC of the VPLMN is the same as that of the HPLMN), then the mobile station periodically checks if the HPLMN of the mobile subscriber could be selected. If possible, registration on the HPLMN of the mobile subscriber is tried. This is specific to the automatic mode operation and is not followed in the manual mode operation of the mobile station.

9.6.1.2 Cell Selection

After selection of the PLMN, the mobile station is required to select a cell and *camp* on it in order to receive services from the PLMN. Camping on a cell has manifold advantages. It enables the MS to:

1. Receive system information from the PLMN.
2. Receive paging messages from the PLMN, e.g. when there is an incoming call for the MS.
3. Initiate call setup for outgoing calls, or any other supplementary service procedures.
4. Receive cell broadcast messages (refer to Chapter 13).

For a mobile station, a suitable cell is defined as a cell that serves the following conditions:

1. It should be a cell of the selected PLMN.
2. The cell should not be 'barred' for the mobile station.

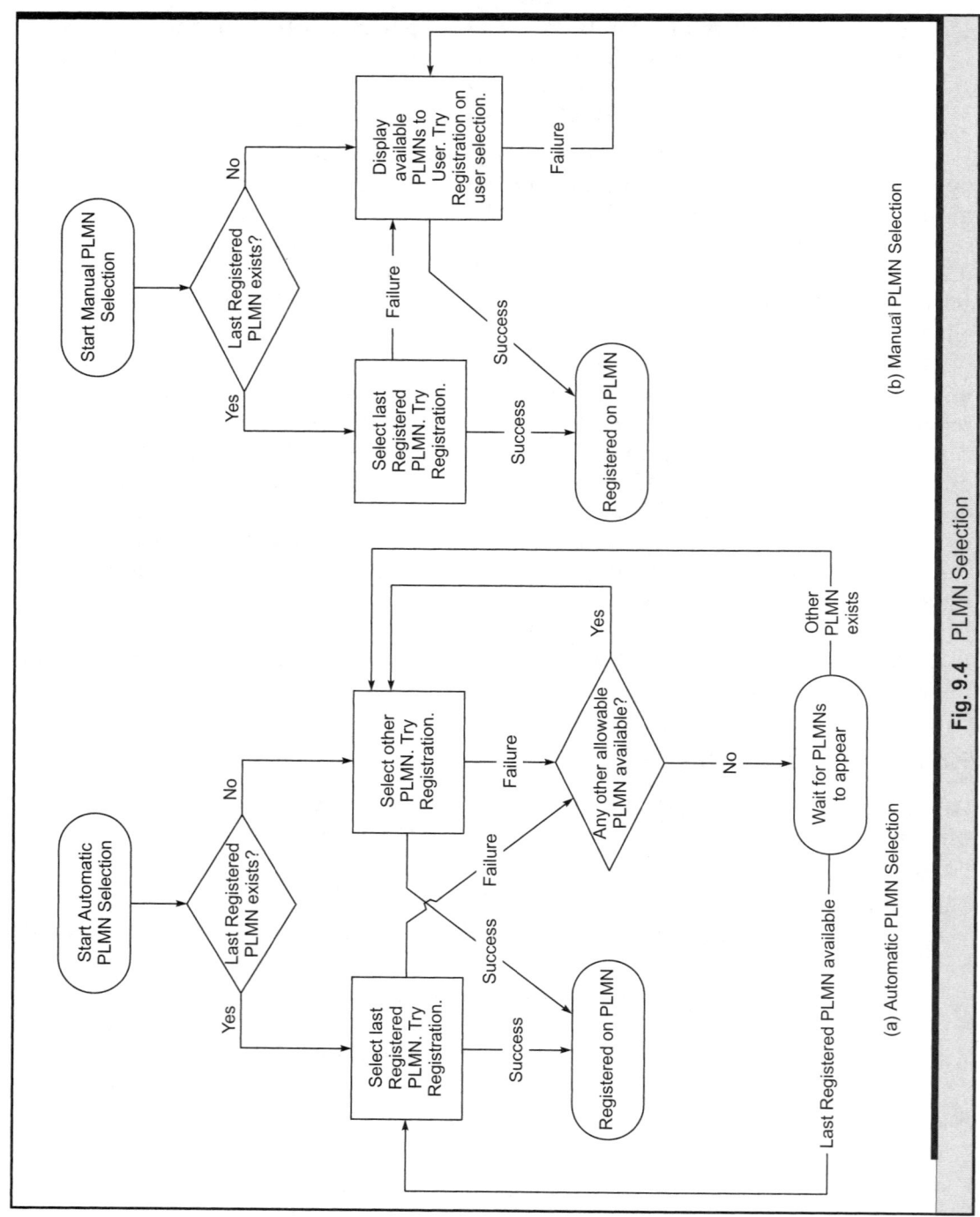

Fig. 9.4 PLMN Selection

(a) Automatic PLMN Selection

(b) Manual PLMN Selection

3. It should not be in a location area that is in the list of 'forbidden LAs' for the mobile station. An LA, for example, may be forbidden for the MS, when the latter is not allowed to roam in it (barring of roaming).

4. The radio path loss between the mobile station and the Node B must be below a threshold.

If such a cell is available, the mobile station selects it for camping. Camping on such a cell is termed as 'Normal Camping' of the mobile station. If however, a suitable cell is not found, then the mobile station camps on any acceptable cell and enters the 'limited service state'. In this state it is only allowed to initiate emergency calls; it cannot participate in any other call/session.

Even after a mobile station has selected a cell and has camped on it, it continues to look out for 'better' and 'more suitable' cells. A better cell in case of normal camping is a cell which has a lower radio path loss between the MS and the Node B. The MS determines this by looking at the received signal strength from the Node B. In case of camping in a limited service state, the 'better' cell is the one on which 'Normal' camping is possible.

9.6.1.3 Location Registration

After selection of a suitable cell, the mobile station is required to register itself with the network using Location Registration (LR) procedures. LR procedures fall under the following categories:

(a) **Normal LR Procedures:** LR request indicating 'Normal Updating' is made under any of the following conditions:

- When the MS changes a cell, and it had not successfully entered the registered state on the last selected cell.
- When the MS enters a new registration area (LA/RA).
- When the Periodic Location Registration timer expires, and the MS was not in a registered state earlier, i.e. the last LR request of the MS was not successful.

(b) **Periodic LR Procedures:** LR request indicating 'Periodic Location Updating' (for CS Domain) or 'Periodic Routing Area Update' (for PS Domain) is made when the Periodic Location Updating timer/Periodic Routing Area Update timer expires, and the MS was earlier in a registered state.

(c) **LR indicating IMSI-Attach:** LR request indicating 'IMSI-attach' is made when the MS is activated (or powered 'on') in the same LA in which it was deactivated (or powered 'off'), and the system information broadcast for the PLMN indicates that IMSI-attach procedures are required. IMSI-attach procedure is a mechanism by which the MS indicates to the system that it has again been activated. At the time of last deactivation, the MS should have carried out an 'IMSI-detach' procedure, indicating that the MS is being deactivated.

(d) **GPRS-Attach procedures:** On activation, a mobile station makes a GPRS- attach, when it requires to avail of the PS domain services. Unlike IMSI-attach, which is carried out conditionally on activation (i.e. only when the LA before detach, and the LA after activation match), GPRS-attach is always done when the MS registers/attaches itself to the PS domain.

On successful completion of location registration, the MS is now ready to receive services from the PLMN. The location of the MS is maintained and registered with the PLMN.

9.6.2   MS 'Connected Mode' Procedures

MS 'Connected Mode' procedures are carried out when the MS is in the 'Connected State' (refer to MM state model). In this state, the MS has an RRC connection established, and hence, is also in the connected mode, as per the RRC state model (refer to RRC protocol state model in Chapter 7).

The 'Connected Mode' procedures of the MS are the same as those in the RRC connected mode. These procedures were discussed in Chapter 7. In the RRC connected mode, the MS is required to carry out a 'Cell Update' on each cell change (while in the CELL_FACH, CELL_DCH and CELL_PCH RRC states), and a 'URA Update' on each URA change (while in the URA_PCH RRC state). The 'Cell Update' and the 'URA Update' message flow diagrams were discussed in Chapter 8. The location of the MS in the connected state is hence tracked by the UTRAN at the Cell/URA level.

At the end of the active session, the MS leaves the connected state and enters the idle state. During this state transition, the MS repeats the Cell Selection/Reselection procedure. After cell selection, if the MS determines that its registration area (LA/RA) has changed in the duration of the active session (i.e. while it was in the connected state), it also repeats the Location Registration procedure. In other words, if the LA/RA of the MS at the end of a call/session is different from its LA/RA at the start of the call/session, the MS registers its new location with the CN at the end of the active session, when moving to the 'Idle State'.

9.7   MM PROCEDURES IN THE ACCESS NETWORK

The role of the UTRAN in Mobility management is limited mostly to tracking the location of the MS when the latter is in the connected state. The UTRAN plays no significant role when the MS is in the idle state. In this case, it is only required to support transparent transfer of mobility management signaling messages between the

MS and the Core Network (CN). The UTRAN is not required to interpret these messages, since these are meant for interpretation at the CN.

Figure 9.2 depicts the mobility management control plane for the UMTS. While the RRC layer provides the communication between the MS and the UTRAN, the Radio Access Network Application Part (RANAP) supports transparent transfer of the signaling messages between the UTRAN and the CN. Before the exchange of Mobility Management Signaling messages between the MS and the CN can begin, a signaling channel is required to be set up between them. This is done by setting up an RRC connection between the MS and the UTRAN, and an instance of RANAP/SCCP serving between the UTRAN and the CN. The message flow diagram for the 'NAS Signaling Connection Setup' was discussed in Chapter 8. After the set up of this signaling connection, the MM messages are exchanged between the MS and the CN. In the idle state of the MS, the role of UTRAN is therefore limited to relay of messages between the MS and the CN. Besides this, the UTRAN also supports 'Paging' of a mobile station in the idle state, when there is an incoming call for the MS. The process of paging was covered in Section 9.4. The message flow diagram for paging support in UTRAN was discussed in Chapter 8.

When the MS participates in a call/session, it enters the 'Connected State'. This enables the UTRAN to track and maintain the location of the MS at the cell/URA level. In the MM connected state, the MS is in the RRC 'Connected Mode', and the RRC procedures of cell update and URA update are used between the UTRAN and the MS for tracking the location of the latter. Besides this, the UTRAN MM procedures include support for SRNS relocation, and soft/hard handovers. These UTRAN procedures for the connected state of the MS were discussed in Chapters 7 and 8.

9.8   MM PROCEDURES IN THE CORE NETWORK

A mobile station can be attached to the CS or the PS domain, or to both. The MM procedures for location management in the CS domain differ slightly from procedures used in the PS domain. Hence, the procedures followed in these domains are discussed separately in the following two sub-sections.

9.8.1   MM Procedures in CS Domain

For a subscriber in the CS domain, the MM procedures in the CN involve interaction between the MSC, VLR and the HLR. An MSC area is defined as an area covered by the entire set of base stations controlled by the MSC. An MSC area may contain multiple location areas. Change in location information is forwarded to the HLR only if the MS has entered a new MSC area. Changes in LA are not propagated to the HLR, but are maintained in the VLR.

Mobility management procedures for the CS domain are discussed in 3GPP TS 23.012 and 3GPP TS 29.002 specifications. This section provides an overview of the MM procedures in the CS domain, which include the MM 'Location Update' and 'Purge MS' procedures. In the case of the former, procedures followed within the VLR and the HLR are also covered in detail. MM procedures for the MSC are not covered in the following sections. This is basically because, firstly, they only involve receipt of messages from the MS and the transfer of these to the VLR, and vice-versa. Hence, the MSC only acts as a bridge or a relay between the VLR and the MS. Secondly, the interface between the MSC and the VLR is not fully standardized. MSC and VLR are considered to co-exist, and also assumed to have an internal proprietary interface.

9.8.1.1   *MM Location Update Procedure*

For registration of location information within the CS domain, the 'MAP_Update_Location' message is used between the VLR and HLR. This message is initiated as a result of receipt of the 'Location_Updating Request' message from the MS (NAS Signaling message) indicating one of the following:

- Normal Location Update
- Periodic Location Update
- IMSI-Attach

As already mentioned, indication of IMSI-attach is only used when the MS is activated in the same LA in which it was last deactivated, which means that the LA of the MS has not changed. The last registered LA is stored in the MS, and on activation of the MS, this is compared with the current LA. If the two match, then using the IMSI-attach, the MS informs the network that it has once again become reachable, within the same LA.

9.8.1.1.1   *Location Update Procedure in the VLR*

The receipt of a 'Location Update' message from the MSC initiates the location management procedure in the VLR. The VLR first checks if there are any missing parameters in this request. In case of any parameter errors in the received message, the VLR sends back an appropriate error response to the MSC. In case the message is correctly received, the handling of the 'Location Update' message in the VLR depends upon whether the received identity of the subscriber in the 'Location Update' message is International Mobile Subscriber Identity (IMSI) or Temporary Mobile Subscriber Identity (TMSI). Recall from Chapter 3 that the IMSI is the permanent and TMSI the temporary identity of the subscriber. The TMSI is used instead of IMSI to hide the permanent identity.

If the received identity is IMSI, then the steps involved in the VLR (see Figure 9.5) are as follows:

1. Check whether the subscriber is known in the VLR (i.e. whether a location update has been previously received for this subscriber). If the subscriber is known, the

VLR verifies whether the previous LA of the subscriber (obtained by the VLR from the 'Location Update' request received from the MSC) is in the area serviced by it.

2. If either of the above conditions are not met (i.e. either the subscriber is new to the VLR, or the previous LA of the subscriber was not in the area controlled by the existing VLR), then the VLR marks a flag indicating the information of the new location, is to be sent to the HLR also. This means that when the subscriber moves into an area serviced by a different VLR, information of the change must also be sent to the HLR.

3. If, however, both the conditions in step 1 are met, then the 'Location Update' message received at the VLR does not trigger the updating of the HLR. This is because the only location information maintained at the HLR is the current MSC/ VLR of the subscriber. Unless this changes, there is no need to inform the HLR.

4. Check if authentication of the subscriber is required. If so, then the subscriber is authenticated. The process of authentication is discussed in detail in Chapter 14.

5. Upon successful authentication, the location information in the VLR is updated, and if HLR updating is required, then the 'MAP_Update_Location' message is sent to the HLR.

If, on the other hand, the subscriber identity received in the 'Location Update' message from the MSC is a TMSI, (Figure 9.5) the sequence of steps involved in the location update procedure in the VLR is as follows:

1. Check if the previous LA of the subscriber is in the area handled by the existing current VLR. It so, then the VLR should be able to identify the subscriber by the TMSI. If, however, the previous LA is not serviced by current VLR, then the previous VLR of the subscriber is contacted to obtain its IMSI. In some cases, if the IMSI of the subscriber cannot be obtained from the previous VLR, then the MS is requested to provide the IMSI of the subscriber to the current VLR.

2. Check whether the MSC area for the subscriber has changed. If it has, and if subscriber tracing is active for the subscriber, then this information is sent to the new MSC. Subscriber tracing is required for certain O&M purposes, as well as for maintaining record of subscriber activities. Subscriber tracing includes maintaining information of subscriber activities, such as the numbers from which terminating calls have been received, numbers to which calls were made, etc.

3. Check if authentication of the subscriber is required. If it is, then the subscriber is authenticated.

4. Upon successful authentication, the location information in the VLR is updated. Also, if the subscriber has moved to this VLR from some other VLR, or if its MSC area has changed, then the HLR is notified accordingly, through a 'MAP_Update_ Location' message.

Note that in the above two cases (where either 'IMSI' or 'TMSI' was received as identity), successful location update in the VLR would only be possible if the subscriber

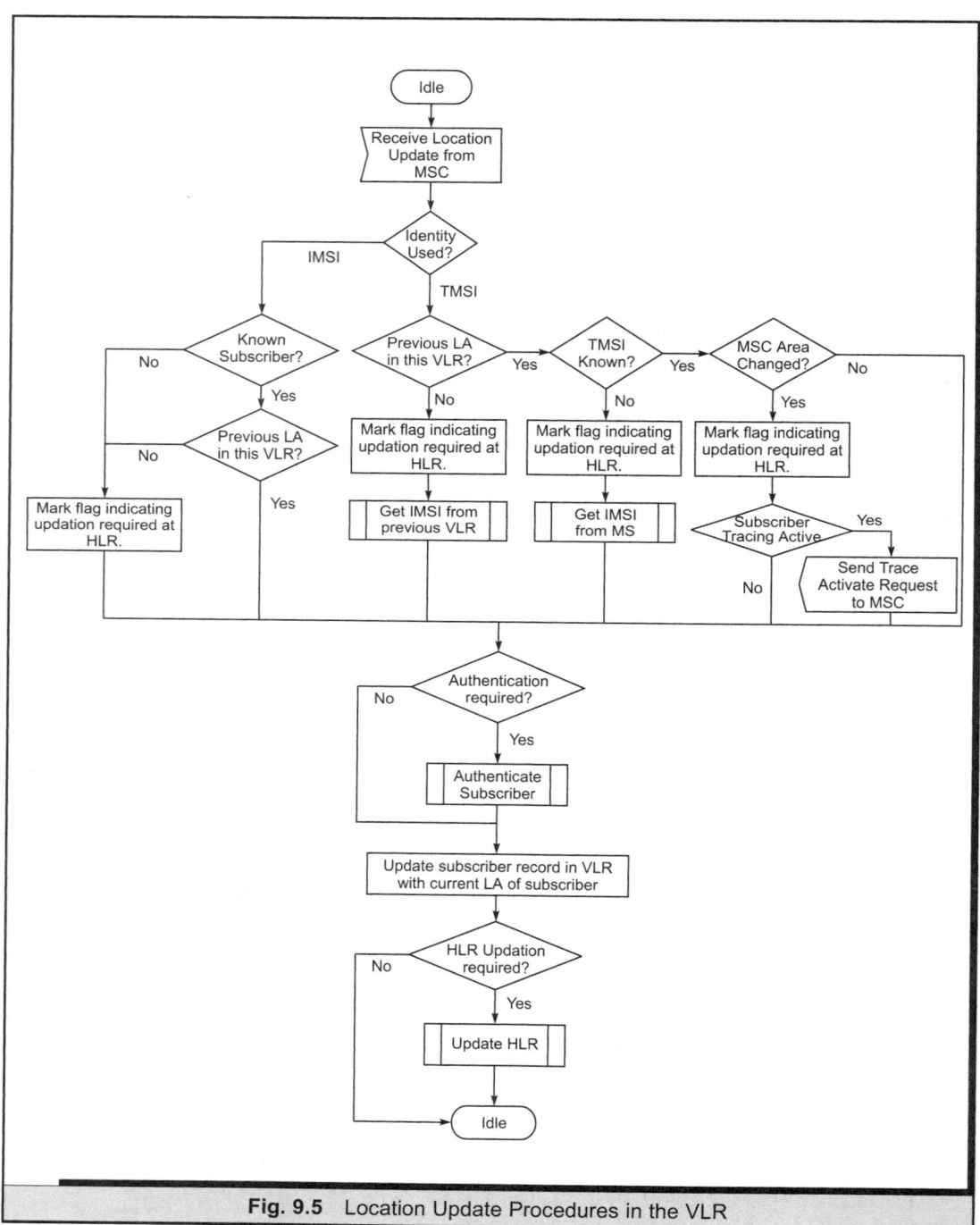

Fig. 9.5 Location Update Procedures in the VLR

is allowed to roam in the network handled by the VLR. If roaming is not allowed in the area serviced by the existing MSC/VLR, then the HLR would indicate this to the VLR (see Section 9.8.1.1.2), and the 'Update HLR' procedure (Figure 9.5) would not be successful.

9.8.1.1.2 *Location Update Procedure in the HLR*

The procedure in the HLR is initiated on receipt of the 'MAP_Update_Location' message from the VLR. During this procedure the location information is updated at the HLR, with the current MSC and the VLR of the subscriber. In this procedure, the subscriber data that is relevant to operations at the VLR is also transferred from the HLR to the VLR.

The sequence of steps involved in location update at the HLR is as follows:

1. Check if the subscriber is known in the HLR. If not known, then a negative response indicating 'unknown subscriber' is sent back to the subscriber.
2. Alternatively, check if the subscriber has access to CS domain services. If not, then a similar response, indicating 'unknown subscriber', is sent back.
3. Else, check if the VLR sending the 'MAP_Update_Location' message is same as the VLR where the subscriber was last known by HLR to exist. This condition is possible because the VLR of the subscriber may not have changed, and the HLR may have received the 'MAP_Update_Location' message as a result of one of the following cases:
   (a) TMSI was not known at VLR
   (b) MSC area has changed (see Figure 9.5)

   If this is a new different VLR, then a "MAP_Cancel_Location" message is sent to the previous VLR. This is a trigger for the previous VLR to delete the subscriber data from its database, since the subscriber has migrated to a new VLR. If the VLR is the same, then this step is skipped.
4. Check if the subscriber is allowed to roam into the visited network. If not, then a negative response is sent to the VLR indicating 'roaming not allowed'. The visited network is identified using the CC-NDC digits of the VLR address.
5. Else, check if subscriber tracing is active. If it is active, then the HLR sends a 'MAP_Activate_Trace_Mode' message to the new VLR, requesting activation of tracing for the subscriber. If not active, then this step is ignored.
6. Next, the relevant subscriber data is sent to the new VLR, which acknowledges this data and stores it in its database for future use.
7. On successful completion of data updation at the VLR, a positive response to the 'MAP_Update_Location' message is sent by HLR to the VLR. This completes the location update procedure in the HLR.

9.8.1.1.3 *Location Update Scenario*

Figure 9.6 depicts the message flow diagram for a location update scenario, which considers the case of a 'Normal Location Update', with change in MSC/VLR Area. The sequence of steps followed is:

1. The MS sends a 'Location_Updating Request' message using the NAS Signaling procedures. The 'Location_Updating' message indicates that the location update is a 'normal location update'. The MS includes the TMSI to identify the subscriber to the MSC/VLR. In addition, the LAI of the previous LA in which the subscriber was registered is also included in this message.

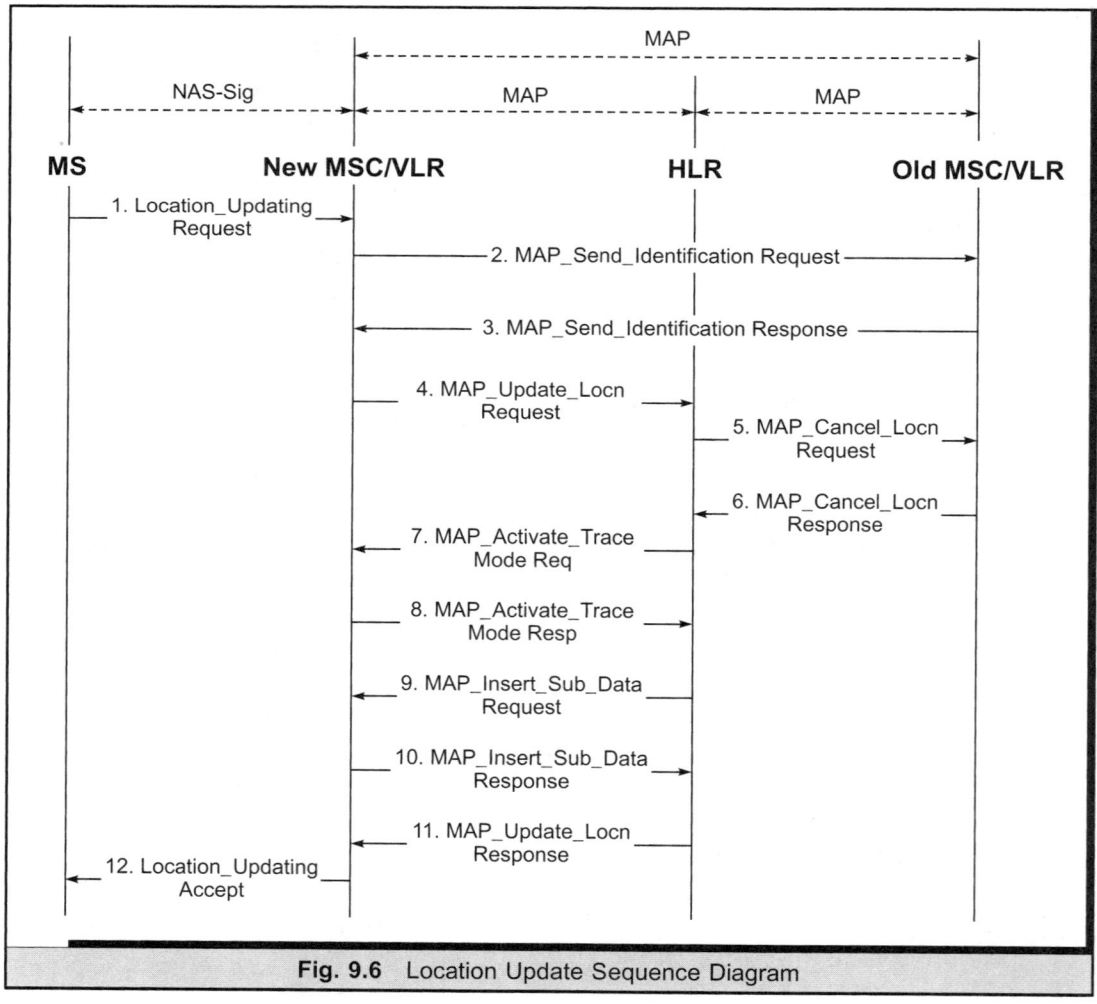

Fig. 9.6 Location Update Sequence Diagram

2. The new VLR checks if the subscriber is known to it. In this scenario, it is assumed that the subscriber is not known to the new VLR to which it has migrated for the first time. The new VLR then sends a 'MAP_Send_ Identification Request' message to the old VLR (which it determined from the LAI of the previous LA), requesting it to send the subscriber's IMSI. The TMSI is sent in this message to identify the subscriber to the old VLR.

3. The old VLR sends the IMSI to the new VLR as part of the 'MAP_Send_Identification Response' message.

4. After completing the sequence of steps described in Figure 9.5 (subscriber authentication, VLR record updation, etc.), the VLR sends a 'MAP_Update_ Location Request' message to the HLR to update the location information of the subscriber in the HLR. The new MSC and VLR numbers are included in this request message. The subscriber is identified in the HLR by using the IMSI received in the message from the VLR.

5. On receipt of this message, the HLR sends a 'MAP_Cancel_Location Request' message to the old VLR to indicate to it that the subscriber has moved out from its service area.

6. The old VLR, on receipt of this message, deletes the subscriber's information stored in its datastore. It then acknowledges the message from HLR by sending it a 'MAP_Cancel_Location Response' message.

7. The HLR then checks if subscriber tracing is active. If so, it sends a 'MAP_ Activate_Trace_Mode Request' message to the new VLR. This message includes information about the tracing records that have to be generated for the subscriber. If, however, tracing is not active for the subscriber at that time, this step is ignored.

8. The new MSC/VLR activate subscriber tracing as per information received from the HLR. The new VLR acknowledges the request from the HLR by sending it a 'MAP_Activate_Trace_Mode Response' message.

9. Next, the HLR sends relevant subscriber data to the new VLR. This data is required by the latter to provide services to the subscriber. The subscriber data is sent as a part of the 'MAP_Insert_Subscriber_Data Request' message to the new VLR.

10. The new VLR updates its datastore with subscriber data, and sends a 'MAP_ Insert_Subscriber_Data Response' message to the HLR.

11. The HLR then acknowledges successful completion of the location update procedure, by sending a 'MAP_Update_Location Response' message to the new VLR.

12. The new VLR in turn confirms the successful completion of the location update procedure by sending a 'Location_Updating Accept' message to the MS, via NAS Signaling, thus completing the location update procedure.

The scenario discussed above is the most exhaustive scenario. A subset of this are the 'IMSI-attach' and 'Periodic Location Update' scenarios. In case of IMSI-attach, the

MS is activated in the same LA in which it was last deactivated, and hence signaling is required only between the MS and the VLR. Similarly, in case of 'Periodic Location Update', signaling is restricted between the MS and the VLR; the HLR is not involved in the process.

9.8.1.2 MM Purge MS Procedure

When a mobile station is deactivated (powered 'off'), there are two mechanisms by which the network gets to know of it. One method involves informing the network through the use of an explicit 'IMSI-Detach message (Section 9.6.1.3). However, though the support for 'IMSI-Attach and Detach' is mandatory in the MS, the implementation of these procedures by the network elements is optional. Information about whether the network supports 'IMSI-Attach and Detach' is obtained by the MS from the system information that is broadcast by the network. In the second method, where the network does not implement the 'IMSI-detach' procedures, a separate mechanism is used to identify a case of potential MS deactivation, as follows.

In case the MS misses a pre-configured number of periodic location updates, the network assumes that it may have been deactivated. This is called Implicit 'IMSI Detach'. In this scenario, the VLR sends a 'Purge MS' request towards the HLR, requesting its permission to delete the subscriber data from the VLR, and to freeze the TMSI that was allocated to the subscriber. The sequence of steps followed at HLR on receipt of a 'Purge MS' message is as follows:

1. Check if the subscriber is known at the HLR. If not known, a negative response is sent indicating 'unknown subscriber'. If, on the other hand, the subscriber is known at the HLR, then the following steps are executed.

2. Check if the Purge MS message is received from the same VLR as the one where the subscriber is currently known by HLR to be registered. If this is not the case, then the HLR simply sends back a successful response indicating that the VLR may delete the subscriber data. The VLR is not required to freeze the TMSI in this case, since as per HLR records, the subscriber has already moved to a new VLR, where it would have been allocated a new TMSI. Such a scenario is possible, for example, when the subscriber moved into a new VLR area, but the old VLR could not process the 'MAP_Cancel_Location' message sent by the HLR as part of the location update scenario.

3. If the 'Purge MS' message is received from the same VLR where the subscriber is registered, then the HLR marks a flag 'MS Purged for non-GPRS' for the subscriber in its database, and sends back a positive response to the VLR. The message indicates to the VLR that it may delete the subscriber data, and freeze the TMSI allocated to the subscriber. The subscriber, from then on, is treated as an unreachable subscriber. The VLR freezes the TMSI for some time, which it does

not allocate to any other subscriber immediately. This is done to safeguard against the original MS coming 'alive', and using the TMSI again.

Once the 'MS Purged for non-GPRS' flag is set at HLR, the flag can be unset as a result of a new location update message received from a VLR, for the subscriber. This means that the subscriber has become reachable and the flag can be unset. The 'Purge MS' sequence diagram between the VLR and HLR is depicted in Figure 9.7.

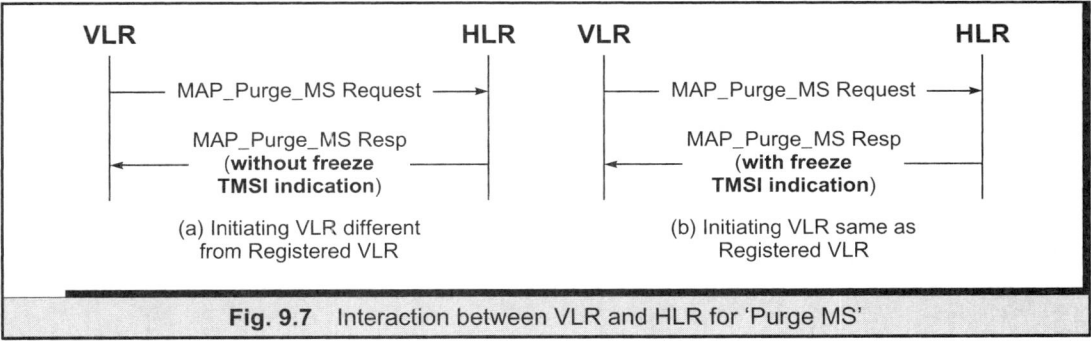

Fig. 9.7 Interaction between VLR and HLR for 'Purge MS'

9.8.2 MM Procedures in PS Domain

This section details the location management procedures for a PS domain subscriber, which slightly differ from those for a CS domain subscriber. MM procedures for the PS domain are discussed in 3GPP TS 23.060 specification. For a PS domain subscriber, location management procedures in the CN involve interaction between the SGSN, GGSN and the HLR. While the SGSN maintains information of the existing Routing Area (RA) of the MS, the HLR maintains information of the subscriber's current SGSN. Hence, the routing updates are sent to the HLR from the SGSN only when the MS enters a new SGSN, and not on each RA change.

The GGSN acts as the gateway between the PLMN and the external packet data networks. A Packet Data Protocol (PDP) context is maintained in the MS, the SGSN, and the GGSN. A PDP context contains information used for routing data packets between the MS and the GGSN, via the SGSN. The PDP context contains information about the PDP address assigned to the MS, the PDP Type (IPv4, IPv6, X.25), Quality of Service (QoS), etc. (refer Chapter 3).

For registration of location information within the CN, the 'MAP_Update_GPRS_Location' message is used between the SGSN and HLR. The location update procedure is triggered on receipt of one of the following messages from the MS:

- 'Attach' message
- 'Routing_Area_Update' message

A GPRS-Attach is carried out by an MS before it can avail the PS Domain services. However, when the MS no longer requires these services, it may carry out a GPRS-detach procedure. On receipt of the 'Attach' message (refer to NAS Signaling messages discussed in Chapter 6) from the MS, the SGSN may initiate location updating towards the HLR if either the SGSN has changed since the last GPRS-Detach, or if it is the first GPRS-attach. Subsequently, each update in the RA of the MS is conveyed to the SGSN through the normal routing area update procedure. Periodic routing area updates are sent by the MS to the SGSN to confirm its presence in the same SGSN area. The receipt of the routing area update indication at the SGSN would initiate location updating towards the HLR only if the SGSN of the MS has changed. The next two sections discuss the interaction between the SGSN and the HLR for both cases: GPRS-Attach and Routing Area Update.

9.8.2.1 GPRS-Attach Procedure

The MS uses the GPRS-attach procedure when it attaches itself to the network. Using this procedure, the MS provides the identity of the subscriber to the network, which may either be the IMSI, or the P-TMSI (Packet-TMSI, similar to the TMSI in CS Domain). In case the MS provides the P-TMSI as the identity, it also includes the Routing Area Identity (RAI) associated with the P-TMSI. Since a P-TMSI is allocated to the subscriber by an SGSN, the RAI is the identity of the RA where the subscriber was registered when it was allocated this P-TMSI.

The steps followed in the GPRS-attach procedure (Figure 9.8) are:

1. The MS sends the 'Attach Request' message to the SGSN using NAS signaling. The message includes the identity of the MS.
2. In case the MS has identified the subscriber by the P-TMSI, the SGSN first checks if the subscriber was previously registered with it. If not, the SGSN contacts the previous SGSN of the MS to request for the IMSI of the subscriber. The new SGSN in this case includes the P-TMSI of the subscriber in the 'Identification Request' message sent to the old SGSN.
3. The old SGSN sends back the IMSI of the subscriber to the new SGSN using the 'Identification Response' message. In case both the old and the new SGSN do not understand the P-TMSI, the MS is requested by the new SGSN to provide its identity—the IMSI. This, however, is not shown in the scenario discussed here.
4. Authentication procedures are followed to authenticate the MS.
5. Optionally, the equipment of the MS (identified by the IMEI) is checked to verify that it is not black or grey-listed.
6. Subsequently, the SGSN checks whether the MS had carried out the GPRS-Detach in its region or not. In case the SGSN has changed since the MS carried out the Detach, or if this is the first GPRS-Attach request from the MS, the SGSN informs

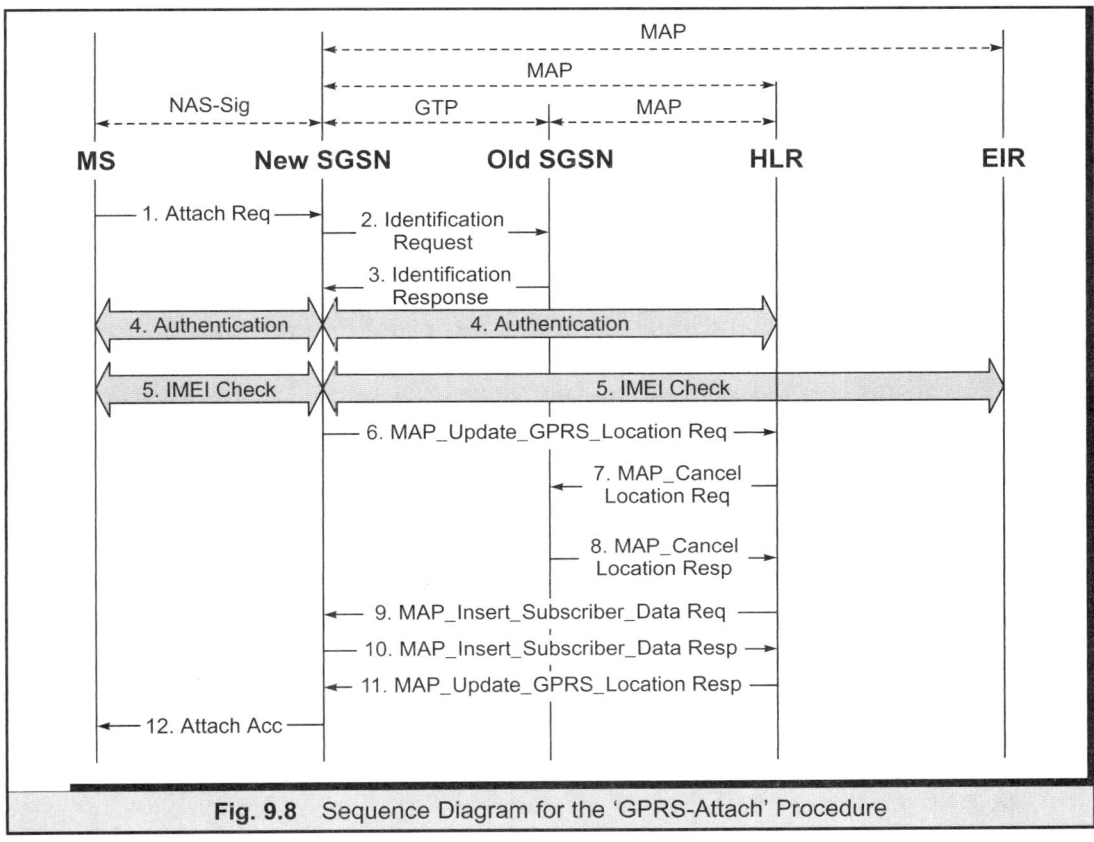

Fig. 9.8   Sequence Diagram for the 'GPRS-Attach' Procedure

the HLR about the location of the MS by sending it a 'MAP_Update_GPRS_ Location Request' message. The location update message to the HLR contains the SGSN number (an E.164 number), the SGSN address (IP address), and the IMSI of the subscriber.

7. The HLR initiates a 'MAP_Cancel_Location Request' message towards the previous SGSN, if any, where the MS was last registered. This is to inform the previous SGSN that the MS has moved to a different SGSN, and that it can purge/delete all data stored for the subscriber.

8. The old SGSN deletes subscriber records for the MS, and acknowledges this by sending a 'MAP_Cancel_Location Response' message to the HLR.

9. The HLR then sends the relevant subscriber data to the existing SGSN through one or more 'MAP_Insert_Subscriber_Data Request' messages. This scenario assumes that only one MAP message is required to transfer the complete subscriber data.

10. The existing SGSN updates its datastore with subscriber data, and sends a 'MAP_ Insert_Subscriber_Data Response' message to the HLR.

11. On successful completion of subscriber data transfer, the HLR sends a 'MAP_ Update_GPRS_Location Response' to the SGSN, indicating successful registration of location information of the MS at the HLR.

12. The SGSN, in turn, conveys this to the MS by sending it an 'Attach Accept' message. This completes the GPRS-attach procedure.

9.8.2.2 Routing Area Update Procedure

The Routing Area Update procedure is executed each time the MS moves into a new RA. Even when within one RA, the MS initiates this procedure periodically, to confirm its presence in the same RA. On receipt of the RA update message, the SGSN checks if the RA or the SGSN of the MS has changed since the last update. In case the RA has changed, the SGSN modifies its information to reflect the current RA of the MS. And if the SGSN has changed, i.e. if it has received an update from the MS for the first time, the SGSN also informs the HLR and the GGSN that the SGSN of the MS has changed.

In case of a change in the RA (but not the SGSN), or in case of the periodic RA updates, the SGSN simply updates its information for the MS. There is no need to inform either the GGSN or the HLR of the MS. In case of change in the SGSN, however, the GGSN and the HLR need to be notified. The GGSN routes incoming packets for the MS through the SGSN, and hence, needs to be informed of updation of the SGSN. The sequence of steps involved in the RA update procedure (Figure 9.9), in case the SGSN has changed, is as follows:

1. The MS sends the 'Routing_Area_Update_Request' message to the SGSN with the old RAI and the old P-TMSI.

2. The SGSN checks if it has received the request from the MS for the first time. If this is the case, the SGSN requests the old SGSN of the MS for information about the MS, by sending a 'SGSN_Context Request' message. The information requested includes information about the PDP contexts of the MS. This step will be executed in case the SGSN for the MS has changed since the last RA update.

3. The old SGSN responds by sending the PDP Context information for the MS in the 'SGSN_Context Response' message.

4. Further, the old SGSN duplicates the buffered messages that it has received for the MS, and starts tunneling them to the new SGSN. This is done to ensure that the messages for the MS are not lost during location update between the two SGSNs.

5. The new SGSN sends a 'Update_PDP_Context Request' to the GGSN to update the PDP context of the MS with the new SGSN address.

6. The GGSN acknowledges the change to the PDP Context by sending a 'Update_ PDP_Context Response' message to the new SGSN.

7. Next, the SGSN informs the HLR of the change in the location (SGSN) of the MS by sending a 'MAP_Update_GPRS_Location Request' message. The location update

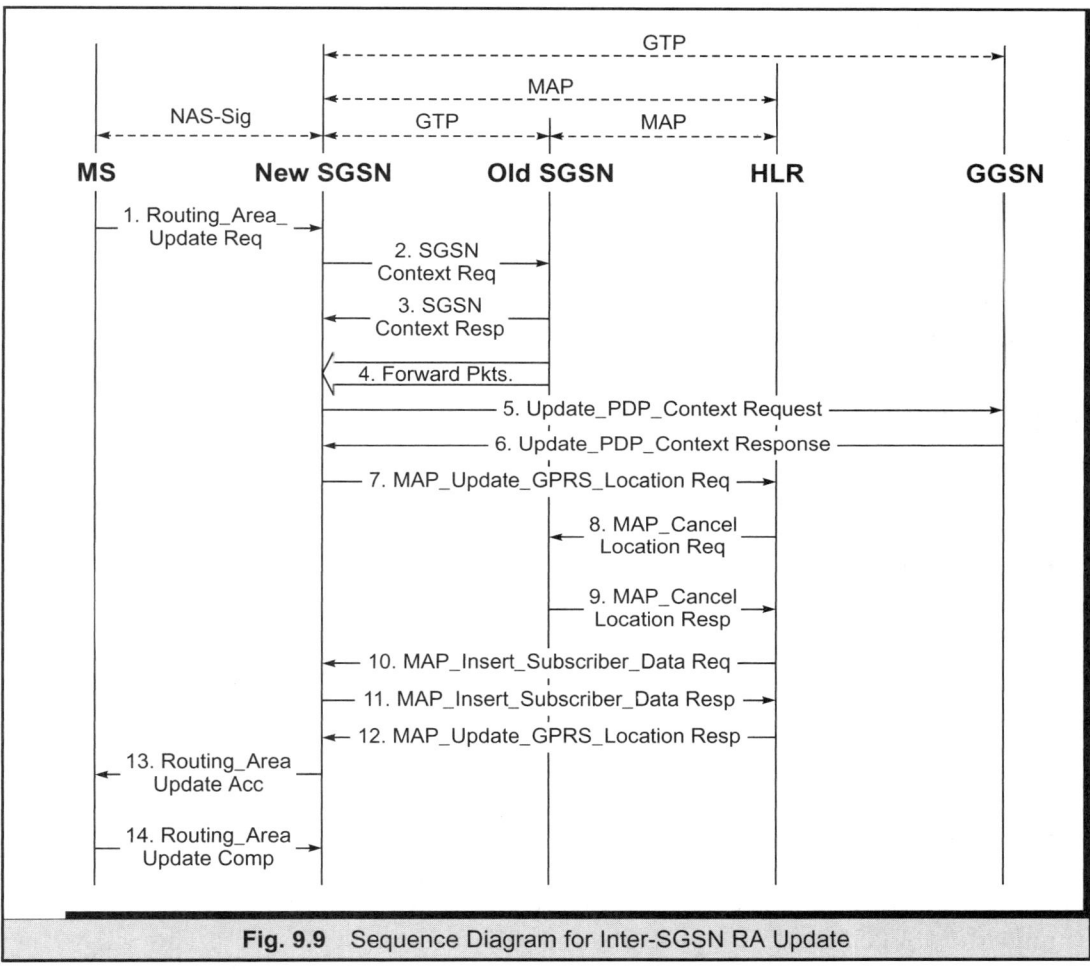

Fig. 9.9 Sequence Diagram for Inter-SGSN RA Update

message to the HLR contains the new SGSN Number (an E.164 number), the SGSN address (IP address), and the IMSI of the MS.

8. The HLR initiates a 'MAP_Cancel_Location Request' message towards the previous SGSN, if any, where the MS was last registered. This is an indication to the previous SGSN that the MS has moved to a different SGSN, and the previous SGSN can purge/delete all data stored for the subscriber.

9. The previous SGSN acknowledges the deletion of subscriber data by sending a 'MAP_Cancel_Location Response' message to the HLR.

10. The HLR then sends the relevant subscriber data to the current SGSN using the 'MAP_Insert_Subscriber_Data Request' message.

11. The SGSN acknowledges receipt of the subscriber data by sending a 'MAP_ Insert_Subscriber_Data Response' message to the HLR.
12. On successful completion of subscriber data transfer, the HLR sends a 'MAP_ Update_GPRS_Location Response' message to the SGSN, indicating successful registration of location information of the MS at the HLR.
13. The SGSN then sends the 'Routing_Area_Update_Accept' message to the MS with the new P-TMSI.
14. The MS confirms the reallocation of the P-TMSI by sending back a 'Routing Area Update Complete' message to the SGSN.

This completes the RA update procedure.

9.8.2.3 Combined RA/LA Update Procedure

The UMTS architecture defines an optional Gs interface between the VLR and the SGSN. The Gs interface was first discussed in Chapter 6. This interface is used to facilitate combined LA/RA update and combined IMSI/GPRS-attach by the MS. This helps in saving the radio resources when the MS is attached wishes to attack to both the CS and the PS domain. Also, the presence of the Gs interface allows the paging of the MS for the CS domain through the SGSN. This section discusses the steps involved in a combined LA/RA update, or a combined IMSI/GPRS-Attach from the MS.

When the MS is both CS and PS-Attached, the LA and RA updating is done in a co-ordinated way to save radio resources. As discussed in Section 9.8.2.2, when an MS enters a new RA, it sends a 'Routing_Area_Update' message to the SGSN. In case the LA of the MS has also changed (i.e. the new RA belongs to a different LA), the 'Routing_ Area_Update' message from the MS contains the LA update as well. The SGSN then forwards the LA update to the MSC/VLR over the GS interface. The latter may return a new TMSI, which is sent to the MS by the SGSN. Similarly, the MS can carry out a combined IMSI/GPRS-attach by sending a request to the SGSN only. The SGSN then forwards a Location Update request to the MSC/VLR, indicating IMSI-attach.

Figure 9.10 depicts the combined LA/RA update or combined 'attach' scenario. Note that the messages exchanged between the SGSN and the VLR are the same, and only the location update request from the SGSN to the MSC/VLR indicates whether it is a case of IMSI-attach, or normal LA update. The BSSAP+ protocol messages are used between the MSC/VLR and the SGSN.

9.8.2.4 Purge MS Procedure

As in the case of a VLR, the Purge procedures are also defined for the SGSN. These procedures are identical to those defined for the CS domain. In case the MS misses a pre-configured number of periodic RA updates, the network assumes that the MS may

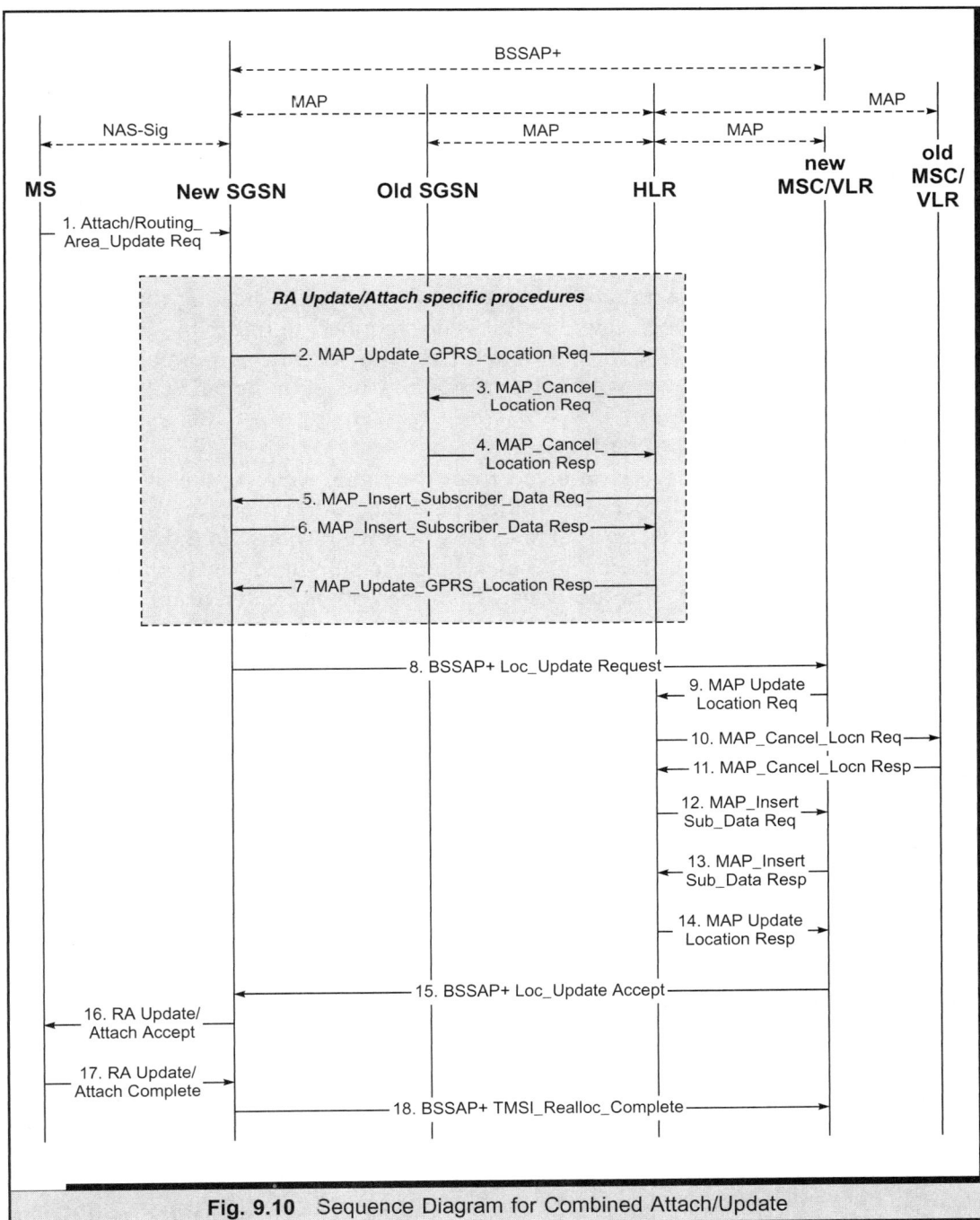

Fig. 9.10 Sequence Diagram for Combined Attach/Update

have been powered 'off', or has moved out of its area. In this scenario, the SGSN sends a 'MAP_Purge_MS' request towards the HLR, requesting its permission to delete the subscriber data from the SGSN, and to freeze the P-TMSI that was allocated to the subscriber. The sequence of steps followed at HLR on receipt of a 'Purge MS' message is:

1. Check if the subscriber is known at the HLR. If not known, a negative response is sent back indicating 'unknown subscriber'. If, on the other hand, the subscriber is known at the HLR, then the following steps are executed.

2. Check if the Purge MS message is received from the same SGSN as the one where the subscriber is currently known by HLR to be registered. If this is not the case, then the HLR simply sends back a successful response indicating that the SGSN may delete the subscriber data. Here, it is not required to freeze the P-TMSI, as in the case of Purge MS for the CS domain. Such a scenario is possible when, for example, the subscriber moved into a new SGSN area, but the old SGSN could not process the 'MAP_Cancel_Location' message sent by the HLR as part of the Routing Area Update scenario.

3. If the Purge MS message is received from the same SGSN as the one where the subscriber is registered, then the HLR marks a flag ('MS Purged for GPRS') for the subscriber in its database, and sends back a positive response to the SGSN. This is an indication to the SGSN that it may delete the subscriber data, and freeze the P-TMSI allocated to it. The subscriber, from then on, is treated as an unreachable subscriber.

Once the 'MS Purged for GPRS' flag is set at HLR the flag is unset as a result of a new location update message received from a SGSN for the subscriber. This means that the subscriber has become reachable, and the flag can be unset. The interaction between the SGSN and the HLR for Purge MS scenario is similar to the interaction between the VLR and the HLR, as depicted in Figure 9.7.

9.8.3 Super-Charger Functionality

When an MS registers in a new VLR/SGSN area, the VLR/SGSN initiates a location update message towards the HLR. As a result, the subscriber data relevant to the VLR/SGSN is transferred from the HLR to the former. In most cases, it is expected that the MS would be shuttling back and forth between only a few VLR/SGSNs. In such a scenario, the VLR/SGSNs can be configured to retain the subscriber data, even after the subscriber has moved out of the area serviced by them (i.e. VLR/SGSN can be configured not to purge subscriber data). If this is the case, the data will not need to be obtained from the HLR each time the subscriber re-registers with the VLR/SGSN. This is the concept behind the '*Super-Charger*' functionality.

The '*Super-Charger*' functionality reduces the number of signaling messages exchanged between network entities at the time of location update. This significantly reduces the time required to update the location of the MS in the network. Further, it also saves on network resources, since the subscriber data is not required to be transferred from the HLR to the VLR/SGSN each time the subscriber registers with the latter.

An important consideration for the success of the super-charger functionality, though, is the assumption that subscriber data does not change very frequently. If this is not the case, then it can be clearly understood that the super-charger would not work, since by the time the MS moves back into a VLR area, the data last retained at the VLR for the MS may have become stale. However, this is normally not the case. Subscriber data modification is a very infrequent activity, much more infrequent when compared to the location update of an MS.

Support for the super-charger at different network entities (HLR, VLR and SGSN) is optional, and if implemented, works as follows:

1. The HLR maintains an 'Age Indicator' for the subscriber data of each subscriber in its data-store. Each time the subscriber data is modified, the HLR updates the age indicator for the data.
2. When the HLR transfers the subscriber data to a VLR/SGSN, which supports the super-charger, it also sends the age indicator, indicating the time when this data was last modified. The VLR/SGSN stores this age indicator along with the subscriber data.
3. When the MS moves into a new VLR/SGSN area, the latter sends a Location Update message towards the HLR. In case the VLR supports the super-charger, this indication is included in the Location Update message sent to the HLR. Also, if the VLR/SGSN already has the subscriber data since the last transfer, the age indicator of this data is also provided.
4. On receipt of the Location Update message with the super-charger support indicator and the age indicator, the HLR compares the age indicator received from the VLR/SGSN with the one stored in its data-store. In case the two match, then the step involving transfer of subscriber data to the VLR/SGSN is skipped. This is done because the VLR/SGSN has a consistent set of subscriber data, and the data at the HLR has not changed since it was last transferred to this VLR/SGSN. However, if the age indicators do not match, normal transfer of data takes place.
5. The super-charger support indicator for this VLR/SGSN is also stored in the HLR data-store for future reference.
6. Next, the HLR checks if the previous VLR/SGSN of the subscriber had indicated support for the super-charger (this information would also have been stored in the HLR, at the time the subscriber registered with the previous VLR/SGSN). If the previous VLR/SGSN of the subscriber also supported the super-charger, then the 'Cancel Location' message is not sent towards it. This is to ensure that the previous VLR/SGSN continues to maintain the subscriber data in its data-store for this subscriber.

7. Hence, even when the MS has moved out of a VLR/SGSN area, the latter continues to hold the subscriber data. However, this means that the VLR/SGSN require a larger data-store than what they would normally have required if the super-charger were not supported. Also, since the size of the data-store cannot be infinite, the subscriber data would have to be purged after some time by the VLR/SGSN, when its data-store becomes full. This mechanism is left open for the VLR/SGSN to implement; no standard mechanism has been defined. As a possible mechanism, the VLR/SGSN may use the 'First-In First-out' or the 'Least Recently Used' principle to purge subscriber data for subscribers not currently in its area.

Figure 9.11 depicts the various scenarios of the super-charger, when the support is available at different network entities. The super-charger is discussed in detail in 3GPP TS 23.116 specification.

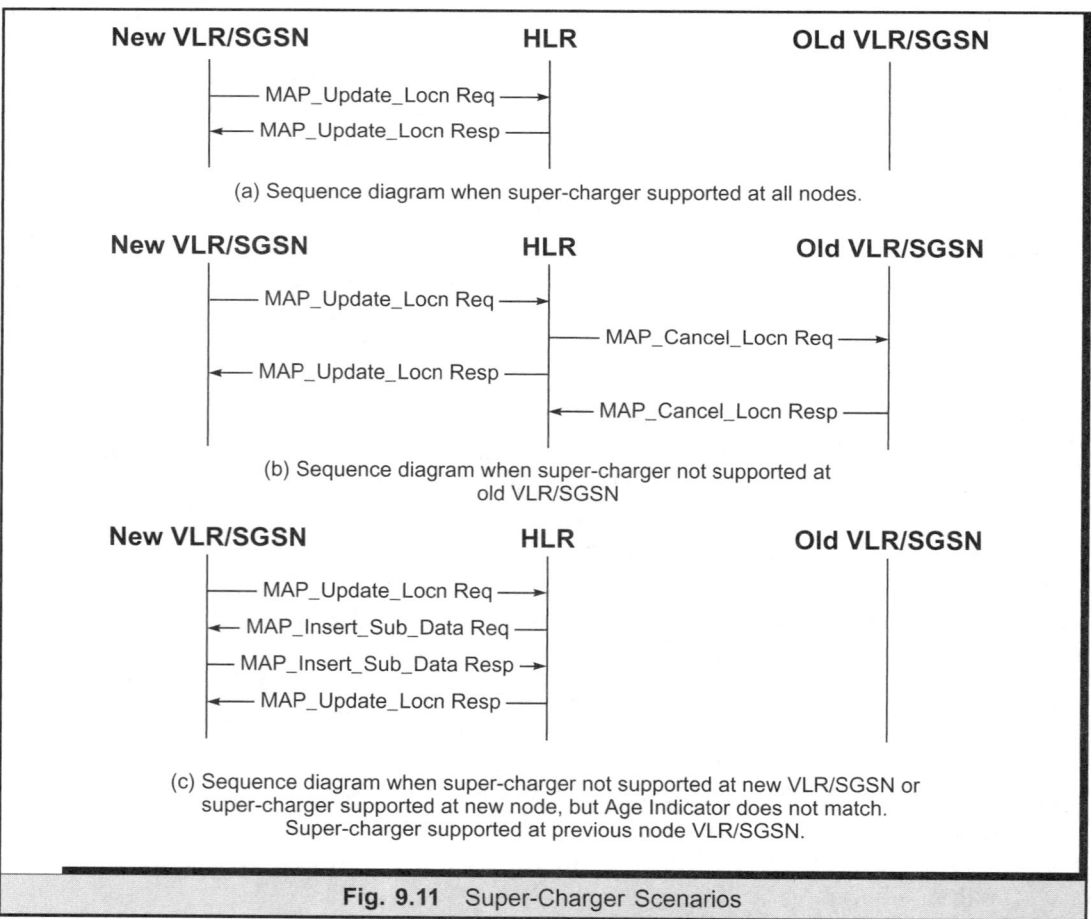

(a) Sequence diagram when super-charger supported at all nodes.

(b) Sequence diagram when super-charger not supported at old VLR/SGSN

(c) Sequence diagram when super-charger not supported at new VLR/SGSN or super-charger supported at new node, but Age Indicator does not match. Super-charger supported at previous node VLR/SGSN.

Fig. 9.11   Super-Charger Scenarios

SUMMARY

The main objective of the Mobility Management (MM) procedures is to maintain and manage the location information of a UE. These procedures also facilitate subscribers to roam into network operated by different service provides. The MM protocol defines a three-state MM state machine, which is implemented in the MS as well as in the CN nodes (i.e. VLR and SGSN), for mobility management activities. The MM state machine defines three states: Detached, Idle and Connected states. The MM state machine has been discussed in detail in Section 9.2. Section 9.3 discussed the location information stored in various network elements and the hierarchical nature of this information.

The MS, Access Network along with the Core Network participate in implementing the MM procedures. MM Procedures for the MS, Access Network and Core Network (both CS and PS domain) have been discussed in the chapter. The important MM procedures include the IMSI Attach/Detach, GPRS Attach/Detach, Location Update, Routing Area Update and Combined Location Area-Routing Area Update procedures. Each of these procedures has been discussed in detail in the chapter.

CALL HANDLING

10.1 INTRODUCTION

The previous chapter covered the Mobility Management procedures, using which the MS registers its presence in the PLMN with the VLR. After registering its location information successfully, the MS is ready to initiate and receive calls. Calls *initiated* by the MS are termed as *Mobile-Originated Calls* (MO-Calls), while those *received* by it are termed as *Mobile-Terminated Calls* (MT-Calls). The procedures involved in handling the MO and MT calls of the MS are clubbed under the 'Call Handling' functionality. This chapter covers the procedures for both the MO and MT calls. A basic mobile-to-mobile (end-to-end) call can be treated as a combination of an MO and an MT call. Hence, the complete end-to-end call establishment is not covered separately in this chapter, but is taken up during the discussion on optimal routing procedures.

Before proceeding with the concepts of call handling, it is important to clarify a few terms like HPLMN, VPLMN and IPLMN, that will be used during the discussion on call handling procedures. The term HPLMN is used to denote the 'Home' PLMN of an MS, which maintains the subscription records and the current location information of the latter. In other words, this HPLMN is the PLMN where the HLR for the MS resides. The VPLMN is the 'Visited' PLMN where the MS is located/registered. While all subscribers are allowed to roam within their HPLMN, only those with the roaming facility are allowed to visit networks (VPLMNs) outside of the HPLMN. Verification of whether or not a subscriber has this facility is done at the time of location registration, as was discussed in the Location Update procedure in Chapter 9. The term IPLMN (or 'Interrogating' PLMN) denotes a PLMN that interrogates the HLR about the current VPLMN of an MS. This information is required whenever the IPLMN needs to route a call towards the MS in question. Normally, the IPLMN and HPLMN of the MS are the same, except when Optimal Routing procedures are used, in which case, the PLMNs may be distinct. The IPLMN, in such cases, is normally the PLMN from where the call originated.

This chapter is organized as follows: The next section discusses the architecture of MO and MT calls, which forms the basis of call handling procedures discussed in Sections 10.3 and 10.4 respectively. Section 10.5 discusses the interaction of the call forwarding and call barring supplementary services with call handling procedures. Optimal routing procedures are discussed in Section 10.6. To conclude, Section 10.7 carries a discussion on the Immediate Service Termination (IST) service.

10.2 ARCHITECTURE OF MO AND MT CALLS

This section describes the various entities involved in establishing an MO and MT call, as they form the basis of the message flow diagrams, discussed in Sections 10.3 and 10.4.

10.2.1 Architecture of Mobile-Originated Call

As mentioned earlier, Mobile-Originated (MO) calls are calls that are initiated by the MS. Figure 10.1 depicts the architecture of a basic MO call, which is handled entirely by the VPLMN of the MS; no interaction is required between the VPLMN and the HPLMN. The IPLMN is not relevant in this case.

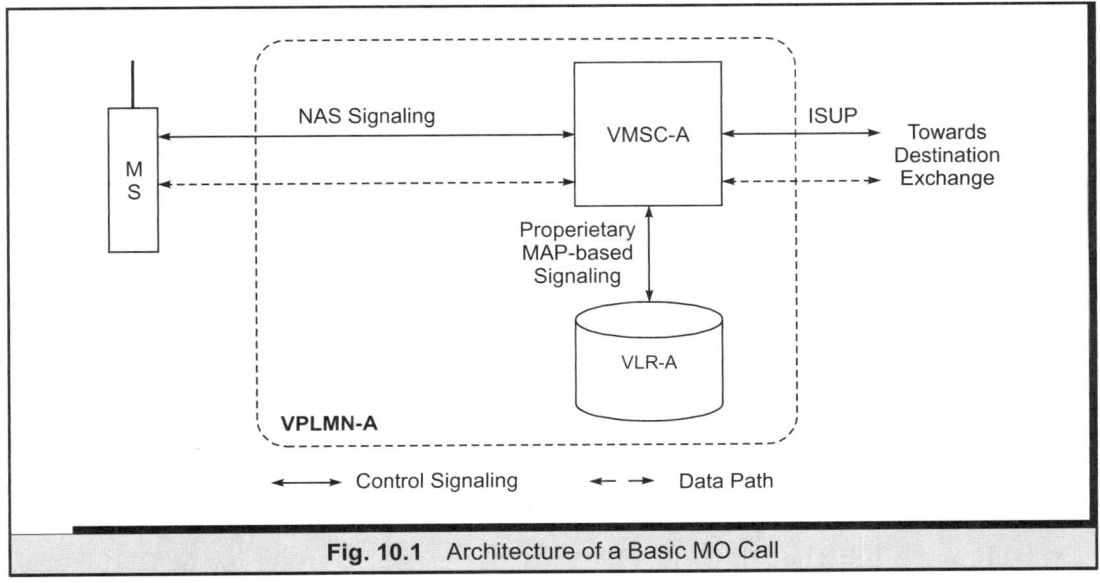

Fig. 10.1 Architecture of a Basic MO Call

A basic MO call involves exchange of signaling messages between the following entities:

- **MS and its VMSC (via the UTRAN):** The MS and VMSC exchange NAS signaling messages for call control over an established MM connection. For call establishment, the signaling messages include information regarding the called and the calling party address, besides other information useful for call setup.

- **VMSC and VLR (using a proprietary/MAP-based interface):** This is used by the VMSC to verify from the VLR that the requested service (here, the outgoing call) is allowed to the subscriber. The VLR maintains the temporary subscription information for the MS, and hence has to be consulted by the VMSC before the outgoing call is allowed. The MAP protocol defines only some of the messages used between the MSC and the VLR. Other signaling messages are considered to be proprietary, and are not standardized. This chapter employs the convention of using the messages defined by MAP, wherever possible. Where MAP messages are not defined, proprietary signaling messages will be used.

- **VMSC and destination exchange:** The VMSC exchanges signaling messages with the destination exchange using the ISUP protocol (discussed in Chapter 6). The ISUP message for call establishment includes address information of the originating and the destination entities, along with information on the dialed digits.

10.2.2 Architecture of Mobile-Terminated Call

Figure 10.2 depicts the architecture of a basic MT call. As shown in the figure, handling an MT call involves interaction between the IPLMN, HPLMN and the VPLMN. Note that though the discussion treats the IPLMN as distinct from the HPLMN, this distinction is evident only when Optimal Routing (OR) procedures are employed to route the MT call to an MS. In most other cases, the IPLMN and the HPLMN will be the same.

A basic MT call involves exchange of signaling messages between the following entities:

- **GMSC and HLR:** The GMSC and HLR exchange MAP signaling messages between them. When the GMSC of the IPLMN has to route a call towards the MS, it requests routing information from the HLR (in the HPLMN). The HLR maintains information of the MSC/VLR where the subscriber is currently registered (refer to Chapter 9), and also provides the GMSC with the routing information to route the call till the VMSC.

- **HLR and VLR:** The HLR and VLR exchange messages using the MAP protocol. On receipt of a request for routing information from the GMSC, the HLR in turn

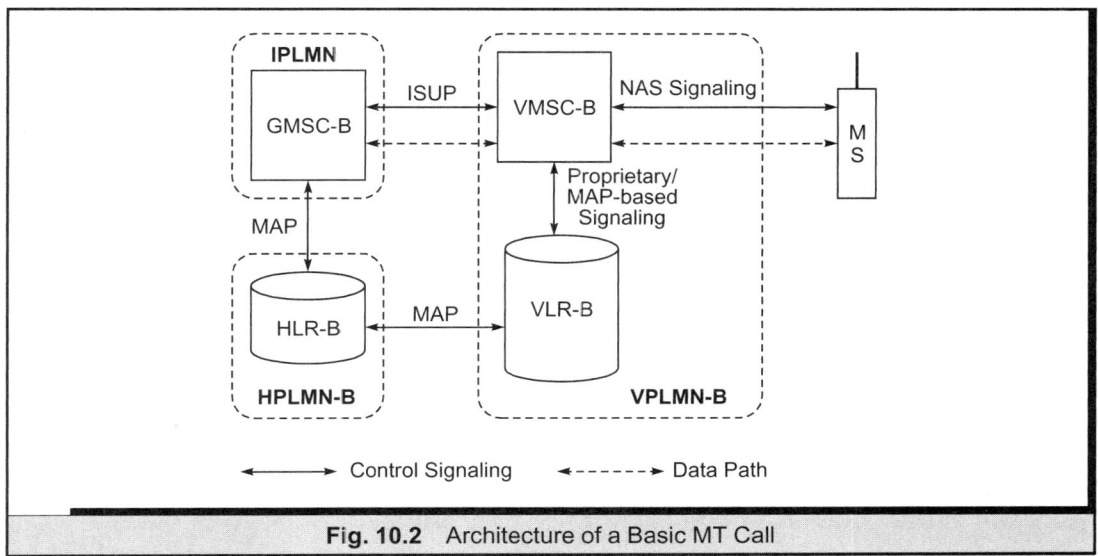

Fig. 10.2 Architecture of a Basic MT Call

requests the VLR (in the VPLMN) of the MS to provide a 'Roaming Number' for this incoming call towards the MS. Each VLR maintains a list of free ISDN numbers that can be dynamically assigned to mobile stations that have an incoming call. This ISDN number can be used to establish the call-leg between the GMSC and the VMSC. The VLR provides one such free ISDN number to the HLR. This number is called a 'Mobile Station Roaming Number, (MSRN)'. It is this number that the HLR returns to the GMSC as routing information, besides the MSC and VLR number. The MSRN was introduced in Chapter 3 as a temporary number allocated to the MS.

- **GMSC and VMSC:** The GMSC and the VMSC exchange ISUP messages for establishing the call-leg between them. The call from the GMSC to the VMSC in the VPLMN is established using the MSRN provided by the HLR to the GMSC.

- **VMSC and VLR:** The VMSC and VLR exchange messages over a non-standardized interface. As in the case of an MO call, the VMSC consults the VLR to verify that the MT call can be delivered to the MS. In certain situations, like the case where the operator has barred incoming calls for the MS, the VLR may decide to disallow the call. This interface is also used by the VLR to direct the VMSC to employ the paging process to determine the exact location of the MS.

- **VMSC and MS:** If the VLR decides that the call can be delivered to the MS, the VMSC uses NAS-layer signaling to inform the MS of the incoming call. NAS-layer signaling is used to establish the call with the MS.

10.2.3 Architecture of a Basic Mobile-to-Mobile Call

As depicted in Figures 10.1 and 10.2, the convention followed in the specifications is to use an 'A' with the entities in the PLMN which initiate the call (MS-A, VMSC-A and VLR-A) and a 'B' with those that terminate the call (VMSC-B, VLR-B and MS-B). Apart from this, the entities in the HPLMN and the IPLMN are also suffixed with 'B' (GMSC-B, HLR-B), especially since they are also participating in the MT call. This convention stems from the fact that the subscriber originating the call is called the 'A' subscriber, while the subscriber terminating the call is called the 'B' subscriber. This terminology will be followed throughout the chapter.

Figure 10.3 clubs together the architecture of the MO and MT call to depict the architecture of a basic mobile-to-mobile call. As is evident, the call from the 'A' subscriber in VPLMN-A to the 'B' subscriber in VPLMN-B always goes through the GMSC of the HPLMN-B. This happens since the HPLMN-B is the only point of contact for VPLMN-A for reaching the 'B' subscriber. HPLMN-B can thus be considered as the fixed part of the mobile subscriber 'B' that is known to VPLMN-A. HPLMN-B itself stores the actual location of the 'B' subscriber. Once the call reaches GMSC-B in HPLMN-B, GMSC-B queries HLR-B as to the location of the 'B' subscriber in order to determine the VPLMN of the latter. This is the normal procedure of routing information retrieval that was covered in discussion on MT call handling. Since it is GMSC-B that normally queries HLR-B as regards the routing information, the IPLMN is normally the same as the HPLMN. Once GMSC-B identifies the VMSC-B where the 'B' subscriber is registered, it connects the call to VMSC-B. Thus, a basic mobile-to-mobile call takes the route as follows: from VMSC-A to GMSC-A, to GMSC-B and then to VMSC-B.

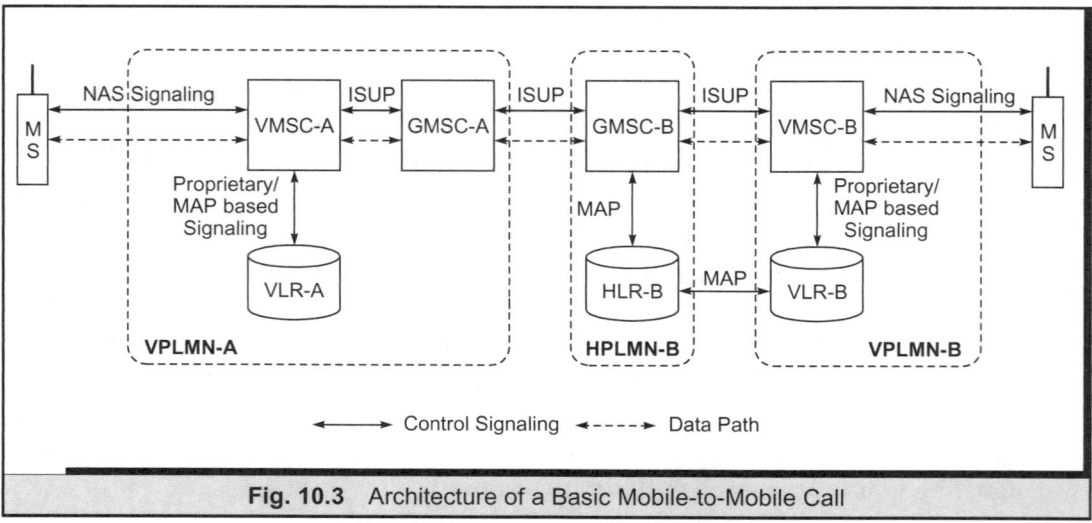

Control Signaling ◄──────►   Data Path ◄----►

Fig. 10.3 Architecture of a Basic Mobile-to-Mobile Call

A few important points that emerge from this description of the MO and MT call handling are as follows:

1. For an MO call, there is no need to involve the HPLMN of the subscriber. This is because all information required to handle the MO call (e.g. checks on whether or not the MS is allowed to initiate the call, etc.) is available at VLR-A. This information is transferred from HLR-A to VLR-A at the time of Location Registration (Location Update procedure), using the MAP_Insert_ Subscriber_Data message (refer to Chapter 9).

2. An MT call always involves interaction with the HPLMN, since the current VPLMN of the MS is available and tracked only at the HPLMN. Hence, for GMSC-B, the HLR-B in the HPLMN is the point of contact to obtain the current location of the MS.

3. For the MT call, the HPLMN has to further interact with the VPLMN, since the call cannot be routed unless the MS is temporarily assigned some fixed ISDN number (within the VMSC-B area) for the duration of the incoming call. This temporary number is called the Mobile Station Roaming Number (MSRN) and is assigned to the MS by VLR-B. The call is connected from GMSC-B to VMSC-B using this MSRN.

4. While the exact location of the MS is known when it initiates an outgoing call, in case of an incoming call, the location of the MS is required to be exactly determined within the MSC area. For an MO call, the location of the MS is known since it is the MS that first establishes contact with the network. However, in case of an MT call, the MS would normally be in the 'Idle State' (refer to MM state model in Chapter 9), and hence its exact cell location would not be known. Paging is used to determine the exact cell location of an MS for which an MT call arrives.

5. Paging may be done at one of the two different stages during call establishment. It may either be done when HLR-B requests the roaming number from VLR-B, or else it may occur after the call is connected from GMSC-B to VMSC-B on the roaming number. The latter is more likely, wherein paging is done after the call-leg between the VMSC-B and the GMSC-B has been established. However, the former is also possible, whereby paging is done at the time the roaming number is requested. This is based on the assumption that a roaming number is requested only when there is an incoming call for the MS, and hence paging could be done earlier in anticipation of the incoming call. This is called *Pre-Paging*. In case pre-paging is employed, both GMSC-B and HLR-B would have to wait longer for their queries as regards the routing information. This is because VLR-B responds to the HLR request for a roaming number after it has successfully determined the cell location of the MS using paging. Thus, pre-paging can only be employed if HLR-B and GMSC-B are ready to wait longer for their respective queries; hence, they are said to be supporting pre-paging.

In the next few sections, the finer details of MO and MT call handling—i.e. information flows between the various entities involved, and the messages exchanged—are discussed.

10.3   MOBILE-ORIGINATED CALL HANDLING

As discussed in the previous section, an MO call requires interaction between entities within the VPLMN. Figure 10.4 depicts the information flow between the MS, VMSC and the VLR for a basic MO call. The sequence of steps involved in the establishment of the MO call is as follows:

1. When MS-A wishes to originate a call, it sends a 'Connection Management Service Request' message to the VMSC-A to establish an MM connection.
2. The VMSC-A forwards the request from MS-A to the VLR-A by sending a 'Process Access Request' message to the latter. The VLR is the central entity that manages the MO call related procedure.
3. On receipt of the 'Process Access Request' message, VLR-A may initiate authentication procedures to authenticate MS-A. As suggested in Note 1 of Figure 10.4, authentication procedures may be initiated at any stage during the establishment of an MO call. This figure gives one possible location of the authentication message in the sequence of message flows. Authentication procedures may be initiated under various circumstances. One such scenario is when the VLR cannot identify the MS.

   The authentication procedure followed between the VLR and the MS uses a Challenge-Response protocol, under which, the VLR first sends a random number to the MS (*Challenge*). The MS then calculates a value, called RES, using the received random number and a secret key as input to an authentication algorithm. The information about the authentication algorithm and the secret key is shared between the MS and the network, and is kept confidential. The RES calculated by the MS is sent back to the VLR (*Response*). The VLR verifies whether the MS has correctly calculated the value of RES. If the value of the RES is correct, then the MS is authenticated. This is because information regarding the secret key and authentication algorithm required to calculate the RES is known only to one MS. Authentication procedures between the MS and the VLR are discussed in detail in Chapter 14.
4. VMSC-A receives the random number from VLR-A in the 'MAP_Authenticate' request message. The received random number is forwarded by VMSC-A to MS-A using the 'Authentication Request' message.
5. On receipt of the 'Authentication' request from VMSC-A, MS-A calculates the RES using the received random number, and responds to the authentication request from VMSC-A with this calculated value of RES.

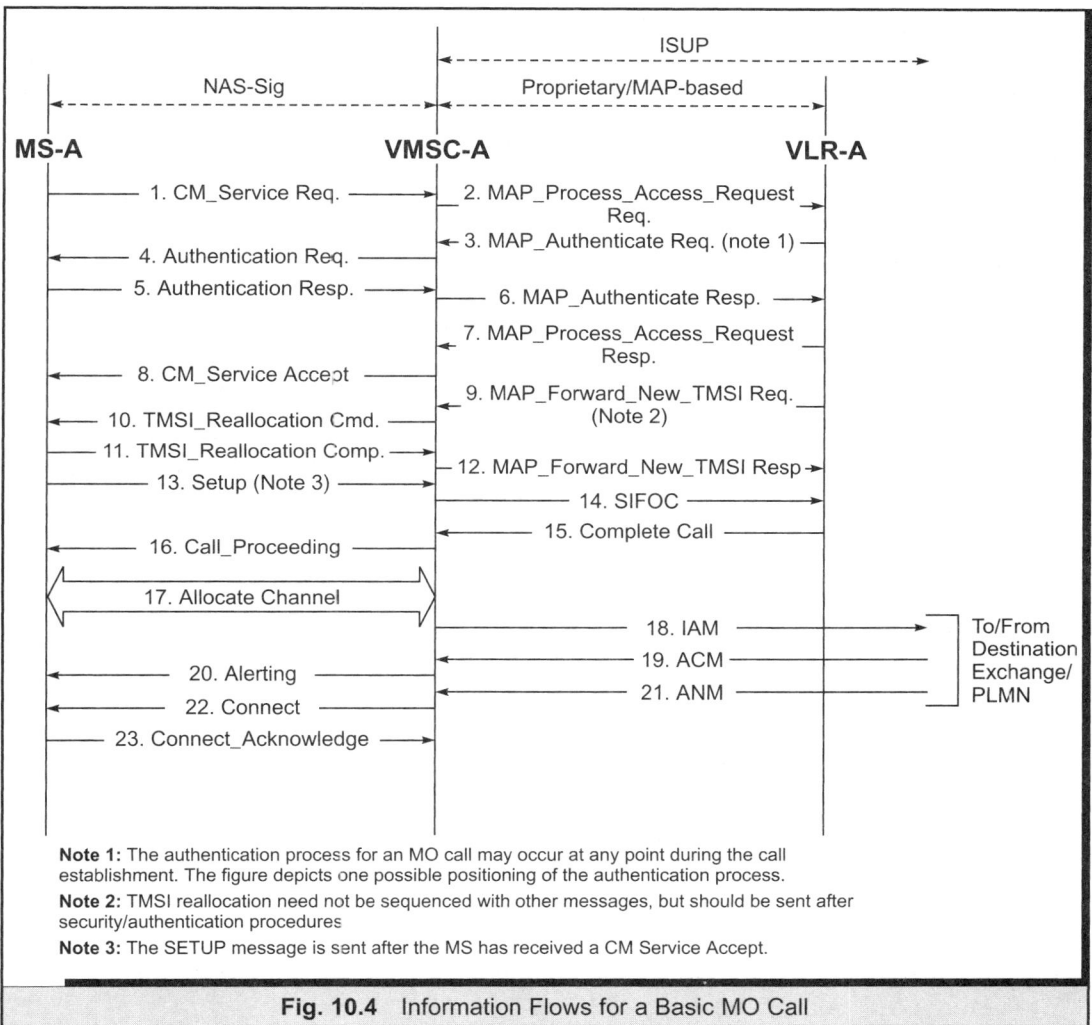

Fig. 10.4 Information Flows for a Basic MO Call

6. On receipt of the authentication response from MS-A, VMSC-A sends the authentication acknowledgement to VLR-A, along with the RES value calculated by MS-A. VLR-A then verifies the value calculated by MS-A, and authenticates it.

7. If VLR-A determines that MS-A is allowed to make the outgoing call, it sends back a positive response to the 'Process Access Request' message received from VMSC-A.

8. On receipt of the 'Process Access Request' response from VLR-A, VMSC-A sends a 'CM_Service Accept' message to MS-A, in response to the request received in step 1. This completes the establishment of the MM connection.

9. VLR-A may also send a new TMSI towards MS-A. TMSI is a temporary identity that is assigned to an MS. The MS uses this TMSI when initiating any further communication with the VLR. This is done to hide the permanent identity (IMSI) of the subscriber.

10. In case VLR-A sends a new TMSI to VMSC-A, the latter forwards this new identity to MS-A. As mentioned in Note 2 of Figure 10.4, TMSI reallocation need not be sequenced with other messages in the information flow, but must be done on completion of authentication/security procedures.

11. MS-A treats this new TMSI as the new subscriber identity, which it uses for all further communication with the network. It sends back an acknowledgement to VMSC-A, indicating that it has accepted the new identity.

12. VMSC-A forwards this acknowledgement from MS-A to VLR-A.

13. After MS-A has received the 'CM_Service_Accept' message, it sends a 'Setup' message to VMSC-A. The 'Setup' message contains the address of the called subscriber (subscriber 'B'). MS-A also indicates the bearer capability required for the call in the 'Setup' message. The bearer capability provides information on the type of data bearer required for the call.

14. On receipt of the 'Setup' message, VMSC-A draws information of the basic service requested by MS-A (e.g. voice telephony or fax) from the bearer capability received in the 'Setup' message. VMSC-A then requests VLR-A to determine if MS-A is allowed to make this outgoing call. For this, it sends a 'Send Information For Outgoing Call (SIFOC)' message to VLR-A, containing the address of the 'B' subscriber.

15. If VLR-A determines that the call is allowed, it responds to the VMSC-A request with a 'Complete Call' message.

16. On receipt of the 'Complete Call' message from VLR-A, VMSC-A sends a 'Call Proceeding' message to MS-A, to indicate that the call has been allowed to go through locally, and is being forwarded towards the destination.

17. Next, VMSC-A sets up a traffic channel over the radio interface towards MS-A for the outgoing call.

18. VMSC-A then sends an 'Initial Address Message (IAM)' towards the destination exchange, with the address of the 'B' subscriber, to set up the end-to-end connection.

19. When the destination exchange has accepted the terminating call, and has initiated alerting of the 'B' subscriber, it sends an 'Address Complete Message (ACM)' to VMSC-A.

20. VMSC-A sends an 'Alerting' message to MS-A to indicate that the 'B' subscriber alerting has been initiated at the destination exchange.

21. When the 'B' subscriber has connected the call (i.e. has accepted the call), the destination exchange sends an 'Answer Message (ANM)' to VMSC-A.

22. On receipt of the 'ANM' from the destination exchange, VMSC-A sends a 'Connect' message to MS-A to instruct it to connect to the traffic channel, and to indicate that the call has been accepted by the 'B' subscriber.

23. MS-A acknowledges the 'Connect' message received from VMSC-A by sending a 'Connect Acknowledge' message. This completes the setup of the MO call.

10.4 MOBILE-TERMINATED CALL HANDLING

Mobile-Terminated Call Handling procedures can be divided into two main sub-procedures:

(a) Retrieval of routing information from the HLR of the 'B' subscriber
(b) Handling of the call at the VPLMN of the 'B' subscriber.

10.4.1 Retrieval of Routing Information

As discussed in Section 10.2.3, an MT call always involves interaction with the HPLMN of the called subscriber. This is because the existing VPLMN of the called MS is available and tracked only by the latter's HPLMN. Thus, to route an MT call to an MS, the HPLMN of the MS is contacted to retrieve routing information required to route the call to its current VPLMN. Figure 10.5 depicts the information flow between the GMSC, HLR, VLR, VMSC and the MS for retrieval of routing information required to route a basic MT call. The sequence of steps involved is as follows:

1. GMSC-B receives the 'Initial Address Message (IAM)' containing the called party address from the originating exchange to set up the connection with the 'B' subscriber.

2. GMSC-B sends a 'Send Routing Information (SRI)' message to HLR-B, requesting it for routing information to route the call to MS-B. GMSC-B indicates in the SRI message whether it supports pre-paging or not.

3. On receipt of the SRI message, HLR-B checks if the subscriber is allowed the incoming call. A number of checks are applied by the HLR to determine this, including verification of whether 'barring of incoming calls' supplementary service (Section 10.5) is applicable for the subscriber. If the call is allowed to be delivered, HLR-B extracts the VLR address for the called subscriber from its database, and sends a 'Provide Roaming Number (PRN)' request to VLR-B. This request is sent to obtain the roaming number (MSRN) from VLR-B. If HLR-B supports pre-paging, and if it has received indication in the 'SRI' request suggesting that GMSC-B also supports pre-paging, then HLR-B indicates its support for the same in the 'PRN' request sent to VLR-B.

4. If HLR-B indicates support for pre-paging, then on receipt of the "PRN" request from HLR-B, VLR-B informs the VMSC-B to page MS-B. If pre-paging support is

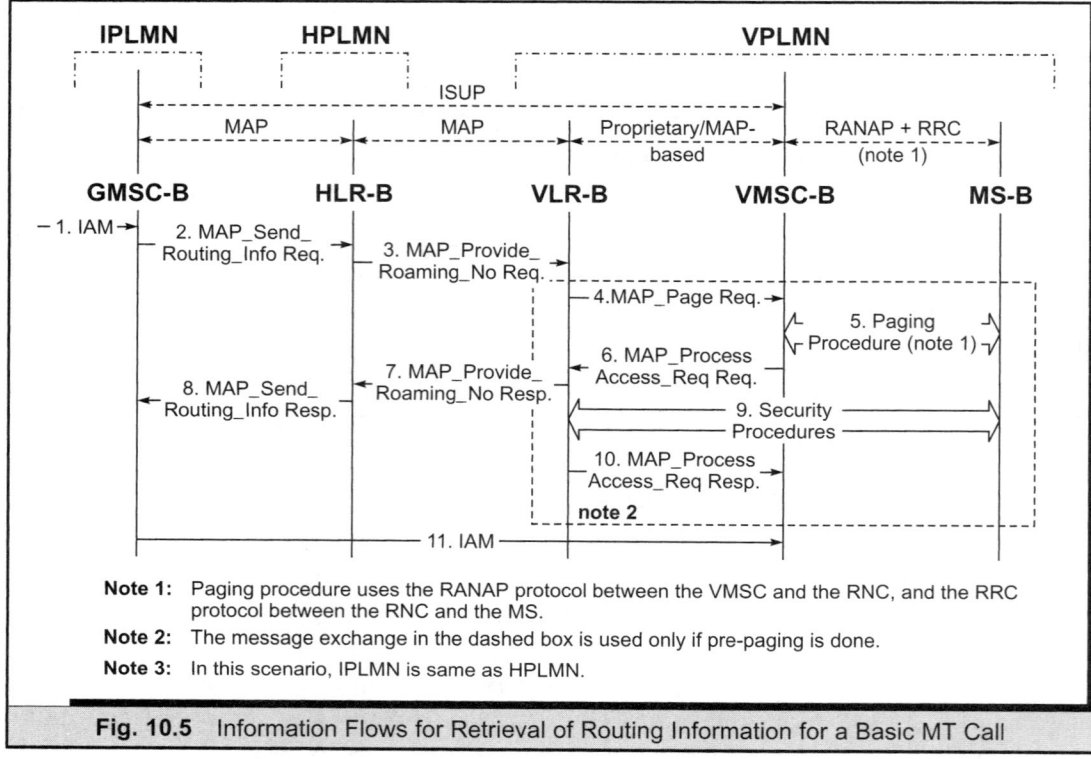

Fig. 10.5 Information Flows for Retrieval of Routing Information for a Basic MT Call

not indicated in the PRN request, then the steps within the dashed box (Figure 10.5) are not executed.

5. On receipt of the 'Page' request from VLR-B, VMSC-B broadcasts the same within its area using the paging channel. The paging request is broadcast in the location area where the MS is known to be located. If the Gs interface between VLR-B and the SGSN is implemented, and there is a valid association between the two for the MS, then this interface can also be used to page the MS. The interaction related to the Gs interface is not depicted in the figure.

When MS-B receives this 'Page' request, it initiates procedures for establishment of a signaling channel over the radio interface. Once this is done, MS-B sends back a 'Page' response to VMSC-B over this channel. Paging procedures between the VMSC and the RNC use the RANAP protocol, and those between the RNC and the MS, use the RRC protocol. The procedures for paging were discussed in Chapters 7 and 8.

6. Next, VMSC-B sends a 'Process Access Request' message to VLR-B indicating that MS-B has responded to paging.

7. VLR-B then identifies a free roaming number from its list of available numbers, and allocates this to MS-B. It sends this roaming number to HLR-B in the 'PRN' response message.

8. HLR-B extracts the roaming number from the 'PRN' response received from VLR-B and sends it to GMSC-B in the 'SRI' response.

9. When VLR-B receives the 'Process Access Request' message from VMSC-B (step 6), it may initiate security procedures with MS-B. Such procedures may be initiated if it is desired that a secure communication channel be set up between the VMSC and the MS (see Chapter 14 for security procedures).

10. VLR-B then acknowledges the receipt and indicates the successful processing of the 'Process Access Request' message by sending a 'Process Accept Request Response' message to VMSC-B.

11. On receipt of the roaming number in the 'SRI' response message (Step 8), GMSC uses this roaming number to construct the 'Initial Address Message (IAM)' and sends it to VMSC-B.

This completes the first stage of the MT call handling procedures. The above steps help in extracting the routing information from the HLR and to use it to route the call till the VPLMN of the called subscriber. The second stage involves handling of the MT call at the VPLMN of the 'B' subscriber, discussed next.

10.4.2 MT Call Handling at VPLMN

The handling of the MT call at the VPLMN of the called subscriber is depicted in Figure 10.6. The sequence of steps involved at this stage is as follows:

1. GMSC-B sends the 'Initial Address Message (IAM)' to VMSC-B to complete the call connection. This message includes the roaming number allocated to MS-B.

2. On receipt of the 'IAM' message from GMSC-B, VMSC-B sends the 'Send Information For Incoming Call (SIFIC)' message to VLR-B. The 'SIFIC' message contains the MS-B roaming number received in the 'IAM' message from GMSC-B.

3. If VLR-B recognizes the roaming number as that allocated to MS-B, it further checks if MS-B is allowed to receive the incoming call. If both conditions are met, VLR-B indicates to the VMSC-B to page the MS. This step, and the subsequent steps till step 11, (see box, Figure 10.6) are followed only if pre-paging was not done at the time of allocating the roaming number.

4. On receipt of the 'Page' request from VLR-B, VMSC-B broadcasts the same within its area using the paging channel. The paging request is broadcast in the location area where the MS is known to be located. This step is similar to the pre-paging related procedure. When MS-B receives this 'Page' request, it initiates the procedures for the establishment of a signaling channel over the radio interface. On successful set up of the channel, MS-B sends back a 'Page' response to VMSC-B

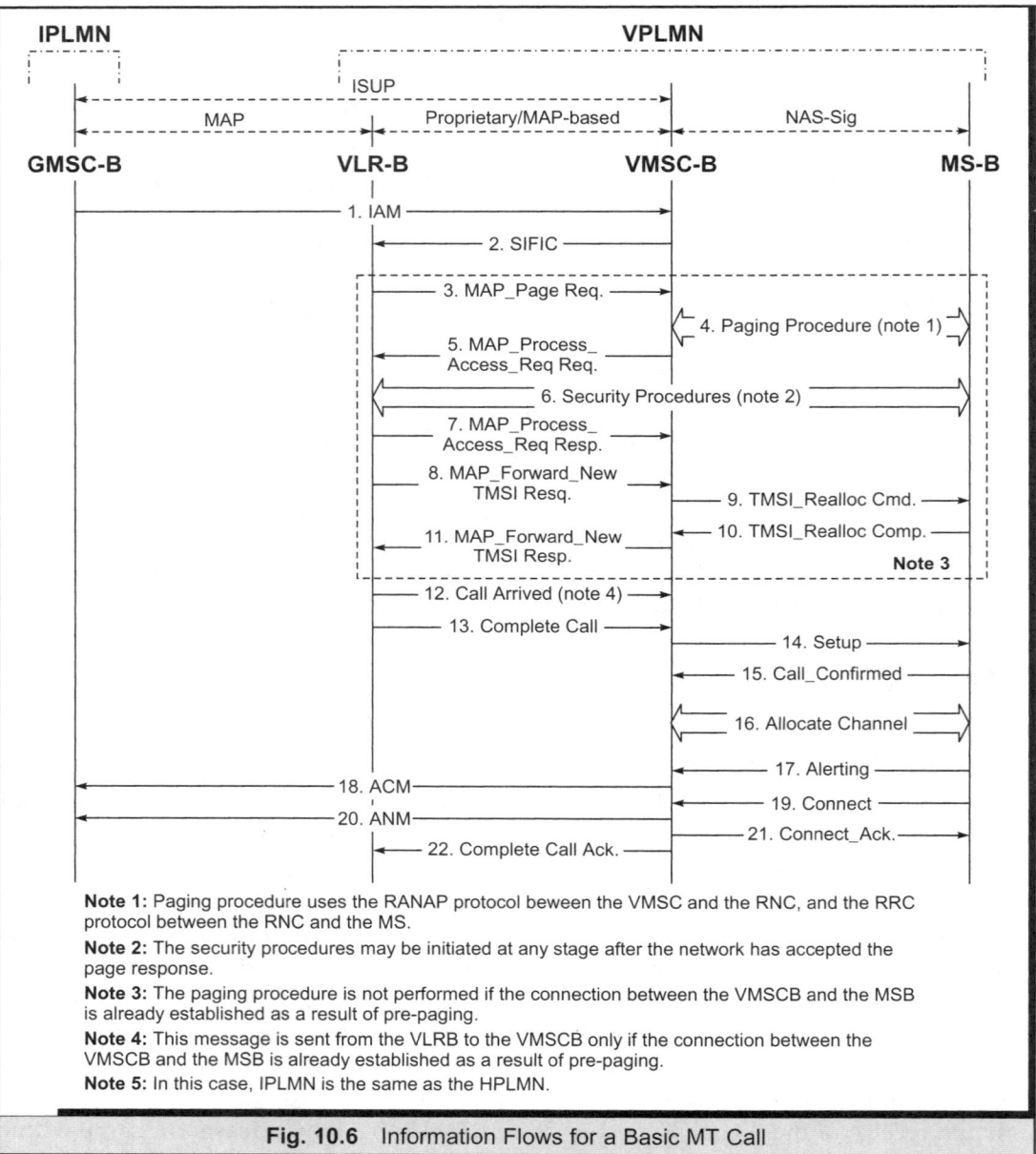

Fig. 10.6 Information Flows for a Basic MT Call

over this signaling channel. Paging procedures between the VMSC and the RNC use the RANAP protocol, and those between the RNC and the MS, use the RRC protocol.

5. On successful completion of the paging procedure, VMSC-B sends a 'Process Access Request' message to VLR-B indicating that MS-B has responded to paging

6. When VLR-B receives the 'Process Access Request' message from VMSC-B, it may initiate security procedures with MS-B. Such procedures may be initiated if it is desired that a secure communication channel be set up with the MS.

7. VLR-B then acknowledges the receipt and indicates the successful processing of the 'Process Access Request' message by sending a 'Process Accept Request Response' message to VMSC-B.

8. VLR-B may also allocate a new TMSI for MS-B, in which case, it will send this new TMSI towards MS-B.

9. In case VLR-B sends a new TMSI to VMSC-B, the latter forwards this new identity to MS-B. The TMSI reallocation procedure need not be sequenced with other messages in the information flow, but must be executed only after security procedures have been completed.

10. MS-B treats this new TMSI as the new subscriber identity, which it needs for all further communication with the network. It sends back an acknowledgement to VMSC-B indicating its acceptance of the new identity.

11. The response to the TMSI reallocation command is forwarded by VMSC-B to VLR-B.

12. If pre-paging was done at the time of allocation of the roaming number, then MS-B would already have an established signaling channel (as discussed in Step 5, Section 10.4.1). In case the channel was established at the time of pre-paging, when the call had not actually reached the VPLMN, the VMSC-B would have started a timer in which duration, it was expecting the call to actually reach the VPLMN. If this timer expires prior to the call reaching the VPLMN, VMSC-B clears the signaling channel established for MS-B. On the other hand, if the call reaches the VPLMN before the expiry of this timer (as is the case here), VLR-B sends a 'Call Arrived' Message to VMSC-B to indicate that the call has arrived, and that VMSC-B should stop the timer. The 'Call Arrived' message is relevant only when pre-paging is used, and not otherwise.

13. On successful completion of the security procedures, VLR-B sends a 'Complete Call' message to VMSC-B, indicating that a call leg be established between the VMSC-B and the MS-B".

14. On receipt of this 'Complete Call' message from VLR-B, VMSC-B sends a 'Setup' message to MS-B. This message may include the bearer capability information for the call.

15. In case the MS-B accepts the call, it responds to this 'Setup' message with a 'Call Confirmed' message.

16. On receipt of the 'Call Confirmed' message the VMSC-B establishes a traffic channel for the MS over the radio interface.

17. The MS-B then synchronizes itself with the specified traffic channel. Subsequently, it sends an 'Alerting' message to VMSC-B to indicate that the called user has been alerted.

18. Next, the VMSC-B sends the 'Address Complete Message (ACM)' to GMSC-B, indicating that the called user has been alerted.

19. When the MS-B user accepts the terminating call, MS-B sends a 'Connect' message to VMSC-B.

20. The VMSC-B, on receipt of the 'Connect' message from MS-B forms and sends the 'ANswer Message (ANM)' to GMSC-B.

21. The acknowledgement of the 'Connect' message is sent from VMSC-B to MS-B.

22. The acknowledgement of the 'Complete Call' message is sent from VMSC-B to VLR-B. The call establishment is complete, and the MS-B is now in the 'Connected State'.

10.5 INTERACTION OF CF AND CB SERVICES WITH CALL HANDLING PROCEDURES

Call Handling procedures and their interaction with multiple supplementary services has been covered in Chapter 12. This section covers only the interaction of the Call Barring (CB) and Call Forwarding (CF) supplementary services with call handling procedures. This is because these supplementary services play significant role in the day-to-day activities of the mobile subscriber.

Consider the following scenario: a mobile subscriber wishes to attend a meeting, undisturbed by any incoming calls, but at the same time does not want to miss any calls. In such a case, the subscriber may wish to forward all incoming calls to his/her residence telephone number, or to any other convenient number where these calls would be attended to on his/her behalf. The call forwarding service would come in handy in such situations. If, however, the subscriber is not too bothered about missing out on incoming calls, but just does not want to be disturbed by them, then he/she could have all incoming calls barred. The call barring service can be put to use in such a scenario. Though this may not be the best example of a call barring service, there are numerous instances in a subscriber's daily routine where Call Forwarding (CF) and Call Barring (CB) services find good use.

While call forwarding services are used only in case of MT calls, the call barring services can be used for both MO and MT calls. The following sections describe the interaction of call forwarding and call barring services with the call handling procedures. For more information on CF and CB supplementary services, the reader is referred to Chapter 12.

10.5.1 Interaction of CF and CB Services with MT Calls

The impact of Call Barring (CB) and Call Forwarding (CF) supplementary service on MT calls is verified at the HLR of the mobile subscriber's HPLMN. This happens when the GMSC requests for routing information from the HLR. Call forwarding checks are also carried out at the VLR of the subscriber's VPLMN, when the MSC in the VPLMN receives the IAM message from the GMSC of the IPLMN. Checks related to call barring are carried out first, followed by checks related to call forwarding. This order seems logical, since it is important to first determine whether the incoming call is allowed at all. If it is allowed (i.e. it is not barred), only then can its interaction with call forwarding be checked to determine if the call is to be forwarded to some other number, or if the mobile subscriber is to be connected to the call. The call barring checks carried out at the HLR of the called mobile subscriber are as follows:

1. Check if 'Barring of all incoming calls' facility is active for the mobile subscriber. If it is active, then the call is not delivered to the subscriber; a negative response to the routing information request is sent to the GMSC indicating that the call is barred by the target subscriber.
2. Check if 'Barring of all incoming calls when roaming outside the HPLMN country' is active for the mobile subscriber. If active, then if the subscriber is roaming out of his/her HPLMN country, the call is not delivered to the subscriber; a negative response to the routing information request is sent to the GMSC indicating that the call is barred by the subscriber. However, if the subscriber is within the same country as his/her HPLMN, then the call is delivered to the subscriber. By comparing the country code of the VLR address of the subscriber's current VLR with the country code of the HPLMN, the HLR can determine whether the subscriber is within or outside the HPLMN country.

If the call barring checks confirm that the call is not barred, then call forwarding checks are carried out to determine whether the call is to be forwarded to a 'Forwarded-To-Number (FTN)', or it is to be connected to the mobile subscriber. The following call forwarding checks are conducted at the HLR of the called mobile subscriber:

1. Check if 'Call Forwarding Unconditional (CFU)' is active for the subscriber. If active, then the call for the subscriber is unconditionally forwarded to the FTN registered by the subscriber.
2. Check if 'Call Forwarding on Mobile subscriber Not Reachable (CFNRc)' is active for the subscriber. If active, and if the HLR is aware of the fact that the mobile subscriber is not reachable (mobile subscriber may be unreachable either because it is not registered in any location area, or the handset is switched off, or has entered a region where service cannot be provided), then the call of the subscriber is forwarded to the FTN registered by the mobile subscriber.

In either of these two cases—and in order to forward the call to the FTN—the HLR responds to the GMSC with the FTN (instead of the roaming number) included in the response to the routing information request message. The GMSC, on receipt of this number in response to the routing information request, connects the call towards this number, just like a normal call.

Call forwarding barring checks may also need to be carried out at the VPLMN (or VLR) of the called subscriber. These are as follows:

1. Check if 'Call Forwarding on mobile subscriber Not Reachable (CFNRc)' is active for the subscriber. If active, and if the mobile subscriber is not reachable, then the call is forwarded to the FTN registered by the mobile subscriber. Note that this may happen when the HLR of the called subscriber has no information of the mobile subscriber being unreachable, and hence allowed the call to proceed to the latter's VPLMN. However, when the VMSC tries to contact the subscriber, it might confirm that the subscriber is actually unreachable.

2. Check if the 'Call Forwarding on mobile subscriber Busy (CFB)' is active for the subscriber. If active, and if the VMSC determines that the mobile is busy and cannot take the incoming call, the call is forwarded to the FTN registered by the mobile subscriber.

3. Check if the 'Call Forwarding on mobile subscriber No Reply (CFNRy)' is active for the subscriber. If active, and if the VMSC determines that the mobile subscriber is not replying to the incoming call, the call is forwarded to the FTN registered by the latter.

Thus, in a nutshell, while the CB-related interactions for an MT call take place at the HLR of the subscriber's HPLMN, the CF-related checks may be carried out at the HLR or the VLR (in the VPLMN of the called subscriber). The 'Call Forwarding Unconditional (CFU)' and 'Call Forwarding on mobile subscriber Not Reachable (CFNRc)' checks are conducted at the HLR, and the 'Call Forwarding on mobile subscriber No Reply (CFNRy)', 'Call Forwarding on mobile subscriber Not Reachable (CFNRc)' and 'Call Forwarding on mobile subscriber Busy (CFB)' checks are carried out at the VLR. Note that CFNRc can be handled at either the HLR or the VLR, depending on which entity confirms that the mobile subscriber is unreachable.

10.5.2 Interaction of CB Service with MO Calls

The CB-related checks for MO calls are carried out at the VLR of the calling subcriber's VPLMN, when the mobile subscriber wishes to place an outgoing call. Call forwarding checks are not applicable in case of MO calls.

The call barring checks include the following:

1. Check if 'Barring of All Outgoing Calls (BAOC)' is active for the mobile subscriber. If it is then the outgoing call is not allowed to go through, and the subscriber is notified accordingly.

2. Check if 'Barring of Outgoing International Calls (BOIC)' is active for the mobile subscriber. If active, then if the subscriber is trying to place an international call,

the call is rejected. If, however, the call is not an international call, it is allowed to go through.

3. Check if 'Barring of Outgoing International Calls Except those directed to the Home PLMN Country (BOIC-exHC)' is active for the mobile subscriber. If active, then if the subscriber is trying to place an international call, the following action is taken:

- If the call is to the subscriber's HPLMN country, the call is allowed to go through.
- If the call is to a country other than the subscriber's HPLMN country, the call is rejected.

If, however, the call is not an international call, it is allowed to go through.

Thus, before a mobile-originated call can be allowed to proceed towards its destination, the VLR of the calling subscriber verifies that the call is not barred by the subscriber as a result of application of the call barring supplementary service. The verification carried out at the VLR depends on the type of CB supplementary service activated by the mobile subscriber (if any activated), and the destination of the outgoing call.

10.6 SUPPORT FOR OPTIMAL ROUTING

The architecture for a basic mobile-to-mobile call was introduced in Section 10.2.3. As discussed therein and as depicted in Figure 10.7, the call from the 'A' subscriber in VPLMN-A to the 'B' subscriber in VPLMN-B always goes through the GMSC of the HPLMN-B. This is because the HPLMN-B is the only point of contact for VPLMN-A for reaching the 'B' subscriber. Thus, a basic mobile-to-mobile call takes the route as follows: from VMSC-A to GMSC-A, to GMSC-B and then to VMSC-B.

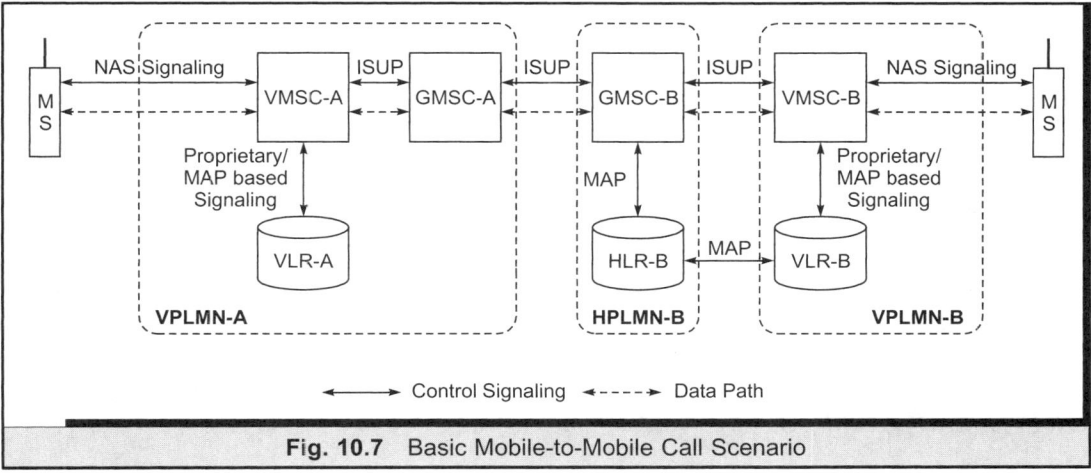

Fig. 10.7 Basic Mobile-to-Mobile Call Scenario

The path taken by the call is fine, as long as the route from VPLMN-A to VPLMN-B through HPLMN-B is of same distance as the direct route from VPLMN-A to VPLMN-B, if it exists. However, consider the case when the VPLMN-A is the same as VPLMN-B, and HPLMN-B is distinct from both VPLMN-A and VPLMN-B. This means that both the 'A' and the 'B' subscriber are in the same PLMN, which is distinct from the HPLMN of the 'B' subscriber. Also, consider the case when VPLMN-A and VPLMN-B is in a different country from HPLMN-B. In this case, the call from the 'A' to the 'B' subscriber would require making two international calls, whereas the direct path between the 'A' and the 'B' subscriber, if used, would have required only an intra-country call. The non-optimally routed call, as it would be referred to in this section, would thus mean extra cost to both the 'A' and the 'B' subscriber, each having to pay for one international call. While the 'A' subscriber pays for this call to reach HPLMN-B from VPLMN-A, the 'B' subscriber pays for this call from the HPLMN-B to VPLMN-B. This problem can, however, be solved if optimal routing is employed to connect the call between subscribers 'A' and 'B'. Optimal routing procedures are discussed in the following sub-sections.

10.6.1   Conditions for Optimal Routing

Optimal routing results in a better route to the destination in multiple situations. One of these scenarios was described in the previous section, where the VPLMN-A and the VPLMN-B were in the same country (but not the HPLMN-B country). However, because of complicated charging mechanisms, it may be difficult at times to gauge whether the route determined by using optimal routing procedures costs less or more. It is, therefore, required that a set of criteria be defined to determine whether optimal routing procedures lead to better routes or not. As per the criteria defined in the 3GPP specification on optimal routing (3GPP TS 23.079), two constraints have been placed on the charging of optimally routed calls. The following text from 3GPP TS 23.079 defines these constraints:

'MoU have imposed two constraints for the charging of optimally routed calls:

- *No subscriber shall pay more for a call, which has been optimally routed than he/she would do under the present routing scheme described in GSM 03.04 in the sub-clauses describing the call cases where the GMSC is in the same PLMN as the HLR.*
- *At least for the first phase of Support of Optimal Routing, the charge for one leg of a call shall be paid for entirely by one subscriber.*

These constraints mean that the direct route for a call cannot always be used. For example, if the calling mobile subscriber (the "A" subscriber) is in Germany, and the "B" subscriber's HPLMN is in Switzerland but he/she has roamed to Finland, the charge payable by the "A" subscriber to route the call by the direct route to Finland would be greater than the charge payable to route the call to HPLMNB, so the HPLMN route must be used.'

These clauses imply that the decision on whether the call should be routed through the optimal or the normal route should take into account the charges payable by the customer in either case. This places the responsibility of calculating these charges, for both the optimal and the normal route, on the GMSC-A. However, for the initial implementation of optimal routing procedures, it is assumed that the GMSC would not be capable of calculating the charges payable for both the routes. Hence, a more stringent, yet simpler, criteria has been defined to decide between the two routes at GMSC-A (In reality, however, GMSC-A relies on the HLR to carry out this simpler verification, as will be discussed later). The criteria is defined as follows:

1. If the country code of both the destination exchange and GMSC-A are the same—in other words, if VPLMN-A and VPLMN-B are in the same country—then optimal routing may be employed at GMSC-A.
2. Optimal route may also be used if the country code of the destination exchange and that of HPLMN-B are the same.
3. Else, the normal route through the HPLMN-B should be used.

The first criterion means that in case the VPLMN-A and VPLMN-B are in the same country, then the call should take the direct route from VPLMN-A to VPLMN-B, by-passing the HPLMN-B. Notice than nothing is said about the location of the HPLMN. This is because irrespective of whether or not the HPLMN is also in the same country, the path obtained using the optimal routing procedures should never be more costly than the path through the HPLMN. The second criterion states that if the VPLMN-B and the HPLMN-B are in the same country, then the path can be obtained using the optimal routing procedures, and the HPLMN-B may be bypassed. Here, the location of VPLMN-A is not important, and irrespective of whether or not it is in the same country, the path obtained through optimal routing procedures should not be costlier than the normally routed path. By clubbing criterion 1 and 2—if the VPLMN-A, HPLMN-B and VPLMN-B are all in the same country—the optimal routing procedures can again be used to connect the call to the destination.

In some cases, however, just checking the country code may not be sufficient to determine whether the optimally routed call would indeed be less costlier than the normally routed call. This is because different costs may be involved in routing the call through different regions of the country. In such a case, the NDC part of the ISDN numbers may also be compared in order to arrive at a decision.

10.6.2 Information Flows for Optimal Routing

Information flows for optimal routing are broadly categorized under three main scenarios:

- Optimal Routing for Basic Mobile-to-Mobile Call
- Optimal Routing for Early Call Forwarding

- Optimal Routing for Late Call Forwarding

10.6.2.1 Optimal Routing for a Basic Mobile-to-Mobile Call

Figure 10.8 depicts the architecture of a basic mobile-to-mobile call that is optimally routed. As shown in the figure, this call is connected directly from VPLMN-A to VPLMN-B, without passing through the HPLMN-B. The HPLMN comes into the picture only when providing the routing information to the VPLMN-A (which in this case is the IPLMN). This information is required by VPLMN-A to route the call to VPLMN-B.

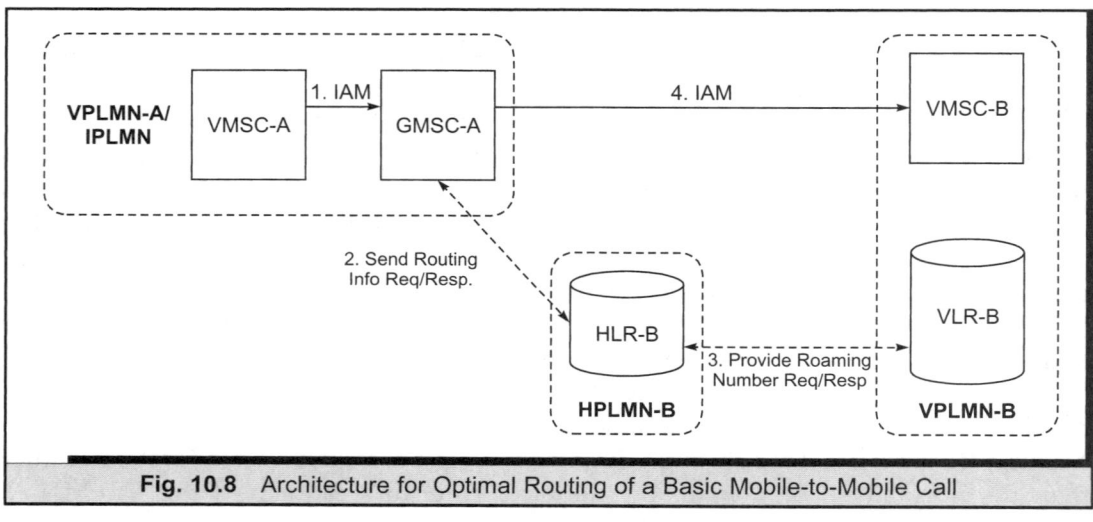

Fig. 10.8 Architecture for Optimal Routing of a Basic Mobile-to-Mobile Call

Figure 10.9 depicts the message flow for optimal routing of a basic mobile-to-mobile call. The sequence of steps followed in this case is identical to that depicted in figure 10.5 for the normal retrieval of routing information for a MT call. The only important point to note in this scenario is that the GMSC interrogating the HLR is GMSC-A, and not GMSC-B. Thus, in case of optimal routing, the distinction between the IPLMN and the HPLMN is clearly evident. In case optimal routing procedures are used, the IPLMN is the PLMN from where the call originated. Both the 'Send Routing Information' message (from GMSC-A to HLR-B) and the 'Provide Roaming Number' message (from HLR-B to VLR-B) include information as to whether the request is for an optimally routed or a normal call.

An important point to note here is that an optimally routed call is possible if the GMSC-A, HLR and the VLR, all three support optimal routing. Further, the optimally routed call can be allowed only if the HLR determines that the conditions discussed in

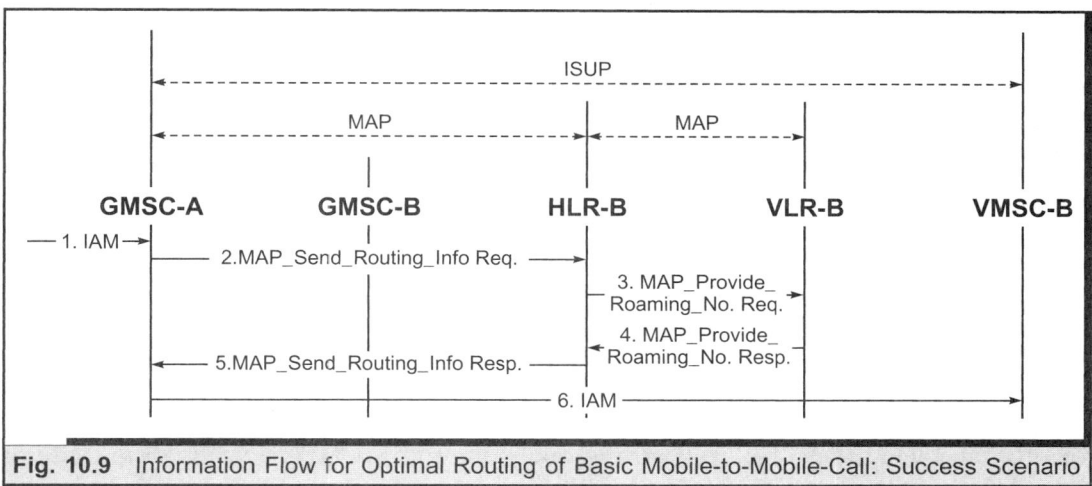

Fig. 10.9   Information Flow for Optimal Routing of Basic Mobile-to-Mobile-Call: Success Scenario

Section 10.6.1 are met. Figure 10.10 depicts the error scenario where the VLR does not support optimal routing procedures. In this case, the VLR sends back an error response to the request from HLR to provide a roaming number (Step 4). The HLR, in turn, sends back a negative response to GMSC-A for the request for routing information retrieval (Step 5). This results in the GMSC-A sending the 'IAM' message to GMSC-B (Step 6)

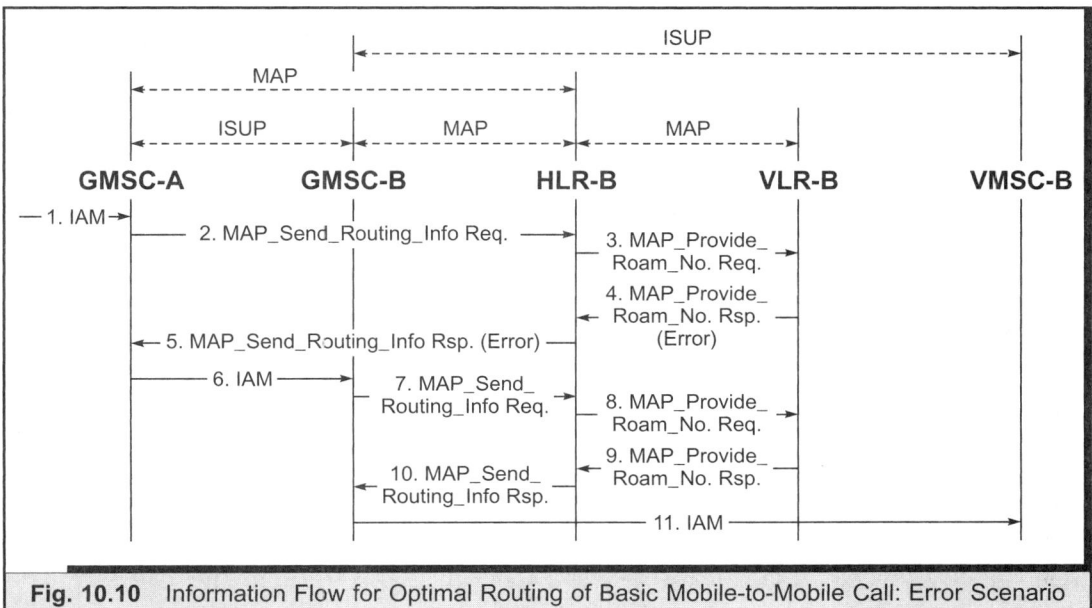

Fig. 10.10   Information Flow for Optimal Routing of Basic Mobile-to-Mobile Call: Error Scenario

and the call is routed and established as a normal call (Steps 7 to 11), through the HPLMN. Similar steps would be followed in case the HLR determines that the conditions for optimal routing are not met, and the call cannot be optimally routed. In this case, the VLR interrogation is not required, and steps 3 and 4 in Figure 10.10 will not apply. The remaining steps remain the same.

10.6.2.2 *Optimal Routing for Early Call Forwarding*

Section 10.5 discussed the interaction of the call forwarding supplementary service with the call handling procedures. As discussed in that Section, the CF supplementary services, when activated, result in the forwarding of the call to a Forwarded-To-Number (FTN). This forwarding can either take place at the HPLMN of the called subscriber (for CFU and CFNRc) or at the VPLMN of the called subscriber (for CFB, CFNRy, and CFNRc). If call forwarding takes place at the HPLMN, then the route of the call is as follows: starting from the IPLMN (GMSC-A) to the HPLMN, and then to the network containing the FTN. On the other hand, if call forwarding takes place at the VPLMN, then the route of the call is from the IPLMN to the HPLMN, to the VPLMN, and then to the FTN network In the latter case, if the original basic call was optimally routed, the route to the FTN would be from the IPLMN to the VPLMN, and then to the FTN network. As in the case of a basic mobile-to-mobile call, the forwarded leg of the call can also be optimally routed, so that the direct path can be taken to the FTN network, bypassing the HPLMN or the VPLMN, as the case may be. If the call forwarding is done at the HPLMN, then optimal routing procedures can be employed to bypass the HPLMN and directly connect the call from the IPLMN to the FTN network. This is termed as Optimal Routing for *Early Call Forwarding,* since the call forwarding is detected early, at the HPLMN itself. In case the call forwarding is done at the VPLMN, then optimal routing procedures can be employed to bypass the VPLMN. This is termed as Optimal Routing for *Late Call Forwarding*.

Figure 10.11 depicts the information flow for optimal routing for Early Call Forwarding. Two different flows are depicted, one which does not require any interrogation of the VLR to employ call forwarding, and the other which requires interrogation of the VLR to determine if call forwarding is to be employed. The term FTNW-LEC stands for the Local Exchange (LEC) in the Forwarded-To-Network (FTNW). In the first information flow (Steps 1 to 4), when the HLR receives the 'Send Routing Information' message for the basic mobile-to-mobile call from GMSC-A, it determines locally that call forwarding is to be employed. This is possible either in case of CFU, or in case of CFNRc, when the HLR can determine locally that the subscriber is currently not reachable. In either case, the HLR sends back the FTN to GMSC-A in the response to the 'Send Routing Information' message. GMSC-A can then directly forward the call to the FTN, without going through the HPLMN. This scenario is similar to the one depicted in Figure 10.9 in case of optimal routing for the

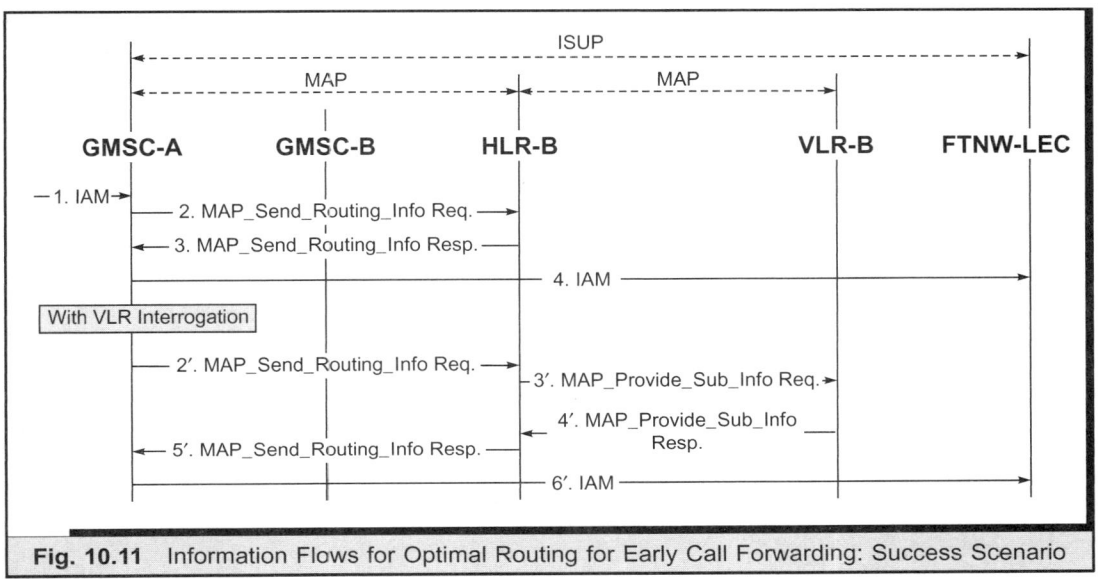

Fig. 10.11 Information Flows for Optimal Routing for Early Call Forwarding: Success Scenario

basic mobile-to-mobile call. The only difference is that in this case, the FTN is sent in the response to the 'Send Routing Information' message, instead of the roaming number. The second information flow (steps 1, 2' to 6' in figure) also considers the case when CFNRc is active for the subscriber. However, unlike the first information flow, the HLR in this case cannot determine locally whether the subscriber is reachable or not (i.e. the HLR has no such information). Here, the HLR interrogates the VLR using the 'MAP_Provide_ Subscriber_Information' message to determine whether the subscriber is reachable or not. Note that in this case, the 'MAP_Provide_Roaming_ Number' message is not used; this message is used only to obtain the roaming number from the VLR, and it does not fetch information as to whether the MS is reachable or not. When the VLR receives the 'MAP_Provide_Subscriber_Information' message from the HLR, it determines the status of the MS (i.e. whether the MS is reachable or not), and sends this status of the MS back to the HLR. The remaining message flow is the same as in the first scenario.

Before the GMSC-A can use the optimally routed path to the destination, the HLR must ensure that the cost of doing this does not exceed the cost incurred by the subscriber on the normal path. If this is not the case, then the normal path through the HPLMN is used and GMSC-A cannot directly route the call to the FTN network. This is as depicted in Figure 10.12, wherein the target subscriber either has CFU or the CFNRc active. The figure shows the case where the VLR interrogation was not required. The scenario in which VLR interrogation is required the similar in nature, and is not shown in the figure.

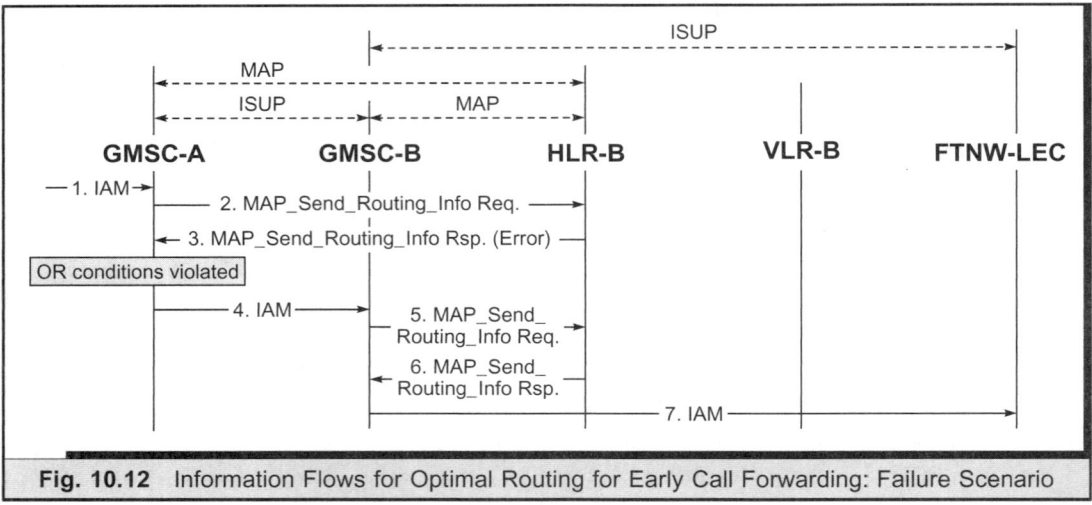

Fig. 10.12   Information Flows for Optimal Routing for Early Call Forwarding: Failure Scenario

In this case, even though CFU/CFNRc is active for the target subscriber, the HLR does not return the FTN to GMSC-A. Instead, it returns an error response to GMSC-A, thus disallowing the optimally routed call. This is because the HLR determines that the conditions introduced in Section 10.6.1 are not met. The call is forwarded to the FTN network by GMSC-B in the HPLMN of the subscriber, as is done in the case of normal call forwarding.

10.6.2.3   *Optimal Routing for Late Call Forwarding*

Call forwarding carried out at the HPLMN of a called subscriber is called *Early Call Forwarding*, and that carried out at the VPLMN of the called subscriber is called *Late Call Forwarding*. In case of late call forwarding, optimal routing procedures can be used to bypass the VPLMN, and call forwarding can be done from the IPLMN network itself. Optimal routing procedures for late call forwarding are discussed in this section.

Figure 10.13 depicts the information flow for optimal routing for late call forwarding; both the success and the failure scenario. Not all steps in the figure need explanation here. Steps 1 to 8 were covered under MT call handling in Section 10.4. Note that the GMSC in the figure is not termed as GMSC-A or GMSC-B. This is because either case is possible, depending upon whether the original call (basic mobile-to-mobile call) was optimally routed or not.

Steps 1 to 8: Refer to Section 10.4.

9. VMSC-B determines that the call to the 'B' subscriber cannot be established, and must be forwarded to a registered FTN. This could be either because the subscriber is not reachable, or else the subscriber is busy (another call in progress), or that

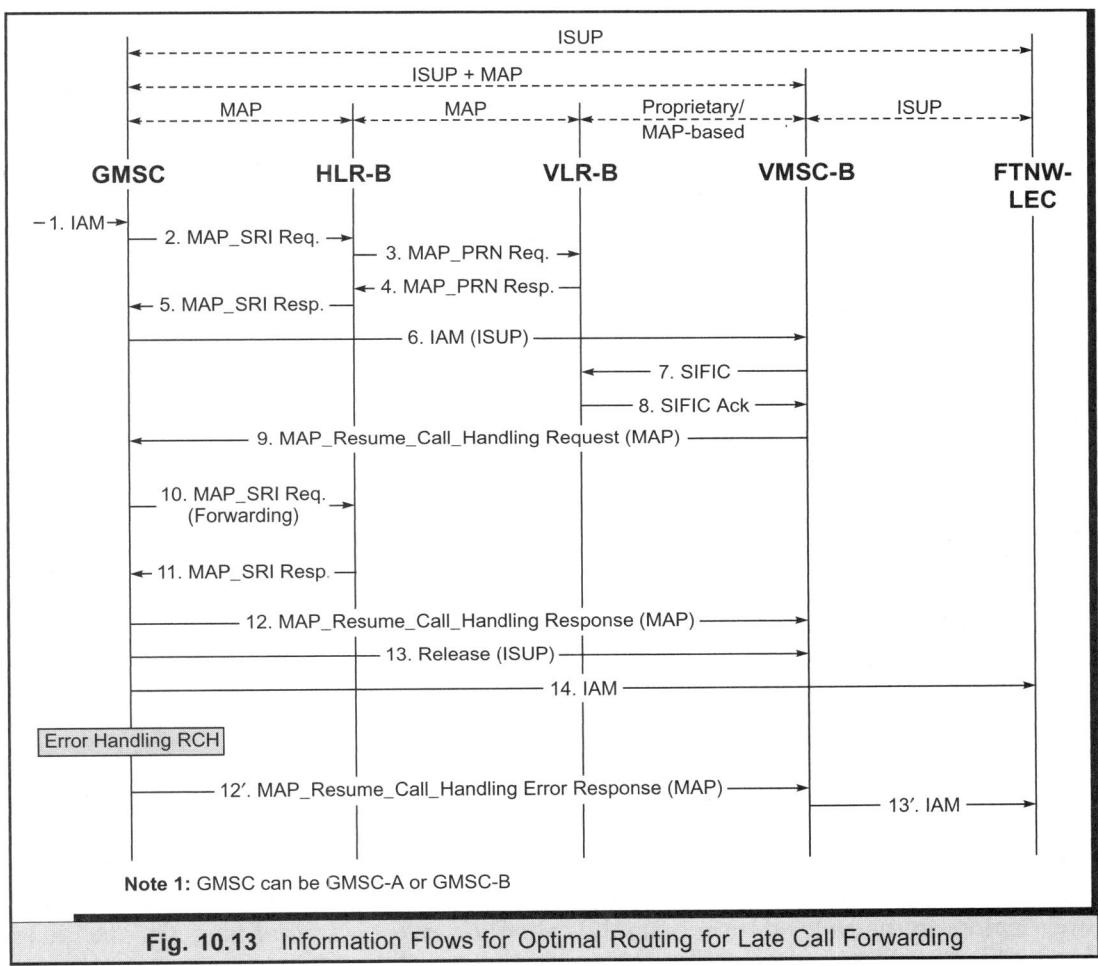

Fig. 10.13 Information Flows for Optimal Routing for Late Call Forwarding

he/she is not replying to the call, and the corresponding CF service is active. In any of the above cases, VMSC-B sends back a 'MAP_ Resume_Call_Handling' message to the GMSC, giving the reason for forwarding the call.

10. The GMSC sends a 'MAP_Send_Routing_Information' message to HLR-B requesting for the FTN registered by the subscriber. The reason for forwarding the call, as received from VMSC-B, is mentioned in the message. This message includes an indication that the GMSC is requesting for the FTN, in case of late call forwarding, and not for a roaming number. This is indicated in the figure by suffixing the 'MAP_Send_Routing_Info' message with the word: 'Forwarding'.

11. HLR-B sends back the FTN in its response to the 'Send Routing Information' message.

12. The GMSC sends a positive response to the 'Resume Call Handling' message from VMSC-B, indicating that the GMSC would forward the call.

13. The GMSC then sends a 'Release' message to VMSC-B to release the already existing connection.

14. The GMSC sends the 'IAM' message directly to the Local Exchange in the Forwarded-to-Network (FTNW-LEC).

On the other hand, if the conditions for using the optimal route to the FTN are not satisfied (refer to Section 10.6.1), the GMSC sends a negative response to the 'Resume Call Handling' message from the VMSC-B. In this case, the VMSC-B constructs and sends an 'IAM' to the FTNW-LEC, (see Steps 12' and 13').

10.7 IMMEDIATE SERVICE TERMINATION (IST)

The IST service enables the service provider to terminate user calls whenever required. The reasons could range from non-payment of bills, to long duration calls when the network is congested. In fact, the IST service is defined to be flexible in that it can be brought to use in a variety of scenarios. This section discusses the IST service as a collection of two distinct procedures: the IST Alert Service, and the IST Command Service. Jointly, the two provide the service provider the control over user calls.

10.7.1 IST Alert Service

The IST Alert Service enables the service provider to continuously monitor the incoming as well as outgoing calls for a particular subscriber. Using this service, the service provider can be notified whenever the subscriber being monitored is involved in any call. The service provider may then either allow the call to continue, or take any other action on the call, e.g. terminate the call. The service provider may also decide to terminate the call if it exceeds a pre-defined duration (e.g. 45 minutes).

To achieve this functionality, the VMSC (for outgoing calls) and the GMSC (for incoming calls) are provided by an IST Alert Timer Value by the HLR of the subscriber's HPLMN. This timer value may be sent to the VLR (and then from the VLR to VMSC) as part of the 'MAP_Insert_Subscriber_Data' message during the location update procedures (refer to Chapter 9). To the GMSC, this timer value is sent in the response to the message for routing information ('MAP_Send_Routing_ Information' response). On receipt of the IST Alert Timer Value, the interpretation of the VLR (or VMSC) and the GMSC is as follows: The service provider wants that the subscriber calls be tracked, and in the event of any subscriber call, the HLR is to be notified. If the HLR allows the call to continue, then as long as the call is active, the HLR would be periodically notified about the call (the period of notification given by the IST Alert

Timer). This provides the HLR the option to terminate the call immediately, on the first alerting from the VMSC/GMSC; to let the call continue; or to terminate the call after more than 'X' alert messages have been received, 'X' being a configurable parameter. The last part of the previous statement means that the service provider may decide to terminate any subscriber calls that exceed the duration of 'X*IST Alert Timer Value'. Figure 10.14 depicts the message flow between the GMSC/VMSC and the HLR.

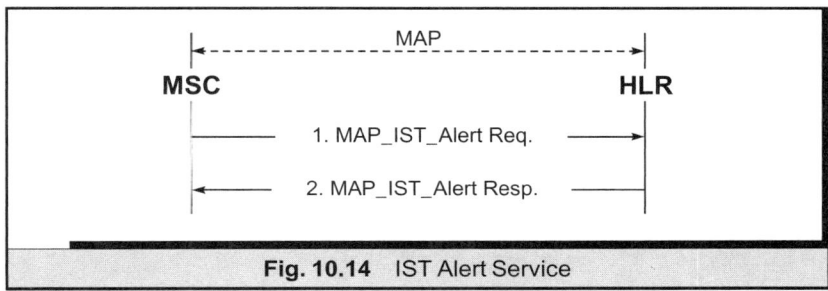

Fig. 10.14 IST Alert Service

The IST Alert message from the VMSC/GMSC is sent to the HLR periodically for each call that the subscriber is involved in. The HLR responds to this message with the action it wants the VMSC/GMSC to take. The action may be any one of the following:

- **Terminate the Subscriber Call:** If this option is used, then the HLR should also mention if it wishes this particular call, or all calls that the subscriber is currently involved in, to be terminated. Using certain supplementary services, a subscriber may be involved in more than one call at a time, as will be discussed in Chapter 12.

- **No Action:** This means that the subscriber call should be allowed to continue for now. However, the VMSC/GMSC should continue to periodically alert the HLR for the ongoing the call. The HLR may also include a new value for the IST Alert Timer, if it so wishes.

- **IST Alert Withdraw:** This option is used to convey to the VMSC/GMSC that the IST service is no longer applicable to the particular subscriber, and that the VMSC/GMSC should discontinue reporting subscriber calls to the HLR.

These three options provide the service provider ample flexibility in choosing the action it wants to take for any subscriber call. The service provider may just monitor the subscriber calls without terminating them, or else, terminate the calls under certain circumstances. The next section discusses the second category of IST service—the IST Command Service, which complements the IST Alert Service to provide complete control to the service provider.

10.7.2 IST Command Service

As discussed in the previous section, the IST Alert Service provides the service provider the flexibility to monitor and control subscriber calls in a variety of ways. However, in this case, the service provider can terminate a call only when notified by the MSC. The IST Alert Service is more of a procedure to monitor subscriber calls, and to terminate these in certain scenarios. But there are times when the service provider is required to immediately terminate all calls of a particular subscriber, without having to wait for the IST Alert message from the MSC. Moreover, the service provider may also wish to immediately bar the subscriber from using the network services any further. The IST Alert Service, whose primary function is to provide monitoring of calls, cannot be used in this scenario. In such situations, the IST Command Service may be used. The message flow for the IST Command Service is depicted in Figure 10.15.

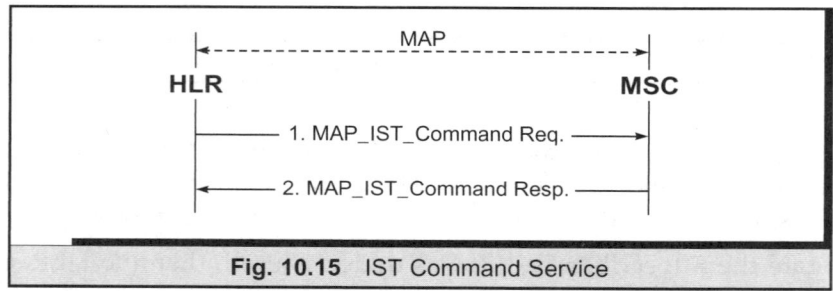

Fig. 10.15 IST Command Service

As depicted in the figure, when the service provider wishes to immediately terminate the subscriber calls, it uses the IST Command message to inform both the VMSC and the GMSCs to terminate all the outgoing and/or incoming calls, respectively, of the subscriber. Note that though there is only one VMSC that is handling all the outgoing calls of the subscriber in question, there could be multiple GMSCs, each handling one or more terminating calls for the subscriber. When the MSCs receive the IST command message, they terminate all call-related activities of the subscriber. Besides, before sending the IST Command message to the MSC, the HLR also sends the 'MAP_Cancel_Location" message (refer to Chapter 9), so that the subscriber cannot commence operations after completion of the IST command procedures.

Thus, while the primary function of the IST Alert Service is to monitor subscriber calls, and to terminate calls only under certain circumstances, the IST Command Service is used for immediate termination of subscriber calls. This service is used along with the 'MAP_Cancel_Location' message to bar the subscriber from commencing operations. Jointly, the IST Alert and Command service provide complete control to the service provider in terms of monitoring and termination of subscriber calls.

SUMMARY

Call Handling (CH) procedures include procedures for handling subscriber calls, which can either be Mobile Originated (MO) or Mobile Terminated (MT) calls. Accordingly, CH procedures are classified into two broad categories: MO CH procedures and MT CH procedures. The MO calls for a subscriber are handled by the VPLMN without involving the HPLMN. On the other hand, MT calls for the subscriber involve interaction between the IPLMN, HPLMN and VPLMN. The architectures for handling the MO and MT calls, as well as the procedures for MO and MT call handling have been discussed in detail in the chapter.

CH procedures are greatly affected by the supplementary services that a subscriber has subscribed to. Important amongst these supplementary services is the Call Forwarding (CF) and the Call Barring (CB) service. Interaction of the CF and the CB supplementary services with the CH procedures has been discussed in Section 10.5. Section 10.6 discussed the Optimal Routing (OR) procedures, which aim at providing a low-cost route for a MT call. The route determined using the OR procedures tends to bypass the HPLMN of the called subscriber, to use the direct path from the IPLMN to the VPLMN.

Immediate Service Termination (IST) procedures, discussed in Section 10.7, are used by service providers to control subscriber calls. IST procedures consist of two distinct sub-procedures: the IST Alert service and the IST Command service. The IST Alert service and the IST Command service together provide great flexibility to the service provider to monitor and terminate subscriber calls.

SESSION MANAGEMENT

11.1 INTRODUCTION

The call handling procedures are covered in the previous chapter, which apply to the CS domain. In the PS domain, the notion of call (or circuits) does not exist. Instead, the concept of 'session' is used. A session is used to avail the services of the PS domain. For each session, a PDP Context is created, which defines its characteristics. The PDP context information is maintained at the MS, SGSN and GGSN. The management of PDP context, which includes the procedures necessary to activate, modify and deactivate a PDP context between the MS, SGSN and GGSN, come under the Session Management (SM) functionality.

To explain the session management procedures, the information in this chapter is organized as follows: First, some of the basic concepts are explained. These include addressing, routing, encapsulation, decapsulation, tunneling and packet filtering. These concepts form the basis of all session management procedures. Thereafter, the PDP activation procedure is explained. The description includes the MS-initiated activation procedure, network-requested activation procedure, and the secondary PDP context activation procedure. After this, the PDP context modification procedure, which can be initiated by various entities including MS, SGSN, GGSN and RNC, is discussed. Only the important procedures are explained. The PDP context deactivation procedure is discussed next. Like the PDP context modification, the deactivation procedure can also be initiated by different entities. Only the important scenarios are discussed.

The discussion in this chapter is based primarily on three 3GPP specifications.

- The 3GPP TS 23.060 'General Packet Radio Service (GPRS)', which describes the SM procedures in significant detail.
- The session management part of 3GPP TS 24.008 'Mobile Radio Interface for Layer 3 specification: Core Network Protocols'.

- The 3GPP TS 29.060 'GPRS Tuneling Protocol', referred for exchanges between SGSN and GGSN.

Apart from these, other protocols like RANAP and MAP also feature in the discussion.

11.2 SESSION MANAGEMENT CONCEPTS

Before delving into the specifics of session management procedures, it is important to explain some of the elementary concepts that are central to understanding them. These concepts are:

- Addressing
- PDP Context Activation and Deactivation
- Packet Routing
- Tunneling and Encapsulation
- Packet Filtering

11.2.1 Addressing

The PS domain enables the MS to communicate with entities of a Packet Data Network (PDN). In order to engage in such communication, the MS must have an address that is valid in the PDN.

It may be noted that the PDN lies outside the PLMN; this implies that the addresses in the PLMN (like IMSI or MSISDN) are alone not sufficient for communication with the PDN entities. Since the most common PDN is based on the Internet Protocol (IP), an MS must have an IP address for communicating with other entities in the IP network. The IP address may be an IPv4 or an IPv6 address. Whatever be its nature, the MS must have an address, called the Packet Data Protocol (PDP) address, to communicate with entities in a PDN.

The MS can obtain the PDP address in different ways. The simplest way is to assign a *static PDP address* to it. This address is allocated by the network operator of the Home PLMN (HPLMN). Since the allocation is static, the address is permanent in nature. However, network addresses are scarce resources. It does not make sense to allocate an address permanently, more so when a subscriber may not be using it all the time. Thus, the addresses are generally allocated *dynamically*, so that a small set of these can be shared between a large number of subscribers.

There are three different ways in which an MS obtains a dynamic PDP address:

- It is assigned to the MS by the Home PLMN (HPLMN) operator.
- It is assigned to it by the Visited PLMN (VPLMN) operator.
- The operator of the PDN assigns a dynamic PDP address to the MS.

As part of subscription information, the home PLMN operator defines how the PDP addresses are assigned. Further, when the PDP address is assigned by the home or the visited PLMN, it is the responsibility of the GGSN to allocate and release this address. The PDP address is allocated by the GGSN during the *activation of PDP context*. A PDP context can be viewed as a set of information maintained by the MS, SGSN and GGSN. It contains a PDP type (that identifies the type of PDN, for example, IPv4), the PDP address (say a dynamically allocated IPv4 address), QoS information and other session information (see Table 11.1).

Activating a PDP context refers to creating the PDP context at the MS, SGSN and GGSN so that the MS can communicate with an entity in PDN using the PDP address maintained in the PDP context. When the communication is over, the *PDP context is deactivated*. During the deactivation of PDP context, the dynamic PDP address is released.

In case the PDP address is assigned by an external PDN, there are again two ways to achieve it. First, the GGSN obtains an address from the external PDP (using protocols like Dynamic Host Configuration Protocol) and provides it to the MS during the activation of the PDP context. Alternatively, the MS may obtain the PDP address directly from the PDN after the PDP context is activated.

Table 11.1   Elements of PDP Context

PDP Type	The PDP type is identified by • the organization that is responsible for the PDP type (e.g. ETSI or IETF) and • a PDP type number (e.g. PPP, IPv4 or IPv6).
PDP Address	This contains the PDP address whose format is governed by the PDP type. There is a length field that defines the PDP address length.
Network Service Access Point Identifier (NSAPI)	NSAPI holds the index of the PDP context. In the MS, this field is used to identify a PDP SAP. In the SGSN/GGSN, it is used to identify the PDP context associated with a Mobility Management (MM) connection.
Quality of Service (QoS)	This defines the QoS for a PDP context. There are three types of QoS: • **Subscribed QoS:** This is the QoS profile maintained at the HLR. A subscriber cannot request QoS greater than the value subscribed for. • **Requested QoS:** This is the QoS profile requested by the MS at the beginning of a session. • **Negotiated QoS:** While a user provides the requested QoS, the network negotiates each QoS attribute to a level that is in accordance with the available network resources.

Contd.

Table 11.1 Contd.

	The important parameters included in the QoS profile are:
	– **Traffic class:** This parameter defines the nature of traffic. Four traffic classes are defined, namely, Conversational class, Streaming class, Interactive class and Background class (refer Chapter 3 for details).
	– **Maximum and Guaranteed bit rate:** This set of parameters defines the maximum and guaranteed number of bits delivered by and to the UTRAN within a period of time, divided by the duration of the period.
	– **Delivery order:** This parameter defines whether the bearer shall provide in-sequence SDU delivery or not.
	– **Maximum SDU size:** This parameter defines the maximum size of SDU that can be carried by the network.
	– **SDU error ratio:** It indicates the fraction of SDUs that are lost or are detected as erroneous.
	– **Residual bit error ratio:** This parameter defines the undetected bit error ratio in the delivered SDUs.
	– **Delivery of erroneous SDU:** This parameter defines whether erroneous SDUs are to be delivered to the application layer or not.
	– **Transfer delay (ms):** It indicates maximum delay for 95% of delivered SDUs during the lifetime of a bearer service.
	– **Traffic handling priority:** This parameter specifies the relative importance for handling of all SDUs belonging to one PDP context compared to the SDUs of other PDP context.
	– **Allocation/Retention Priority:** This parameter specifies the relative importance of a RAB compared to other RAB for allocation and retention of the RAB.
Tunnel Endpoint Identifier (TEID)	This field is used between SGSN and GGSN and between RNC and SGSN to identify a tunnel endpoint in the receiving GTP-C or GTP-U protocol endpoint and to identify a PDP context. Note that GTP-C tunnels do not exist between RNC and SGSN.
Access Point Name (APN)	An APN is used to access a service associated with a GGSN. The name is translated by SGSN using the Domain Name System (DNS) to obtain the IP address of GGSN that can provide the requested service. An APN consists of an APN Network Identifier (mandatory) and an APN Operational Identifier (optional). The APN Network Identifier is a Fully Qualified Domain Name (FQDN) (for example, 'service.company.com'). This identifier is used to identity the external network that the GGSN is connected to and optionally, a requested service by the MS. The APN Operational Identifier is also an FQDN ending with '.gprs'. This is used to identify the PLMN GPRS backbone in which the GGSN is located.

11.2.2   PDP Context Activation and Deactivation

As mentioned above, an MS obtains the dynamic PDP address during the activation of PDP context. Even if it is a static PDP address, it is ready for use only after the PDP context is activated. During the activation of the PDP context, the GGSN creates a mapping between the IMSI and the PDP address. The GGSN also keeps other information (e.g. QoS, TEID, etc.) in the PDP context. When the GGSN receives a packet from an external network, it uses the PDP context (that has the TEID-PDP address mapping) to route the packet to the MS. The SGSN also keeps the PDP context information to route packets from the RNC to GGSN and vice-versa.

The PDP context is always activated by the MS. Even if a PDP context does not exist and GGSN receives a packet with a static PDP address, it requests the MS to activate the PDP context. This is done through the network-requested PDP context activation procedure. Upon receiving the request from the network, the MS activates the PDP context.

While the PDP context is always activated by the MS, it can be deactivated either by the MS or by the network (i.e. SGSN or GGSN). A PDP context is deactivated when data transfer is over and the PDP address is not required any more. Apart from activation and deactivation, there are also procedures to modify the attributes of the PDP context (e.g. QoS). The PDP context activation, modification and deactivation procedures are explained in Sections 11.4, 11.5 and 11.6, respectively.

11.2.3   Packet Routing

After a PDP context is activated, the PDP address is available for data transfer. The PDP PDUs are routed between the SGSN and GGSN over the UDP/IP protocols. The exact path depends upon whether the MS is in HPLMN or VPLMN and whether the address is allocated by the GGSN of HPLMN or GGSN of VPLMN. Figure 11.1 depicts an example where the MS is roaming in a VPLMN. Two scenarios are shown in the figure. These are explained as follows.

In case the GGSN of HPLMN assigns a PDP address to the MS, the packets are routed from the SGSN of the VPLMN to the GGSN of HPLMN. During the activation of the PDP context, the SGSN stores the address of the GGSN that provides connectivity to the external PDN. This address is used to tunnel the packets to the GGSN using the GTP-U. Since the exchange takes place between GSNs of two different PLMNs, an inter-PLMN backbone is used for communica-tion. Further, the Border Gateway (BG) of the respective PLMN performs security functions and prevents unauthorized access. In reverse direction, when the GGSN of HPLMN receives a packet with a PDP address, it uses the PDP context to determine the SGSN to which this packet is to be forwarded. Note that as the PDP address is allocated by the GGSN

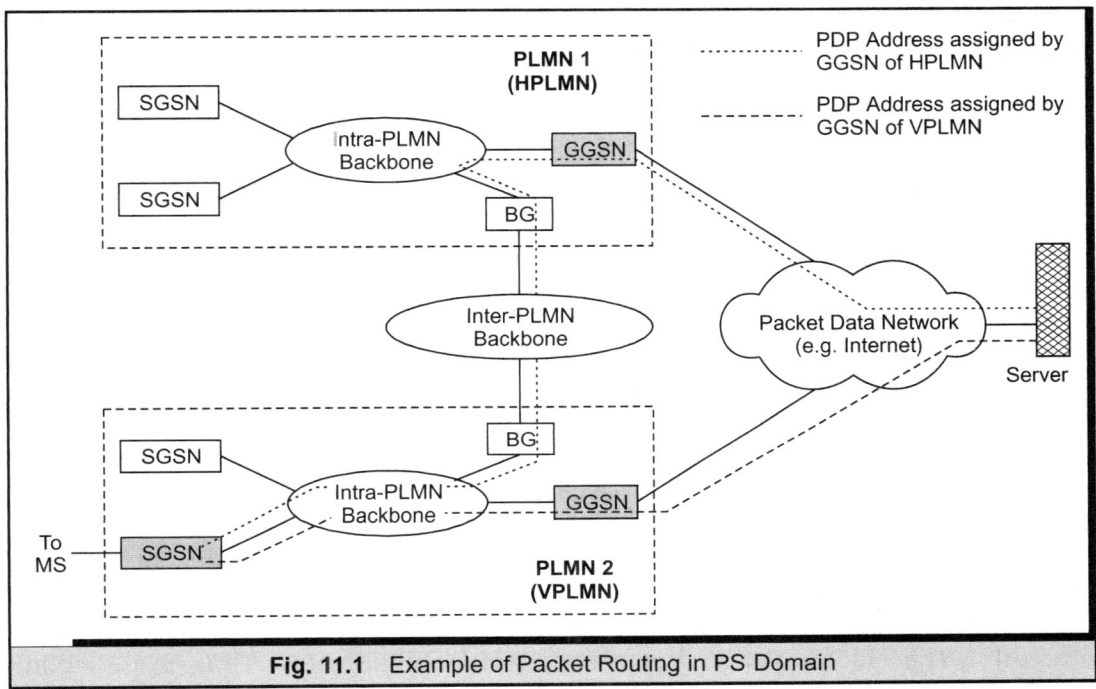

Fig. 11.1 Example of Packet Routing in PS Domain

of HPLMN, all packets destined for that PDP address are directed to that GGSN. The GGSN of HPLMN then takes further action.

It is also possible that the GGSN of VPLMN allocates a PDP address. Whether this is allowed or not is known by the SGSN when it queries the subscriber information from HLR. In case this is possible, the GGSN of VPLMN assigns a dynamic PDP address. All packets are then routed over the intra-PLMN backbone.

11.2.4 Encapsulation and Tunneling

The twin concepts of *encapsulation* and *tunneling* are central to the way data is exchanged in the PS domain. These concepts are explained as follows:

Encapsulation is formally defined as *the addition of address and control information to a data unit for routing packets within and between the network(s)*. In the context of session management, encapsulation enables packets of any packet data protocol to be exchanged between the MS and the PDN. Encapsulation makes the PLMN transparent to the data protocols carried by it. This provides it the flexibility to transport packets of any protocol across the PLMN, without having to define protocol-specific mechanisms for such a transport. The encapsulation function is performed by the entity sending an encapsulated packet. The receiving entity decapsulates the packet (i.e. removes the address and control information) and obtains the original packet.

In the PS domain, the PDP PDU is encapsulated/decapsulated at the MS, RNC, SGSN and GGSN. Between the MS and RNC, the PDP packet is encapsulated using Packet Data Convergence Protocol (PDCP). Between RNC and SSGN, and between SGSN and GGSN, the PDP PDU is encapsulated using the GPRS Tuneling Protocol-User Plane (GTP-U) Protocol.

Along with encapsulation, tunneling is used to exchange messages between the MS and GGSN. Formally, tunneling is defined as *the transfer of encapsulated data units within and between the network(s) from the point of encapsulation to the point of decapsulation.* The tunnel is a two-way, point-to-point logical path used for information exchange. In PS domain, the tunnels (referred to as GTP-U and GTP-C tunnels) are identified using Tunnel Endpoint Identifier (TEID). The RANAP and GPRS Tunneling Protocol-Control Plane (GTP-C) is used for establishing GTP-U tunnels and exchanging TEID.

The GTP-U tunnel exists between RNC and SSGN, and between SGSN and GGSN. This is depicted in Figure 11.2. Between RNC and SSGN, the TEID is exchanged during the RAB establishment using the RANAP protocol. The RAB establishment is part of PDP context activation procedure.

Between SSGN and GGSN, the TEID is exchanged as part of PDP context activation procedure. The TEID is always assigned by the downlink entity. Thus, in the uplink direction (i.e. SGSN to GGSN direction), the TEID is allocated by GGSN. In the downlink direction (i.e. GGSN to SGSN direction), it is allocated by SGSN. The TEID is sent with every message as part of the GTP header. A PDP PDU is encapsulated along with a GTP header. This encapsulated PDU, called the GTP PDU, is carried as a UDP PDU, which in turn is carried as an IP PDU. The TEID of the GTP header, along with the GSN address and port number, is used by the receiving entity to identify the GTP tunnel to which the PDU belongs. This information then gives the PDP context,

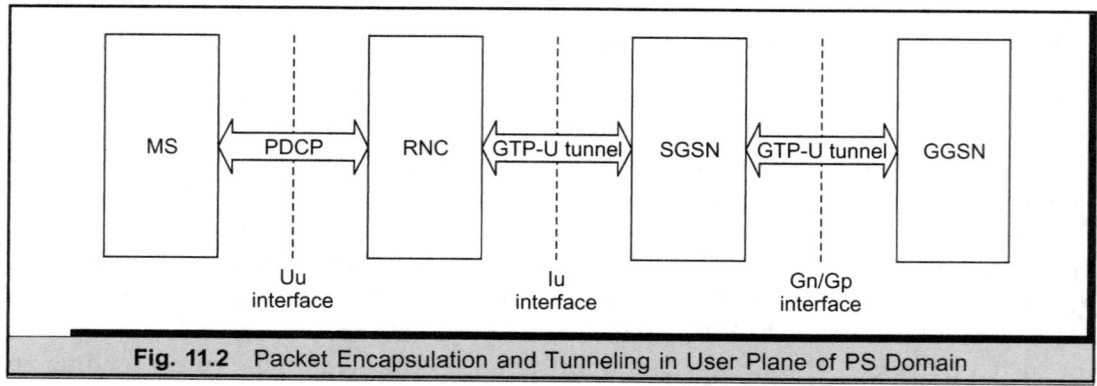

Fig. 11.2 Packet Encapsulation and Tunneling in User Plane of PS Domain

with which the PDP PDU is associated. The packet is then handled as per the parameters of the PDP context.

11.2.5 Packet Filtering

It is possible that a PDP address is associated with multiple PDP contexts. For example, when the UE is participating in many application layer sessions, it will be getting different flows for different applications, all with the same PDP address. Under such circumstances, the GGSN routes downlink PDUs based on the Traffic Flow Template (TFT) assigned to different PDP contexts. The TFT is used to classify packets at the GGSN and police incoming packets for downlink direction (i.e. GGSN to MS direction). A TFT consists of one to eight packet filters, each uniquely identified by a packet filter identifier. A packet filter also has an evaluation precedence (similar to priority) that is unique within all TFTs associated with the PDP context that share the same PDP address. Besides the packet filter identifier and evaluation precedence, there is at least one attribute (used for filtering) from among the following:

- Source Address and Subnet Mask
- Protocol Number (IPv4)/ Next Header (IPv6)
- Destination Port Range
- Source Port Range
- IPSec Security Parameter Index (SPI)
- Type of Service (TOS) for IPv4/Traffic Class and Mask for IPv6
- Flow Label for IPv6

Based on the settings of the packet filter attributes and the contents of the received PDP PDU, the GGSN does packet filtering and classification. Upon receiving a PDP PDU, the GGSN uses the packet filters with the smallest evaluation precedence to find a match. If no match is found, the packet filter with the next smallest evaluation precedence is inspected. When a match is found, the PDP PDU is associated with the corresponding PDP Context. In case no match is found, the PDP context without any TFT is used to associate the PDP PDU. In case all PDP contexts have a TFT and no match is found, the PDP PDU is discarded.

11.3 PDP PROTOCOL STATES

To manage the PDP addresses, the MS, SGSN and GGSN implement a PDP state machine. Figure 11.3 depicts a simplified view of the PDP state machine with two PDP protocol states (INACTIVE and ACTIVE) and the state transitions between these two states. Apart from these two states, there are intermediate states (like PDP-INACTIVE-PENDING and PDP-ACTIVE-PENDING). However, for sake of

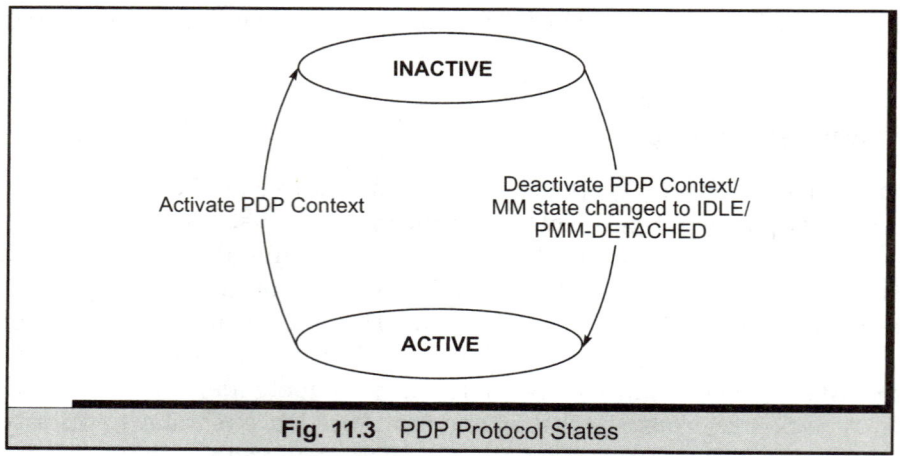

Fig. 11.3 PDP Protocol States

simplicity, these intermediate states are not shown in the figure. The PDP states are explained as follows:

- **INACTIVE State:** This state indicates that the PDP address is currently not available for use (i.e. PDP context for the PDP address is not active). In this state, the PDP data cannot be exchanged by the MS. In case the GGSN receives a PDP PDU for a PDP address that is in an INACTIVE state, it may initiate a network-requested PDP context activation (see Section 11.4.3). This procedure applies to static PDP addresses. If GGSN is unable to serve the request (e.g. when it does not have information about the PDP address), it returns an error (e.g. ICMP unreachable error). An MS makes transition to the ACTIVE state by activating the PDP context.

- **ACTIVE State:** In this state, the PDP address is available for data transfer. The GGSN contains a PDP context for the PDP address in ACTIVE state. The PDP context contains the necessary information to route a PDP PDU from and to the MS.

11.4 PDP CONTEXT ACTIVATION PROCEDURES

In order to communicate with PDNs, the MS activates a PDP context. The MS may initiate the PDP context activation procedure by itself or it may initiate the procedure after receiving a request from the network.

A PDP context applies to a specific QoS and a specific NSAPI. A given PDP address can be associated with one or more PDP contexts. In case more than one PDP context exists for a given PDP address, the first PDP context is established using the PDP context activation procedure while the additional contexts are activated using the *secondary*

PDP context activation procedure. The secondary PDP context may be used to activate a PDP context while reusing the PDP address and other PDP context information from an already active PDP context, but with a different QoS profile.

Both normal and secondary PDP context activation procedures are explained in this section.

Further, if more than one PDP context exists for a PDP address, there is a Traffic Flow Template (TFT) for the additional contexts. The TFT is sent transparently by the MS to the GGSN. As mentioned earlier, TFT is used for packet filtering and classification at GGSN.

11.4.1 PDP Context Activation Procedure

The PDP context can be activated by the MS provided the MM connection (i.e. RRC and Iu connection) exists. Figure 11.4 shows the messages exchanged during the PDP context activation procedure. The steps involved in this procedure are summarized as follows:

1. In order to request PDP context activation, the MS sends 'Activate PDP Context Request' message to the SGSN. The message contains the NSAPI, PDP type, PDP address, APN and the requested QoS. The PDP address is sent only if the MS has a static address. In case the MS wants a dynamic address, the address field is left empty.
2. Upon receiving the 'Activate PDP Context Request', the SGSN validates the request. After this, the SGSN uses the APN to determine the GGSN that can handle the request. If a GGSN cannot be determined, an error response is sent to the MS.

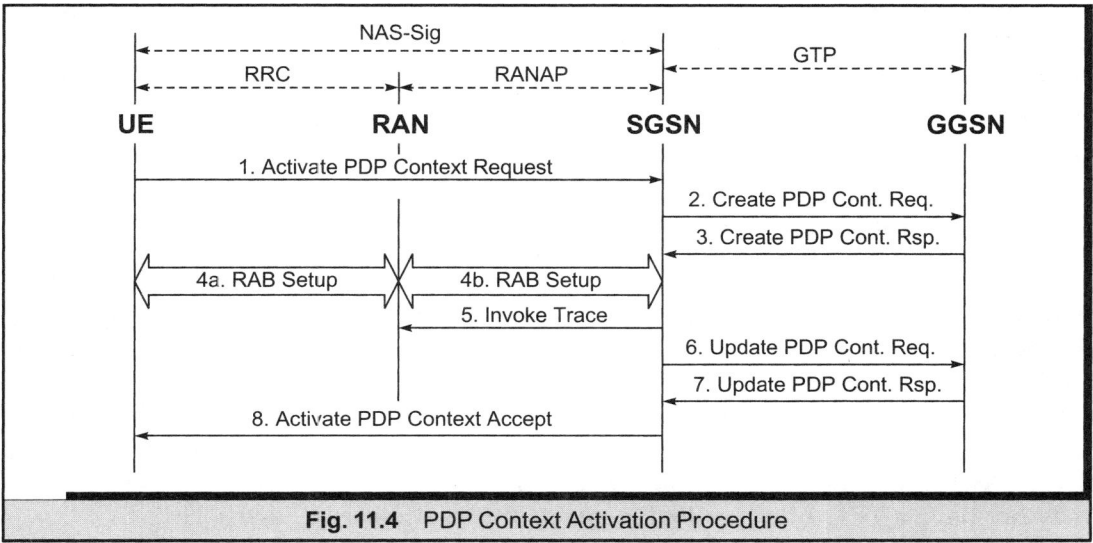

Fig. 11.4 PDP Context Activation Procedure

Else, the SGSN sends a 'Create PDP Context Request' to GGSN. In this message, the SGSN sends the information received from the MS along with the TEID to be used by the GGSN in the downlink direction. The SGSN, based on the availability of resources, may restrict the QoS. Thus, the QoS sent by SGSN to GGSN is called the 'negotiated QoS' (and not the 'requested QoS').

3. If the GGSN successfully processes the request, it creates an entry in the PDP context table. This entry allows the routing of packets between the SGSN and the external PDN. GGSN then creates the 'Create PDP Context Response' message and sends it to SGSN. The response message has a PDP address if it is dynamically allocated by the GGSN. In case the address is to be obtained from an external PDN, then a value of 0.0.0.0 is set to indicate that the MS shall get a PDP address after the completion of the PDP context activation procedure. The GGSN also sends a TEID to be used by SGSN in uplink direction.

4. On receiving a success the SGSN initiates the RAB setup procedure (refer to Chapter 8 for details of RAB setup procedure).

5. The SGSN also checks whether tracing is activated or not. If it is activated, then it sends an 'Invoke Trace' message to RNC.

6. If during the RAB setup the QoS attributes are downgraded, the SGSN updates the PDP context information at the GGSN using the 'Update PDP Context Request'.

7. The GGSN responds with 'Update PDP Context Response' message.

8. The SGSN sends 'Activate PDP Context Accept' message with NSAPI, GGSN address, QoS negotiated and other parameters to the MS. The procedure is complete when the MS receives this message.

Depending upon the nature of the application, different QoS profiles can be requested for by the MS. For example, for E-mail the QoS profile indicates delay-insensitive application. Similarly, for real-time applications the QoS profile indicates so. In case the network cannot provide the requested QoS profile, it can negotiate the QoS profile to a value close to the requested one. The MS has the option to accept the negotiated profile or deactivate the PDP context.

11.4.2 Secondary PDP Context Activation Procedure

The Secondary PDP Context is activated when the MS wishes to reuse the PDP address and some other PDP context information from an active PDP context. The secondary PDP context must have a Traffic Flow Template (TFT). The TFT is used by the GGSN to differentiate between two PDUs, each with the same PDP address but associated with different PDP contexts. In case all previously activated PDP contexts already have a TFT associated with them, a secondary PDP context activation may choose not to send a TFT. Note that at most, only one PDP context associated with a PDP address may not have a TFT. All other PDP contexts must necessarily have a TFT.

The secondary PDP context activation procedure is similar to the steps depicted in Figure 11.4. Some of the differences are as follows: First, the APN selection is not done. Note that since a GGSN is already selected, there is no need for the SGSN to select a GGSN that can provide a given service. Moreover, as the PDP address is already obtained, the dynamic PDP address allocation procedure by GGSN is not carried out; nor is the invoke trace procedure.

11.4.3 Network-Requested PDP Context Activation Procedure

As discussed in the previous sub-sections, the PDP context activation procedure is always initiated by the MS. However, it is possible that the GGSN may request the MS to initiate this procedure. This happens when a PDP PDU is received for a PDP address for which the GGSN of the home PLMN has some static PDP information (i.e. static PDP address). Note that even for network-requested PDP context activation, the MS actually initiates the PDP context activation procedure after receiving a request to this effect from the network.

Figure 11.5 shows the messages exchanged during the network-requested PDP context activation procedure. The steps involved in this procedure are summarized as follows:

1. The GGSN receives a PDP PDU for which there is no active PDP context. It then determines whether it has to initiate a network-requested PDP context activation procedure. This procedure is initiated when the received PDP address is a static address and the GGSN is the home PLMN of the MS under consideration.

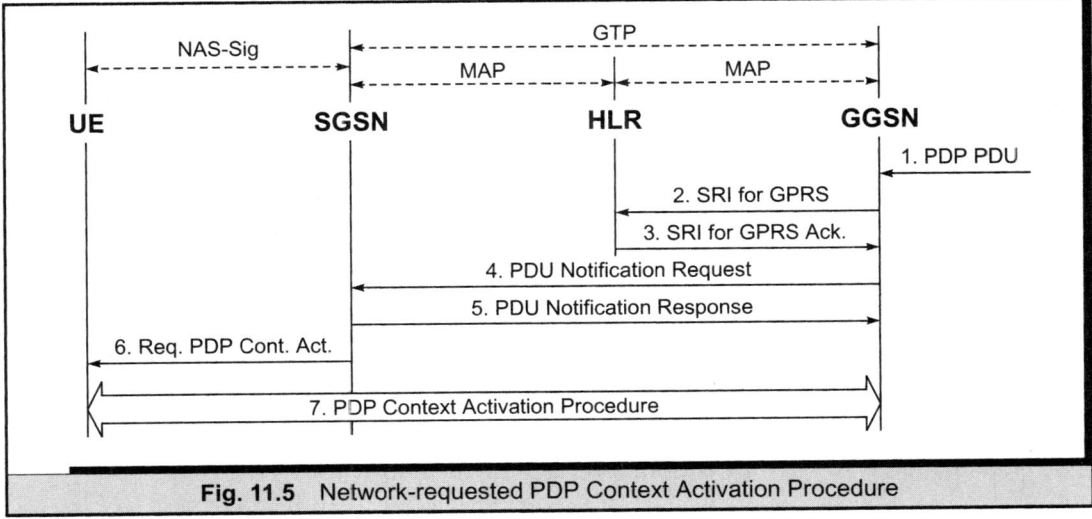

Fig. 11.5 Network-requested PDP Context Activation Procedure

2. If the GGSN so decides, it sends the 'Send Routing Information for GPRS' message to HLR (using MAP protocol). Note that while the GGSN receives a PDP PDU containing a PDP address, it sends a request to HLR using the IMSI. This implies that the GGSN must have static information that maps the PDP address to IMSI.

3. In case the HLR has the SGSN address, it sends this in the response message. Moreover, if the flag 'MS Not Reachable for GPRS (MNRG)' is set in HLR, the HLR also sends 'MS Not Reachable Reason (MNRR)' to GGSN. The MNRR indicates the reason why the MS is not reachable.

4. The GGSN takes further action based on information received from the HLR. In case the response contains the SGSN address and the 'MS Not Reachable Reason (MNRR)' is missing, the GGSN sends a 'PDU Notification Request' to the SGSN indicated by the HLR. The 'PDU Notification Request' contains the IMSI, PDP type, PDP address and APN. Even if the MNRR is present and indicates 'No Paging Response', the GGSN still sends the 'PDU Notification Request' to the SGSN, provided that the HLR has furnished the SGSN address. In all other cases (i.e. if the MNRR is different from 'No Paging Response' or SGSN was not provided by HLR), the GGSN marks the 'MS Not Reachable for GPRS (MNRG)' flag for that MS.

5. The SGSN acknowledges the request by sending a response.

6. The SGSN also sends 'Request PDP Context Activation' message to MS with the information received from the GGSN. Before this message can be sent from the SGSN to the UE, the paging procedure is used to determine the latter's exact location. A NAS signaling connection is also set up between the UE and the SGSN. The paging and the NAS signaling connection establishment procedures are not depicted in the figure. The same were discussed in Chapter 8.

7. If the MS accepts the invitation, it follows the PDP context activation procedure, mentioned in Section 11.4.1.

11.4.3.1 *Unsuccessful Network-Requested PDP Context Activation Procedure*

It is possible that the network-requested PDP context activation procedure may not succeed. This may be because of an unreachable MS or due to other reason. In such a case, the specifications define different mechanisms that could ward off unnecessary enquiries to HLR. Among these, the mechanism using 'MS Not Reachable for GPRS (MNRG)' flag is most important and is explained below:

- When GGSN sends a 'PDU Notification Request' message to SGSN and SGSN detects that the MS is GPRS-detached or if the IMSI is unknown, it send a 'PDU Notification Response' message to the GGSN with the appropriate cause.

- The subsequent action at GGSN depends upon the cause received from SGSN. Two different course of actions are explained below:

  1. In case the cause indicated by SGSN is 'MS is GPRS-detached', the GGSN sends a 'Failure Report' message to the HLR. The HLR then sets the MNRG

flag. HLR also stores the GGSN number and address. This is done to contact the GGSN in case some MS-related activity is later detected by the HLR. Now, when HLR receives another request from a different GGSN and it sees that the MNRG flag is set, the HLR sends a response containing the 'Mobile Not Reachable Reason (MNRR)'. The HLR also adds the number and address of this GGSN in a list of GGSNs that have to be contacted when some MS-related activity is detected. Upon receiving the response, the GGSN can avoid contacting the SGSN because the presence of MNRR indicates that the MS is not reachable for GPRS.

2. In case the SGSN had reported 'IMSI is not known', the GGSN may again query HLR requesting it to provide the SGSN address again. If the SGSN address is same as that obtained earlier, the GGSN sends 'Failure Report' message to HLR. If the SGSN address is different, the GGSN again tries to perform network-requested PDP Context Activation Procedure in the anticipation that the SGSN has changed since the previous attempt.

- Subsequently, when activity for MS is detected (e.g. when HLR receives a 'Ready for SM' message or 'Update Location' message), the HLR indicates this to all the GGSNs maintained for the given subscriber using 'Note MS Present for GPRS' message. This allows the GGSN to know that MS is reachable again and that network-requested PDP context activation should be attempted.

The procedure mentioned above has the advantage that a failure of network-requested PDP Context Activation Procedure due to lack of activity is notified to the HLR. On subsequent queries, the HLR can then inform that the MS is not reachable for GPRS. This prevents the GGSNs from making unnecessary requests to SGSN once they detect that MNRG is set. Since a GGSN may also maintain the MNRG flag, it can avoid attempting network-requested PDP context activation procedure for an MS that has the MNRG set.

11.5 PDP CONTEXT MODIFICATION PROCEDURES

After a PDP context is activated, the session management procedures provide the means to modify the PDP context parameters. The modification procedure can be initiated by the MS, SGSN, GGSN and RNC. The modification of subscriber data at HLR can also trigger the initiation of the PDP context modification procedure. Not all parameters of the PDP context are modifiable. Those that can be modified are:

- QoS negotiated
- Radio Priority
- Packet Flow Identifier
- PDP Address (in case of GGSN-initiated modification procedure)
- Traffic Flow Template (in case of MS-initiated modification procedure)

The exact modification procedure followed depends upon the entity that initiates the request, and also on the nature of data that is modified. The following sections provide details of the PDP context modification procedure initiated by various entities.

11.5.1 MS-Initiated PDP Context Modification Procedure

The MS can initiate the PDP context modification procedure to modify the negotiated QoS and/or TFT. The negotiated QoS indicates the desired QoS profile. The TFT can be added, modified or deleted from a PDP context.

Figure 11.6 shows the messages exchanged during the MS-Initiated PDP context modification procedure. The steps involved in this procedure are summarized as follows:

1. In order to modify the parameter(s) of a PDP context, the MS sends 'Modify PDP Context Request' message to the SGSN. The message contains the NSAPI, QoS Requested and/or TFT.

2. Upon receiving the 'Modify PDP Context Request', the SGSN validates the request and sends the 'Update PDP Context Request' to GGSN. In this message, the SGSN sends the information received from the MS.

3. The GGSN determines whether the QoS negotiated and/or TFT are compatible with the PDP context being modified. The GGSN may further restrict QoS negotiated as per its capabilities and the load of the system. The GGSN then stores the QoS Negotiated. It also stores, modifies or deletes the TFT of the PDP context as per the TFT in the request message. The GGSN then sends 'Update PDP Context Response' to SGSN.

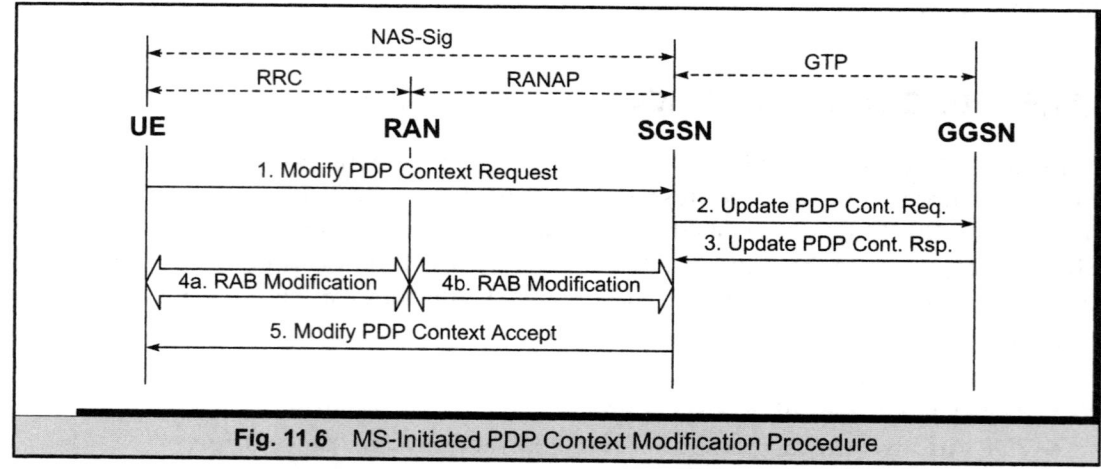

Fig. 11.6 MS-Initiated PDP Context Modification Procedure

4. On receiving a success response, the SGSN initiates the RAB modification procedure (refer to Chapter 8 for details of RAB modification procedure).

5. The SGSN then sends 'Modify PDP Context Accept' with QoS negotiated, and other parameters to the MS.

11.5.2 SGSN-Initiated PDP Context Modification Procedure

The SGSN can also initiate the PDP context modification procedure. This is done primarily to change the QoS of the PDP context.

Figure 11.7 shows the messages exchanged during the SGSN-initiated PDP context modification procedure. The steps involved in this procedure are summarized as follows:

1. In order to modify the parameter(s) of a PDP Context, the SGSN sends 'Update PDP Context Request' message to the GGSN. The message contains the TEID, NSAPI, QoS Negotiated and other parameters.

2. The GGSN determines whether the QoS Negotiated is compatible with the PDP context being modified. The GGSN may further restrict QoS negotiated as per its capabilities and the load of the system. It then stores the QoS negotiated and sends the 'Update PDP Context Response' to SGSN.

3. On receiving a success response, the SGSN selects the radio priority and packet flow identifies based on the negotiated QoS. The SGSN then sends the 'Modify PDP Context Request' to the MS.

4. The MS acknowledges this by sending an accept message.

5. The SGSN initiates the RAB modification procedure (refer to Chapter 8 for details of RAB modification procedure).

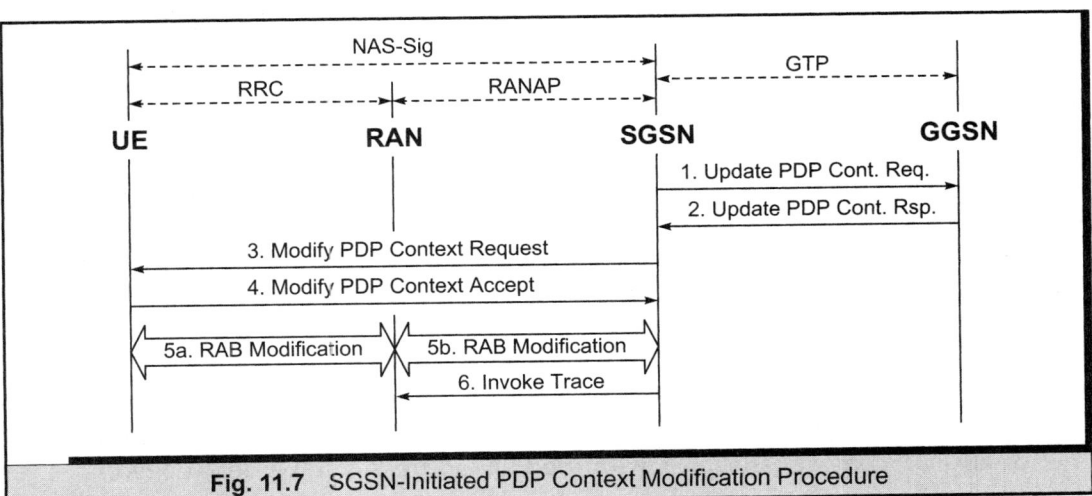

Fig. 11.7   SGSN-Initiated PDP Context Modification Procedure

6. Further, the SGSN checks whether tracing is activated or not. If it is activated, then it sends an 'Invoke Trace' message to RNC.

11.5.3 Other PDP Context Modification Procedures

The GGSN can also initiate the PDP context modification procedure. This is done primarily to change the QoS of the PDP context or to change the PDP address. The procedure for this is similar to the SGSN-Initiated PDP context modification procedure described in Section 11.5.2. The only difference is that in this case the GGSN sends the 'Update PDP Context Request' message to the SGSN, and SGSN responds with 'Update PDP Context Response' message after the modification procedure is complete.

Apart from the MS, SGSN and GGSN, the RNC can also initiate the PDP context modification procedure. This happens when an Iu connection is released (e.g. due to break of the radio connection or due to user inactivity). In such cases, the *preservation procedure* is used. The preservation procedure allows the active PDP contexts associated with the released RABs to be preserved. The RABs can then be re-established at a later stage.

The PDP context modification also takes place when a RAB is released by the RNC. For details of the preservation procedures, the reader is referred to 3GPP TS 23.060 'General Packet Radio Service (GPRS)'.

11.6 PDP CONTEXT DEACTIVATION PROCEDURES

When a PDP context is not required any more (e.g. when data exchange is complete), it is deactivated. The deactivation procedure can be initiated by the MS or by the network (i.e. SGSN or GGSN). The following section describes the PDP context deactivation procedure in detail.

11.6.1 MS-Initiated PDP Context Deactivation Procedure

The MS carries out the MS-initiated PDP context deactivation procedure to tear down an existing PDP context.

Figure 11.8 shows the messages exchanged during the MS-initiated PDP context deactivation procedure. The steps involved in this procedure are summarized as follows:

1. For PDP context deactivation, the MS sends a 'Deactivate PDP Context Request' message to the SGSN. The message may contain a Tear-down Indicator, which indicates whether only the PDP context under consideration is to be deactivated or all PDP context associated with the PDP address is to be deactivated.

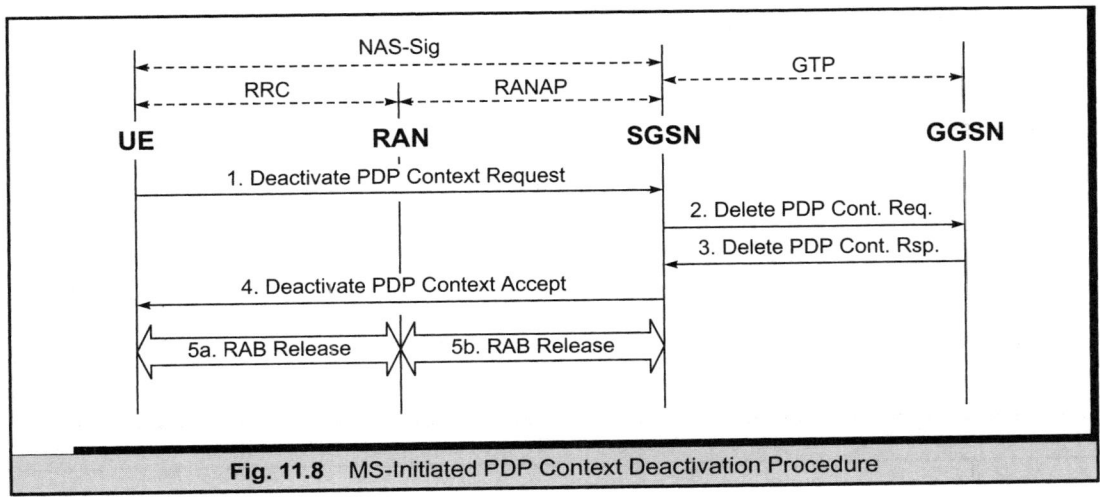

Fig. 11.8 MS-Initiated PDP Context Deactivation Procedure

2. Upon receiving 'Deactivate PDP Context Request', the SGSN validates the request. After this, the SGSN sends a 'Delete PDP Context Request' to GGSN. In this message, the SGSN sends the TEID, NSAPI and the tear-down indicator.

3. If the MS has included the tear-down indicator in the request message, the GGSN deletes all PDP context associated with the PDP address. If the GGSN has assigned a dynamic PDP address and the PDP context was the last context associated with the PDP address, then the GGSN releases this PDP address and makes it available. The GGSN then sends 'Delete PDP Context Response' to SGSN.

4. Thereafter, the SGSN sends 'Deactivate PDP Context Accept' message to the MS.

5. Further, the SGSN initiates the release of the RAB associated with the PDP context (refer to Chapter 8 for details of RAB release procedure).

11.6.2 SGSN-Initiated PDP Context Deactivation Procedure

The SGSN can also deactivate an active PDP context. Figure 11.9 shows the messages exchanged during the SGSN-initiated PDP context deactivation procedure. The steps involved in this procedure are summarized as follows:

1. The SGSN sends a 'Delete PDP Context Request' to GGSN. In this message, it sends the TEID, NSAPI, and optionally, the tear-down indicator.

2. If the SGSN has included the tear down indicator in the request message, the GGSN deletes all PDP context associated with the PDP address. If the GGSN has assigned a dynamic PDP address and the PDP context was the last context associated with the PDP address, then the GGSN releases this PDP address and makes it available. The GGSN then sends 'Delete PDP Context Response' to SGSN.

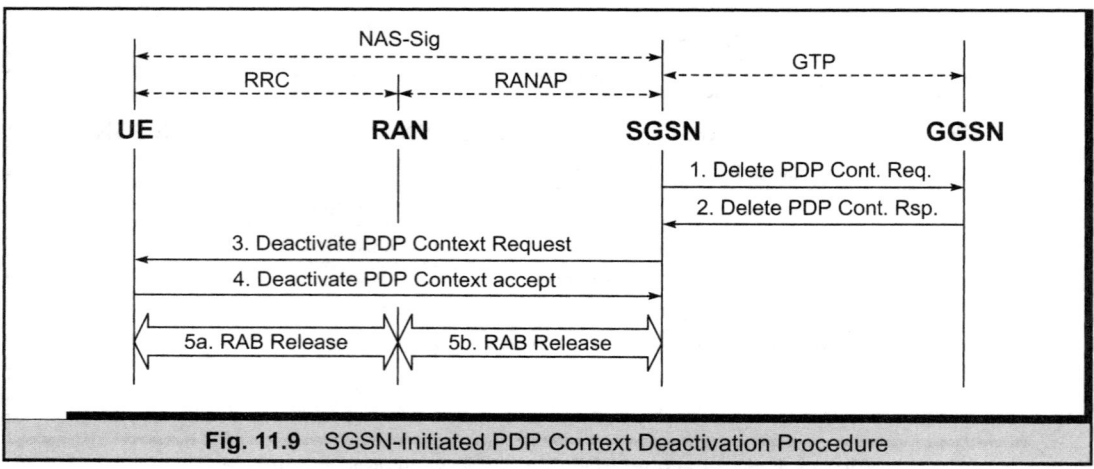

Fig. 11.9   SGSN-Initiated PDP Context Deactivation Procedure

3. Thereafter, the SGSN sends the 'Deactivate PDP Context Request' to the MS. In case the tear down indicator is included in the request message, the MS deletes all PDP context associated with the PDP address.
4. The MS responds with 'Deactivate PDP Context Accept' message.
5. The SGSN also initiates the release of the RAB associated with the PDP context (refer to Chapter 8 for details of RAB release procedure).

11.6.3 Other PDP Context Deactivation Procedure

Like the SGSN, the GGSN can also deactivate the PDP context. The procedure for this is quite similar to the SGSN-initiated PDP context deactivation procedure, described in Section 11.6.2.

SUMMARY

This chapter explained the basics of Session Management, the heart of which is the concept of PDP Context. A PDP Context is a set of information maintained at MS, SGSN and GGSN. The PDP context contains the PDP address, which is used to communicate with entities of a PDN. The PDP address is obtained during activation of the PDP context. The PDP context activation, modification and deactivation procedures have been elaborated in the chapter.

After PDP context is activated, the data packets are encapsulated using GTP-U and are tunneled between RNC and GGSN. When data transfer is over, the PDP context is deactivated and the associated resources are freed.

12

SUPPLEMENTARY SERVICES

12.1 INTRODUCTION

A classification of the services offered in a UMTS network was provided in Chapter 3. These included the Bearer Services, Teleservices, and Supplementary Services (SS). Collectively, the Bearer Services and Teleservices are also referred to as Basic Services (BS).

As the name suggests, basic services are the primary services offered by a mobile network. They include voice telephony, fax, SMS, circuit-switched data service and GPRS-based packet services. Supplementary services, on the other hand, modify or supplement the basic services and do not have any significance independently. Take, for example, the Call Forwarding (CF) supplementary service that was introduced in Chapter 10. This service does not have any significance by itself, since it is not known what calls/messages are to be forwarded (voice calls, fax messages or SMS messages). In other words, a supplementary service, like CF, has always to be associated with a basic service. Also, from the perspective of a service provider, while the network may not offer supplementary services to the subscribers, it must offer them the basic services.

The network operator has total control over the provisioning of basic services to a mobile subscriber, who has to request the former each time a new basic service is required, or if some existing service is to be removed. On the other hand, in case of supplementary services, the mobile subscriber is offered some flexibility, in that he has control over activation/deactivation of these services. Take, for instance, a Call Barring (CB) supplementary service. A subscriber may wish to bar all incoming voice calls (using CB) while going for a meeting, but may remove this barring once the meeting is over.

This chapter discusses the supplementary services provided in UMTS, and the procedures involved in the control or management of these services. Section 12.2 discusses some of the important concepts related to supplementary services; association of supplementary services with basic services; and supplementary service operations and state information. Section 12.3 covers the procedures related to call independent management of supplementary services by a mobile subscriber as well as Man-Machine Interface for supplementary service management. Section 12.4 discusses the various supplementary services offered in a UMTS network, and the call-related management of these services.

Network operators do not necessarily have to offer all (or for that matter, any) of the services discussed in Section 12.4. It is their prerogative to decide what supplementary services are to be offered to the subscribers. Further, network operators may also offer some non-standardized supplementary services to their subscribers. Unstructured Supplementary Services Data (USSD), discussed in Section 12.5, provides a means to offer such non-standardized services.

12.2 SUPPLEMENTARY SERVICE CONCEPTS

Supplementary services are used to modify/supplement the basic services offered in a UMTS network. Each supplementary service is associated with a set of basic services, and each has some associated state information. This section discusses the association of supplementary services with basic services, and the supplementary service related operations and state information.

12.2.1 Association with Basic Services

Supplementary services are offered to mobile subscribers as a supplement to their basic services. However, not all supplementary services can be associated with a particular basic service; only a subset of these is relevant/applicable to each basic service. Table 12.1 depicts the association of supplementary services with basic services. A 'Yes' against a supplementary service indicates its applicability to the basic service in the relevant column, while a blank entry implies the opposite. An example, is the combination of call deflection supplementary service and telephony, which is a basic service. A 'Yes' in the space corresponding to this combination implies that the call deflection supplementary service is applicable to telephony—a basic service.

The association of Supplementary Services (SS) with the Basic Services (BS) of a subscriber can either be explicit or implicit, depending upon the requirement of the supplementary service in question. In case of explicit association, separate record entries are made for each SS–BS pair indicating the basic services on which the supplementary service is applied. These records are stored as part of the subscription data in the HLR of the subscriber's HPLMN. Thus, it is not essential to apply the supplementary service on all applicable basic services simultaneously. The association

Table 12.1   Applicability of Supplementary Services (SS) to Basic Services (BS)

Basic Services → *Supplementary Services?* ↓	Telephony	Emergency Call	SMS-PTP	SMS-CB	Fax	CCT Data	GPRS	VGCS TS 91	VGCS TS 92
Enhanced Multi-level Precedence and Preemption	Yes	Yes			Yes	Yes		Yes	Yes
Call Deflection	Yes				Yes	Yes		Yes	Yes
Calling Line Identification Presentation	Yes	Yes			Yes	Yes		Yes	Yes
Calling Line Identification Restriction	Yes				Yes	Yes		Yes	Yes
Connected Line Identification Presentation	Yes				Yes	Yes		Yes	Yes
Connected Line Identification Restriction	Yes				Yes	Yes		Yes	Yes
Call Forwarding Unconditional	Yes				Yes	Yes		Yes	Yes
Call Forwarding on Mobile Subscriber Busy	Yes				Yes	Yes		Yes	Yes
Call Forwarding on No Reply	Yes				Yes	Yes		Yes	Yes
Call Forwarding on Mobile Subscriber Not Reachable	Yes				Yes	Yes		Yes	Yes
Call Waiting	Yes				Yes	Yes		Yes	Yes
Call Hold	Yes							Yes	Yes
Multi-party	Yes							Yes	Yes
Closed User Group	Yes				Yes	Yes		Yes	Yes
Advice of Charge	Yes				Yes	Yes		Yes	Yes
User-to-User Signaling	Yes				Yes	Yes			
Barring of All Outgoing Calls	Yes		Yes		Yes	Yes		Yes	Yes
Barring of Outgoing International Calls (BOIC)	Yes		Yes		Yes	Yes		Yes	Yes
BOIC except those directed to the Home PLMN Country	Yes		Yes		Yes	Yes		Yes	Yes
Barring of Incoming Calls (BIC)	Yes		Yes		Yes	Yes		Yes	Yes

Contd.

Table 12.1 Contd.

Basic Services → / Supplementary Services ? ↓	Telephony	Emergency Call	SMS-PTP	SMS-CB	Fax	CCT Data	GPRS	VGCS TS 91	VGCS TS 92
BIC when Roaming outside HPLMN Country	Yes		Yes		Yes	Yes		Yes	Yes
Explicit Call Transfer	Yes							Yes	Yes
Completion of Calls to Busy Subscriber	Yes				Yes	Yes		Yes	Yes
Calling Name Presentation	Yes							Yes	Yes
Multiple Subscriber Profile	Yes		Yes		Yes	Yes		Yes	Yes
Multi-Call	Yes	Yes			Yes	Yes			

SMS-PTP: Short Message Service-Point to Point

Fax: Facsimile Services

GPRS: General Packet Radio Service

SMS-CB: Short Message Service – Cell Broadcast

CCT Data: Circuit-Switched Data Service

VGCS: Voice Group Call Service

can be judiciously made as per the requirements of the subscriber. An example of such a supplementary service is the 'Barring of All Outgoing Calls (BAOC)' service, wherein the BAOC service can be selectively applied on one or more applicable basic services.

On the other hand, certain supplementary services have the property that they can either be applied on all or none of the applicable basic services subscribed to by the subscriber. In other words, as soon as the supplementary service is 'switched-on', it is automatically applied on all applicable basic services. In such a case, a separate record entry is not required for each SS–BS pair. A single record entry in the subscriber data can indicate whether the supplementary service is currently applied on the subscriber's basic services or not. This is called implicit association, wherein the supplementary service is not explicitly associated with any particular basic service. An example of such a supplementary service is the 'Call Deflection' service. For more details on association of supplementary services with basic services, the reader is referred to 3GPP TS 22.004.

12.2.2 SS Operations

Each supplementary service has some *state information* associated with it. *The state information* associated with each supplementary service, provides information on whether the service is subscribed by the subscriber, and whether it is currently active for a basic service or is dormant. However, before proceeding with the discussion on this topic, it is important to understand the operations that can be carried out on a supplementary service. These operations are as follows:

- **Provisioning:** This is the mechanism by which a supplementary service is subscribed to. A supplementary service is set to be provisioned for a subscriber provided it is available to the latter. The service provider may automatically provision certain supplementary services, for example, the 'Call Waiting' service, to all its subscribers free of cost. On the other hand, certain supplementary services may require an arrangement with the service provider before they can be provided to the subscriber.

- **Withdrawal:** Withdrawal is the opposite of provisioning. It makes the supplementary service unavailable to the subscriber.

- **Registration:** Some supplementary services require that certain information related to a particular service be registered with the network, before the use of this service can be used. For example, the 'Forwarded-to-Number (FTN)' must be registered with the service provider before the subscriber can use the call forwarding service. The process of providing this extra information required for the supplementary service is termed as Registration. However, not all supplementary services require additional data, and hence, nor the registration.

- **Erasure:** The process of removal of information registered for a supplementary service is termed as Erasure. Hence, erasure is a reversal of the registration operation.

- **Activation:** A supplementary service which is provisioned for a subscriber, and for which, registration (if required) has already taken place, can be activated and made operational. This is called the active phase of a supplementary service, wherein this service becomes operational. Again, consider the case of Call Forwarding Unconditional (CFU) supplementary service. Once the CFU service is provisioned and registered, the subscriber can activate it when going for a meeting, so that all calls are forwarded to the FTN. There are certain supplementary services, however, which do not require explicit activation, and are implicitly activated as soon as the service is provisioned. An example of such a service is the Calling Line Identification Presentation (CLIP) service, discussed in Section 12.4.3

- **Deactivation:** The reverse of the activation process is deactivation. Using deactivation, the supplementary service is brought back into a non-active state. Con-tinuing with the example of the Call Forwarding Unconditional (CFU) service—when the mobile subscriber finishes with the meeting, he/she may deactivate the CFU supplementary service. The incoming calls can now be received normally by the subscriber.

- **Interrogation:** Interrogation is the process by which a subscriber is allowed to query the state of a particular supplementary service (P, Q, R, A bits discussed in Section 12.2.3). The subscriber may also query the data registered for a particular supplementary service—e.g. the 'Forwarder-to-Number (FTN)' for the call forwarding service—using the Interrogation mechanism.

- **Invocation:** As the name suggests, invocation is the process of invoking a supplementary service. The invocation of a supplementary service can either be the result of a mobile subscriber request, or it can happen automatically on fulfillment of certain conditions.

12.2.3 SS State Information

With the background of the SS operations, the state information associated with a supplementary service can be easily defined. This information, maintained for each supplementary service, consists of 4 bits as follows:

(a) **'P'-Bit:** This bit signifies whether the supplementary service is provisioned or not.

(b) **'R'-Bit:** This indicates whether or not the extra information required for the supplementary service has been registered. The registration can be done only after the service is provisioned, and hence, the 'R'-Bit can be set only if the 'P'-Bit is set. For services that do not require registration, the 'R'-Bit is not applicable.

(c) 'A'-Bit: The 'A'-Bit indicates whether the activation for the supplementary service has taken place or not. This Bit can be set only after the 'P'-Bit and the 'R'-Bit (if applicable) has been set, i.e. the service has been provisioned and registered. For services like CLIP that do not require explicit activation, the 'A'-Bit is not applicable. In fact, both the 'R'-Bit and the 'A'-Bit are not applicable in this case.

(d) 'Q'-Bit: The 'Q'-Bit complements the 'A'-Bit. 'Q' stands for Quiescent, and it specifies if the service is dormant or active. If the 'A'-Bit and the 'Q'-Bit are both set, it indicates that the supplementary service is 'Active, but Quiescent'. This means that though the service is active, it is currently dormant. If the 'Q'-Bit is not set, then it indicates that the service is active, and not dormant. There are many reasons for a service being quiescent/dormant. A typical example is that of the Call Forwarding (CF) and the Call Barring (CB) supplementary services. Interaction of the CF and the CB services with MT Call handling procedures was discussed in Chapter 10. Consider the case when the mobile subscriber has barred all incoming calls using the 'Barring of All Incoming Calls (BAIC)' category of the CB service. In this case, the call forwarding service, if active, would remain dormant. This is because CB-related checks on the MT call would be carried out first, whereby the CF-related checks would never be required in this case.

For more details on the SS state information, the reader is referred to 3GPP TS 23.011.

The SS operations discussed in Section 12.2.2 are used to update or interrogate the state information maintained with a supplementary service. Of the eight SS operations discussed in Section 12.2.2, the provisioning and withdrawal operations are controlled only by the service provider, and not by the mobile subscriber. However, both the service provider as well as the mobile subscriber can control the registration and erasure, and activation and deactivation of a supplementary service. The interrogation facility is specifically for the mobile subscriber, and is not used by the service provider. Invocation of a supplementary service can either be network-controlled or subscriber-controlled. The control or management of the supplementary services by the MS or the network is grouped under the Supplementary Services Management Procedures.

12.3 CALL INDEPENDENT SS MANAGEMENT

Management of supplementary services has been broadly classified under two main categories: *Call Related SS Management and Call Independent SS Management*. The call related SS management procedures revolve around the invocation operation of the supplementary services, where the invocation is a result of some call related activity which might take place either during the call establishment phase within the call-connected state, or during the call release phase. On the other hand, the call independent SS management procedures involve management of supplementary service state information as a result of registration, activation, erasure and/or deactivation operations by the MS. Call independent SS management procedures also involve the interrogation (by the MS) of the state information associated with a supplementary service. This category of SS management has no relation with any ongoing call. The MS can independently modify/interrogate the SS states, irrespective of whether or not there is an ongoing call.

The next section discusses the Call Independent SS Management Procedures. These procedures are generic in the sense that they are applicable to almost all supplementary services. The Call Related SS Management procedures, on the other hand, are specific to each supplementary service, and are covered in Section 12.4, along with a discussion on supplementary services.

12.3.1 Call Independent SS Management Procedures

This section discusses the call independent SS management procedures for registration, erasure, activation, deactivation and interrogation of supplementary services. These procedures allow a mobile subscriber to control and interrogate the supplementary service states and the associated information. The MS and the HLR are the primary entities involved in this interaction.

Further, the service provider also has some control over the supplementary service states and the associated information. The service provider may control the information associated with supplementary services through an operator inter-face. However, the focus of SS management procedures is mainly on procedures for the mobile subscriber, and therefore, procedures related to control of supplementary services by the service provider are not discussed in this section.

The supplementary service related data for a subscriber is maintained at the HLR of the subscriber's HPLMN, and is transferred to the VLR when the subscriber registers in the area serviced by the VLR. Transfer of subscriber data from the HLR to the VLR was discussed in Chapter 9. This section discusses the procedures used by the mobile subscriber to manage the supplementary service data stored at the HLR.

The mobile subscriber can initiate the following SS-related procedures:

- Registration and erasure of the SS-related data.
- Activation and deactivation of the SS.
- Interrogation of the state and SS-associated information stored at the HLR for an SS.

Though the above is not an exhaustive list of the SS-related procedures, it includes the ones that are most important. For details on the other procedures, the reader is referred to 3GPP specifications TS 29.011 and TS 24.080.

Call independent SS management procedures take place over a dedicated MM connection between the MS and the MSC. The concept of MM connections was discussed in Chapter 9. Once an MM connection is established, the signaling messages for SS management procedures flow over this connection. Figure 12.1 depicts the generic procedure for Call Independent SS Management. The sequence of steps involved in this procedure is as follows:

1. The MS initiates the procedure by transferring either a 'Register' or a 'Facility' message across the radio interface. To convey the supplementary service operation, the *Facility* information element present in the 'Register'/'Facility' message is used. Normally, the 'Register' message is the first message exchanged between the MS and the MSC, for particular SS management procedure. Subsequently, the 'Facility' message is used.

   (*Note:* 'Facility' message is different from the *Facility* information element. The latter is a part of the 'Facility' and 'Register' messages).

2. On receipt of the 'Register'/'Facility' message from the MS, the MSC/VLR sends a MAP message to the HLR. The MAP message used depends on the supplementary service operation indicated by the *Facility* information element received from the MS. Table 12.2 provides a mapping between the contents of the *Facility* information element, and the MAP message used.

3. The HLR carries out the requested SS operation (Registration, Erasure, Activation, Deactivation or Interrogation), and conveys the result of the same in response to the MAP message from the MSC/VLR.

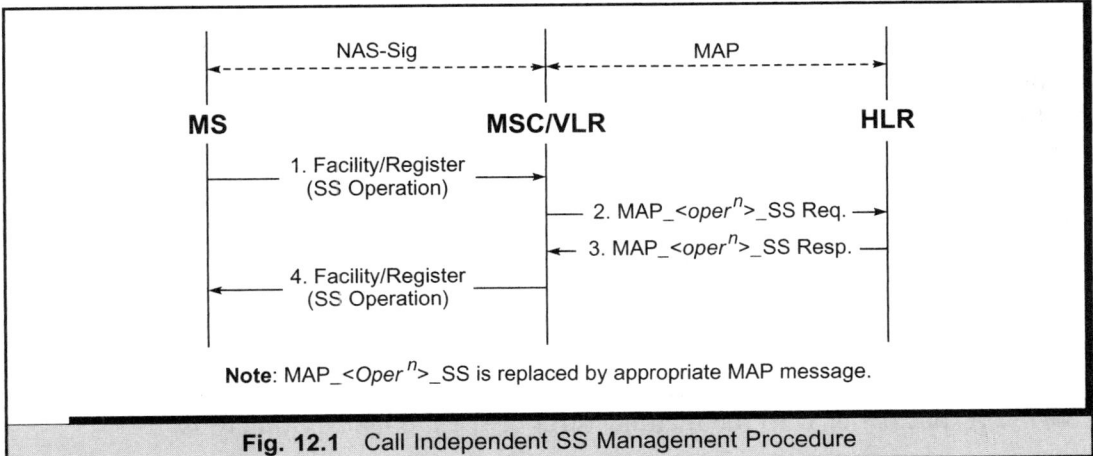

Fig. 12.1 Call Independent SS Management Procedure

Table 12.2 Mapping between SS Operation indicated in Facility Information Element and MAP messages

SS Operation in Facility Information Element	MAP Message
RegisterSS	MAP_REGISTER_SS
EraseSS	MAP_ERASE_SS
ActivateSS	MAP_ACTIVATE_SS
DeactivateSS	MAP_DEACTIVATE_SS
InterrogateSS	MAP_INTERROGATE_SS

4. The MSC/VLR then sends back the response of performing the SS operation to the MS. The response is sent back as a part of the 'Register'/'Facility' message.

For more details on the SS management procedures, the reader is referred to 3GPP TS 29.011.

As already mentioned, supplementary service operations may either be carried out by the network operator through an operator interface at the HLR, or by the mobile subscriber. Further, not all SS operations are applicable to all supplementary services, since many services may not require a registration/activation operation. Table 12.3 presents a consolidated report on these aspects of SS management. For each supplementary service, the table provides information on the SS operations that are valid for the service, as well as how these operations are carried out.

12.3.2 Man-Machine Interface for SS Management

The 'Man-Machine Interface (MMI)' is the interface between the mobile subscriber and the mobile equipment. The MMI has been standardized by the 3GPP to prevent proprietary schemes from being employed by different equipment manufactures on the interface between the mobile subscriber and the mobile equipment. The MMI defines how user-input is provided by the mobile subscriber to carry out certain control activities. This input is provided in the form of standardized *MMI codes*.

Though the MMI has been standardized by the 3GPP, equipment manufacturers are free to provide added features, such as a user-friendly interface, which provides the mobile subscriber with an easier option than having to deal with the more complex MMI codes. Nevertheless, all mobile equipment are required to support the MMI codes on the MMI; hence, the mobile subscriber can always use the standardized MMI codes, irrespective of who the manufacturer of the mobile equipment is.

Table 12.3	Supplementary Service Management

Operations → *Supplementary Services* ↓	*Regist-* *ration*	*Erasure*	*Acti-* *vation*	*Deacti-* *vation*	*Interro-* *gation*
Enhanced Multi-level Precedence and Pre-emption	A/S	W/R	–	–	DR
Call Deflection	–	–	P	W	–
Calling Line Identification Presentation	–	–	P	W	SC
Calling Line Identification Restriction	–	–	P	W	DR
Connected Line Identification Presentation	–	–	P	W	SC
Connected Line Identification Restriction	–	–	P	W	SC
Call Forwarding Unconditional	A/S	W/R/S	R/S	E/S	DR
Call Forwarding on Mobile Subscriber Busy	A/S	W/R/S	R/S	E/S	DR
Call Forwarding on No Reply	A/S	W/R/S	R/S	E/S	DR
Call Forwarding on Mobile Subscriber Not Reachable	A/S	W/R/S	R/S	E/S	DR
Call Waiting	–	–	A/S	A/S	SC
Call Hold	–	–	P	W	–
Multi-Party	–	–	–	–	–
Closed User Group	–	–	P	W	–
Advice of Charge	–	–	P	W	–
User-to-User Signaling	–	–	S	C	–
Barring of All Outgoing Calls	A/S	W/R	A/S	S/A	DR
Barring of Outgoing International Calls (BOIC)	A/S	W/R	A/S	S/A	DR
BOIC except those directed to the Home PLMN Country	A/S	W/R	A/S	S/A	DR
Barring of Incoming Calls (BIC)	A/S	W/R	A/S	S/A	DR
BIC when roaming outside HPLMN Country	A/S	W/R	A/S	S/A	DR
Explicit Call Transfer	–	–	P	W	–
Completion of Calls to Busy Subscriber	–	–	P	W	–
Calling Name Presentation	–	–	P	W	SC
Multi-Call	A/S	W	P	W	DR

P – As a result of Provision
E – As a result of Erasure
SC – As a result of Status Check
DR – As a result of Data Request
C – When conditions in subscription option are met (for Activation); At the end of call in case of per call activation (for Deactivation)

– – Not Applicable
A – Service Provider Controlled Procedure
S – Subscriber Controlled Procedure
W – As a result of Withdrawal
R – As a result of Registration (for Activation); As a result of New Registration (for Erasure)

12.3.2.1 MMI Codes for SS Management

MMI codes have been defined to provide a standardized interface for the MMI. Table 12.4 depicts the MMI codes as defined for the management of supplementary services.

Table 12.4 MMI Codes for SS Management

Supplementary Service Operation	MMI Code
Activation	*SC*SI#
Deactivation	#SC*SI#
Interrogation	*#SC*SI#
Registration	*SC*SI# and **SC*SI#
Erasure	##SC*SI#

The MMI codes consist of the following parts:

- **Service Code (SC):** The service code consists of 2–3 digits, and it uniquely defines the supplementary service that is being controlled.

- **Supplementary Information (SI):** The supplementary information is of variable length, and defines the associated information for an SS control operation.

Each MMI code always begins with a *, #, **, ## or *# and ends with a #. Each sub-code within the MMI code is separated by a *. The Service Code (SC) uniquely defines the supplementary service that is being controlled. The supplementary service can either be a standardized supplementary service (as discussed in Section 12.4), or an operator specific supplementary service.

As depicted in Table 12.4, both activation and registration operations may use the same MMI code. This is possible because the mobile equipment can determine the intended SS operation from the Supplementary Information (SI) provided in the MMI code. For example, in case of call forwarding supplementary service, if the FTN is provided as a part of the SI, then the mobile equipment can determine that the registration operation was intended. If, however, the FTN is not provided in the SI, then the mobile equipment can determine that the activation operation was intended.

Supplementary Information (SI) may not be required for all SS operations. For SS operations that do not require any additional information, the SI need not be provided as part of the MMI code. Similarly, some SS operations require multiple (possibly optional) supplementary information to be entered. In this case, the SI is represented as SIA, SIB and SIC, indicating the order of the SI in the MMI code.

Table 12.5 provides examples of valid SI inputs in an MMI code. Note that the placeholder is left blank, for each optional SI information that is missing.

| **Table 12.5** | Examples of valid SI Inputs in an MMI Code |

SI Input String	Interpretation
*SIA*SIB*SIC#	All three SI (SIA, SIB and SIC) are present in the MMI Code
*SIA**SIC#	The optional SIB is missing. SIA and SIC are present
*SIA*SIB#	Only SIA and SIB are present. SIC may be either optional, or may not be defined for the SS operation.
*SIA#	Only SIA is present. SIB and SIC may either be optional, or may not be defined for the SS operation.
**SIB*SIC#	The optional SIA is missing. SIB and SIC are present
***SIC#	The optional SIA and SIB are missing. Only SIC is present

12.3.2.2 MMI Codes for Standardized Supplementary Services

The 3GPP specification TS 22.030 defines the MMI codes for standardized supplementary services. The Service Codes (SC) and the Supplementary Information (SI) for these standardized supplementary services is provided in Table 12.6.

Besides the standardized supplementary services, a network operator may also offer certain non-standardized supplementary services to its subscribers. These non-standardized supplementary services are provided a Service Code (SC) that is distinct from the service codes of standardized supplementary services. Upon input of the MMI code by the mobile subscriber, the USIM would decipher the code to determine whether the SC maps to a standardized, or an operator-specific supplementary service. If the supplementary service is a defined supplementary service, the USIM shall use the procedures discussed in Section 12.3.1 for SS management. If, however, the USIM determines that the supplementary service is not standardized, it shall use the Unstructured SS Data (USSD) procedures, which are covered in detail in Section 12.5.

12.4 SUPPLEMENTARY SERVICES IN UMTS

A variety of supplementary services have been standardized for the UMTS networks. This section provides a description of these services, which are as follows:

- Enhanced Multilevel Precedence and Pre-emption (eMLPP)
- Call Deflection (CD)
- Line Identification
- Call Forwarding (CF)
- Call Barring (CB)
- Call Waiting (CW)
- Call Hold (CH)

Table 12.6 Codes for Supplementary Services

Supplementary Services	Service Code	SIA	SIB	SIC
Enhanced Multi-level Precedence and Pre-emption	75 and 75n (n = 0–4)	–	–	–
Call Deflection	66	–	–	–
Calling Line Identification Presentation	30	–	–	–
Calling Line Identification Restriction	31	–	–	–
Connected Line Identification Presentation	76	–	–	–
Connected Line Identification Restriction	77	–	–	–
Call Forwarding Unconditional	21	DN	BSG	–
Call Forwarding on mobile subscriber Busy	67	DN	BSG	–
Call Forwarding on No Reply	61	DN	BSG	T
Call Forwarding on Mobile Subscriber Not Reachable	62	DN	BSG	–
Call Waiting	43	BS	–	–
User-to-User Signaling 1	361	R	–	–
User-to-User Signaling 2	362	R	–	–
User-to-User Signaling 3	363	R	–	–
Barring of All Outgoing Calls	33	PW	BSG	–
Barring of Outgoing International Calls (BOIC)	331	PW	BSG	–
BOIC except those directed to the Home HPLMN country	332	PW	BSG	–
Barring of Incoming Calls (BIC)	35	PW	BSG	–
BIC when Roaming outside HPLMN country	351	PW	BSG	–
Explicit Call Transfer	96			
Completion of Calls to Busy Subscriber	37	n (n = 1–5)	–	–
Calling Name Presentation	300	–	–	–
Multiple Subscriber Profile	59n (n = 1–4)	PW	–	–
Multi-Call	88	Nbr_User	–	–

R–UUS Required Option   **BSG**–Basic Service Group   **PW**–Password
Nbr-User–Maximum Number of Simultaneous   **T**–No Reply Condition Timer   **DN**–Directory Number
    CS Bearers

- Multiparty (MPTY)
- Closed User Group (CUG)
- Advice of Charge (AoC)
- User-to-User Signaling (UUS)
- Explicit Call Transfer (ECT)
- Multi-Call (MC)

Further, the call related SS management procedures for the supplementary services are also discussed in this section. For a better understanding of these procedures, it may be beneficial to cross-refer to the call handling procedures discussed in Chapter 10. The reader may also refer to 3GPP TS 29.011 for more information on Call Related SS Management procedures.

12.4.1 Enhanced Multilevel Precedence and Pre-emption (eMLPP)

The eMLPP service enables the subscriber to assign different levels of priorities to each of his/her calls. Further, this service can also be used by the network to differentiate between the calls of different subscribers, by assigning different levels of priorities to calls of different subscribers. In the former case, the eMLPP provides a mechanism for a higher priority incoming call to a subscriber, to pre-empt his/her existing lower priority call. In the latter case, the eMLPP service provides priority-based handling of calls within the network, especially if the network is congested. Thus, eMLPP facilitates the prioritization of calls within the network, as well as the calls of a particular subscriber.

The eMLPP service is described in 3GPP TS 23.067.

The eMLPP service defines seven different priority levels. Of these, the two highest levels are reserved for network administrative purposes. The other five are available to subscribers for use. A subscriber with an eMLPP service can assign a priority to each call originated by him/her at the time of call setup. The subscriber may also register a default priority with the HLR using the SS management procedures for registration. The default priority is assigned by the network to any outgoing call that has not been explicitly assigned a priority by the subscriber.

Priority treatment for point-to-point calls is carried out as follows:

- **Mobile-Originated (MO) Point-to-Point calls:** The priority of the MO point-to-point call depends on the priority assigned to it by the calling subscriber. If the calling subscriber does not have an eMLPP subscription, the network assigns a pre-defined default priority to the call. This default priority is usually the lowest priority level. If, however, the subscriber has an eMLPP subscription, the priority of the call depends on whether or not he/she has explicitly specified a priority level for it. If the subscriber has specified a priority for the call at setup time, then the specified priority level is used. On the other hand, if the subscriber has not explicitly specified a priority level for the call, the default priority level registered by the subscriber is used.

- **For Mobile-Terminated (MT) Point-to-Point calls:** For MT point-to-point calls, the priority level of the call depends on the priority assigned by the calling party. In this case, if the calling party is outside the mobile network, interworking with the ISDN MLPP (Multi-level Precedence and Pre-emption) service is required. Thus, the priority of the call depends on the priority level indicated by the interfacing network. If the call is not an MLPP call, then it is treated with a default priority level in the mobile network.

- **Mobile-to-Mobile Point-to-Point calls:** For a mobile-to-mobile point-to-point call, the calling subscriber's priority is defined as described in the mobile-originated call case. For the called subscriber, the priority is as defined in the mobile-terminated call case.

For more details on the eMLPP service, the reader is referred to 3GPP TS 23.067.

12.4.2   Call Deflection (CD)

The Call Deflection (CD) supplementary service enables the subscriber to deflect an incoming call to another destination. The option to deflect an incoming call may either be pre-configured in the mobile station by the subscriber, or selected by him/her at the time of receiving an incoming call. The destination to which the call is deflected may further deflect the call to another destination. The maximum number of times that a call can be deflected is determined by the network operator, and is generally in the range of one to five deflections per call.

On receipt of a call, a subscriber using a call deflection service can request the deflection of the call by providing a 'Deflected-To-Number (DTN)'. On receipt of the request for call deflection, the network deflects the call towards the DTN specified by the subscriber.

Figure 12.2 depicts the message flow between the MS, VMSC and the VLR for the invocation of the call deflection supplementary service. The following is the sequence of steps involved in the procedure:

1. As discussed in Chapter 10, a mobile terminating call is initiated by the VMSC by sending a 'Setup' message to the MS.
2. In case the mobile subscriber wishes to deflect the incoming call to a 'Deflected-To-Number (DTN)', he/she would reject the call. The MS would then send a 'Disconnect' message to the VMSC, with an indication to deflect the incoming call to a DTN.

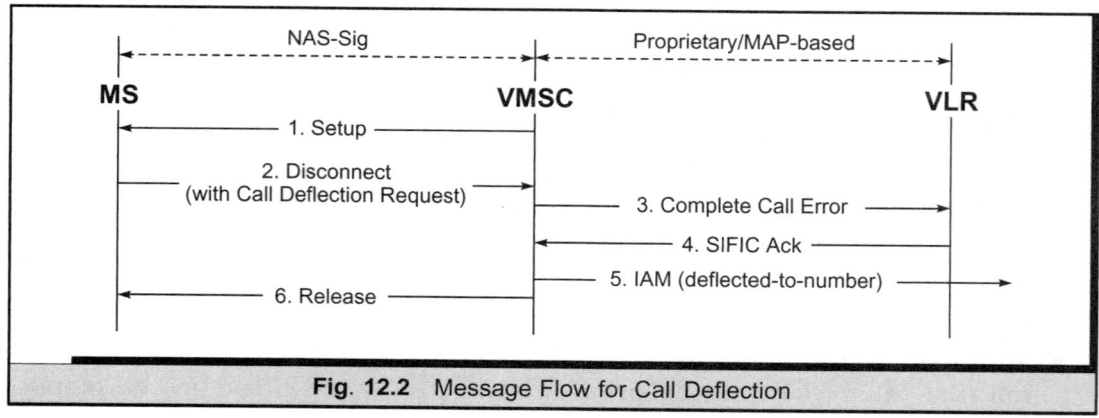

Fig. 12.2   Message Flow for Call Deflection

3. On receipt of this request from the MS, the VMSC sends an error response to the 'Complete Call' request received from the VLR.
4. The VLR then verifies whether the call is allowed to be deflected to the DTN indicated by the mobile subscriber. If the VLR determines that the call can be deflected, it sends an acknowledgement to the 'Send Information for Incoming Call (SIFIC)' request from the VMSC (**Note:** The SIFIC request message is sent by the VMSC to request the VLR to provide information for handling the incoming call. Refer to Chapter 10 for details).
5. On receipt of the 'SIFIC' acknowledgement from the VLR, the VMSC sends an IAM message towards the network containing the DTN.
6. Subsequently, the VMSC releases the call towards the MS by sending a 'Release' message.

For more information on the call deflection supplementary service, the reader is referred to 3GPP TS 23.072.

12.4.3 Line Identification

The Line Identification supplementary services provide features that can present or restrict the identity of the parties involved in the call. The line identity services group consists of four related supplementary services; each may be subscribed/unsubscribed to individually, by a mobile subscriber.

These four services are:

- Calling Line Identification Presentation (CLIP)
- Calling Line Identification Restriction (CLIR)
- Connected Line Identification Presentation (COLP)
- Connected Line Identification Restriction (COLR)

The Calling Line Identification Presentation (CLIP) supplementary service enables the subscriber to receive the ISDN number of the calling party along with the call setup request. The CLIP service can be used by the subscriber to screen incoming calls before accepting them. Based on the ISDN number of the calling subscriber, the called subscriber may either accept the call or reject it. Further, in case the mobile subscriber does not answer the incoming call, the ISDN number of the calling party may be stored by the mobile-equipment to indicate a 'missed call'.

The Calling Line Identification Restriction (CLIR) supplementary service, on the other hand, prevents the calling subscriber's identity from being presented to the called subscriber, by not providing the former's ISDN number to the called subscriber at call setup time. Further, in case the calling subscriber has CLIR subscription and the called subscriber has CLIP subscription, the CLIR service over-rides the CLIP service, and the called subscriber is not presented the identity of the calling subscriber. An exception exists to this rule though, whereby national regulations may permit a category of users (e.g. police), to over-ride the CLIR service with the CLIP service.

The Calling Line Identification services (CLIP and CLIR) can allow/restrict the presentation of the calling subscriber ISDN number to the called subscriber. On the other hand, the Connected Line Identification services (COLP and COLR) provide the facility to allow/restrict the presentation of the connected subscriber ISDN number to the calling subscriber. If the calling subscriber has subscribed to the Connected Line Identification Presentation (COLP) service, then the ISDN number of the connected party is provided to the calling subscriber. This is especially useful in cases where the call may be diverted/forwarded by the called subscriber to some other destination (the FTN or the DTN). In such cases, the connected party is different from the original called party, and the calling subscriber may wish to know of the final destination of the call. The COLP service is depicted in Figure 12.3.

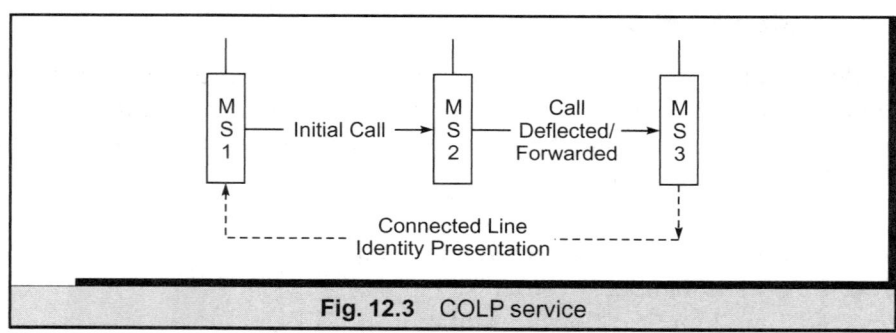

Fig. 12.3   COLP service

The COLR service is similar to the CLIR service. The Connected Line Identification Restriction (COLR) service prevents the identity of the connected party to be provided to the calling party. As in the case of calling line identification services, if the calling subscriber has subscribed to the COLP service, and the connected party has subscribed to the COLR service, the COLR service would over-ride the COLP service. Hence, in such a case, the connected party identification would not be provided to the calling party. However, an exception similar to the one in the CLIR case exists here as well. As a national concern option, certain categories of subscribers (e.g. police) may be provided the facility to over-ride the COLR service of the connected party, and thereby, may be able to receive the connected party identification.

Figure 12.4 depicts the message flows for Line Identification Services. In each case, the 'Complete Call' message from the VLR to the VMSC indicates whether the corresponding supplementary service (CLIP, CLIR, COLP, COLR) is active for the

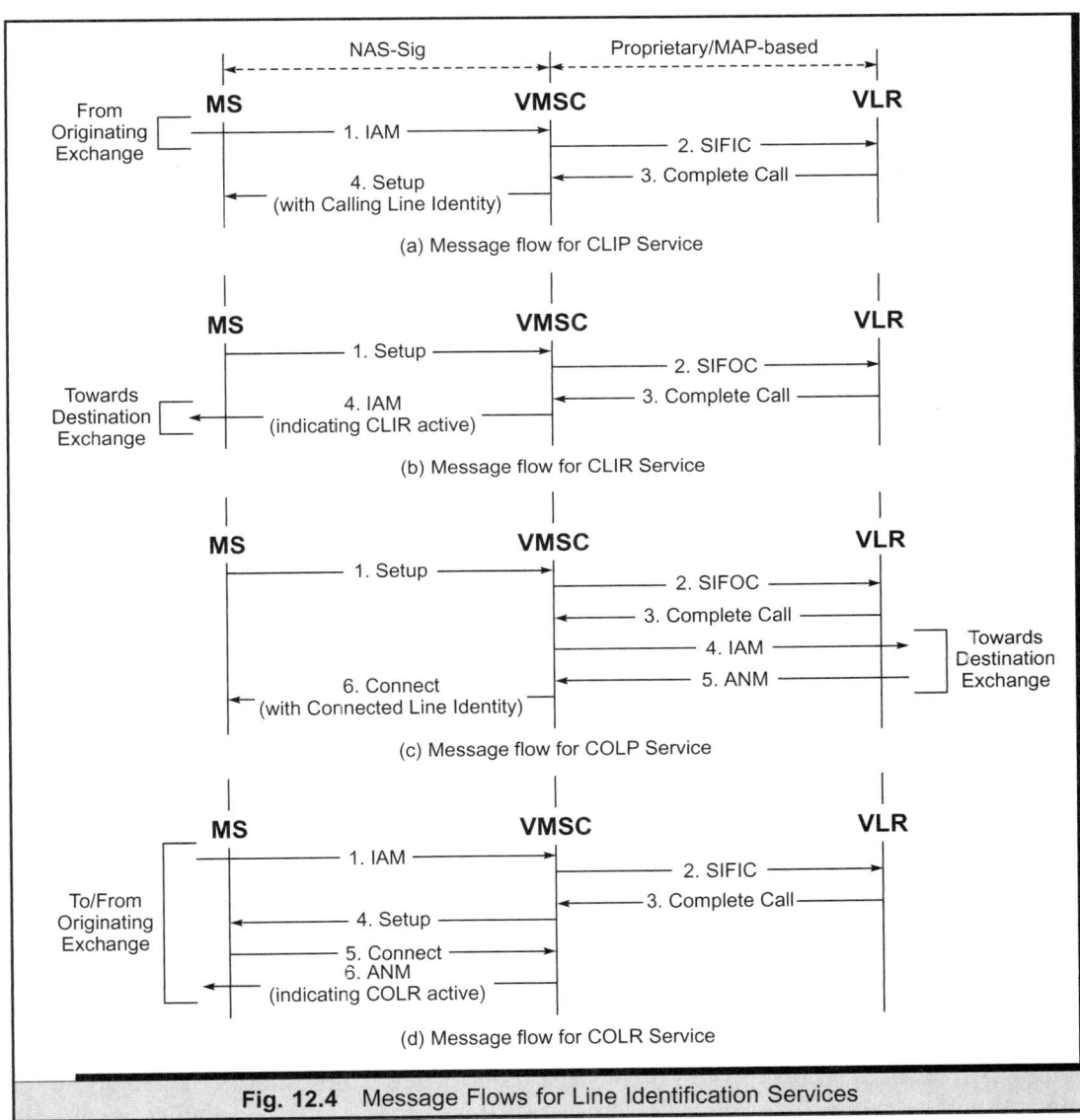

Fig. 12.4 Message Flows for Line Identification Services

calling/called subscriber, or not. The following are the salient points to note for each of the services:

(a) CLIP: In case the called subscriber has the CLIP service, the 'Setup' message sent to the MS includes the calling line identification. This happens only if CLIR is not indicated in the incoming call setup message, and the calling subscriber number is present in the incoming call setup message.

(b) CLIR: If the calling subscriber has the CLIR service, this is indicated in the Initial Address Message sent to the destination network.

(c) COLP: In case the calling subscriber has the COLP service, the connected identity is sent to the calling party in the 'Connect' message. This, however, is possible only if the connected line identification is present in the response received from the destination network.

(d) COLR: If the called subscriber has the COLR service, this information is included in the response to the calling subscriber network (originating network).

The rest of the message flow is similar to the message flow in MO and MT call handling, discussed in Chapter 10. The line identification supplementary services are defined in 3GPP TS 23.081.

12.4.4   Call Forwarding (CF)

The Call Forwarding (CF) supplementary services enable the subscriber to forward his/her incoming calls to another number called the *Forwarded-To-Number* (FTN). The CF supplementary services group consists of four independent supplementary services. A mobile subscriber may individually subscribe/un-subscribe to each of these services, which are as follows:

- Call Forwarding Unconditional (CFU)
- Call Forwarding on mobile subscriber Busy (CFB)
- Call Forwarding on No Reply (CFNRy)
- Call Forwarding on mobile subscriber Not Reachable (CFNRc)

The Call Forwarding Unconditional (CFU) supplementary service enables the subscriber to unconditionally forward all incoming calls to a 'Forwarded-To-Number (FTN)'. Thus, if CFU is active, the network forwards all calls—addressed to the called subscriber's directory number—to the FTN. This supplementary service is useful in various scenarios. For example, a subscriber can use the CFU service when he/she is busy in a meeting and wishes to forward all incoming calls to his/her secretary.

The Call Forwarding on mobile subscriber Busy (CFB) service enables the subscriber to forward an incoming call to a 'Forwarded-To-Number (FTN)' when he/she is busy in another ongoing call. Thus, if CFB is active and the served subscriber is busy in another call, the network forwards all incoming calls—addressed to the called subscriber's directory number—to the FTN. This supplementary service is useful in scenarios when a subscriber does not want to miss a call just because he/she is busy talking on the line.

The Call Forwarding on No Reply (CFNRy) service enables the subscriber to forward an incoming call to a 'Forwarded-To-Number (FTN)' if it does not get answered. Thus, if CFNRy service is active and the served subscriber does not answer/reply to an incoming call, the network forwards this call to the FTN. This supplementary service is useful in scenarios where the subscriber cannot answer an incoming call due to various reasons, yet does not wish to miss the call. The reasons could be that the subscriber is busy in another activity (e.g. in a meeting, or is driving) or because he/she does not hear the ring of the incoming call. In such a case, the incoming call may, for example, be forwarded to the subscriber's secretary, or to his/her residential PSTN number.

The Call Forwarding on mobile subscriber Not Reachable (CFNRc) service enables the subscriber to forward an incoming call to a 'Forwarded-To-Number (FTN)' if the user equipment is not reachable to an incoming call. Thus, if the CFNRc service is active and the served subscriber is not reachable at the time of an incoming call, the network forwards this call to the FTN. This supplementary service is useful in various scenarios, like when the subscriber is in a region where it cannot receive the signal from the base station, and is hence unreachable. Alternatively, the subscriber may have switched-off his/her user equipment. In such cases, the incoming calls may be forwarded to an FTN as desired by the subscriber.

For the call forwarding supplementary services, the mobile subscriber is required to register the FTN with the network using the SS management procedures for the registration operation. The CF supplementary services only affect the handling of a subscriber's incoming calls; outgoing calls are not affected. Interaction of call forwarding supplementary services with call handling procedures is discussed in Chapter 10. For more details on the call forwarding supplementary service, the reader is referred to 3GPP TS 23.082.

12.4.5 Call Barring

The Call Barring (CB) supplementary services enable the subscriber to bar incoming or outgoing calls to/from the subscriber. The CB supplementary services group consists of five independent supplementary services. A mobile subscriber may individually subscribe/unsubscribe to each of these services, which are as follows:

- Barring of All Outgoing Calls (BAOC)
- Barring of Outgoing International Calls (BOIC)
- Barring of Outgoing International Calls except those directed to the Home PLMN Country (BOIC-exHC)
- Barring of All Incoming Calls (BAIC)
- Barring of all Incoming Calls when Roaming outside the HPLMN country (BIC-Roam)

The CB services are password-protected. This requires the subscriber to provide the network with the correct password before performing any operation on the supplementary services.

The Barring of All Outgoing Calls (BAOC) supplementary service enables the subscriber to unconditionally forbid all outgoing calls. Thus, if BAOC is active, the network does not allow any call originated by the mobile subscriber. This supplementary service may be useful in various scenarios. For example, when the subscriber wishes to leave the user equipment unattended, and does not want any other person to make outgoing calls in his/her absence.

The Barring of Outgoing International Calls (BOIC) service enables the subscriber to bar all outgoing international calls. Thus, if BOIC is active, the network would reject any such call that is being made. The BOIC supplementary service may also be useful in various scenarios, for example, the scenario discussed for the BAOC service.

The Barring of Outgoing International Calls except those directed to the Home PLMN Country (BOIC-exHC) service enables the subscriber to forbid all outgoing international calls, except those made to the country of the subscriber's Home PLMN. For a subscriber traveling out of his/her Home PLMN country, the difference between BOIC-exHC and BOIC is that while the latter would not allow the subscriber to call back his/her Home PLMN country, the BOIC-exHC would allow such a call. For a subscriber located within his/her Home PLMN country, BOIC and BOIC-exHC offer similar services.

The Barring of All Incoming Calls (BAIC) service enables the subscriber to unconditionally bar all incoming calls. Thus, if the BAIC service is active, the network does not allow any incoming call to reach the mobile subscriber. A subscriber may use the BAIC service when he/she does not wish to receive any calls.

The Barring of All Incoming Calls when Roaming outside the HPLMN country (BIC-Roam) service allows the subscriber to bar all incoming calls when he/she is roaming out of the Home PLMN country. Thus, if the BIC-Roam service is active and the subscriber is roaming out of the Home PLMN country, the network does not allow any incoming call to reach the mobile subscriber. The subscriber may decide to use the BIC-Roam supplementary service in case he/she does not want to receive any incoming calls when roaming, and thereby reduce the roaming charges.

While the first three CB services (BAOC, BOIC and BOIC-exHC) affect the handling of a subscriber's outgoing calls, the latter two (BAIC and BIC-Roam) affect the handling of the incoming calls. Interaction of the call barring supplementary services with call handling procedures is discussed in Chapter 10. For more details on the call barring supplementary service, the reader is referred to 3GPP TS 23.088.

12.4.6 Call Waiting (CW) and Call Hold (CH)

The Call Waiting (CW) and the Call Hold (CH) services are distinct supplementary services. However, they are covered together in this section. This is done because the CW and CH may be used together in certain scenarios. Here, we discuss the scenario when both call waiting and call hold supplementary services are used together.

The call waiting supplementary service is used to indicate an incoming call when the subscriber is already busy with another call. If subscribed to, the call waiting service notifies the subscriber of the new incoming call, which is in the *wait-state*, till he/she decides to accept, reject or ignore it.

The call hold supplementary service is used when a mobile subscriber wishes to place an ongoing call on hold, in order to originate or receive another call. Subsequently, the mobile subscriber may toggle between these two calls. The call hold service can be used in conjunction with the call waiting service. A typical scenario is when the subscriber wishes to receive an incoming call, while already involved in an ongoing call. In this case, he/she may place the ongoing call on hold, and accept the new incoming call which is in *wait-state*.

The CW and CH supplementary services are defined in 3GPP TS 23.083. As per the specifications of the CW service, the maximum number of calls that can be allowed to be in the *wait-state* is limited to one. However, a network operator may decide to allow more than one waiting call for a subscriber at its own discretion. Similarly, as per the specifications, the maximum number of calls that a subscriber may put on hold is limited to one.

Figure 12.5 depicts the message flow for invocation of call waiting and call hold supplementary services. The following points describe the scenario that is highlighted in the figure:

- Subscriber 'B' has both CW and CH service.
- Subscriber 'B' has an ongoing call with subscriber 'A'. This call is referred to as call A-B.
- Subscriber 'B' receives an incoming call from subscriber 'C'. This call is referred to as call C-B.

The sequence of steps involved is as follows:

1. In the initial state, the Call A-B is in active state, i.e. there is an ongoing call between subscriber 'A' and subscriber 'B'. The VMSC receives the 'Initial Address Message (IAM)' from the originating network, containing the called address of subscriber 'B' and the calling address of subscriber 'C'.
2. VMSC requests information for the incoming call using 'Send Information For Incoming Call (SIFIC)' message from the VLR to determine if the call can be accepted.
3. VLR determines that the call can be accepted, and sends the 'Complete Call' message to the VMSC.
4. VMSC then sends a 'Setup' message to subscriber 'B' indicating the incoming call C-B.
5. Since subscriber 'B' is involved in a call with subscriber 'A', this 'Setup' message may be accepted, ignored or rejected by subscriber 'B'. Here, it is assumed that the call is accepted by subscriber 'B', and is put on wait. A 'Call Confirmed' message is sent back to the VMSC.

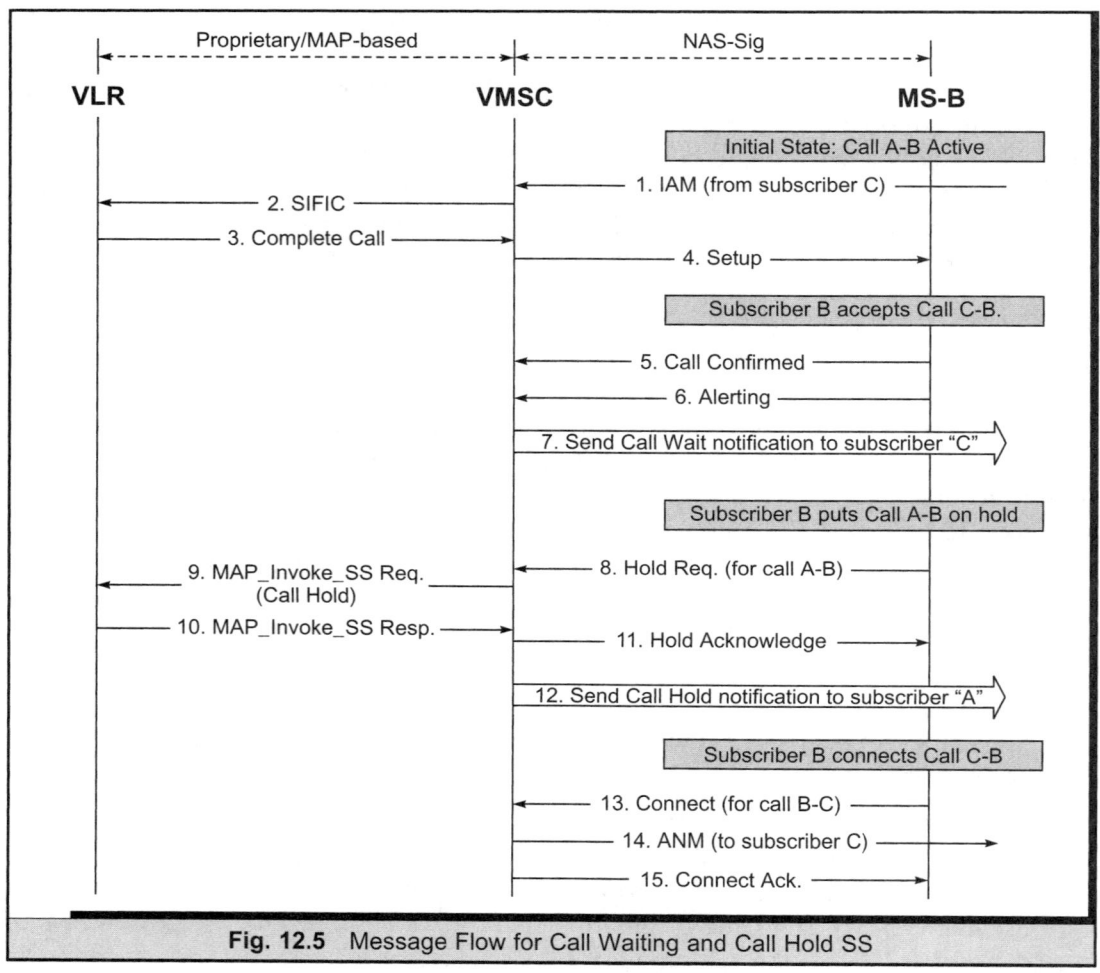

Fig. 12.5 Message Flow for Call Waiting and Call Hold SS

6. Subscriber 'B' next sends an 'Alert' message to the VMSC. Note that no new traffic channel is allocated to subscriber 'B' for the new call (call C-B), since only one call can be active at a time.

7. On receipt of the 'Alert' message, VMSC sends a notification to subscriber 'C' indicating that the call is put in *wait-state*. Normally, the calling subscriber (here subscriber 'C') receives notification of the waiting call using the *facility* information element in the 'Alerting' message delivered to him/her. This step completes the procedure for putting a call on wait using the CW procedures.

8. Subsequent to accepting the call from subscriber 'C' and putting it in *wait-state*, subscriber 'B' may wish to put the call with subscriber 'A' on hold, and connect to

call C-B. The 'Call Hold' procedures are used for this purpose. To achieve this, subscriber 'B' sends a 'Hold' request message for call A-B to the VMSC.

9. VMSC then requests VLR to verify if call hold supplementary service can be invoked for subscriber 'B'. For this, it sends a 'MAP_Invoke_SS' message to the VLR.

10. The VLR determines that subscriber 'B' is allowed to invoke the call hold service, and sends a positive response to the 'MAP_Invoke_SS' message from the VMSC.

11. The VMSC then sends a 'Hold Acknowledgement' message to subscriber 'B', indicating that the call has been put on hold.

12. The VMSC also sends a notification towards subscriber 'A' indicating that the call A-B has been placed on hold. This completes the CH procedure for putting a call on hold.

13. Subscriber 'B' now requests the VMSC to connect call C-B. The traffic channel allocated to subscriber 'B' for call A-B is now used for call C-B.

14. The VMSC sends an 'Answer Message (ANM)' to subscriber 'C' indicating that the call has been connected.

15. The VMSC also sends a 'Connect Acknowledgement' message to subscriber 'B' indicating that the call C-B is connected. Subscriber 'B' can now communicate with subscriber 'C', while call A-B remains on hold.

12.4.7 Multiparty

The Multiparty supplementary service makes it possible for a mobile subscriber to organize a conference call, which allows simultaneous communication with more than one party. Initiating a multiparty call involves the following sequence of steps:

1. The subscriber (let us say 'A') with multiparty subscription establishes a call with the first party (say 'B').

2. The subscriber then puts this call on hold, and establishes another call with a second party (let us say 'C'),

3. In this state, the subscriber has one ongoing call, and one call on hold. The subscriber can now request the network to build a multiparty call.

4. After the multiparty call has been established between 'A', 'B' and 'C', the served subscriber 'A' can now add new parties to the multiparty call. Subscriber 'A' can also drop parties from the call.

Subscriber 'A' can add five other remote parties to the multiparty call. These five parties can further add another five members each, thus involving a seemingly unlimited number of participants in the multiparty call.

Figure 12.6 depicts the message flow for invocation of a multiparty call. Subscriber 'A' has an ongoing call with subscriber 'C' and a call on hold with subscriber 'B'. In

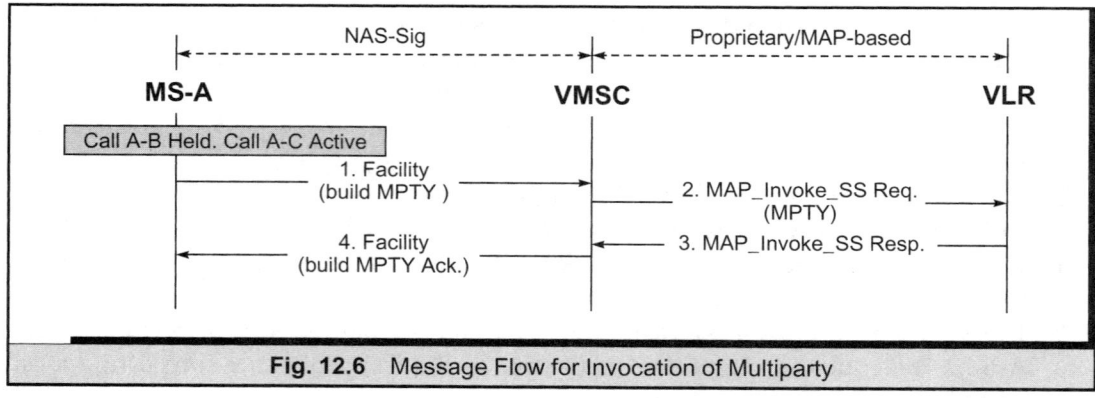

Fig. 12.6   Message Flow for Invocation of Multiparty

this state, subscriber 'A' can request the network to establish a multiparty call. The sequence of steps involved in this procedure is an follows:

1. Subscriber 'A' requests the VMSC to build a multiparty call by including the 'build MPTY' operation in the 'Facility' request message sent to the VMSC.
2. The VMSC requests the VLR to verify if the multiparty service can be provided to subscriber 'A'. For this, the VMSC sends a 'MAP_Invoke_SS' request message to the VLR. This message contains the indication that the multiparty service has been requested by the subscriber.
3. If the VLR can verify that the multiparty service can be offered to subscriber 'A', it sends a positive acknowledgement to VMSC's 'MAP_Invoke_SS' request message.
4. The VMSC then sends a 'Facility' message to subscriber 'A', acknowledging the request for a multiparty call.

The multiparty call is set-up by the network between subscriber 'A'. 'B' and 'C'. Subscriber 'A' may subsequently add or remove parties to/from the multiparty call, and may also put the entire multiparty call on hold using procedures similar to those discussed in Section 12.4.6. For details on the multiparty supplementary service, the reader is referred to 3GPP TS 23.084.

12.4.8   Closed User Group (CUG)

The Closed User Group (CUG) supplementary service provides a means for mobile subscribers to form user groups to which access is restricted. To understand the concept of the CUG service, consider the group of all employees working in a company. Suppose that the company management decides to provide mobile equipment to all employees, but wishes to restrict communication to only between its employees. In this case, the employees of the company can be placed within one CUG, and access rules for the CUG can be defined to restrict communication to only the employees of the company. Access to this CUG can be barred to outsiders who are not employees of the company. In other words, while members of this CUG can place calls to other members of the CUG, they are restricted from placing calls to people outside the CUG. Similarly, incoming calls are generally restricted to calls from members within the CUG.

There are many other scenarios where the CUG service can be used. Based on different requirements, different access rules can be defined for a CUG. In another example, consider the scenario when all incoming calls to mobile subscribers are provided free-of-cost by the PLMN service provider. In this case, restriction on the incoming calls to members of the CUG may be relaxed to allow all incoming calls, whether from within or outside of the CUG. However, outgoing calls may still be allowed only to members within the CUG. Thus, the CUG supplementary service provides means for defining various combinations of access rules to offer an extremely flexible service.

A mobile subscriber is allowed to be a member of multiple CUGs. This may be necessary, for instance, to meet the requirements of a subscriber who works as a consultant to more than one organization. As an employee of more than one organization, the subscriber could be a part of multiple CUGs. *CUG subscription data* is maintained for a subscriber who has a CUG subscription. This data is maintained at the HLR of the subscriber, and consists of a list of CUGs that the subscriber is a member of, as well as access rules, both for Intra-CUG and Inter-CUG calls.

The CUG subscription data maintained per subscriber includes the following information:

- **CUG-List:** This is the list of CUGs of which the subscriber is a member. A subscriber can be a member of a maximum of 10 CUGs.

- **Intra-CUG Access Rules:** Intra-CUG rules are maintained for each CUG that the subscriber is a member of. These rules are applied when the mobile subscriber places a call to another member of its CUG. The Intra-CUG access rules for a subscriber consist of one of the following access rights:
  - Both Incoming and Outgoing Calls allowed
  - Only Incoming Calls allowed
  - Only Outgoing Calls allowed

- **Inter-CUG Access Rules:** Inter-CUG access rules are used when the mobile subscriber places a call to a called party that does not belong to any of the CUGs that he/she is a member of. These rules are defined per-subscriber, and not per-CUG. The Inter-CUG access rules for a subscriber consist of one of the following access rights:
  - No Inter-CUG calls allowed
  - Only Outgoing Inter-CUG calls allowed

> – Only Incoming Inter-CUG calls allowed
> – Both Incoming and Outgoing Inter-CUG calls allowed.

To ensure that CUG access rules are not violated, CUG related checks are applied before an incoming or outgoing call is allowed to the subscriber. These checks are applied at both the VLR of the calling subscriber, and at the HLR of the called subscriber. Figure 12.7 depicts the message flow for the CUG service. The following is the sequence of steps involved:

1. Subscriber 'A' initiates an outgoing call to subscriber 'B' by sending a 'Setup' message to VMSC-A. Optionally, subscriber 'A' includes information about the CUG that it wishes to use while making the call. This is required when the subscriber is a member of multiple CUGs. The CUG is identified by a CUG Index.

2. VMSC-A sends a 'Send Information For Outgoing Call (SIFOC)' message to VLR-A. If the CUG Index is received from MS-A, this is also included in the SIFOC message.

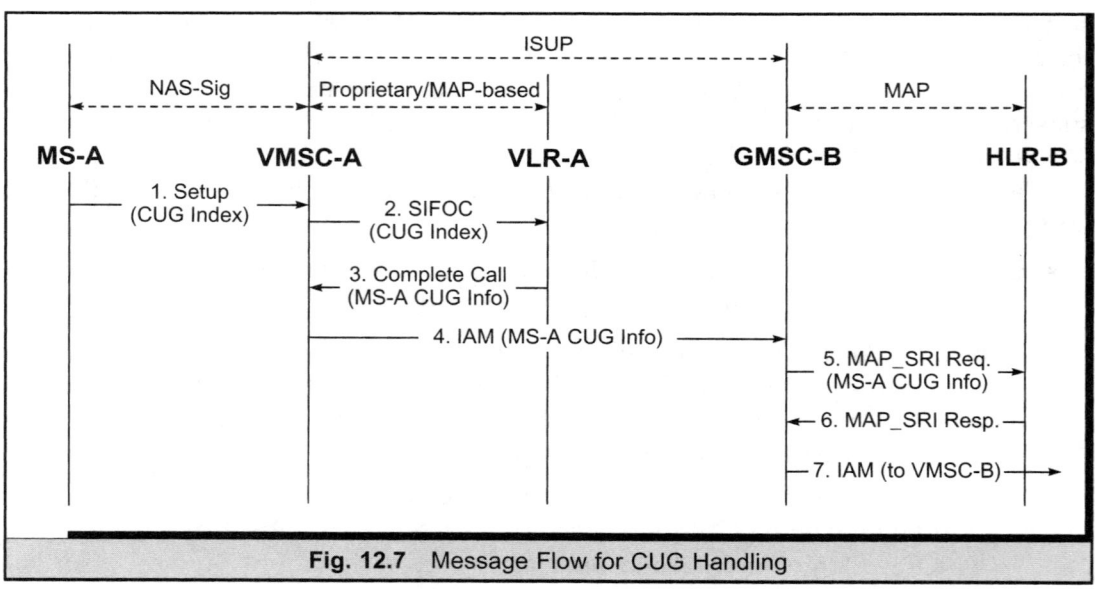

Fig. 12.7   Message Flow for CUG Handling

3. VLR-A applies the CUG related checks to determine if the call is allowed to the subscriber. Since VLR-A has no information on the CUG of the called party, or the access rules defined for the called party, it can carry out only a part of the verification process for the call. Simply put, VLR-A conducts the following checks:

   • If neither the Intra-CUG nor the Inter-CUG rules define outgoing access for the subscriber, the outgoing call is rejected.

- If the Intra-CUG rules for the subscriber do not allow outgoing access, but the Inter-CUG rules allow this, then the call is treated as a *normal call* (i.e. non-CUG call) by VLR-A.
- If the Intra-CUG rules for the subscriber allow outgoing access, then the call is allowed to go through. However, VLR-A indicates to VMSC-A to treat this call as a *CUG call* and not as a normal call. This means that the CUG related information of the calling subscriber is to be sent by VMSC-A to the destination network along with the call setup request.

If VLR-A determines that the outgoing call can be allowed to proceed, it sends back an acknowledgement to the SIFOC request from VMSC-A. If it determines that the call can be treated as a normal call, then no CUG information is sent in the response to the SIFOC message. Else, VLR-A includes the calling subscriber CUG information in its response to the SIFOC message. This CUG information includes the CUG identity of the calling subscriber, and optionally, its Inter-CUG access rules. The information is used by the HLR of the called subscriber to apply further CUG-related checks.

4. VMSC-A sends an 'Initial Address Message (IAM)' to GMSC-B. If VMSC-A has received the optional CUG information for the calling subscriber from VLR-A, this information is also included in the IAM message to GMSC-B.
5. GMSC-B sends a 'MAP_Send_Routing_Information' request message to HLR-B, with the optional CUG information for MS-A, if it is received in the IAM message.
6. HLR-B applies further CUG related checks to determine if the incoming call is allowed. To carry out these checks, HLR-B uses the CUG information received for the calling subscriber from GMSC-B. The HLR also uses the CUG information of the called subscriber stored in its database, to conduct CUG-related checks.

The following checks are carry out at HLR-B:

- If the CUG Index for subscriber 'A' does not match any of the CUG Index for subscriber 'B', the call is treated as if it is an Inter-CUG call. In this case, if subscriber 'A' does not have outgoing access as per Inter-CUG access rules, the call is rejected. On the other hand, if subscriber 'A' is allowed outgoing access as per Inter-CUG access rules, then the call is accepted or rejected on the basis of whether or not subscriber 'B' has incoming access as per the Inter-CUG access rules.
- If the CUG Index for subscriber 'A' matches a CUG Index for subscriber 'B', the call is treated as if it is an Intra-CUG call. In this case, if subscriber 'B' does not have incoming access as per Intra-CUG access rules, the call is rejected. On the other hand, if subscriber 'B' is allowed incoming access as per Intra-CUG access rules, the call is accepted.

- If no CUG information is received for subscriber 'A' (i.e. the call is a normal call), but subscriber 'B' has CUG subscription, then the call is treated as an Inter-CUG call. In this case, the call is accepted or rejected on the basis of whether or not subscriber 'B' has incoming access as per the Inter-CUG access rules.
- If subscriber 'B' is a normal (non-CUG) subscriber, but subscriber 'A' has CUG subscription, then the call is treated as an Inter-CUG call. In this case, the call is accepted or rejected depending on whether or not subscriber 'A' has outgoing access as per the Inter-CUG access rules.
- If the HLR determines that the call is allowed, it sends back a positive response to the 'MAP_Send_Routing_Information' request message from GMSC-B. The roaming number is included in the response to GMSC-B. The procedure to obtain the roaming number from VLR-B is not shown in the figure.

7. GMSC-B forms an 'Initial Address Message (IAM)' and sends it towards VMSC-B, where the called subscriber is registered.

CUG related procedures also include checks carried out for call forwarding. This is required when HLR-B determines that the call cannot be delivered to the called subscriber, and is to be forwarded to a FTN. However, these procedures are beyond the scope of this section. For more details on the CUG supplementary service, the reader is referred to 3GPP TS 23.085.

12.4.9   Advice of Charge (AoC)

The Advice of Charge (AoC) supplementary service provides the MS with the information to produce an estimate of the charge levied to a subscriber for the services used. Charges are indicated for the call(s) in progress. Any charges for non-call related transactions, and for certain supplementary services, such as call forwarding are not indicated.

To offer the AoC service, the network provides the MS with charging information required to calculate the approximate cost of each call. This information includes the charge applicable per unit of usage, time dependence, and unit increments. The charging information is available to the MS at the beginning of the call, and if required, during the call. The MS maintains information about the duration of the call and uses this information, along with the charging information provided by the network, to calculate the charge for the call. To indicate the charge for a call, the MS displays the units consumed so far during the present call and maintains this value until the MS is switched off or a new call setup is attempted. The MS may also indicate the total accumulated charge for all calls by storing the running cumulative unit charge in the USIM.

Figure 12.8 depicts the message flow for the AoC service for both the Mobile Originated (MO) call and the Mobile Terminated (MT) call. In both cases, the VMSC gets to know that the Advice of Charge supplementary service is applicable to the mobile subscriber, through an indication present in the 'Complete Call' message from the VLR to the VMSC.

In case of a MO call, the charging information may be provided to the calling subscriber as part of a 'Facility' message. Alternatively, the charging information may be included within the 'Connect' message that is sent to the calling subscriber after the called subscriber has answered the call. In case of a MT call, the charging information is provided to the called subscriber as part of a 'Facility' message after the latter has answered the call. The remaining steps in the figure are similar to those followed for an MO/MT call as discussed in Chapter 10. For more details on the AoC service, the reader is referred to 3GPP TS 23.086.

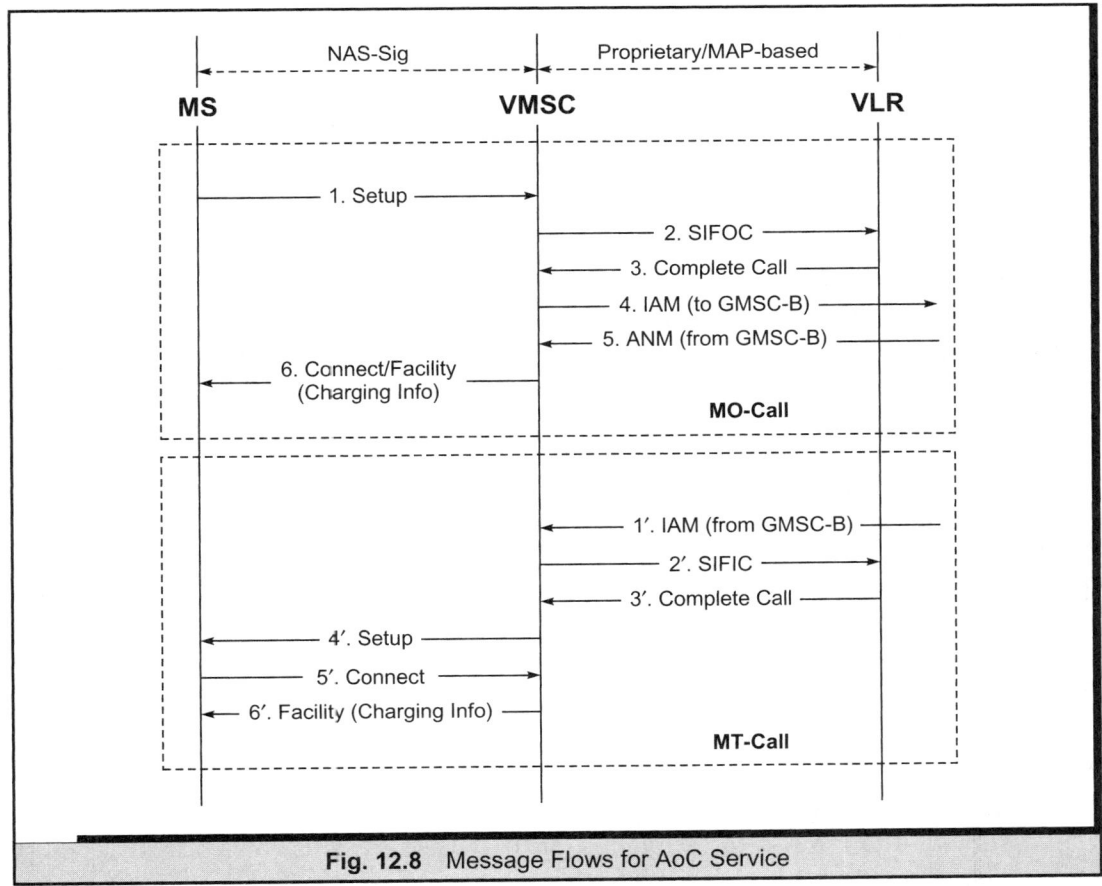

Fig. 12.8 Message Flows for AoC Service

12.4.10 User-to-User Signaling (UUS)

The User-to-User Signaling (UUS) supplementary service allows a subscriber to send/receive a limited amount of subscriber generated information to/from another user in the call. This information is carried transparently through the network without any modification of the contents. The network does not try to interpret or act upon this information.

The UUS service is available in three flavors: UUS1, UUS2 and UUS3, as follows:

- **UUS1:** Using the UUS1 service, the User-to-User Information (UUI) can be sent/received during the originating and terminating phases of a call. In this case, the UUI is sent embedded within call control messages. The UUS1 supplementary service may be implicitly activated for a call by simply sending the UUI in call control messages. Alternatively, the UUS1 service may be explicitly activated using the procedures described below.

- **UUS2:** Using the UUS2 service, UUI can be sent/received only after the served subscriber receives an indication that the remote party has being alerted to the call. Further, the UUI is allowed to be sent/received until the time the connection establishment is pending. After the connection establishment is complete, no UUI information is exchanged. The UUS2 supplementary service requires explicit activation.

- **UUS3:** Using the UUS3 service, UUI can be sent/received only after the connection is established, and before the connection release is initiated. The UUS3 supplementary service requires explicit activation.

The UUS1 service may be implicitly activated by including the UUI in the call setup request from the MS. If the calling subscriber has UUS1 provisioned and if this UUS1 is implicitly requested/activated, the network shall transparently transfer all UUI contained in the call control messages.

The served subscriber can explicitly activate any of the UUS services by including an activation request within the call setup request message for a MO call. In addition, the UUS3 service may alternatively be activated during an established call by using a 'Facility' message. The network verifies the availability of UUS capabilities for the call by passing the UUS request to the remote side. If the UUS service is available for the call, an appropriate *UUS provided* indication is sent within the first message from the remote side.

Figures 12.9, 12.10 and 12.11 depict the message flow for the different UUS services. The following steps in the figures need explanation:

- **Request for UUS service:** For all flows, explicit activation of the UUS service is assumed. The request for activation of the UUS service is carried in the 'Setup' message sent to initiate the MO Call.

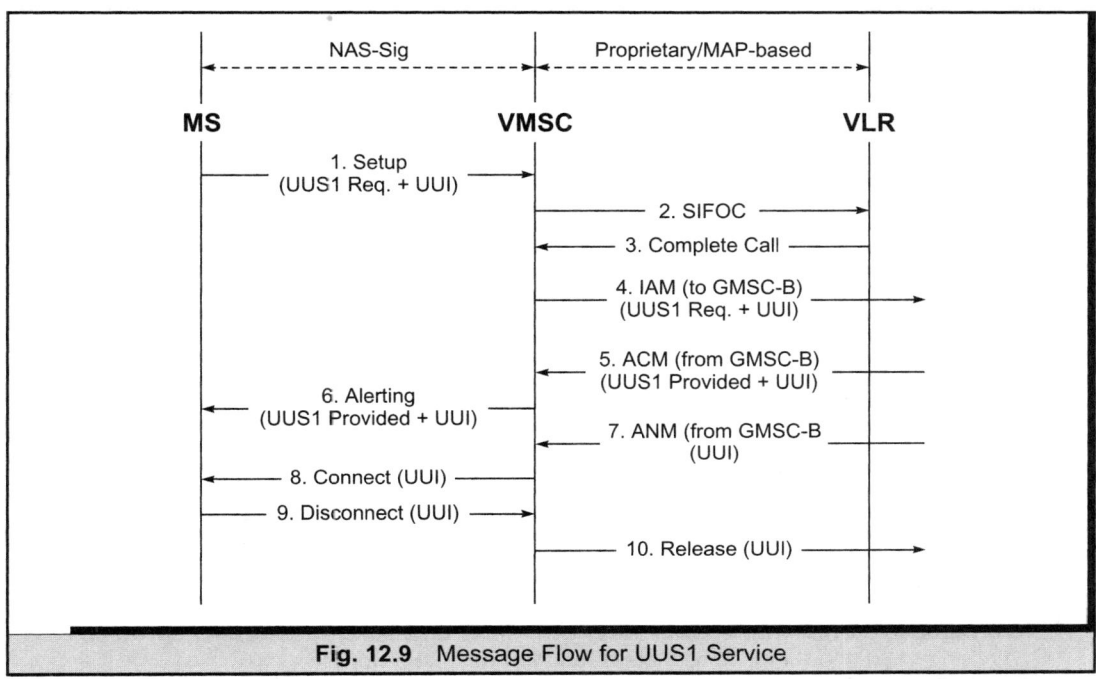

Fig. 12.9   Message Flow for UUS1 Service

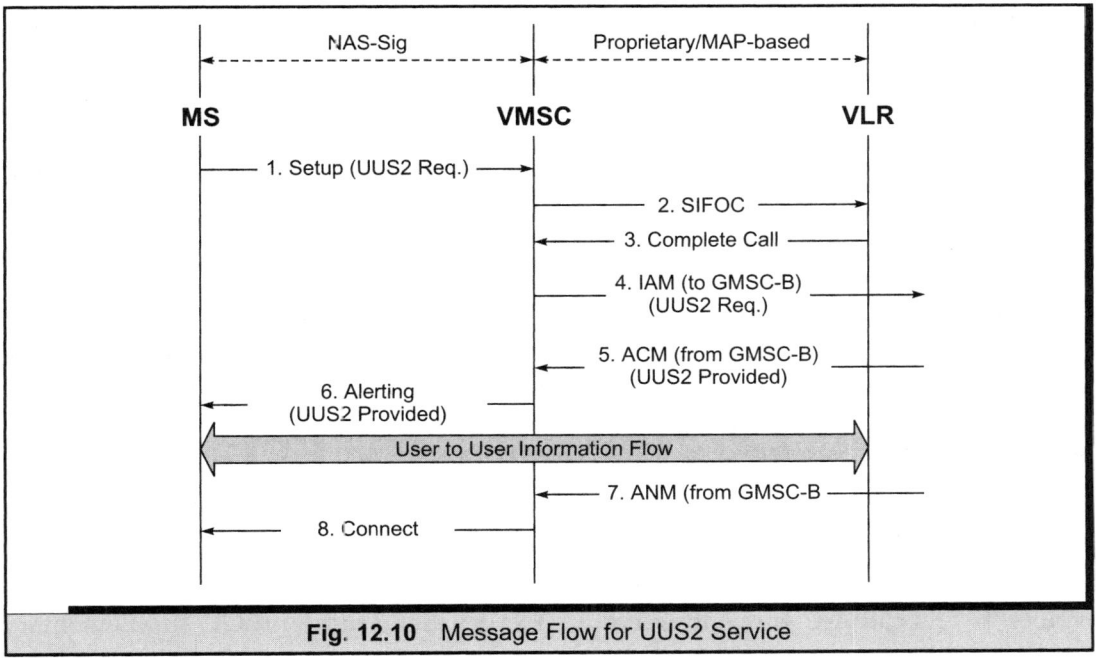

Fig. 12.10   Message Flow for UUS2 Service

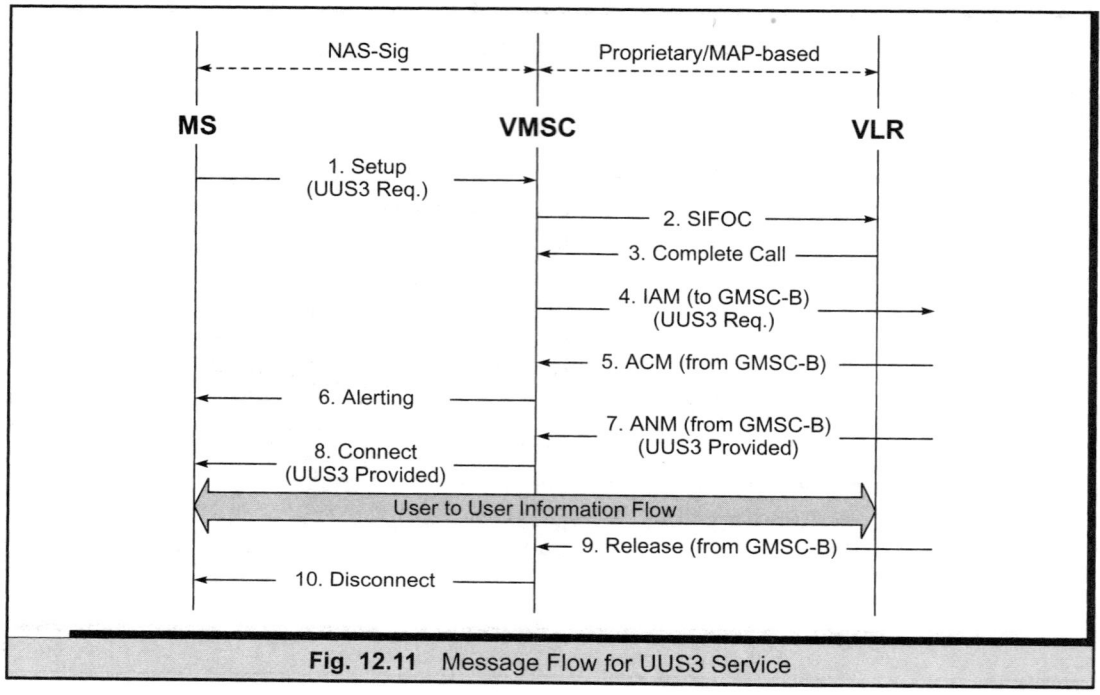

Fig. 12.11   Message Flow for UUS3 Service

- **Response to UUS service request:** The response to the UUS request is included in either the 'Connect' message, or the 'Alerting' message sent to the calling subscriber.

- **Transfer of User-to-User Information (UUI):** The UUI for UUS1 is transferred embedded within the call control messages. For UUS2, the UUI is transferred after the remote party is alerted but before the connection is established. UUI for UUS3 is transferred after the connection is established, and before it is released.

The rest of the steps in the figures are similar to the steps for MO call establishment discussed in Chapter 10. Between the three UUS services, the main difference is the phase during which the User-to-User Information is exchanged. The procedure for explicit activation is almost identical for all services. For more details on the UUS service, the reader is referred to 3GPP TS 23.087.

12.4.11   Explicit Call Transfer (ECT)

Consider a mobile subscriber 'A' handling two simultaneous calls: call A-B with subscriber 'B' and call A-C with subscriber 'C'. Explicit Call Transfer (ECT) supplementary service enables the served subscriber 'A', to connect together the other parties in the two calls ('B' and 'C'), and release its own connections (for calls A-B and A-C).

The following rules govern the use of the ECT service:

(a) Before the ECT service can be used, the connection for at least one of the two calls (let us say call A-B) should be established.

(b) For the other call (call A-C), either the connection should be established prior to call transfer, or as a network option, call transfer may occur while the other party (subscriber 'C') is being informed of the call (i.e. the connection has not yet been established).

After transfer of the call, the resources reserved for the calls of subscriber 'A' shall be cleared, and the call between subscribers 'B' and 'C' shall proceed as a normal call.

Figure 12.12 depicts the message flow for the ECT service. The steps in the figure are explained as follows:

1. In the initial stage, the call A-B is assumed to be in *held-state* due to the application of the call hold service. The connection for call A-C is also assumed to be established, and the call is in *idle-state* (i.e. subscribers 'A' and 'C' are not currently communicating). In this state, subscriber 'A' may wish to invoke the ECT service. If so, subscriber 'A' sends a request for invocation of the ECT service to the VMSC by using the 'Facility' message.

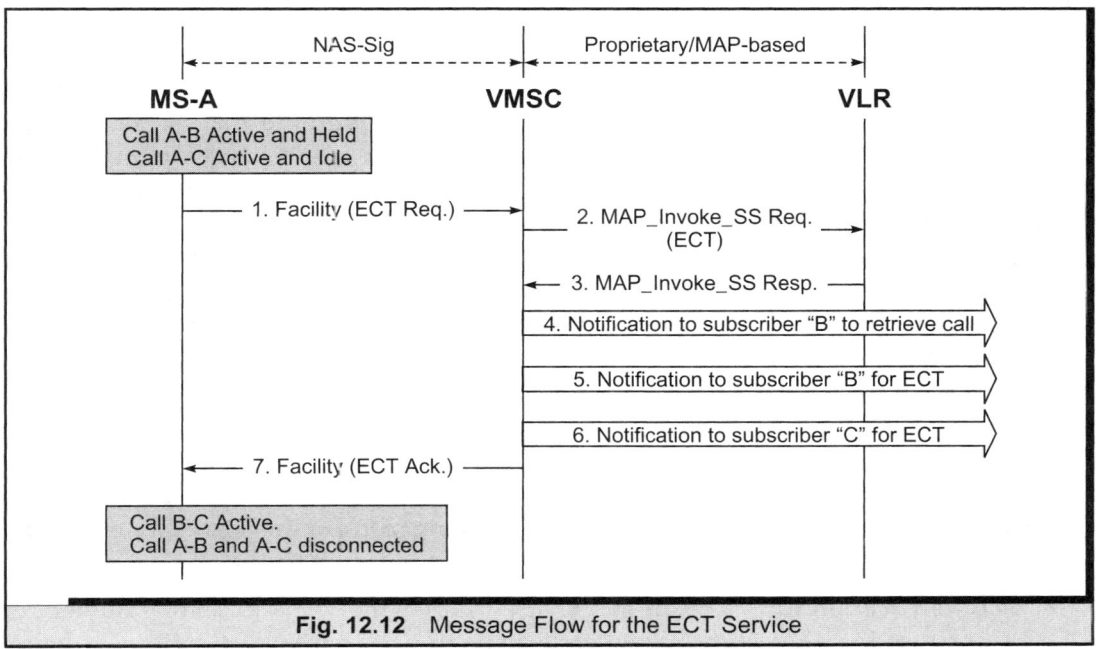

Fig. 12.12 Message Flow for the ECT Service

2. The VMSC subsequently sends a 'MAP_Invoke_SS' request message to the VLR indicating that subscriber 'A' has requested for invocation of the ECT service.

3. The VLR determines whether subscriber 'A' can be offered the ECT service. It then sends a response to the VMSC indicating whether or not the ECT service can be offered.

4-6. If the ECT service can be offered, the VMSC sends a notification message towards subscriber 'B' to indicate retrieval of the held call. Also, notification messages are sent towards both subscriber 'B' and 'C', indicating call transfer.

7. An acknowledgement is sent back to subscriber 'A' indicating that the call transfer was successful. This is followed by disconnection of the calls A-B and A-C, and the resources for these calls are freed.

Subsequent to the application of the ECT service, call B-C is treated as a normal call; as if it were explicitly set-up between the two subscribers. Normal call clearing procedures are followed for tearing down the call B-C, when either of the subscribers, 'B' or 'C', disconnects the call.

The ECT service is very useful in scenarios where each incoming call is first received by a telephone operator and then transferred to another party, based on the request of the calling subscriber. For more details on the ECT service, the reader is referred to 3GPP TS 23.091.

12.4.12 Multi-Call (MC)

The Multi-Call (MC) supplementary service allows a mobile subscriber to have multiple circuit-switched (CS) calls active at the same time, with each call using a dedicated bearer connection. For a mobile subscriber with a subscription to the MC service, the maximum number of bearers that can be simultaneously set-up is pre-defined by the service provider as part of the subscription information. As per 3GPP specification TS 23.135 for multi-call supplementary services, the currently defined range for the number of simultaneous bearers is from 2 to 7. However, the subscriber may decide to use fewer bearers than what is offered in the subscription. If this is the case, then the subscriber may register a self-imposed maximum value with the network. This value is denoted in the specifications as *Nbr_User*. The *Nbr_User* is registered with the network using the SS management procedures for the registration operation. The registered value for Nbr_User must be less than or equal to the maximum value defined by the service provider in the subscriber's subscription data.

Figure 12.13 depicts the message flow for the multi-call service for both the mobile-originated and mobile-terminated calls. The following steps in the figure are explained further:

- **VLR verification for MO call:** The VMSC informs the VLR of the total number of bearers used by the subscriber when it sends the 'Send Information For Outgoing

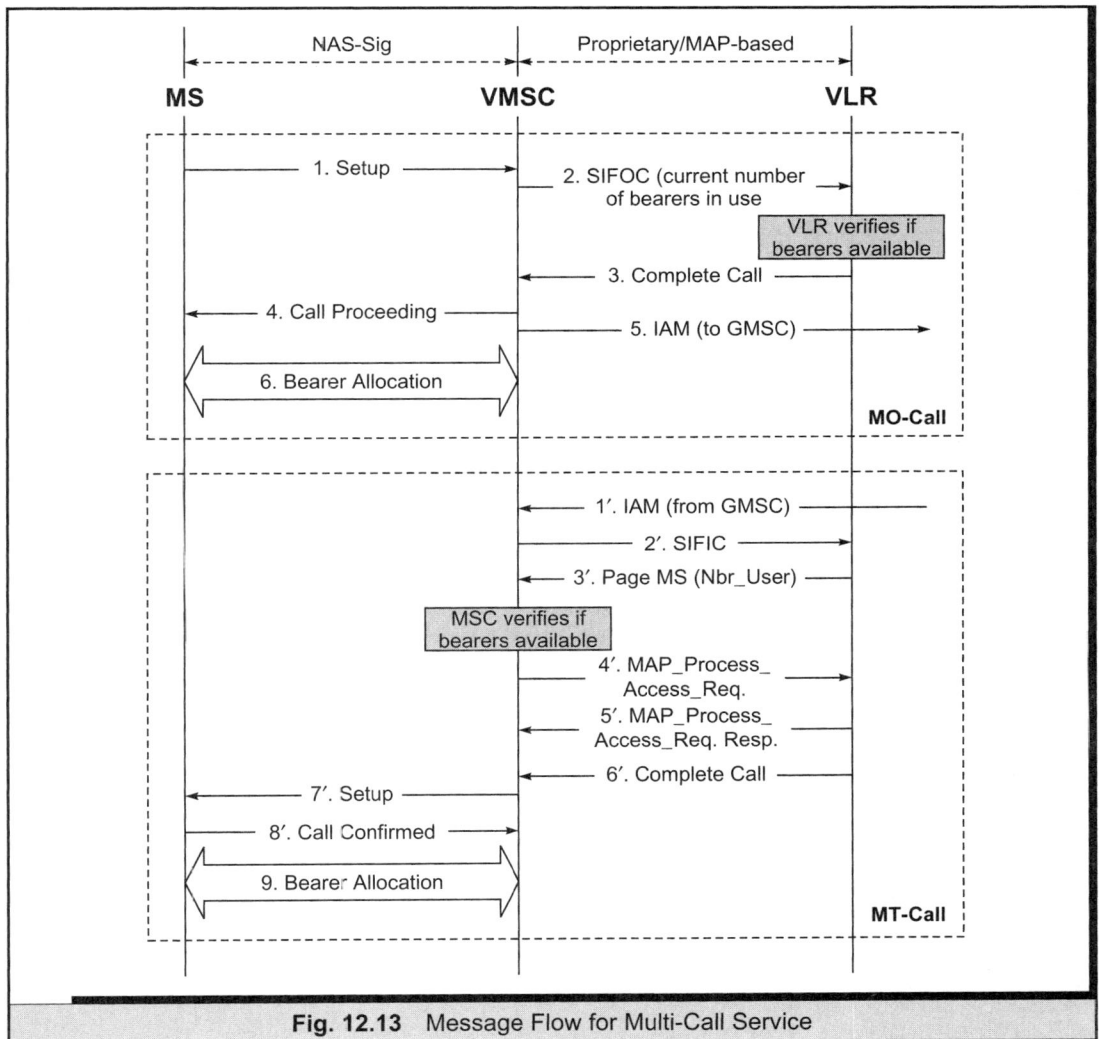

Fig. 12.13 Message Flow for Multi-Call Service

Call (SIFOC)' message to the VLR. The VLR determines if the number of bearers in use by the subscriber has reached the limit as defined by *Nbr_User*. If so, it sends a negative response to the SIFIC message from the VMSC. If, on the other hand, the number of bearers used by the subscriber is less than that defined by *Nbr_User*, it sends a 'Complete Call' message to the VMSC. The rest of the steps in the figure are similar to the steps for MO call establishment discussed in Chapter 10.

- **MSC verification for MT call:** For MT call handling, when the VLR sends a 'Page MS' message in response to the VMSC's SIFIC message, it includes the registered value of the *Nbr_User* in the message. The VMSC verifies if the number of bearers currently being used by the subscriber is less than the *Nbr_User* value. If so, the call proceeds as a normal call. Else, the call is rejected. The rest of the steps in the figure are similar to the steps for MT call handling discussed in Chapter 10.

For more details on the multi-call supplementary service, the reader is referred to 3GPP TS 23.135.

12.4.13 Other Supplementary Services

Apart form the supplementary services already discussed, there are a few other supplementary services, as follows:

- Calling Name Presentation (CNAP)
- Multiple Subscriber Profile (MSP)
- Completion of Calls to Busy Subscriber (CCBS)

12.4.13.1 *Calling Name Presentation (CNAP)*

The Calling Name Presentation (CNAP) supplementary service provides information about the calling party name to the called party at the time of call setup. Normally, mobile subscribers find it difficult to identify the calling user by just looking at the calling party number. It is easier to recognize the calling user by looking at the name of the calling party. CNAP supplementary service can therefore be considered as a variation of the CLIP service, wherein the calling party name is provided to the called party, instead of the calling party number.

Like the CLIP service, the delivery of the calling party name to the called party may be affected by the CLIR service. If the calling party has subscribed to Calling Line Identification Restriction (CLIR), then the calling party line identity as well as its name is not presented to the called party. For more information on the CNAP service, the reader is referred to 3GPP TS 23.096.

12.4.13.2 *Multiple Subscriber Profile (MSP)*

The Multiple Subscriber Profile (MSP) service allows mobile subscribers to have several profiles associated with a single IMSI. Each profile of the subscriber has separate subscription data. Thus, the MSP service provides subscribers the flexibility to categorize their service requirements according to their need. For example, a subscriber may wish to separate his/her service requirements into two categories, one for office-use (e.g. fax and voice-telephony) and the other for personal-use (voice-telephony only). The subscriber may subsequently use any of the profiles for both MO and MT calls.

The MSP service provides the option to define up to four different profiles per subscriber. Separate charging information is maintained for each profile for the services used. Thus, in the above example, the mobile subscriber may charge a part of his total bill to his office, for charges accrued on the profile defined for office-use. For more information on the MSP service, the reader is referred to 3GPP TS 23.097.

12.4.13.3 *Completion of Calls to Busy Subscriber (CCBS)*

Consider a scenario where subscriber 'A' places a call to destination 'B' and the latter is already busy in another call. In this scenario, if the Completion of Calls to Busy Subscriber (CCBS) service is used, it allows the calling subscriber 'A' to be notified when destination 'B' becomes idle. Further, it also allows the network to automatically generate a CCBS call to destination 'B', when it becomes idle, if subscriber 'A' so desires.

When subscriber 'A' determines that destination 'B' is busy, subscriber 'A' can request for the activation of the CCBS supplementary service for destination 'B'. If so, the network monitors destination 'B' to determine when it becomes idle. When destination 'B' becomes idle, the network indicates this condition to subscriber 'A'. After receiving indication, that subscriber 'B' is now idle, subscriber 'A' may respond within a pre-defined time indicating that it wishes to place a call to destination 'B'. The network will then automatically generate a CCBS call to destination "B".

For the CCBS service to be useful, it has to be supported by both the originating as well as the terminating network. For more information on the CCBS service, the reader is referred to 3GPP TS 23.093.

12.5 UNSTRUCTURED SUPPLEMENTARY SERVICE DATA

As mentioned in Section 12.3.2.2, a network operator may offer certain non-standardized supplementary services to its subscribers. These services are provided a Service Code (SC) that is distinct from the service codes of standardized supplementary services. Upon input of the MMI code by the mobile subscriber, the USIM deciphers the code to determine whether the SC maps to a standardized or an operator-specific supplementary service. If it is a defined supplementary service, the USIM uses the procedures discussed in Section 12.3.1 for SS management. If, however, the USIM determines that the supplementary service is not standardized, it uses the Unstructured Supplementary Services Data (USSD) procedures.

The next section discusses the USSD Architecture for the support of non-standardized supplementary services, followed by a discussion on MS-Initiated and Network-Initiated USSD procedures in Section 12.5.2.

12.5.1 USSD Architecture

The USSD procedures provide a mechanism whereby mobile subscribers and service providers can communicate with each other using means that are transparent to the intermediate network entities. Since the intermediate network elements are transparent to the information exchange, this mechanism allows for the development of services that are specific to a particular service provider. These services are also called *Unstructured Supplementary Services*, or USSD services in short. Figure 12.14 depicts the architecture for the handling of USSD services. Each network entity has the following logical components for the support of the USSD:

- **USSD Handler:** The USSD handler is a mandatory component that must be present in all intermediate network entities to support the USSD. Its main function is to route a USSD message, carrying information on the management of non-standardized supplementary services, to the appropriate destination. The USSD handler in each network element maintains a table indicating the network entity that hosts the USSD application, which can service the USSD message.

- **USSD Application:** Each non-standardized supplementary service is handled by a USSD application, which may reside in any of the network elements depicted in Figure 12.14. These network elements include the MSC or the VLR of the subscriber's VPLMN, or the HLR of the subscriber's HPLMN, or else a gsmSCF determined by the HLR of the HPLMN. The USSD application is not a mandatory component of each network element, which may or may not be hosting this application. A network element hosting a USSD application for a particular supplementary service is said to be handling that supplementary service.

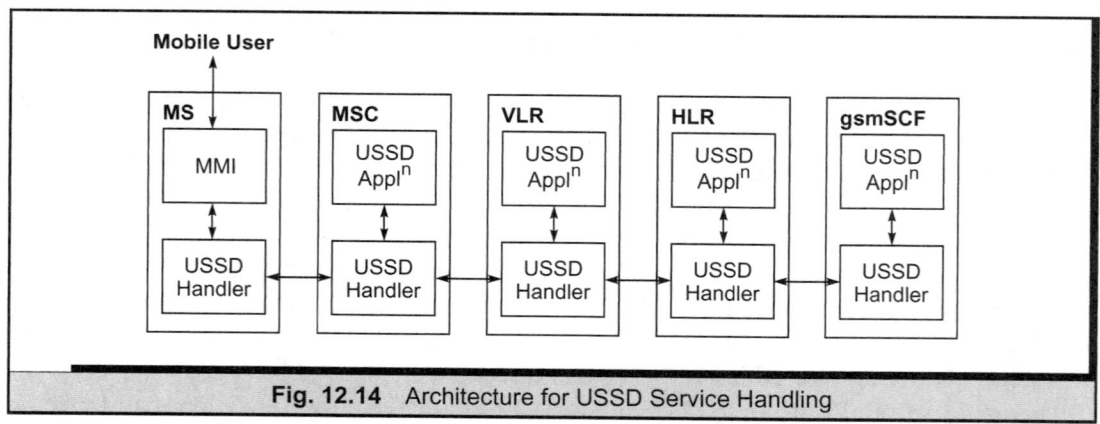

Fig. 12.14   Architecture for USSD Service Handling

A mobile subscriber wishing to carry out an SS operation for a non-standardized SS follows these steps:

1. The mobile subscriber includes information on the SS operation to be carried out, and the service code of the particular supplementary service in the USSD message.

2. The USSD handler at the MS services the USSD message and routes it towards the USSD application hosted on a particular network element (e.g. VLR or HLR).
3. The USSD handler on each intermediate network element transparently forwards the USSD message towards the network element handling the particular SS.
4. When the USSD message reaches the network element handling the particular SS, the USSD handler on that network element forwards the message towards the USSD application.
5. The USSD application on the network element carries out the requested SS operation. It then sends back a response to the mobile subscriber by including it in a USSD message. This message is then routed back to the mobile station in a similar manner.

USSD services can be handled at the MSC, the VLR of the subscriber's VPLMN, the HLR of the subscriber's HPLMN, or at a gsmSCF determined by the HLR of the HPLMN. The USSD handler at each network node decides where to route the USSD message—to a local application, or to the next node. The next section discusses the USSD messages defined by the MAP protocol and the message flows for MS-Initiated and Network-Initiated USSD procedures.

12.5.2 USSD Message Flows

The MAP protocol defines three messages for the exchange of USSD information between the network elements. These messages are as follows:

- **MAP_Process_Unstructured_SS_Data Message:** This message is used for MS-initiated USSD service operation. It is exchanged between the MSC and the VLR, the VLR and the HLR and between the HLR and gsmSCF to relay information related to the unstructured supplementary service operation.

- **MAP_Unstructured_SS_Data_Request Message:** This message is used for the network-initiated USSD service operation. It is exchanged between the gsmSCF and the HLR, the HLR and the VLR and between the VLR and the MSC when the invoking entity requires information from the mobile user in connection with unstructured supplementary service handling.

- **MAP_Unstructured_SS_Data_Notify Message:** This message is also used for the network-initiated USSD service operation. It is exchanged between the gsmSCF and the HLR, the HLR and the VLR and between the VLR and the MSC. The message is used when the invoking entity requires a notification to be sent to the mobile user, in connection with unstructured supplementary services handling.

The existing 'Facility' message is used between the MS and the MSC. In case of USSD, the 'Facility' message is augmented to indicate that it is carrying information for a USSD operation.

Figure 12.15 depicts the message flow for the network-initiated USSD service operation. This operation is initiated by the network element hosting the USSD application for the particular USSD service. The figure depicts the message flow when the USSD service operation is initiated by the HLR, the VLR or the MSC.

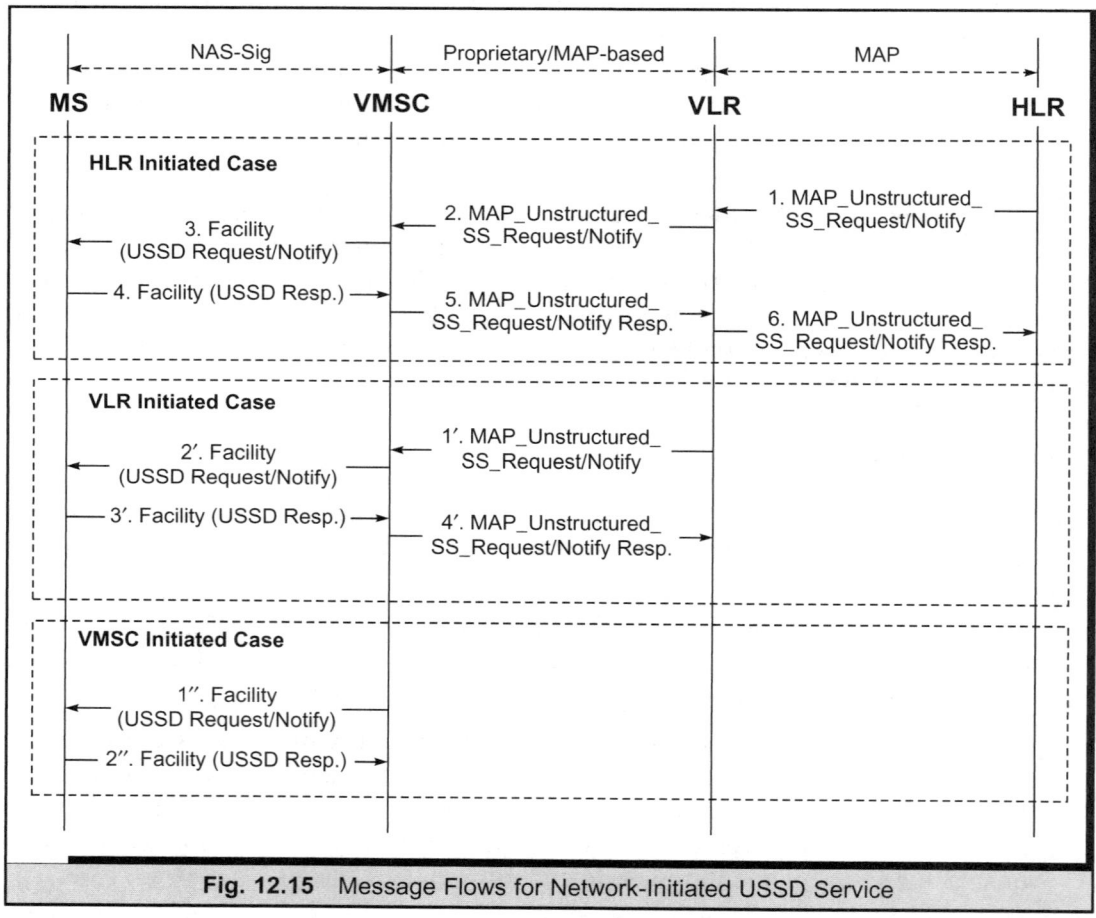

Fig. 12.15   Message Flows for Network-Initiated USSD Service

Figure 12.16 depicts the message flow for the MS-initiated USSD service operation. The MS initiates this operation by sending a 'Process Unstructured SS (PUSSD) Request' message towards the network entity handing the USSD service. On receipt of this request, the network entity may require some additional information from the MS to service this USSD service operation. In this case, the USSD application on the network entity may initiate USSD procedures to fetch the additional information from

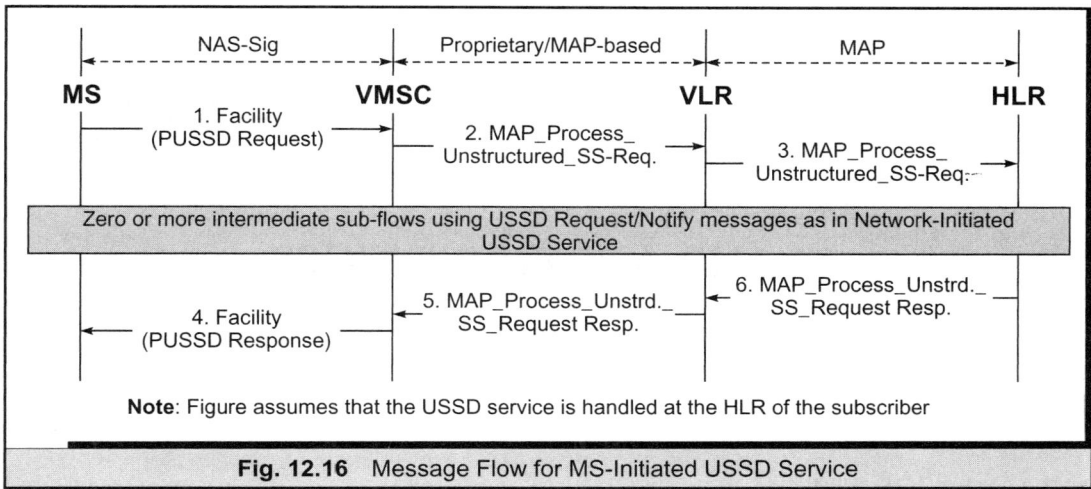

Fig. 12.16 Message Flow for MS-Initiated USSD Service

the MS. These procedures are similar to those used for the network-initiated USSD service operation, as depicted in Figure 12.15.

For more information on the USSD service, the reader is referred to 3GPP TS 23.090.

SUMMARY

Supplementary Services modify or supplement the basic services offered to subscribers in a UMTS network. All the supplementary services cannot be associated with a particular basic service, and only a subset of the supplementary service will be applicable to each basic service. The association of supplementary services with basic services, and the supplementary service operations and state information has been discussed in Section 12.2. The chapter also discussed in detail the supplementary services provided in UMTS, and the procedures for the control/management of these supplementary services. Supplementary service management procedures are classified into two broad categories: Call-independent and Call-related management of supplementary services. Both these categories of procedures have been discussed in the chapter.

Network operators do not necessarily have to offer all supplementary services and it is their prerogative to decide what supplementary services are to be offered to the subscribers. Network operators may also offer some non-standardized supplementary services to their subscribers. Unstructured Supplementary Services Data (USSD) provides a means to offer such non-standardized services, as discussed in the chapter.

VALUE-ADDED SERVICES

13.1 INTRODUCTION

In a UMTS network, service providers are expected to provide various value-added services, which include the Short Message Service (SMS), the Cell Broadcast Service (CBS), Multimedia Messaging Service (MMS) and Location Services (LCS), among others. Some of these services are already provided in 2G networks (e.g. SMS). It is expected that these services will be supported in the UMTS networks as well. This chapter discusses some of the value-added services offered in a UMTS network.

13.2 SHORT MESSAGE SERVICE

The Short Message Service (or SMS, as is commonly known) provides a means for sending limited size text messages to mobile subscribers. An SMS can carry up to 140 octets of text message. At times, SMS is preferred over voice calls, especially because SMS is non-intrusive in nature. The sender can send a text message to another subscriber, and the receiver may read the message at his/her own leisure. In fact, Short Message Service has proved to be a boon for 2G service providers, as it has found wide acceptability amongst 2G mobile subscribers.

The next section introduces the network architecture for supporting the Short Message Service. This is followed by a discussion on Mobile-Originated (MO) and Mobile-Terminated (MT) SMS handling procedures.

13.2.1 SMS Network Architecture

The SMS network architecture (Figure 13.1) includes new network entities defined for the support of SMS. These entities (briefly discussed in Chapter 6), are as follows:

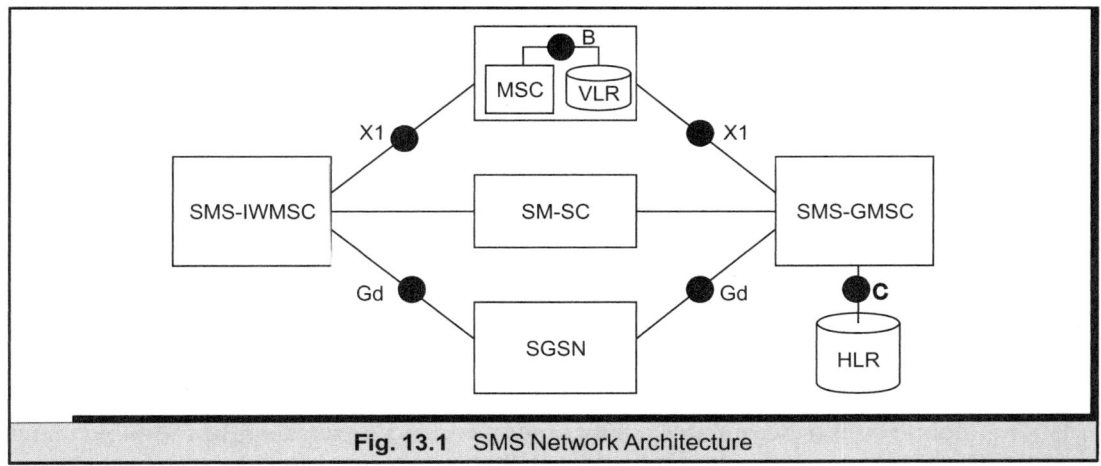

Fig. 13.1 SMS Network Architecture

- **Short Message Service Center (SM-SC):** The SM-SC is a data-store for storing all short messages until they are delivered to the recipient.

- **SMS Inter-Working Mobile Switching Center (SMS-IWMSC):** The SMS-IWMSC receives the mobile-originated short message from the MSC/VLR and/or the SGSN. It then sends the short message to the SM-SC that stores it for delivery to the final recipient.

- **SMS Gateway Mobile Switching Center (SMS-GMSC):** The SMS-GMSC interfaces with the MSC/VLR and/or the SGSN to deliver a short message, stored in the SM-SC, to the recipient mobile subscriber.

The interfaces defined for these new network entities were briefly discussed in Chapter 6. The interface between the SMS-GMSC/SMS-IWMSC and the MSC/VLR has not been assigned any name by the 3GPP specifications. Here, it is referred to as X1 interface. The X1 interface is based on the MAP protocol specified in 3GPP TS 29.002. Similarly, the SGSN interfaces with the SMS-GMSC/SMS-IWMSC using the Gd interface. The Gd interface is also based on the MAP protocol.

Further, the SMS-GMSC interfaces with the HLR over the C interface to retrieve information related to the location of the destination mobile subscriber. This is similar to the MT call handling procedure, wherein the C interface is used by the GMSC to retrieve the routing information from the HLR. The C interface is based on the MAP specification.

The interface between the SMS-GMSC/SMS-IWMSC and the SM-SC is a proprietary interface.

A GSM/UMTS network supports the SMS service by transferring short messages between the mobile stations and the SM-Service Center. Mobile-originated short messages are transferred from the mobile stations to the SM-SC, via the MSC/VLR or SGSN and the SMS-IWMSC. At the SM-SC, the short messages are stored until they are delivered. Mobile-terminated short messages are transferred from the SM-SC to the destination mobile station via the SMS-GMSC and the MSC/VLR or SGSN. Thus, the SM-SC acts as an intermediate storage arrangement for the short messages, while the SMS-GMSC and SMS-IWMSC are involved in the actual transfer of the short messages.

13.2.2   Mobile-Originated SMS Procedures

The Mobile-Originated (MO) SMS handling involves procedures for the transfer of a short message from the mobile station to the SM-Service Center (Figure 13.2). The steps involved in the handling of Mobile-Originated (MO) SMS are sumarized as follows:

1. The mobile subscriber creates a short message, and forwards it to the MSC using the 'CP_Data' message defined by the NAS Signaling protocols.
2. On receipt of the 'CP_Data' message, the MSC requests the VLR for subscriber-related information required to forward the short message towards the SM-SC. The information provided by the VLR to the MSC includes the mobile-subscriber ISDN number (MSISDN).
3. If the VLR determines that the mobile subscriber is allowed to initiate the short message, it responds to the MSC indicating the same.

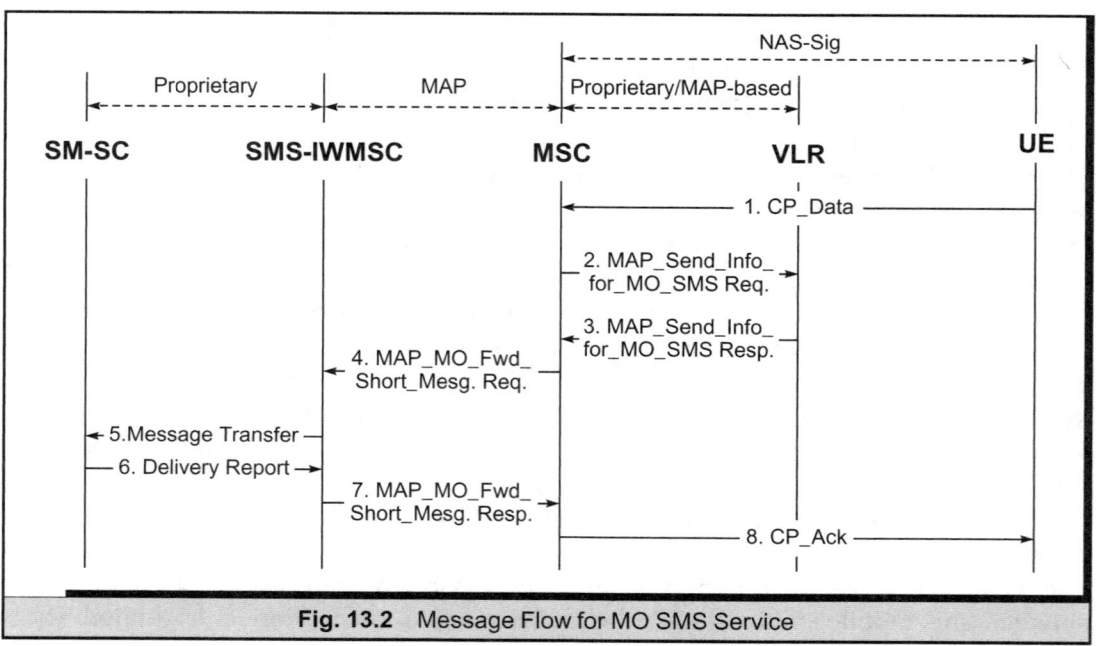

Fig. 13.2   Message Flow for MO SMS Service

4. The MSC then forwards the short message to the SMS-IWMSC, to be sent to the SM-SC.

5. The SMS-IWMSC transfers the short message to the SM-SC, which stores the message until it is delivered.

6-8. The delivery status is sent back from the SM-SC to the SMS-IWMSC, which sends it to the mobile subscriber via the MSC.

The short message can also be transferred from the mobile subscriber to the SM-SC via the SGSN, instead of the MSC/VLR. The followed steps are similar to those discussed above. For more details, the reader is referred to 3GPP specifications TS 23.040 and TS 24.011.

13.2.3 Mobile-Terminated SMS Procedures

The Mobile-Terminated (MT) SMS handling involves procedures for the transfer of a short message from the SM-Service Center to the destination mobile subscriber. Figure 13.3 depicts the message flow for the delivery of the mobile-terminated short message. The steps involved in the handling of Mobile-Terminated (MT) SMS are summarized as follows:

1. The SM-SC sends the short message to the SMS-GMSC, to be delivered to the destination mobile station.

2. The SMS-GMSC requests the HLR of the destination mobile subscriber for information on the serving MSC/VLR of the mobile subscriber. For this, it uses the 'MAP_Send_Routing_Information_for_SM' request message. This step is similar to the step followed for delivery of a MT call to the mobile subscriber, as discussed in Chapter 10.

3. The HLR provides information on the current MSC of the destination mobile subscriber in its response to the SMS-GMSC.

4. The SMS-GMSC then forwards the short message to the serving MSC of the destination mobile subscriber for delivery.

5. Next, the MSC requests the VLR for subscriber-related information required to forward the short message towards the destination mobile subscriber.

6. If the VLR determines that the mobile subscriber is allowed to receive the short message, it responds to the MSC indicating the same.

7. The MSC then forwards the short message to the destination subscriber. For this, it uses the 'CP_Data' message defined by the NAS signaling protocols.

8. On successful receipt of the short message, the destination mobile station responds with a 'CP_Ack' message to the MSC.

9. The MSC then forwards the delivery status for the short message to the SMS-GMSC.

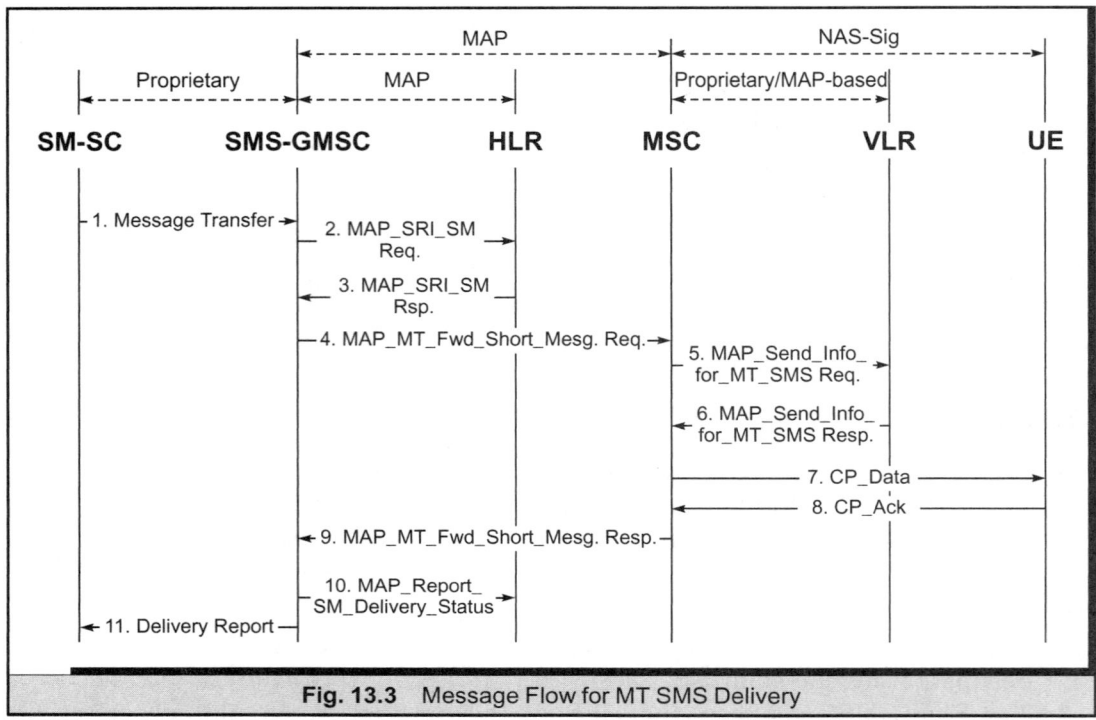

Fig. 13.3   Message Flow for MT SMS Delivery

10. The SMS-GMSC informs the HLR of successful delivery of the short message. This is done to indicate to the HLR that the destination mobile subscriber is currently reachable and is ready to receive the short message. Using this message, the HLR gets to know of the reachability status of the mobile subscriber.

11. The delivery status for the short message is also provided by the SMS-GMSC to the SM-SC. The SM-SC can now delete the short message from its data-store.

The transfer of the short message from the SM-SC to the mobile subscriber is also possible via the SGSN, instead of the MSC/VLR. The sequence of steps followed is similar to that discussed above. For more details, the reader is referred to 3GPP specifications TS 23.040 and TS 24.011.

13.3 CELL BROADCAST SERVICE

The Cell Broadcast Service (CBS) enables the broadcast of a number of unacknowledged CBS messages to all receivers within a particular region. The CBS service is in some way similar to the SMS service, except that in CBS, messages are broadcast to all users within a particular geographical area. This geographical area is also known as a *cell broadcast area*, which may comprise of one or more cells, or even the entire PLMN.

A CBS message consists of multiple CBS *pages*. A CBS *page* comprises of 82 octets. Up to 15 of these *pages* may be concatenated to form a CBS message. Each CBS message is assigned its own geographical coverage area by mutual agreement between the information provider and the PLMN operator.

Both the GSM and UMTS networks provide the CBS service. However, the network architecture for CBS differs for GSM and UMTS. The next section introduces the network architecture for providing the CBS service in UMTS. For details of the GSM network architecture, the reader is referred to 3GPP TS 23.041.

13.3.1 CBS Network Architecture

The UMTS network architecture for the support of CBS is depicted in Figure 13.4. The following new network entities are defined for the CBS service:

- **Cell Broadcast Center (CBC):** The CBC manages the delivery of the CBS messages. It delivers the messages to the RNS, from where they are broadcast within the specified cell broadcast area. A CBC may be connected to more than one RNS.

- **Cell Broadcast Entity (CBE):** The CBE is the originator of the CBS message. The CBS messages broadcast within a network may originate from a number of Cell Broadcast Entities (CBEs).

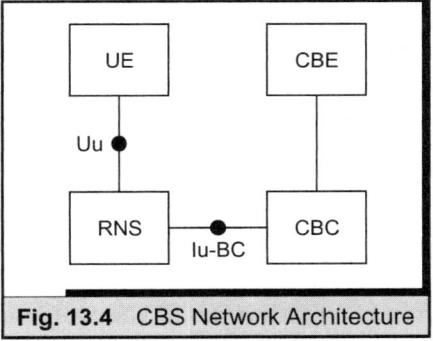

Fig. 13.4 CBS Network Architecture

The interface between the CBS and the RNS is referred to as the Iu-BC interface. This interface was introduced in Chapter 6. The interface between the CBS and the CBE is outside the scope of the 3GPP specifications. A proprietary interface may exist between the CBE and the CBS. The only assumption that the 3GPP specifications have made for this interface is that the CBE is responsible for all aspects of formatting the CBS. This includes the splitting of a CBS message into a number of pages. In other words, it is assumed by the 3GPP specifications that the CBC will not be involved in the formatting of the CBS.

CBS messages may originate from a number of Cell Broadcast Entities (CBEs), which are connected to the Cell Broadcast Center. CBS messages are then sent from the CBC to the cells, in accordance with the CBS's coverage requirements.

13.3.2 CBS Message Transfer Procedures

The CBS message transfer procedures involve the following protocols for the transfer of the CBS message:

- **Service Area Broadcast Protocol:** The Service Area Broadcast Protocol (SABP) is used on the Iu-BC interface for the transfer of the CBS message from the CBC to the RNS. The SABP protocol (covered in Chapter 5) is defined 3GPP TS 25.419.

- **Broadcast/Multicast Control:** The Broadcast/Multicast Control (BMC) protocol is used on the Uu interface for the broadcast of a CBS message within a cell broadcast area. The BMC protocol was covered in Chapter 5, and is defined in 3GPP TS 25.324.

The procedures for the broadcast of a CBS message within the cell broadcast area using the SABP and the BMC protocols have already been covered in Chapter 8.

13.4   MULTIMEDIA MESSAGING SERVICE

The SMS service has proved to be a major success for the current GSM service providers. However, the service offered by SMS is limited to transfer of short text messages between subscribers. Service providers are toying with the idea of enhancing the SMS service to offer a Multimedia Messaging Service (MMS), which can enable the transfer of multimedia messages between subscribers. MMS messages can contain multiple components such as voice, video and pictures, besides text.

This section discusses some of the MMS concepts covered in the MMS specification. The reference architecture for MMS is discussed in the next section. This is followed by a discussion on the MMS protocol framework in Section 13.4.2. MMS message transfer procedures are discussed in Section 13.4.3.

Multimedia Messaging Service is defined in 3GPP TS 23.140.

13.4.1 MMS Reference Architecture

The 3GPP TS 23.140 specification defines the reference architecture for the Multimedia Messaging Service. This reference architecture is depicted in Figure 13.5. The following entities are involved in offering the MMS service:

- **MMS User Agent:** The MMS User Agent resides either on the UE, or on some external entity connected to the UE. The MMS User Agent includes the application layer functionality, which includes support for the following functions:
    - Retrieval of MMS messages
    - Composition of the MMS message

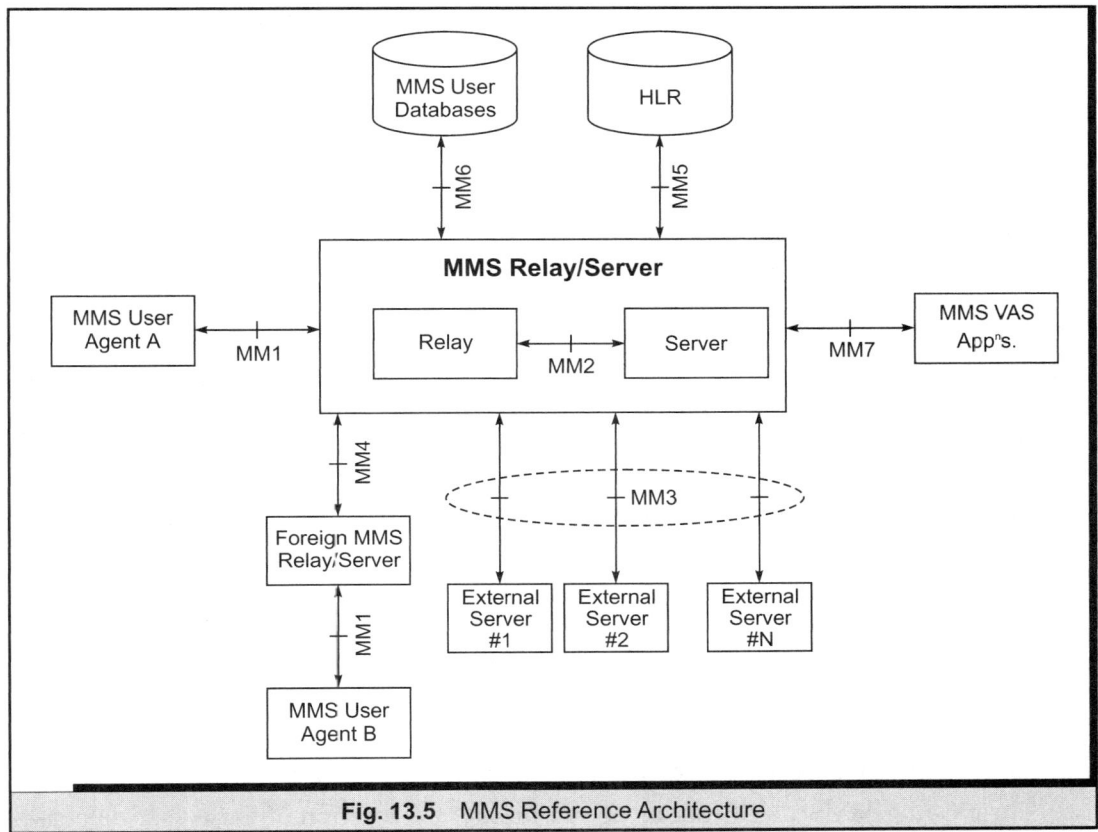

Fig. 13.5 MMS Reference Architecture

- Submission of the MMS message for transfer
- Presentation of the MMS message to the User

Besides these functions, an MMS User Agent may include support for various media formats for text, audio, video and images embedded within the MMS messages.

- **MMS Relay/Server:** The MMS Relay/Server is responsible for the storage and handling of MMS messages. It may also provide convergence functionality between External Servers and MMS User Agents. This function enables the integration of different server types across different networks. An example of an external server may include an SM-SC for sending/receiving SMS messages to/from an MMS User Agent.

Though the MMS Relay and MMS Server are depicted as separate entities in the MMS reference architecture, the allocation of functions to each of these entities is not separately defined in the 3GPP specifications. Instead, the 3GPP specifications define the joint functionality required from an MMS Relay/Server. The functions of an MMS Relay/Server include:

- Receiving and sending MMS messages
- Personalising the MMS message based on the receiver user profile information
- Deletion of MMS message based on the user profile and/or the filtering information
- Media type and format conversion
- Conversion of messages from legacy messaging systems to MMS message formats and vice versa

- **External Server:** External servers provide service to MMS User Agents via MMS Relay/Servers. This enables the integration of different server types across different networks. External servers include e-mail servers and SM-SC (SMS server), among others.

- **MMS User Databases and HLR:** The MMS Relay/Server may have access to several user databases. These databases may include a user profile database, a subscription database, and the database maintained at the HLR, among others. The databases provide an MMS Relay/Server the following information:

  - MMS user subscription information
  - Information to control access to MMS
  - A set of rules on how to handle incoming messages and their delivery
  - Information on the capabilities of the users terminal

  The existing MMS specifications do not discuss the location of these user databases nor how they will be accessed.

- **MMS VAS Applications:** The MMS VAS applications provide value-added services to MMS users. The existing MMS specifications do not discuss the kind of VAS applications that may be available. Neither do they specify how the MMS VAS application will provide the value-added services. MMS VAS applications may be introduced in an MMS network in future.

Table 13.1 provides a brief description of the reference points defined between the entities in the MMS reference architecture as depicted in Figure 13.5.

13.4.2 MMS Protocol Framework

MMS User Agents may exchange MMS messages with each other via the MMS Relay/Server. Figure 13.6 depicts the protocol framework for exchange of MMS messages between two MMS User Agents connected to different MMS Relays/Servers.

Table 13.1 MMS Reference Points

Interface	Between	Description
MM1	MMS User Agent— MMS Relay/Server	Used between the MMS User Agent and the MMS Relay/Server for the transfer of MMS messages and MMS delivery reports. Also used by the MMS Relay/Server to notify the MMS User Agent of the presence of an MMS message.
MM2	MMS Relay—MMS Server	Used between the MMS Relay and the MMS Server. The actual information exchanged on this interface has not been detailed by the MMS specifications.
MM3	MMS Relay/Server— External Server	Used by the MMS Relay/Server to send and retrieve multimedia messages from servers of external messaging systems.
MM4	MMS Relay/Server— MMS Relay/Server	Used for the exchange of MMS messages between two MMS Relays/Servers.
MM5	MMS Relay/Server— HLR	Used by the MMS Relay/Server to access MMS User Agent related information stored in the HLR.
MM6	MMS Relay/Server— MMS User Databases	Used by the MMS Relay/Server to access MMS User Agent related information stored in the MMS User Databases.
MM7	MMS Relay/Server— MMS VAS Applications	Used by the MMS Relay/Server to interface with MMS VAS applications that provide value-added services to the MMS users.

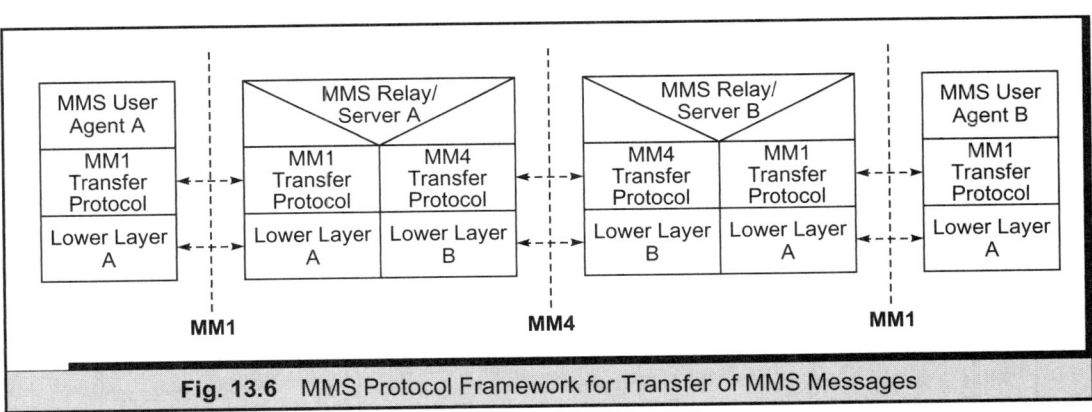

Fig. 13.6 MMS Protocol Framework for Transfer of MMS Messages

The two protocols required for the transfer of MMS messages, as depicted in the figure, are:

- **MM1 Transfer Protocol:** The MM1 transfer protocol is used for the exchange of MMS messages and MMS delivery reports between the MMS User Agent and the MMS Relay/Server. The MMS Relay/Server also uses this protocol to notify the MMS User Agent of the presence of an MMS message.

Multiple options exist for the implementation of MM1 transfer protocol, including the one based on the WAP protocols. Another implementation option is based on the applications using the Mobile Execution Environment (MExE). For more details on the implementation options for the MM1 transfer protocol, the reader is referred to 3GPP TS 23.140.

- **MM4 Transfer Protocol:** This protocol is used on the MM4 interface for the exchange of MMS messages between two MMS Relays/Servers. The MM4 transfer protocol is based on the Simple Mail Transfer Protocol (SMTP) defined by the IETF in its specification STD 10 (RFC 821).

MMS User Agents may also be served by external servers. Figure 13.7 depicts the protocol framework for interworking between an MMS User Agent and an external server. Besides the MM1 transfer protocol already discussed, the following protocol is additionally required for interworking between the MMS User Agent and the external servers:

- **MM3 Transfer Protocol:** The MM3 transfer protocol is used by the MMS Relay/Server to send and retrieve multimedia messages from servers of external messaging systems that are connected to the service provider's MMS Relay/Server. Different implementations of the MM3 transfer protocol is possible in several scenarios, depending upon the type of external server. For example, in case of an Internet Email server, the MM3 transfer protocol could be based on the SMTP, POP3 or the IMAP protocol, or a combination of these. In most cases, it is assumed that the lower layers for the MM3 transfer protocol would be IP-based. For more details on the implementation of the MM3 transfer protocol, the reader is referred to 3GPP TS 23.140.

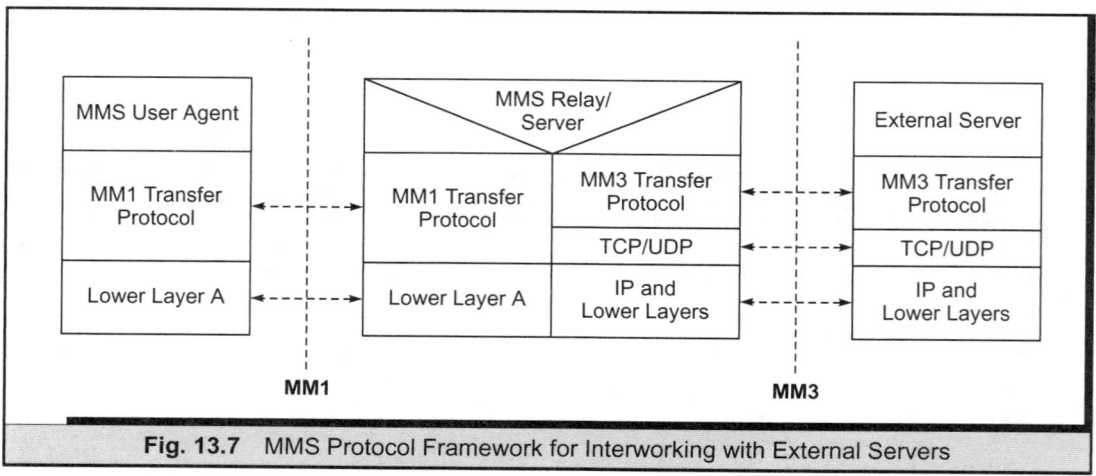

Fig. 13.7 MMS Protocol Framework for Interworking with External Servers

13.4.3 MMS Message Transfer Procedures

The MM1 and MM4 transfer protocols are used for the exchange of multimedia messages between two MMS User Agents. Figure 13.8 depicts the abstract message flow for the transfer of a multimedia message from one MMS User Agent to the other. The actual message flow would depend upon the MM transfer protocols used.

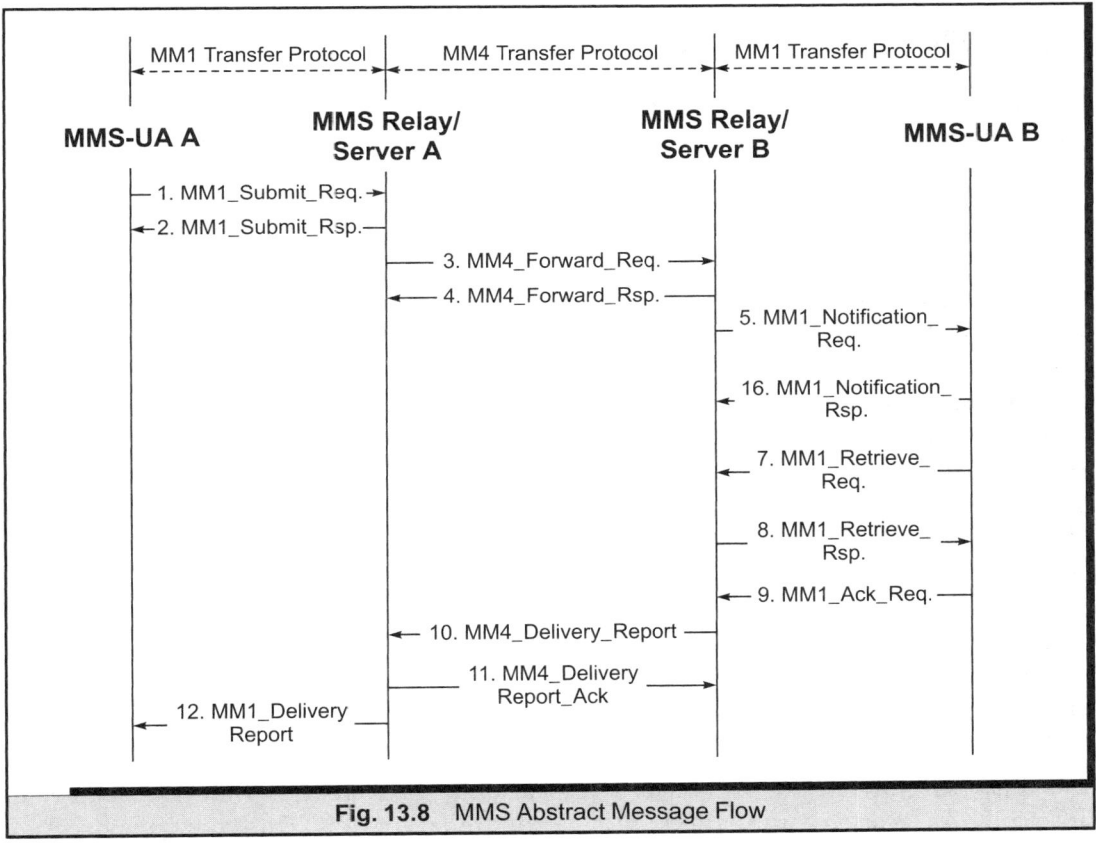

Fig. 13.8 MMS Abstract Message Flow

The salient points of the MMS message flow are:

- **Notification of MMS message:** The MM1 transfer protocol provides a mechanism whereby the MMS Relay/Server can notify the arrival of an MMS message to the MMS User Agent. Once the MMS User Agent receives this notification, it *pulls* the message from the MMS Relay/Server (see Steps 5 to 9 in Figure). Thus, from the perspective of an MMS User Agent, the delivery of an MMS message is based on a *pull* rather than a *push* mechanism.

- **Delivery Report of MMS message:** The MM4 and MM1 transfer protocols include a mechanism to report whether the delivery of the MMS message was successful or not. This is similar to the delivery report mechanism provided in the SMS service.

For more details on the MMS message transfer procedures, the reader is referred to 3GPP TS 23.140.

13.5   LOCATION SERVICES

Location Services (LCS) constitute a category of services that uses the positional information (or geographical location) of the UE to offer certain *specialized* services. These specialized services include emergency services, location-based charging services, tracking services and location-based information services. The geographical location of a UE is expressed in terms of its latitude and longitude co-ordinates. Different location services have varying requirements as regards the accuracy of the location information and the time within which the location information is provided to them. Multiple techniques exist for the determination of location information. As is to be expected, techniques that provide higher accuracy *vis-a-vis* the location information require more processing time. In other words, the more the time that the network has in determining the location of the UE, the more precise is the location information.

The following subsections discuss the reference model for Location Services (LCS), mechanisms used to determine location information, and the various services that are offered under LCS. For more details on location services, the reader is referred to 3GPP TS 23.271.

13.5.1   LCS Logical Reference Model

Figure 13.9 depicts the reference model for location services. The following logical entities are involved in this model:

- **LCS Client:** The LCS client is a logical entity that requests the LCS server to provide location information on one or more target UEs. This entity may require location information with different Quality of Service (QoS), where QoS includes the response time, and the accuracy of the location information. Being a logical entity, the LCS client may also reside within a UE. Alternatively, the LCS client may reside within a PLMN, or in an entity that is outside the PLMN.

- **LCS Server:** An LCS server provides a platform from which location services can be offered and consists of a number of location service components. It responds to requests received from authorized LCS clients, with the location information of the target UEs.

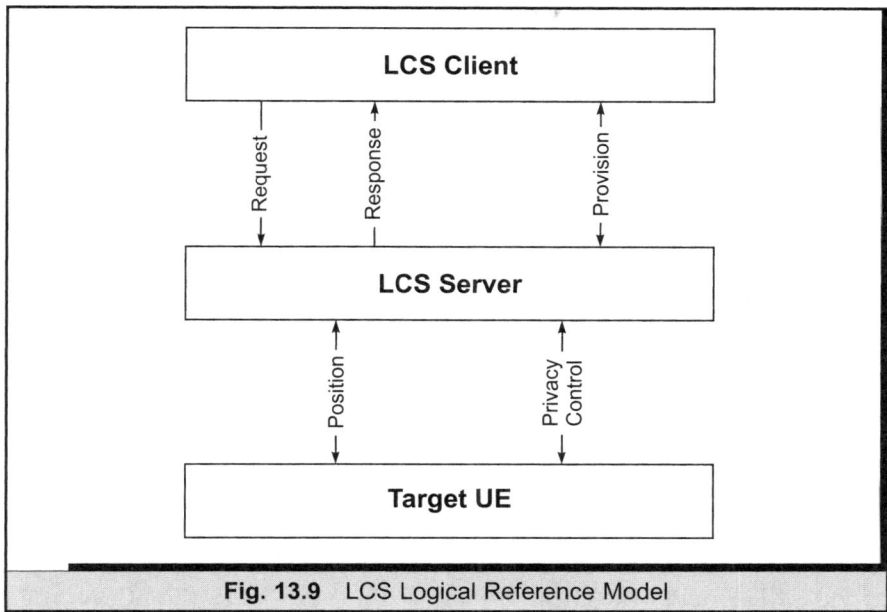

Fig. 13.9 LCS Logical Reference Model

- **Target UE:** The target UE is the logical entity whose position is to be determined by the LCS server. Multiple techniques for determination of location information exist; these include *network-based, mobile-assisted,* and *mobile-based positioning methods.* In case of mobile-assisted or mobile-based positioning methods, the target UE also includes components to actively support determination of location information.

13.5.2 LCS Control Procedures

The following control procedures are defined on the interfaces between the logical entities in the LCS reference model:

- **Request-Response procedures between LCS Client and Server:** The LCS client requests the LCS server for location information on one or more target UEs. The LCS server employs a positioning technique to obtain the location information and furnishes it to the LCS client.

- **Provisioning procedures between LCS Client and Server:** The LCS server must authorize the LCS client before it can obtain location information. The provisioning procedures are used to identify the LCS client, along with its characteristics, to the LCS server.

- **Positioning procedures between the LCS Server and Target UE:** These procedures are used to determine the geographical location of the UE. As mentioned earlier, the mechanisms used to identify the position of a UE include network-based, mobile-assisted, and mobile-based positioning methods. In case of mobile-assisted and mobile-based positioning methods, the positioning procedures between the LCS server and the target UE are used.

- **Privacy Control procedures between the LCS Server and Target UE:** The target UE may be allowed to determine which LCS clients are allowed to obtain its location information. This information is stored at the LCS server in the form of *associated privacy subscription options* for the target UE, and enables a UE to determine which LCS clients are allowed to request for its location information. In the absence of this information, it is assumed that the target UE has restricted all LCS clients from seeking information on its location.

13.5.3 LCS Network Architecture

The network architecture for the LCS services is depicted in Figure 13.10. By comparing the LCS Network Architecture with the LCS Reference Model, it can be seen that the functionality of the LCS server entity of the LCS RM is spread across multiple entities of the UMTS network. These consist of the MSC/VLR, the SGSN, the HLR, the RNS and a new CN entity—the GMLC. The Gateway Mobile Location Center (GMLC) is a new CN entity that has been defined to interface with the external LCS clients. Thus, it can be considered to be the gateway for all location services.

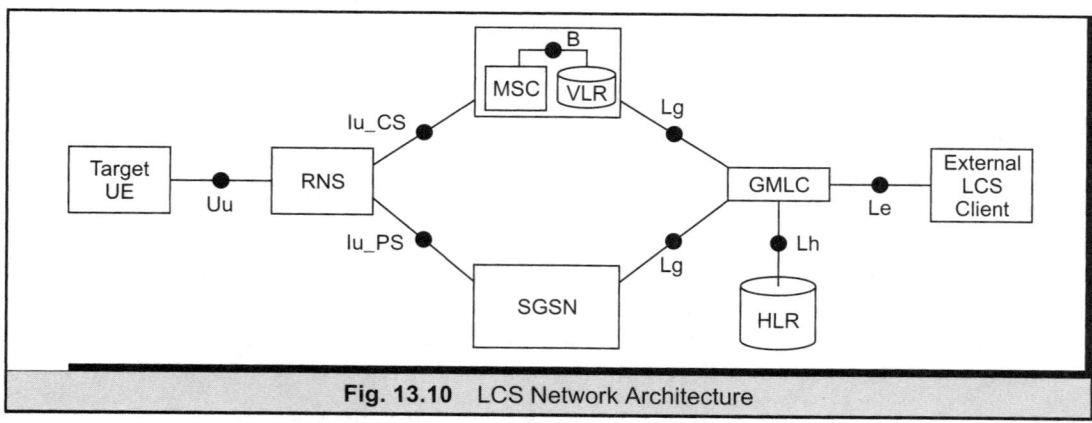

Fig. 13.10 LCS Network Architecture

The interfaces of the GMLC with other CN entities and external LCS clients are depicted in Figure 13.10. The following interfaces are defined:

- **Le Interface:** The Le interface is defined between the GMLC and the external LCS clients. The GMLC receives requests for location information from the external

LCS clients over this interface. It verifies whether the client is allowed to query the location information for the target UE. If so, then the GMLC sends this location information over the Le interface.

- **Lh Interface:** To handle the requests for location information, the GMLC interfaces with the HLR to determine the current MSC/VLR area and/or SGSN area where the subscriber is currently located. LCS-related information for a target UE is also retrieved by the GMLC from the HLR of the target UE over the Lh interface.

- **Lg Interface:** The main function of determining the location information of the target UE is performed by the RNS. The GMLC routes the query for location information to the Radio Network Subsystem (RNS) where the target UE is registered, via the MSC/VLR or SGSN for the target UE. The Lg interface of the GMLC with the MSC/VLR or SGSN is used for this purpose. The MSC/VLR or the SGSN then forward this request to the RNS over the Iu-CS or Iu-PS interface, respectively.

The RNS uses multiple techniques to determine the location information of the target UE. These mechanisms are discussed in the next section.

13.5.4 Mechanisms for Determination of Location Information

The mechanisms for determination of location information fall under three broad categories, as follows:

- **Network-based Positioning Methods:** In network-based positioning methods, the network determines the geographical location of the UE without any support from the latter. Thus, in this case, no additional hardware/software support is required in the UE. Cell-coverage based technique, explained later in this section, is an example of the network-based positioning methods.

- **Mobile-assisted Positioning Methods:** In mobile-assisted positioning methods, the network determines the geographical location of the UE using location measurements reported by the UE. The UE itself does not perform the location information calculation. However, it assists the network by providing certain measurements required for this purpose.

- **Mobile-based Positioning Methods:** In mobile-based positioning methods, the UE collects the measurements required for calculation of the location information. However, unlike in case of mobile-assisted methods, in this case, the UE also analyses the collected measurements to calculate its location information. Thus, the collection of measurements, and the calculation of the geographical location is done by the UE.

Three basic techniques are defined for calculation of location information. These are:

- **Cell Coverage-based Method:** The cell coverage-based method is the easiest technique used to gauge the geographical location of the UE. In this technique, the current cell location of the UE is determined. This can be done either when the UE initiates a location update procedure in the cell, or as a result of the paging procedure. Once the current cell of the UE is determined, the geographical location of the UE is given by the co-ordinates of the Node B in this cell.
  While the cell coverage-based method provides poor accuracy, it is the easiest mechanism since it does not require any measurements or calculations to be carried out by the UE or the network. Thus, the cell coverage-based method can be supported with no change to the network infrastructure.

- **Observed Time Difference of Arrival (OTDOA):** In the OTDOA technique, a UE is required to carry out measurements on the Common Pilot Channel (CPICH) signals received from a Node B. The UE measures the propagation delay in the transmission of information on the CPICH from the Node B to itself. If the exact co-ordinates of the Node B are known, then the propagation delay in transmission of information from Node B to the UE can be used to estimate the position of the UE with respect to the Node B. However, this provides a very rough estimation of the UE location. To obtain a more accurate estimation, the UE conducts this measurement on all CPICH signals that it is currently receiving from all the Node Bs in its vicinity.
  Once the measurements are done, the calculation of the geographical location can either be carried out by the UE itself (mobile-based method) or by the network, based on the measurements reported by the UE (mobile-assisted method). The exact details of the OTDOA method are beyond the scope of this section.

- **Global Positioning System (GPS) with UMTS:** The GPS uses a constellation of satellites to determine the location of an entity. This is done by measuring the propagation delay of satellite transmission to the entity whose geographical location is to be determined. This technique, in some ways, is similar to the OTDOA technique, except that in this case the propagation delay of satellite transmission is measured, rather than the propagation delay of the Node B CPICH signal. Like the OTDOA technique, the GPS-based technique may either be used as a UE-based, or as a UE-assisted method. In case of a UE-based method, a full GPS receiver is required in the UE, while in case of the latter, the UE requires only a reduced complexity GPS receiver. However, the UE-assisted method increases the signaling load on the air interface. A detailed discussion of the GPS-based techniques is beyond the scope of this section.

13.5.5 Location-based Services

These services include various types of services. Table 13.2 depicts some of the service types that can be offered under location services. Besides these, the service providers are free to offer any additional location-based services. The following sections provide a brief description of some of the services that can be offered under LCS.

Table 13.2 LCS Service Types

Location-based Services Categories	LCS Service Types
Public Safety Services	Emergency Services
	Emergency Alert Services
Location-based Charging	Charging on basis of Location
Tracking Services	Person Tracking
	Fleet Management
	Asset Management
Traffic Monitoring	Traffic Congestion Reporting
Enhanced Call Routing	Roadside Assistance
	Routing to Nearest Commercial Enterprise
Location-based Information Services	Navigation
	City Sightseeing
	Localized Advertising
	Mobile Yellow Pages

13.5.5.1 Public Safety Services

Public Safety Services are offered for the safety of the public and are available to all users without subscription. One of the services that can possibly be offered under this category is the Emergency Call Tracking. Under this service, the location information of a UE, from which an emergency call has been received, can be determined. Help can then reach the required location on receipt of the emergency call. Emergency alert services are also provided under the public safety services category and are used to notify mobile subscribers, within a specific geographic location, of emergency alerts like tornado warnings, volcano eruptions, etc.

13.5.5.2 Location-based Charging

Location-based charging enables a network operator to charge the subscriber according to its location or geographic zone. This service may be offered to subscribers either on a per-subscriber basis, or on a group basis. Though many mechanisms could exist for offering such services, a few are discussed in this section.

Examples of per-subscriber services include a service that offers a reduced rate to subscribers when they are roaming in a region that they select as their *home zone*. The *home zone* can be defined by the subscriber as the geographical region around his/her home, place of work, place of travel, and so on. Besides this, location-based charging may be used to provide a service whereby the subscriber can be charged differently when he/she is calling from places like country clubs, golf courses and other similar places.

Examples of group-based services include a service that offers special call rates to business groups. This means that when subscribers in a business group communicate with each other within a particular location, they are charged under special rates. The location, in this case, can be defined to be corporate campuses, business zones or other similar places.

13.5.5.3 Tracking Services

Tracking services include a host of services like person tracking, fleet management and asset management. Person tracking involves tracking the location and movement of a mobile subscriber. This service may be very useful for legal purposes. Fleet management allows an enterprise to track the location of its vehicles (cars, trucks, etc.). Asset management services provide the facility to track an *asset* when it enters or leaves a defined zone. Overall, the scope of tracking services is vast.

13.5.5.4 Enhanced Call Routing

Enhanced Call Routing (ECR) services include services like Emergency Roadside Services. Consider a scenario where a car breaks down in the middle of a journey, and the driver requires the services of a service station. In such a case, the distress call of the driver can be routed to the closest service station, based on his/her location. Thus, ECR can help in routing the calls of a subscriber to the nearest commercial center, based on the location of the subscriber.

13.5.5.5 Location-based Information Services

Location-based Information Services provide access to specialized information, after filtering the same according to the location of the subscriber. These services can be used to provide:

- **Navigation service:** This can be used to guide the subscriber to his/her destination. The input to the service is the current location of the subscriber, and the destination where the subscriber wishes to reach. The service can then be used to display a map showing the route to the destination.

- **City Sightseeing:** The city sight seeing service enables the delivery of location-specific information to a sightseer. When at a particular location, the sightseer can be provided information about the places of interest, and the distance and map showing how to reach there.

- **Location-dependent Content Broadcast:** This service offers the network the flexibility to broadcast information to subscribers within a particular geographical location. One possible usage of location-dependent services can be visualized as follows: Take a scenario where there are multiple commercial establishments, ranging from gas stations to eateries, running along the sides of a highway. Using the location-based services, commercial establishments can post the so-called *messages-in-the-air*, so that when a vehicle reaches the location where the message is posted, the same gets flashed on the handset of the vehicle owner. The message could read something like 'Gas Station, 0.5 miles ahead, Right'. This service of posting *messages-in-the-air* is simple to implement with LCS. Firstly, a database would be required to store a mapping between geographical co-ordinates and messages posted by commercial organizations. Then, handsets with LCS subscription can continuously send their location information to the network, which can in turn consult the database and post messages back to the handset, based on its geographical coordinates.

13.6 SERVICE CAPABILITY FEATURES

The bearer services, teleservices and supplementary services offered by a UMTS network have been standardized by the 3GPP. This means that a service provider offering any of these services will do so as per the 3GPP specifications. Though standardization of services brings uniformity to the services offered by various service providers, it restricts development of proprietary services that are specific to a particular service provider. It may be useful to allow service providers the flexibility to offer proprietary services that may become their Unique Selling Points (USPs). This helps them to retain mobile subscribers by offering them special services. To achieve this objective, 3GPP TS 22.105 defines the concept of *Service Capability Features*.

Services Capability Features (SCF) are open, technology independent building blocks for developing new value-added services. These services are accessible via a standardized application interface. The service capability features enable value-added applications to make use of the service capabilities of the underlying network in an open and secure way.

The Service Capability Features (SCF) are categorized into two types, as discussed in the following section. Some of the toolkits available for development of new value-added services using the SCF concept are discussed in Section 13.6.2.

13.6.1 SCF Types

The two types of Service Capability Features are:

- **Framework Service Capability Features:** These features define a framework for the development of new value-added services. The framework includes commonly used utilities that are necessary for making the value-added services accessible, secure, resilient and manageable. The framework services provide functionality that is independent of any particular type of service. Examples of such services include authentication, service registration and service discovery, among others.

- **Non-Framework Service Capability Features:** The Non-Framework Service Capability Features represent the collection of service capability features that are not included in the framework. These are the value-added services that are offered by service providers. Examples of non-framework services include messaging services and location services.

A detailed discussion on the Framework and Non-Framework Service Capability Features is beyond the scope of this section. For more details, the reader is referred to 3GPP TS 22.121.

13.6.2 SCF Toolkits

Value-added services are accessible to subscribers via a standardized application interface. The standardized interface is provided to subscribers using various toolkits, some of which are:

- **Customized Applications for Mobile network Enhanced Logic (CAMEL):** The CAMEL service enables the use of Operator Specific Services (OSS) by a subscriber even when roaming outside the HPLMN. CAMEL subscribers have one or more CAMEL Subscription Information (CSI) elements stored as part of their subscription data. The CSI is used to offer value-added services to subscribers. The CSI itself is provided by the HPLMN operator using administrative mechanisms. For more information on the CAMEL service, the reader is referred to 3GPP TS 22.078.

- **Open Service Access (OSA):** OSA enables the applications to make use of network functionality through an open standardized interface. This interface is referred to as the OSA API. OSA provides the glue between applications and the network functionality. Using OSA API, applications implementing the services become independent of the underlying network technology. For more details on OSA, the reader is referred to 3GPP TS 22.127.

- **USIM Application Toolkit (USAT):** USAT provides mechanisms to allow applications existing in the USIM to interact with any ME. This provides interoperability between a USIM and a ME, irrespective of who the equipment manufacturer or network operator is. For more details on USAT, the reader is referred to 3GPP TS 22.038.

- **Mobile Execution Environment (MExE):** MExE provides a standardized mobile execution environment in a UE. It enables the UE to negotiate its supported capabilities with a MExE service provider. This allows applications to be developed independent of the UE platform. For more details on MExE, the reader is referred to 3GPP TS 22.057.

The 3GPP TS 22.105 specification provides examples of services built using Service Capability Features. These include services like *Call Filtering* and *Call Back When Free* service. Besides these services, the above mentioned toolkits have also been proposed for realizing the Virtual Home Environment (VHE) concept, which is defined as a concept for personal service environment portability across network boundaries and between terminals. Using this concept, the mobile subscribers are always presented with the same personalized features and services, irrespective of their current location and terminal equipment. The VHE service is defined in 3GPP TS 22.121.

SUMMARY

This chapter discussed the various value-added services that are supported in a UMTS network. These services include the Short Message Service (SMS), the Cell Broadcast Service (CBS), Multimedia Messaging Service (MMS) and Location Services (LCS). While some of these services are already provided in 2G networks (e.g. SMS), it is expected that the UMTS network will continue to support these services. SMS is one of the most commonly used services and has been discussed in Section 13.2. It provides a means for sending limited size text messages to mobile subscribers, with each SMS capable of carrying up to 140 octets of text message. On the other hand, the CBS enables the broadcast of a number of unacknowledged CBS messages to all receivers within a particular region. The CBS service is similar to the SMS service, in some way except that in CBS, messages are broadcast to all the users in a particular geographical area.

Even though the SMS service has proved to be successful, the service is limited to the transfer of short text messages only. Service Providers are toying with the idea of enhancing the SMS service to offer MMS, which can enable the transfer of multimedia messages between subscribers. MMS service has been discussed in detail in Section 13.4. Another value-added service offered in UMTS is the LCS. LCS constitutes a category of services that uses the positional information (or geographical location) of the UE to offer certain specialized services. These specialized services include emergency services, location-based charging services, tracking services and location-based information services. LCS was discussed in detail in Section 13.5. Section 13.6 discussed the Services Capability Features (SCF), which define open, technology independent building blocks for developing new value-added services.

SECURITY MANAGEMENT

14.1 INTRODUCTION

The security requirements in any given network are limited to few key aspects. These include *data integrity*, which means that the contents of packets are not altered in an unauthorized manner. Next is *origin authentication*, which refers to the corroboration that the source of data received is as claimed. Typically, data integrity and origin authentication is achieved using a signature or a message digest of the complete or partial message. The message digest is also known as the Message Authentication Code (MAC). It is customary to club data integrity and origin authentication together, and simply refer to it as *authentication*. Next comes the *confidentiality* of data, which refers to the property that information is not made available or disclosed to unauthorized entities. Confidentiality is achieved by encrypting the data using ciphering techniques. Authentication and confidentiality are two key concepts as regards the security. Apart from these, there are security aspects like *replay-protection*. To understand this concept consider a case where the attacker can neither breach the contents of the packet, nor impersonate someone else. In such a scenario, it may be possible for the attacker to hold a valid packet and replay it after sometime. This replay of a stale packet may cause limited or significant disruption of service. A protocol is replay-protected if a captured packet loses its relevance if stored over a long period of time. Replay-protection is ensured through the use of sequence numbers and/or time-varying parameters.

In Second Generation (2G) networks, there was limited support for the security aspects. One of the major limitations of 2G networks was the user's inability to check whether the network providing the service was authorized to do so or not. Apart from this limitation, there was no support for data integrity in 2G. The cipher keys and authentication data were transmitted in cleartext between and within the network.

Moreover, the security features between two network elements were also non-existent.

When the Third Generation (3G) networks were designed, it was decided that the security architecture would be built upon the 2G architecture. Those aspects that were found to be robust would be retained. The areas requiring improvement would be looked into. Thus, the use of Subscriber Identity Module (SIM), as a removable hardware was retained. The subscriber identity confidentiality on the radio interface was also retained. The use of *shared secret key* between the user and the network was also continued. The shared secret key forms the basis for all authentication procedures.

To strengthen security features, the notion of *mutual authentication* was introduced. This allowed the user to authenticate the network. Note that provision for the network to authenticate the user already existed in 2G. By allowing the user to authenticate the network, mutual authentication was possible. Then, the possibility of brute-force attack was also reduced by increasing the length of ciphering key from 8 bytes to 16 bytes. Apart from this, *network domain security* was introduced to provide security features between different nodes of a network.

Figure 14.1 shows the four basic functional blocks of UMTS security, explained in the following four sections. The first block—*user domain security*—provides users a secure access to mobile stations. This is achieved through the use of PIN codes at the interface between the user and USIM.

The next functional block—*network access security*—encompasses security features to facilitate secure access to UMTS services. Network access security is provided through user identity confidentiality, mutual authentication, confidentiality and data integrity.

The third functional block is of *network domain security*. This security feature enables the network nodes to securely exchange signaling data. Network domain security can be provided at two levels: *application layer* and *network layer*. Application layer security in UMTS networks is provided through MAPsec. MAPsec is used between two entities (e.g. between HLR and VLR) employing MAP protocol to securely communicate with each other. Here, note that the term 'application' is seen in relation to the application layer signaling protocol and not in the context of user applications.

At network layer, the network domain security is provided through IPSec. The IPSec protocol suite includes a key distribution protocol, e.g. Internet Key Exchange (IKE) protocol, and a protocol that provides authentication and confidentiality, e.g. Encapsulating Security Payload (ESP) protocol. Various architectures are possible to deploy IPSec; in the case of UMTS, the IPSec is used to provide secured communication between two signaling gateways. Thus, two entities of different domains use the services of signaling gateways to ensure secured communication. Within a domain, the exchange may or may not be secure. Further, network domain security applies only to control plane, not for user plane.

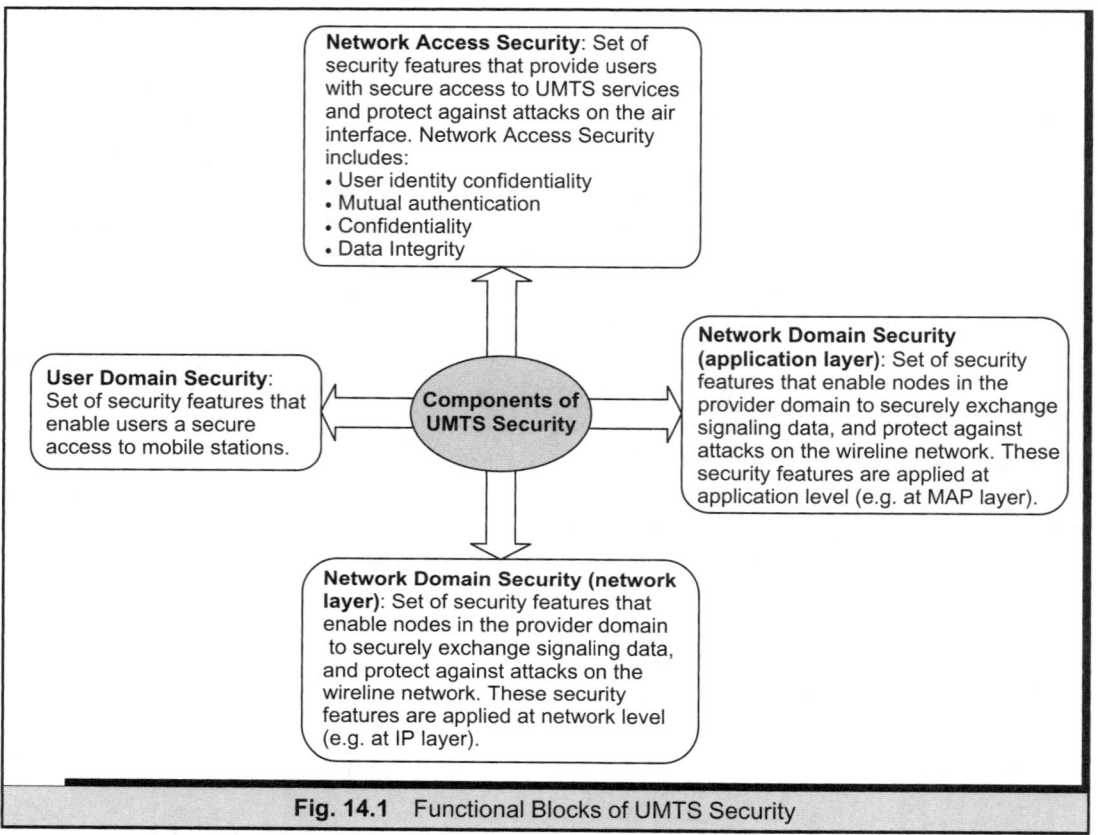

Fig. 14.1 Functional Blocks of UMTS Security

Note that the security features of IP Multimedia Subsystem (IMS) are not covered specifically in this chapter. The IMS which has been introduced for the first time in Rel5 standards, is discussed in Chapter 16. The IP Multimedia Subsystem brings with it new security features, which can be broadly divided into two categories, namely *access security* features and *network domain security* features. The access security features of IMS are, to a large extent, based on access security mechanisms of CP/PS domain as discussed in this chapter. The specific modifications for access security in IMS are discussed in Chapter 16. The network domain security in IMS is based on IPSec and is explained later in this chapter.

14.2 USER DOMAIN SECURITY

The *user domain security* function ensures that only an authenticated user can gain access to a mobile station. This functionality is achieved by restricting access to USIM until the user has been authenticated. For this, a secret password (called the PIN) is shared between the user and the USIM. The password resides in the USIM and the user has the option to modify it. Access to USIM is granted only when the correct password is provided by the user. To ensure that brute force is not applied to obtain the pin, the USIM is rendered unusable after a given number of unsuccessful attempts.

Among the various security features, the user domain security is simplest. The user domain security function is specified in 3GPP TS 31.101.

14.3   NETWORK ACCESS SECURITY

In the UMTS, the phrase 'Access Security' encompasses a number of security aspects. The term *access* refers to accessing or utilizing the services provided by the network. Access security principles are applied to ensure that only legitimate users are allowed access to the network. In this regard, two concepts are defined. The aspect, whereby the serving network corroborates the identity of the user, is referred to as *user authentication*. The second aspect, whereby the user corroborates that the serving network is authorized by the user's home network to provide service, is known as *network authentication*. The user authentication and network authentication are collectively referred to as *mutual authentication*, which is one of the most important components of access security.

A scheme called the UMTS Authentication and Key Agreement (UMTS AKA) provides the means to achieve mutual authentication. UMTS AKA also provides the means whereby security keys are generated as part of the authentication procedure. These security keys are of two types: Integrity Key (IK) and Cipher Key (CK). The integrity key, IK, is used for maintaining the *integrity* of signaling data between the mobile terminal and RNC. The integrity protection principles are not applied for user plane due to performance reasons. The ciphering key, CK, is used to *encrypt* the user and signaling data. The encryption provides confidentiality over the air interface.

Apart from mutual authentication, data integrity, and confidentiality, another important feature of access security is user *identity confidentiality*, which refers to the property that the permanent identity, IMSI, of a user is not eavesdropped on along the radio access link. To achieve this objective, the user is allocated—and is identified by— a temporary identity called the TMSI. However, in some rare cases, it is required that the user provides its permanent identity, IMSI. Such scenarios occur when the VLR/ SSSN does not have knowledge of this permanent identity, nor can it obtain this information from other VLRs/SGSNs. Apart from such specific scenarios, a user is always identified by the TMSI.

The *mobile equipment identification* function is also considered to be part of access security. In this, the serving network seeks the International Mobile Equipment Identity (IMEI), from the MS to verify its legitimacy. The Equipment Identity Register

(EIR) is used to verify the IMEI. Recall from Chapter 6 that the EIR maintains three lists of IMEI, namely the white list, black list and the gray list. These lists are used to check the legitimacy of the mobile equipment. The mobile equipment identification function is not elaborated upon in this chapter.

To summarize, the access security includes four important functions:

- Mutual Authentication
- Data Confidentiality
- Data Integrity
- User Identity Confidentiality

14.3.1 Mutual Authentication

In the UMTS, mutual authentication is accomplished by using a secret key K that is shared by the Authentication Center (AuC) of the Home Environment (HE) and the USIM of the Mobile Station (MS). This secret key is never exchanged between the user and the network. Rather, it is a part of the USIM at the time it is made. The AuC keeps a mapping between the user identity IMSI and the secret key K. Thus, AuC knows the secret key K of a given user identified by IMSI. The knowledge of secret key K and that of security algorithm used for mutual authentication is proof of an authentic user.

Apart from the secret key K, the USIM and AuC also share a sequence counter called SQN. The SQN maintained at AuC is called the SQN_{HE} (referring to SQN maintained by HE). The SQN_{HE} is essentially an individual counter that AuC maintains for each user identified by its permanent identity IMSI. At the MS, the USIM also maintains a sequence number SQN_{MS}. The SQN_{MS} denotes the highest sequence number accepted by the USIM. The use of SQN_{MS} and SQN_{HE} is explained in the following sub-sections.

Using the secret key K and sequence counter SQN, the UMTS Authentication and Key Agreement (UMTS AKA) works as follows: To authenticate a user, the VLR or SGSN (depending upon whether the domain is circuit-switched or packet-switched) requests AuC to send an array of authentication vectors. The authentication vectors are a set of five elements (generated using the secret key K and sequence counter SQN). The elements of authentication vectors are as follows:

- **RAND:** 16 byte field, refers to a random number computed using pseudo random number generators

- **XRES:** 4 to 16 byte field, refers to expected response from USIM

- **CK:** 16 byte field, refers to cipher key; it is used for ciphering purposes

- **IK:** 16 byte field, refers to integrity key; it is used for integrity protection

- **AUTN:** 16 byte field, refers to authentication token; AUTN is used by the serving network to authenticate itself to the MS. The SQN used to generate an authentication vector forms a part of the AUTN. To maintain the confidentiality of SQN, an Anonymity Key (AK) may be used.

Each authentication vector is good for one authentication and key agreement between the VLR/SGSN and the USIM. Each set of vectors achieves the same function. The vectors are used by the VLR/SGSN in the order in which they are received. The VLR/SGNS pre-fetches an array of vectors from AuC to expedite authentication when it takes place again.

From a given authentication vector (which is the first in the available set of vectors), the VLR/SGSN only picks the elements RAND and AUTN and sends them to USIM. On receiving these fields, USIM runs the same algorithm that was earlier used by AuC to generate the authentication vectors. During the execution, of the authentication algorithm the USIM uses the RAND and a part of AUTN as input. The output includes RES, XMAC, CK and IK. The XMAC is compared with the last 8 bytes of AUTN (which is called the MAC). In case XMAC is equal to MAC, the network is said to be authenticated by the user. The RES is sent by the USIM to VLR/SGSN. The CK and IK generated during the process are identical to the CK and IK received by VLR/SGSN from AuC. This CK and IK are later used for encryption and data integrity.

When the VLR/SGSN receives the RES, it checks whether the latter is equal to the XRES (belonging to the authentication vector used for authentication). In case RES is equal to XRES, the user is said to be authenticated by the network. This completes the process of mutual authentication. The CK and IK are now ready to be used at USIM and RNC for encryption and data integrity. The subsequent sections provide details of mutual authentication and the key generation process.

14.3.1.1 *Scenarios Requiring Initiation of Authentication Procedure*

The process of authentication begins when the VLR/SGSN determines that a given scenario requires the MS to be authenticated. Such scenarios are listed by the 3GPP specifications. Associated with each scenario, the 3GPP specifications also mention whether authentication *must*, *should* or *may* be performed. Since the scenarios for SGSN and VLR are on similar lines, only the scenarios in which a SGSN initiates the authentication procedure are listed below (refer 3GPP TS 23.060):

- **First 'GPRS attach' procedure:** In case the MS performs 'GPRS attach' for the very first time, the SGSN *must* initiate the authentication procedure.

- **Subsequent 'GPRS attach' procedure:** In case the MS performs 'GPRS attach' and the SGSN has changed since the last detach and the MS is unknown in both the old and new SGSN (i.e. there is no Mobility Management (MM) context for the MS in the network), the new SGSN *must* initiate the authentication procedure.

- **'GPRS detach' procedure:** In case the MS performs 'GPRS detach' and (1) the P-TMSI signature[1] is not included in the 'GPRS detach' message or (2) the P-TMSI signature sent by MS is not valid, the SGSN *should* initiate the authentication procedure. In case the P-TMSI signature sent by MS in 'GPRS detach' is valid, the SGSN *may* initiate the authentication procedure.

- **'Routing Area Update' procedure:** In case the MS performs an inter-SGSN RA update procedure and the P-TMSI signature provided by MS is not valid, the new SGSN *should* initiate the authentication procedure. Otherwise, the new SGSN *may* initiate the authentication procedure.

- **'Service Request' procedure:** This procedure is used by MS to establish a connection with the SGSN in order to send uplink signaling messages (e.g. Activate PDP Context Request), user data, or as paging response, or after the MS has regained radio coverage. The 'Service Request' procedure is a NAS procedure performed between the MS and SGSN using the 'CM Service Request' message (refer Chapter 6 for NAS signaling messages). During 'Service Request' procedure, if the MS is in idle state, the SGSN *must* initiate the authentication procedure.

- **Intersystem change:** During intersystem change (e.g. the MS changes the radio system from GSM to UMTS), the SGSN *may* initiate the authentication procedure.

14.3.1.2 *Exchange of Authentication Vectors*

Once it is decided to perform authentication, the VLR/SGSN request AuC of the Home Network to send authentication vectors. The way the whole authentication procedure is defined, the VLR/SGSN does not take part in the *generation* of authentication vectors. Rather, VLR/SGSN acts like an agent of the AuC (i.e. AuC provides VLR/SGSN with the information necessary to carry out mutual authentication). The reason behind this is as follows: the secret key K, which is an essential component of the authentication process, cannot be exchanged between any entity. This is to prevent any likelihood of the secret key K becoming vulnerable to external attacks. Given this, it is imperative that the owner of this key, AuC, generates authentication vectors. This is indeed what actually happens. AuC does all the calculation. What is left for VLR/SGSN to do is to communicate with USIM and ensure that the information received from AuC (i.e. XRES) and that received from USIM (i.e. RES) match. AuC also provides VLR/SGSN the information required by the serving network to authenticate itself (i.e. AUTN).

---

[1]The P-TMSI signature is sent by the SGSN to the MS in 'Attach Accept' and 'RA Update Accept' messages. The MS includes the P-TMSI Signature in the next 'RA Update Request', 'Detach Request' and 'Attach Request' for identification checking purposes.

Figure 14.2 shows the exchange of authentication vectors between VLR/SGSN and AuC. The exchange takes place using the MAP_SEND_AUTHENTICATION_ INFO request/response message of Mobile Application Part (MAP) protocol. In the request message, VLR/SGSN sends the user identity IMSI and the number of vectors it wants from AuC. The AuC may already have pre-computed vectors, in which case it sends them to VLR/SGSN immediately. Otherwise, AuC computes fresh vectors using secret key K and sequence number SQN_{HE}. The response message contains an array of vectors comprising of the quintet {RAND, XRES, CK, IK and AUTN}.

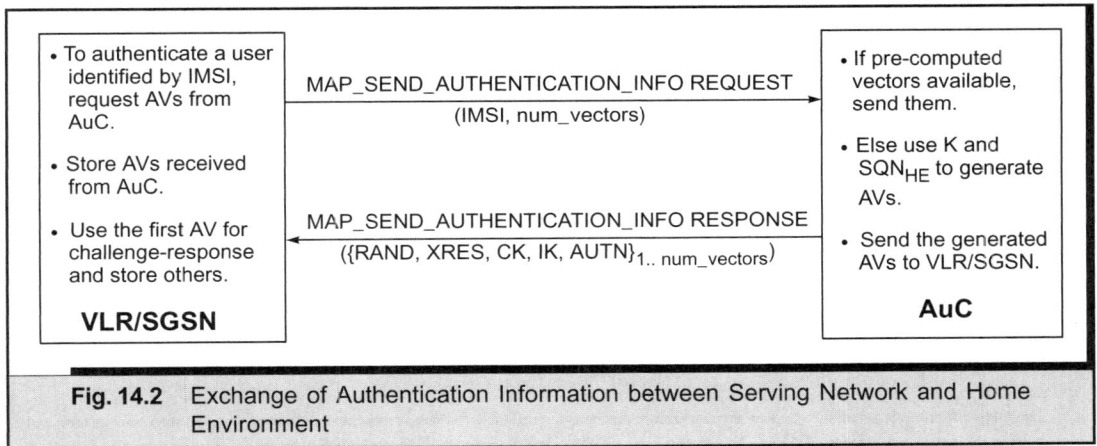

Fig. 14.2 Exchange of Authentication Information between Serving Network and Home Environment

14.3.1.3 *Generation of Authentication Vectors*

The AuC generates authentication vectors using the secret key K and sequence number SQN_{HE}. It maintains separate records, containing K and SQN_{HE}, for each subscriber identified by the IMSI. The authentication vectors are generated using an authentication algorithm, which is specific to a Home Network. The authentication algorithm is not standardized as per as 3GPP specification. However, 3GPP does provide a sample algorithm MILENAGE that can be used by operators who do not wish to develop their own algorithm.

Even though the algorithm for generation of authentication vectors is not standardized, the framework for vector generation is standardized. This framework defines seven functions, namely f1, f1*, f2, f3, f4, f5, and f5*. How these functions are implemented is an operator issue. However, their input and output are well defined. Moreover, the USIM and Home Network use the same algorithm to implement UMTS AKA.

Figure 14.3 shows the framework for generation of authentication vectors. The inputs to the process are SQN_{HE}, K and AMF. While the use of first two fields is already explained, the third—Authentication and Key Management Field (AMF)—is a field

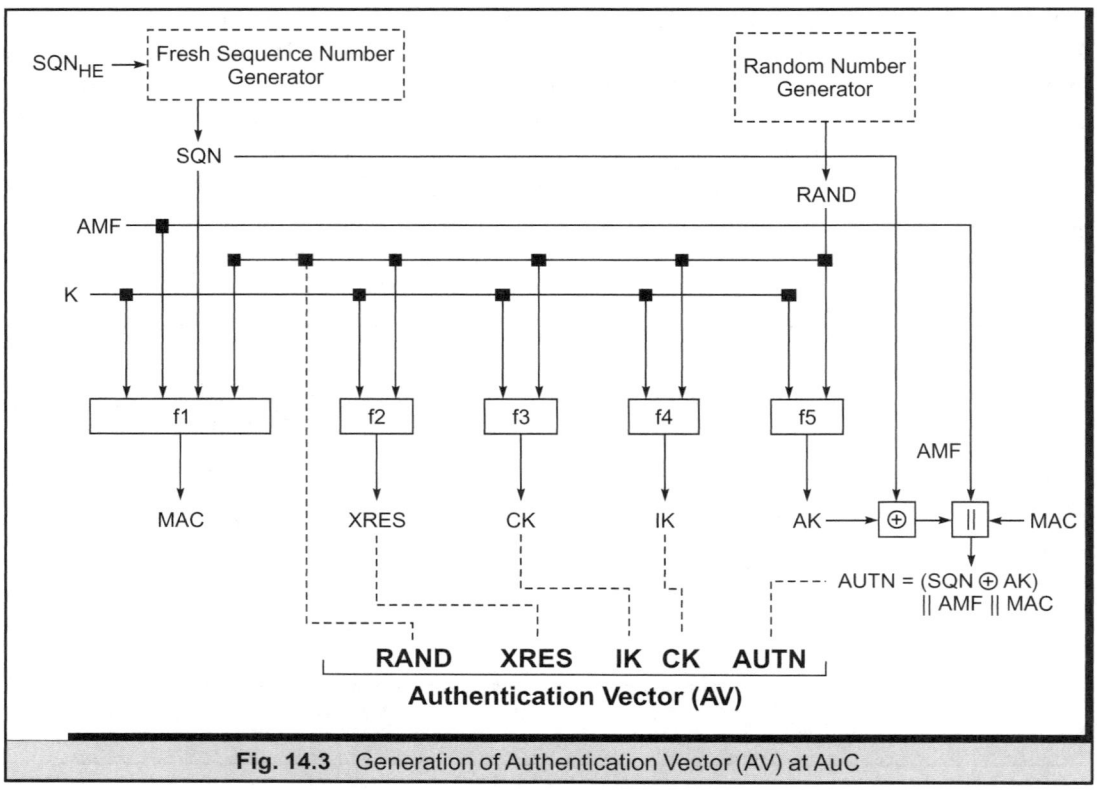

Fig. 14.3   Generation of Authentication Vector (AV) at AuC

that is generated by AuC and is sent to USIM as part of the Authentication Token (AUTN). The use of AMF is not standardized. However, 3GPP TS 33.102 Annexure F provides examples of how this field can be used, one of which is to play a part in the verification of SQN_{HE}. For details of this field, and its possible uses, the reader is referred to Annexure F of 3GPP TS 33.102.

The authentication vectors are generated using the three inputs, (i.e SQN_{HE}, K and AMF) five functions (f1, f2, f3, f4 and f5) and a random number generator. The SQN_{HE} is used to generate a fresh sequence number. For each vector (i.e. quintet), a new sequence number SQN is used. The random number generator gives RAND, which along with the secret key K, forms an input to all the five functions. The function f1 uses RAND, K, SQN, and AMF as input to produce Message Authentication Code (MAC). USIM validates the MAC to authenticate the network. The function f2 uses RAND and K to produce XRES. A similar computation is done at USIM to give RES. The RES and XRES are compared to authenticate the user. Functions f3 and f4 provide CK and IK respectively. Function f5 produces Anonymity Key (AK), which is used to conceal the SQN. This is required in case SQN reveals information that must be kept confidential (user identity IMSI for example). The AK is then XORed with SQN. The AUTN is formed by the concatenation of (AK ⊕ SQN), AMF and MAC. The authentication vector is the quintet {RAND, XRES, CK, IK and AUTN}.

14.3.1.4 Challenge-Response Protocol using Authentication Vectors

The previous two sub-sections explained how AuC generates authentication vectors and how VLR/SGSN obtains these vectors. This sub-section explains the mutual authentication procedures that follow. The procedure is also known as challenge-response protocol (Figure 14.4).

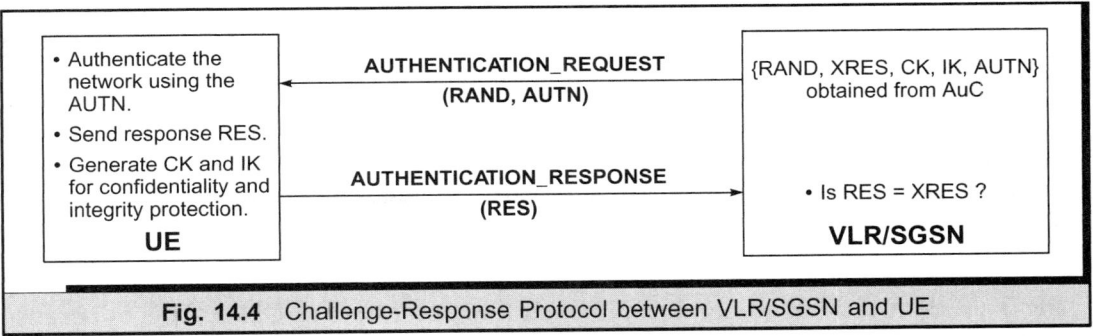

Fig. 14.4 Challenge-Response Protocol between VLR/SGSN and UE

After receiving the authentication vectors, VLR/SGSN challenges USIM to establish its authenticity. For this, it sends the RAND and AUTN from the authentication vector to USIM. The remaining parameters of the vector (i.e. XRES, CK, and IK) are not sent to USIM.

Upon receiving the challenge request from VLR/SGSN, the USIM carries out the following operations. It first uses the AUTN to verify whether the serving network challenging it is authentic or not. If it is authentic, the USIM responds with RES field. The USIM also calculates the CK and IK for future use.

Upon receiving the response, VLR/SGSN verifies whether the received RES matches the XRES sent by AuC. If the two match, the user is authenticated. The CK and IK are then ready for further use.

14.3.1.5 Detailed Procedure at USIM

The previous section briefly describes the authentication procedures at USIM. This section elaborates these procedures, as depicted in Figure 14.5. If this figure is compared with Figure 14.3 that shows the generation of authentication vectors at AuC, certain differences can be noted. First, at AuC, the random number is generated

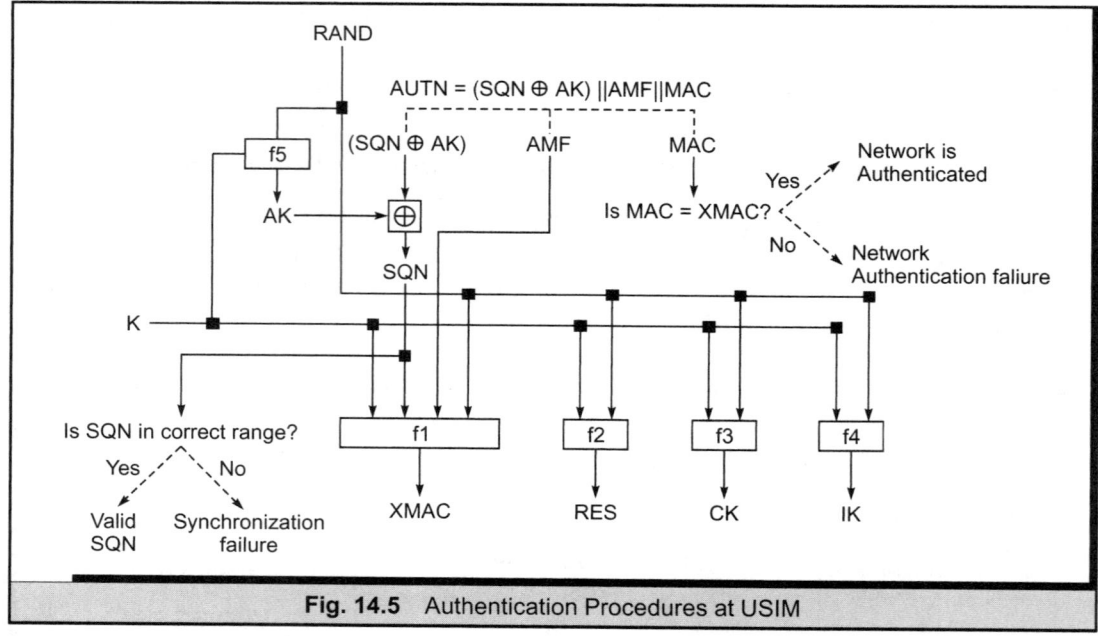

Fig. 14.5 Authentication Procedures at USIM

using a random number generator. No such function exists at USIM. This is because USIM receives the RAND from AuC (it does not require to generate it). This is done because the RAND forms the input to all security functions, and it is necessary that both AuC and USIM use the same value of RAND. Second, the SQN too is received from AuC; this is not generated at USIM. USIM checks whether the SQN received is fresh or not. This check is done to prevent reply attacks, using the SQN_{MS} stored in the USIM. 3GPP TS 33.102 provides guidelines to check freshness of SQN numbers. If the sequence numbers are not fresh, the USIM returns a synchronization failure message to VLR/SGSN. Handling synchronization failures is explained in the next sub-section.

Assuming that the SQN is fresh, USIM uses the function f5 to calculate the Anonymity Key (AK). The AK is XORed with the first six bytes of AUTN to give the SQN (It may be mathematically proven that $(A \oplus (B \oplus A)) = B$). The SQN then forms the input to function f1, along with the received RAND, the secret key K (maintained by USIM), and the AMF (which is the seventh and eighth byte of AUTN). This operation gives XMAC. The XMAC thus generated is compared with the received MAC (the last eight bytes of AUTN). If this match succeeds, the network is authenticated. The function f2 is then used to generate RES, which is sent to VLR/SGSN for user authentication. The functions f3 and f4 give CK and IK respectively. This completes the authentication procedures at USIM.

14.3.1.6 *Handling Synchronization Failure*

As mentioned, besides the secret key K, the entities AuC and USIM maintain another counter called the Sequence Number (SQN). Whenever the USIM receives the challenge message, it first checks the freshness of the SQN. This provides protection against the replay of stored authentication vectors. To provide this protection, SQN can be of three types. In the first case, it can be a simple counter that is incremented by one for each vector generated by AuC. In another case, the SQN can be derived from a global counter (e.g. clock giving universal time). There can also be a hybrid of these two schemes.

Whatever the technique used to generate SQN, the USIM maintains the highest SQN accepted so far. In case the received SQN and maintained SQN differ by more than a configurable value, the synchronization is said to be lost. To minimize chances of inadvertent synchronization failures, gaps between SQN are permitted.

In case synchronization failure occurs, the USIM calculates the AUTS using the function f1* and f5*, as depicted in Figure 14.6. AUTS is a parameter sent by USIM to VLR/SGSN, which in turns sends it to AuC, for resynchronization. The generation of AUTS requires the SQN stored by USIM, the secret key K, and the received RAND and AMF field. The AUTS has two components, the concealed SQN (by XORing SQN with AK), and a MAC-S field. When AuC receives the AUTS field, it first verifies the MAC-S field using the f1* function. The f1* is the same as that used by USIM. If the user is authenticated using MAC-S, the received SQN is inspected to determine whether synchronization failure has actually occured. In case it has, the received SQN is used to overwrite the SQN_{HE} stored at AuC.

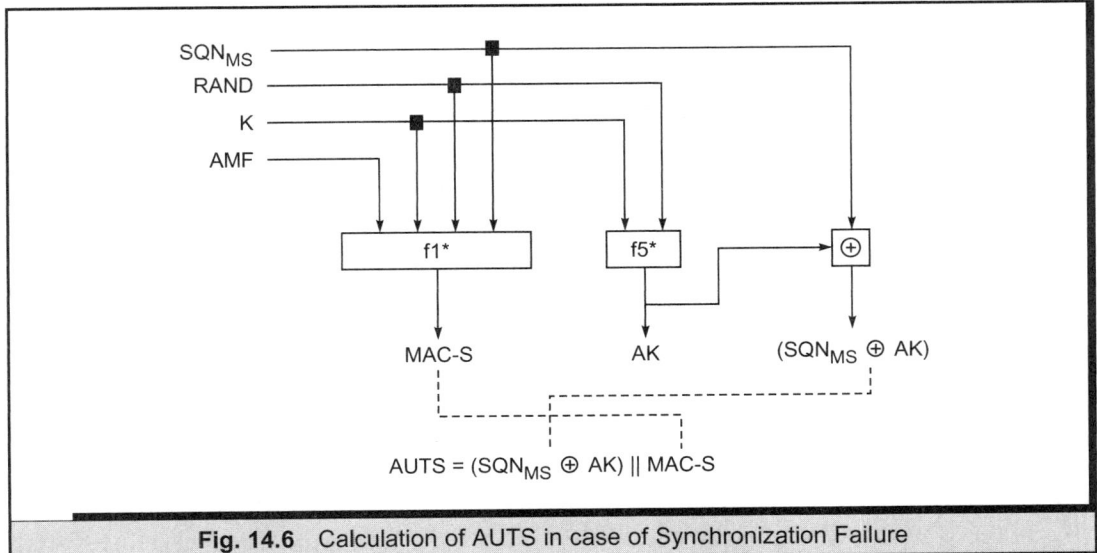

Fig. 14.6 Calculation of AUTS in case of Synchronization Failure

14.3.1.7 MILENAGE

The UMTS AKA procedure standardized as per 3GPP TS 33.102, and discussed in previous sections, defines a framework for mutual authentication. This UMTS AKA framework, besides mutual authentication, is also used to generate Cipher Key (CK) and Integrity Key (IK). This framework uses seven functions, namely f1, f1*, f2, f3, f4, f5, and f5*. While the use of these functions is defined by 3GPP TS 33.102, the standard does not specify what exactly these functions are or how they are implemented, the reason being that 3GPP has decided not to standardize these functions, thereby providing operators the freedom to develop their own authentication and key generation functions.

However, it is possible that not all operators may wish to use their own authentication and key generation functions. For such operators, it would make sense to have an example algorithm, which is generic enough for large-scale use. Keeping this objective in mind, a task force of security experts, called the Security Algorithms Group of Experts (SAGE) was formed. The goal of this task force was to come up with an example set of these algorithms, with the intent that it shall be offered to the UMTS operators, who can utilise it. Table 14.1 shows the functions developed by the ETSI SAGE task force.

Table 14.1 Cryptographic Functions Developed by ETSI SAGE Task Force

Name	Description
f0	The random challenge generating function (Note 1).
f1	The network authentication function.
f1*	The re-synchronisation message authentication function.
f2	The user authentication function.
f3	The Cipher Key (CK) derivation function.
f4	The Integrity Key (IK) derivation function.
f5	The Anonymity Key (AK) derivation function.
f5*	The Anonymity Key (AK) derivation function for the re-synchronisation message.

Note 1: It was decided by the task force not to propose function f0.

The example algorithm, called MILENAGE, is based on the block cipher Rijndael. Rijndael was also chosen as the candidate algorithm for Advanced Encryption Standard (AES). The AES standard replaces the old Data Encryption Standard (DES) and triple-DES standards.

The MILENAGE algorithm is defined in multiple specifications. These specifications are 3GPP TS 35.205 to 35.208. The 3GPP TS 35.205 specification provides general aspects of MILENAGE; 3GPP TS 35.206 specifies the actual MILENAGE algorithm including a C program of the actual algorithm. Figure 14.7 provides an extract from 3GPP TS 35.206, showing the variable definitions and function prototypes of MILENAGE algorithm. The 3GPP TS 35.207 specification provides Implementors' Test Data. This specification is a useful document to verify the MILENAGE output for operators implementing the MILENAGE algorithm. The 3GPP TS 35.208 provides design conformance Test Data. It supplements the test data provided by 3GPP TS 35.207. Apart from these four specifications, there is a technical report 3GPP TR 35.909 that provides MILENAGE summary and results of design and evaluation.

While details of MILENAGE are beyond the scope of this section, one important point needs a mention. As shown in Figure 14.7, MILENAGE provides a configurable field called the Operator Variant Algorithm Configuration Field. This 128-bit field provides the means to distinguish between the functionality of the algorithms when used by different operators. Each operator has the freedom to choose a value of OP. Even if the operator uses a publicly known OP, the MILENAGE algorithm is secure. In the C program provided in TS 35.206, the OP is used as a global variable and hence is not passed as an argument to the seven functions.

```
typedef  unsigned  char  u8;

/*--------- Operator  Variant  Algorithm  Configuration  Field  --------*/

           /*------- Insert  your  value  of  OP  here  -------*/
u8  OP[16]  = {0x63,  0xbf,  0xa5,  0x0e,  0xe6,  0x52,  0x33,  0x65,
               0xff,  0x14,  0xc1,  0xf4,  0x5f,  0x88,  0x73,  0x7d};
           /*------- Insert  your  value  of  OP  here  -------*/

/*------------------------- prototypes  -------------------------*/

void  f1  ( u8  k[16],  u8  rand[16],  u8  sqn[6],  u8  amf[2],  u8  mac_a[8]  );
void  f2345  ( u8  k[16],  u8  rand[16],  u8  res[8],  u8  ck[16],  u8  ik[16],  u8  ak[6]
);
void  f1star( u8  k[16],  u8  rand[16],  u8  sqn[6],  u8  amf[2],  u8  mac_s[8]  );
void  f5star( u8  k[16],  u8  rand[16],  u8  ak[6]  );
void  ComputeOPc( u8  op_c[16]  );
void  RijndaelKeySchedule( u8  key[16]  );
void  RijndaelEncrypt( u8  input[16],  u8  output[16]  );
```

Fig. 14.7 Variable Definitions and Function Prototypes for Milenage Algorithm

14.3.2 Data Confidentiality

The previous sub-sections explained how mutual authentication takes place and how CK and IK are generated. This sub-section explains how CK is used for encryption in UTRAN. The next sub-section explains the use of IK for integrity protection in the UTRAN.

After successful AKA, the VLR/SGSN determines which UMTS Encryption Algorithm (UEA) and which UMTS Integrity Algorithm (UIA) is to be used. In Section

14.3.3.1, the possible values of UEA and UIA are provided. Once this is decided, the UIAs, UEAs, IK and CK are sent to the serving RNC (SRNC) by VLR/SGSN using the RANAP command 'Security Mode Command'. The SRNC uses this security information for encryption and integrity protection.

Encryption is applied at the RLC or the MAC layer, using the f8 function. Figure 14.8 shows how the f8 function is used for encryption. The main input parameter to f8 function is the 128-bit CK. Apart from CK, there are four other parameters, namely COUNT-C, BEARER, DIRECTION, and LENGTH. The COUNT-C is a 32-bit field, composed of two logical parts: Connection Frame Number (CFN) and Hyper Frame Number (HFN). The CFN acts as a sequence number. The HFN is incremented whenever the CFN wraps around. The most significant bits of HFN are initialized using the START field. The START field has a very specific function. It is stored in the USIM and is sent to UE (by USIM) when the MS is powered-on. The MS then sends the field to SRNC during the initial connection setup. If the START field reaches a THRESHOLD value, the MS requests the SRNC to stop using the CK any further. The SRNC uses a new set of keys from then on.

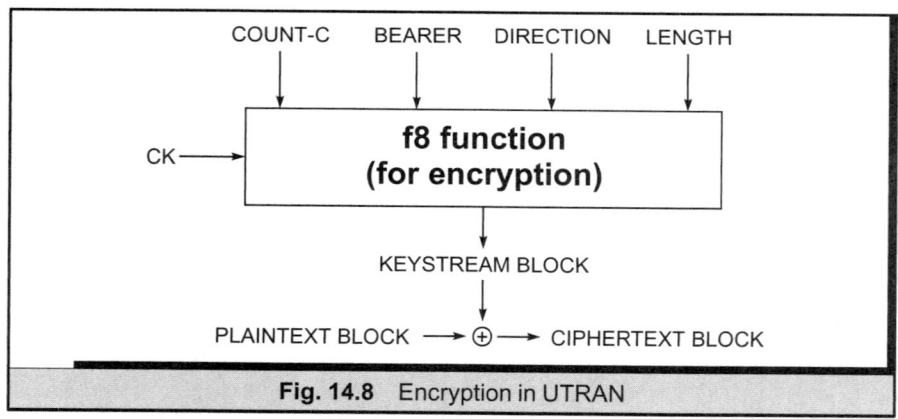

Fig. 14.8 Encryption in UTRAN

Apart from COUNT-C, the other fields used in encryption are as follows. The BEARER is a 5-bit field, used to identify the radio bearer. This is required to have different keystream blocks for different radio bearer channels. The DIRECTION is a 1-bit field, whose value is 0 for messages flowing from UE to RNC, and 1 for reverse direction. The LENGTH is a 16-bit field that determines the length of the required keystream block.

The five input parameters are used as input to the f8 function to give a keystream block. The keystream block is then XORed with the plaintext to give the ciphertext. At the receiving end, the plaintext is recovered by generating the same keystream and applying a bit by bit binary addition with the ciphertext.

14.3.3 Data Integrity

The basics of integrity protection in UTRAN are similar to the basics of encryption. The IK generated during AKA is used for integrity protection. This, along with the UMTS Integrity Algorithm (UIA), is sent to the serving RNC (SRNC) by VLR/SGSN using the RANAP command 'Security Mode Command'. Integrity protection is applied at the RRC layer.

Integrity protection is done using a Message Authentication Code (MAC). This MAC is generated for the input RRC message using the IK. Apart from IK, there are three other parameters, namely COUNT-1, DIRECTION and FRESH. The COUNT-I, like COUNT-C, is divided into two logical components: A 4-bit sequence number called the RRC SN, and a 28-bit long sequence number, called the RRC Hyper Frame Number (RRC HFN). The HFN is incremented whenever the sequence number wraps around. The most significant bits of HFN are initialized using the START field.

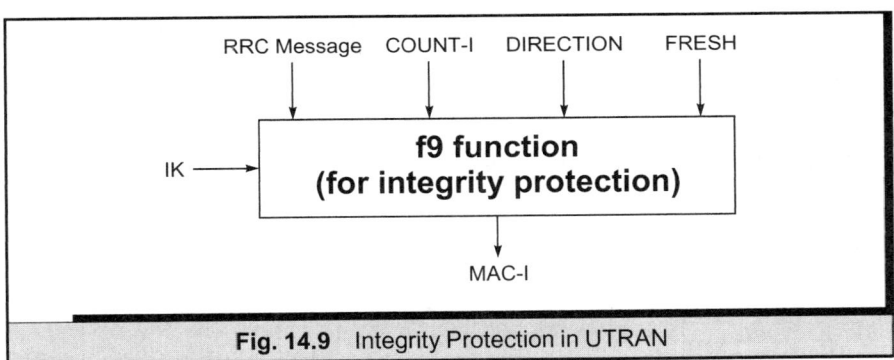

Fig. 14.9 Integrity Protection in UTRAN

Apart from COUNT-I, the DIRECTION field specifies the direction of the message. Its value is the same as that defined for encryption in the previous sub-section. The FRESH field is used for replay protection (i.e. it prevents a malicious user replaying stored signaling messages). At the connection set up, the SRNC generates a random FRESH value and sends it to the user. The value of FRESH stays constant through the entire duration of a connection. The input parameters, when applied to the RRC message, give a 32-bit MAC-I. This MAC-I is appended to the message when sent over the radio link. The receiver ensures that the received MAC-I is the same as the expected MAC (XMAC-I).

14.3.3.1 f8 and f9 Specification (based on Kasumi)

After security mode setup, the functions f8 and f9 are used for encryption and integrity protection respectively. Unlike the seven AKA functions (namely f1, f1*, f2, f3, f4, f5, and f5*), the f8 and f9 functions are well defined and standardized. For encryption, two possibilities exist: UMTS Encryption Algorithm 0 (UEA0) and UMTS Encryption Algorithm 1(UEA1). UEA0 or no encryption implies that the encrypted text is same as the cleartext. UEA1 is a KASUMI based f8 function.

For integrity protection through the f9 function, only one algorithm is defined, which is the UMTS Integrity Algorithm 1 (UIA1). The UIA1, like UEA1 is based on Kasumi.

The Kasumi based f8 and f9 functions are defined in multiple specifications. These specifications are 3GPP TS 35.201 to 35.204. The specification 3GPP TS 35.201 defines the normative part for f8 and f9 algorithm. It also includes an informative C program for the actual implementation of f8 and f9 functions. Both the f8 and f9 functions use Kasumi algorithm as their core part. The Kasumi algorithm is defined in 3GPP TS 35.202. This specification also includes an informative section of a C program of the Kasumi implementation. The 3GPP TS 35.203 provides the implementors' test data, which is a useful document to verify the f8 and f9 output for operators implementing these functions. The 3GPP TS 35.204 specification provides design conformance test data. It supplements the test data provided by 3GPP TS 35.203. For details of Kasumi based f8 and f9 functions, the reader is referred to 3GPP TS 35.201 to 35.204.

14.3.4 User Identity Confidentiality

To maintain the confidentiality of a user's permanent identity IMSI, a temporary identity, called the Temporary Mobile Subscriber Identity (TMSI), is used on the radio access link. The TMSI has local significance only in the location area or routing area in which the user is registered. The association between the permanent identity IMSI and temporary identity TMSI is kept by the VLR/SGSN in which the user is registered. In PS domain, the TMSI is referred to as P-TMSI (or packet-TMSI). In this chapter, for the sake of simplicity, both TMSI and P-TMSI will be referred to as TMSI).

A new TMSI is allocated by the VLR/SGSN. To do this, the new TMSI, along with the Location Area Identifier (LAI) or Routing Area Identifier (RAI), is sent to the user. Upon receiving such a request, the user clears the previous mapping and acknowledges the receipt of the message. The VLR/SGSN, upon receiving the acknowledgement message, deletes the old mapping. The new TMSI is now ready for use.

In case the user does not acknowledge the message, the network maintains the association between the new temporary identity TMSIn ('n' stands for new) and the IMSI, as well as the mapping between the old temporary identity TMSIo (if there is any) and the IMSI. From here on, depending on whether it is the user or the network that initiates a transaction, two possibilities arise. In case the user initiates a transaction, it is allowed the liberty to identify itself using its old TMSI (TMSIo) or new TMSI (TMSIn). Depending on what is received, the network keeps one of the mappings, and deletes the other. The freed TMSI is available for allocation to other users. In case the network initiates a transaction, it uses the permanent identity IMSI, instead of temporary identity TMSI (This is because the network does not know which temporary identity, TMSIo or TMSIn, to use). When the network-initiated transaction is successful and radio contact is established, the user is instructed to delete any stored TMSI so that a fresh one can be allocated.

In case the user roams, as a result of which its VLR/SGSN changes, it is required that the use of old mapping between TMSI/IMSI as well as of the old CK/IK is continued. To maintain this continuity, the new VLR/SGSN, upon receipt of the location-update/ routing-update message from the user, sends a request to the old VLR/SGSN asking it to return the IMSI and the unused vectors. The request message contains TMSIo and LAIo/RAIo that the user had sent with the location-update/routing-update message. If the old VLR/SGSN has the information of the user in its database, it responds with the IMSI. Optionally, the old VLR/SGSN may include unused authentication vectors and other security data like CK, IK and KSI (While CK and IK are used for confidentiality and integrity protection, the Key Set Identifier (KSI) is used to identify a particular authentication vector). In case the old VLR/SGSN cannot identify the user, it responds stating that the user identity cannot be retrieved.

The new VLR/SGSN takes further action based on the action of the old VLR/SGSN. If the response is positive, the new VLR/SGSN creates an entry for the user and stores any authentication-vectors/security-data that was sent by the old VLR/SGSN. If the response is negative, the VLR/SGSN sends a User Identity Request to the user, requesting the latter to identify itself using its permanent identity IMSI. The user responds with its permanent identity IMSI. Note that in this case, the user's IMSI is transmitted in cleartext. This represents a violation of user identity confidentiality.

14.3.5 Access Security Flow Diagram

This section summarizes the message flows for network access security based on the UMTS Authentication and Key Agreement (UMTS AKA) protocol. The flow diagram is depicted in Figure 14.10. The steps involved are as follows:

1. The procedure starts when the VLR/SGSN has to authenticate a UE, and it does not have authentication vectors to do so. So it requests the AuC to provide authentication vectors for a given user (identified through IMSI).

2. The AuC returns authentication vectors to VLR/SGSN. The vectors are either pre-computed or are dynamically generated. The message exchange between VLR/SGSN and AuC takes place using the MAP protocol.

3. After obtaining the authentication vectors, the VLR/SGSN sends RAND and AUTN from the first authentication vector to USIM. The remaining parameters of the vector (i.e. XRES, CK, and IK) are not sent to USIM.

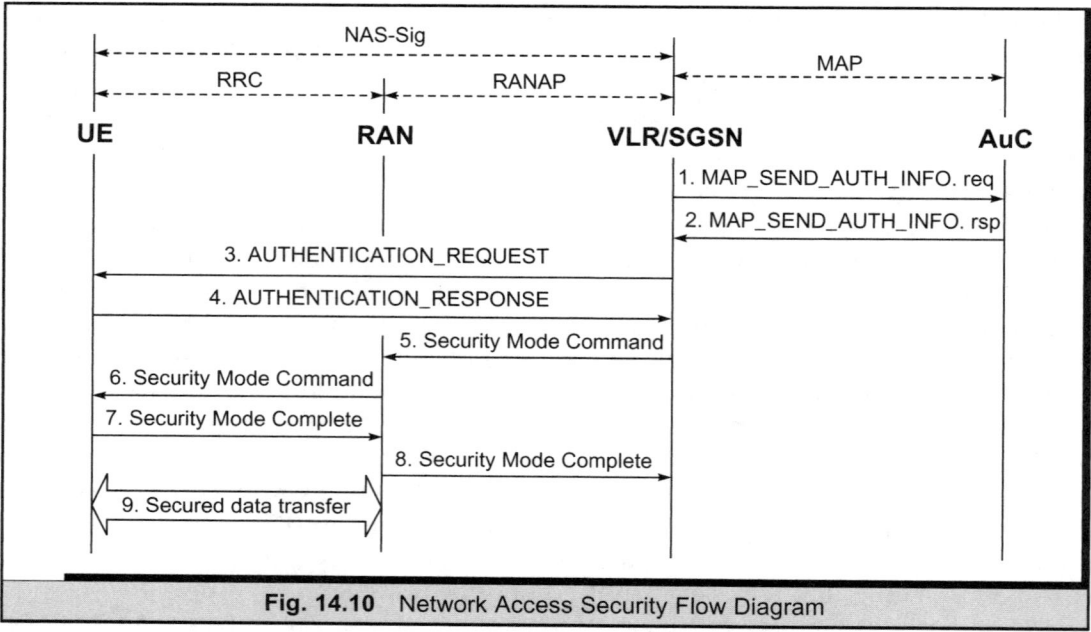

Fig. 14.10 Network Access Security Flow Diagram

4. USIM uses the AUTN to verify whether the serving network challenging it is authentic or not. If it is authenticated, then the USIM responds with RES field. The USIM also calculates the CK and IK for future use. The message exchange takes place using NAS-Signaling messages as listed in Chapter 6.

5. When mutual authentication is over, the VLR/SGSN determines the UMTS Encryption Algorithm (UEA) and UMTS Integrity Algorithm (UIA) that are to be used. The VLR/SGSN then initiates integrity and ciphering by sending the RANAP message 'Security Mode Command' to SRNC. The message contains the UEAs, CK, UIAs and IK, and a few other parameters.

6. SRNC then determines the UEA and UIA to be used. Thereafter, it sends the RRC message 'Security Mode Command' to UE. The message contains the selected UEA and UIA, the random value FRESH, a MAC and other parameters.

7. The UE verifies the MAC. If this check, and a few other checks are successful, it sends the RRC message 'Security Mode Complete' to SRNC. The UE also includes a MAC in the message.

8. The SRNC verifies the received MAC and if the verification is successful, sends RANAP message 'Security Mode Complete' to VLR/SGSN. This completes the setting of security mode between the UE and RNC.

9. Subsequently, data is ciphered/integrity protected as per the parameters of the security mode. The ciphering function is applied at the RLC layer (both for

Acknowledged and Unacknowledged mode) or at the MAC layer (in case the Transparent Mode of RLC is used). The data integrity function is applied at the RRC layer.

14.4 NETWORK DOMAIN SECURITY USING MAPsec

In the 2G networks, the absence of security features in SS7 was viewed as an important security weakness. However, this weakness was justified is view of the fact that the SS7 networks were controlled by a small number of large companies. Thus, security threats arising out of the interoperability between networks were not seen as a prime threat. With the large-scale deployment of UMTS networks, and burgeoning demand for interoperability and global roaming, the security requirements have drastically altered. Thus, in addition to the *access security* mechanism (using AKA), Release 4 onwards specification of 3GPP mandates the use of *network domain security* mechanisms for *control plane signaling* within and between the core networks.

At the MAP application level, the security protocol used to protect MAP messages is referred to as the MAPsec protocol. Formally, '*the complete set of enhancements and extensions to facilitate security protection for the MAP protocol is termed MAPsec and it covers transport security in the MAP protocol itself and the security management procedures*'.

The stage 2 specification for the MAPsec protocol is described in 3GPP TS 33.200, while the stage 3 specification is described in 3GPP TS 29.002. As per the former, MAPsec provides four security services. First, it provides *data integrity*. The integrity of MAP message is protected by using a 32-bit Message Authentication Code (MAC). The MAC is also used for *origin authentication* MAPsec also provides *anti-replay protection*. This is done by including a 32-bit time-stamp in the MAP message header. The presence of a time-stamp ensures that the MAP messages cannot be stored and replayed (if the MAP messages are actually stored and replayed, the presence of a stale time-stamp indicates a possible violation). Lastly, MAPsec optionally provides *confidentiality*. Confidentiality is achieved by encrypting the original MAP message.

To provide these four security services, MAP peers establish a Security Association (SA) between them. The SA contains the parameters (i.e. security keys) necessary for the operations of MAPsec. The following sections provide details of the MAPsec protocol.

14.4.1 Protection Modes and Message Formats

MAPsec messages use a special security header and a protected payload to provide security services. The format of MAPsec message is shown in Figure 14.11. The security header contains five fields: The Security Parameters Index (SPI) is an arbitrary 32-bit value used in combination with the Destination PLMN-Id to uniquely identify a

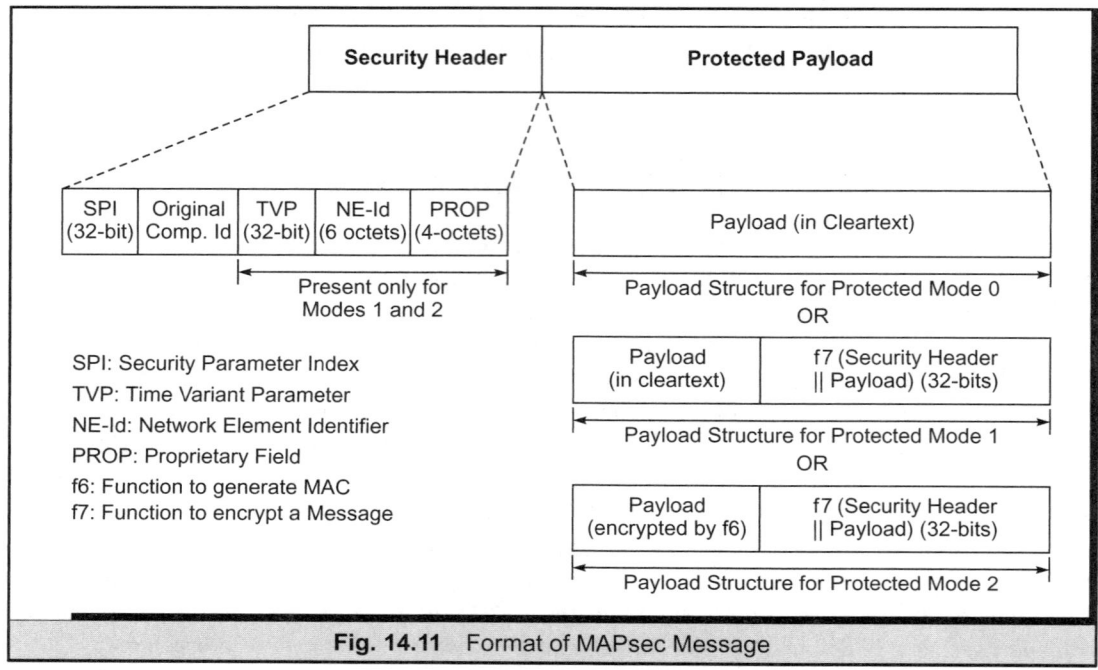

Fig. 14.11 Format of MAPsec Message

Security Association (SA). SA is explained in greater detail in the next section. The SPI is followed by Original Component Id that identifies the type of component (invoke, result or error) within the MAP operation that is being securely transported. This is followed by the Time Variant Parameter (TVP) field. The TVP is used for anti-replay protection. The TVP is essentially a time-stamp that renders a message useless if stored and replayed after a configurable interval. Then comes the Network Entity Identifier (NE-Id), a 6-octet field. The last field is the 4-octet Proprietary field (PROP), used to create different Initialization Vectors (IV) values for different protected MAP messages within the same TVP period for one NE. The use of the proprietary field is not standardized.

The security header is followed by a protected payload, whose structure depends upon the protection mode. MAPsec provides three protection modes, namely Protection Mode 0, 1 and 2. Protection Mode 0 offers no protection at all. Thus, the payload in this protection mode is identical to the original MAP message payload in cleartext (see Figure 14.11). For this mode, the security header only contains the SPI and the Original Component Id.

Protection Mode 1 provides integrity and authenticity. To provide this, the protected payload contains the payload (in cleartext) and a 32-bit MAC. The MAC is computed using function f7 that is applied to the concatenation of security header and the payload (in cleartext). This 32-bit MAC is computed using an integrity key agreed upon by the communicating NEs. The MAC acts as a signature of the message. If the payload changes in transit, the expected MAC and the received MAC differ, thereby indicating an integrity violation. The authenticity of the message is assured by the fact that only the sender of the message has the integrity key to compute the correct MAC. The algorithms used to compute MAC are based on Advanced Encryption Standard (AES), the details of which are available in 3GPP TS 33.200.

Protection Mode 2 provides integrity, authenticity and confidentiality. To provide this, the protected payload contains the encrypted payload and a 32-bit MAC. The payload is encrypted using function f6. This function uses a confidentiality key agreed upon by the communicating NEs. As the message is not visible in cleartext, its confidentiality is maintained. The algorithm used to encrypt the payload is based on AES. The MAC is computed using function f7 that is applied to the concatenation of security header and the encrypted payload. The handling of MAC for integrity protection in this mode is the same as explained in case of Protection Mode 1.

In short, Protection Mode 0 provides no protection, Protection Mode 1 provides integrity and authenticity, and Protection Mode 2 provides confidentiality, integrity, and authenticity. Each mode uses a different message format.

14.4.2 Components of MAPsec Protocol

To implement MAPsec, a Network Entity (NE) maintains two different databases, namely Security Policy Database (SPD) and Security Association Database (SAD). The SPD contains general rules that govern how MAPsec is used for communication between two NEs of different PLMNs. The policy rules given in SPD are not local to an individual NE. Instead, SPD entries of different NE within the same PLMN are identical.

Figure 14.12 shows the attributes contained in the SPD. As depicted, the SPD has three main entries: The 'fallback to unprotected mode' is a Boolean variable that determines how unprotected messages are handled, and whether an NE can fall back to unprotected mode in case it receives an error from peer entity. The specifications recommend that MAPsec security features are useful for a particular PLMN if it disallows fall back to unprotected mode for MAP messages received from any other PLMN. Providing the option to fall back to unprotected mode makes a PLMN vulnerable to external security attacks. The 'table of MAPsec operation components' defines whether or not MAPsec has to be applied to an incoming message. The last attribute of the SPD is 'Explicit policy configuration'. This is a table, which contains an entry for each PLMN the network entity is allowed to communicate with.

Apart from SPD, another database, called the Security Association Database (SAD), is maintained by an NE. The SAD contains the Security Association (SA) information.

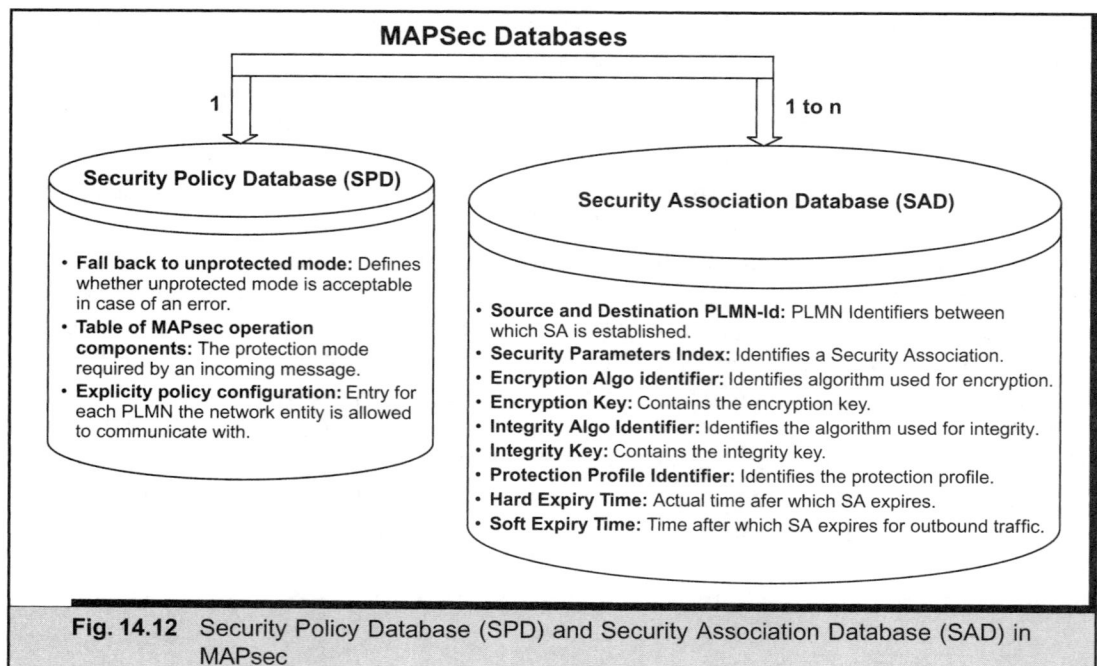

Fig. 14.12   Security Policy Database (SPD) and Security Association Database (SAD) in MAPsec

All communication at MAPsec level is tied to a SA, which means that the encoding and decoding of the MAPsec message is based on the parameters defined by the SA. All the SA applicable to an NE is contained in the SAD. The difference between SAD and SPD is that while the latter defines high-level policies, SAD defines the actual parameters applicable to a communication. Thus, the SPD contains one set of attributes, while there are many records in the SAD.

An SA is established between a sending-PLMN and destination-PLMN. A PLMN is identified by PLMN-Id, which is a concatenation of the Mobile Country Code (MCC) and Mobile Network Code (MNC). Thus, the first two parameters of an SA are the sending-PLMN-Identifier and destination-PLMN-Identifier.

Since it is possible that two PLMNs may have multiple associations between them, there must therefore be a mechanism to distinguish between different associations. For this, a 32-bit identifier, called the Security Parameters Index (SPI) is used in combination with the destination-PLMN-Identifier to identify the SA.

Apart from the identifiers, the SA contains the core information, which includes the encryption/authentication algorithm identifiers, and the associated keys. Encryption is optional; an SA may or may not have the encryption algorithm identifier and encryption key. In case encryption is used, the encryption identifier and the encryption key is used by the sender to encrypt the MAP payload. The recipient uses the SPI of the security header to identify the SA for which the MAP message is received. The SPI is used as an index to the identified SA to obtain the encryption identifier and the encryption key. This key is then used to decrypt the MAP payload. Similarly, the authentication algorithm identifier and the authentication keys are used to provide data integrity and origin authentication. The operations of MAPsec are detailed in the next sub-section.

The next element of an SA is the 16-bit Protection Profile Identifier (PPI). In simple terms, the PPI defines the level of protection applicable to a message. For example, a certain value of PPI may define that for 'Send Authentication Info' MAP message, the request message will have Protection Mode 1; the answer message (which contains the authentication vectors) will have Protection Mode 2 (implying greater protection with encrypted payload); and an error response will have Protection Mode 0 (implying lower level of protection). It may be noted that the SPD element 'table of MAPsec operation components' only defines whether MAPsec must or must not be applied to an incoming message. The actual protection mode applied to a message is governed by the PPI.

Every SA is valid for a definite period of time. After this, the SA expires and cannot be used any further. There are two types of expiry times. The first is the *hard expiry time*, which is the actual time after which an SA expires. There is another expiry time, called the *soft expiry time*. This variable provides an expiry period for outbound traffic. In case the soft expiry time of a SA is crossed, then this SA should only be used when there is no other valid SA for the same PLMN. However, no circumstance can a SA be used after the hard expiry time is crossed.

14.4.3 Operations of MAPsec

In the previous sections, various aspects of MAPsec were explained. The discussion included the different security services provided by MAPsec, the protection modes, and the SAD/SPD databases. This section describes how these aspects apply to MAPsec for its normal operations.

Assume that a network element of PLMN-A, say NE-A, has to send MAP message to another network element of PLMN-B, say NE-B, and that NE-A knows the PLMN-Id of PLMN-B. To achieve this, the operations of MAPsec at NE-A and NE-B can be summarized in the following steps:

1. NE-A inspects the SPD to check whether NE-A is allowed to communicate with an entity of PLMN-B (using PLMN-Id of PLMN-B). If this is not allowed, an error is returned.

2. NE-A then looks for a valid SA in the SAD that can be used for PLMN-B. If no SA exists, or if the SA has expired, an error is returned. If only one valid SA exists, it is chosen. In case more than one valid SA are available, the one that expires first is chosen.

3. The chosen SA provides the protection mode to be applied to the message. The mode may be 0, 1 or 2. The SA also provides the encryption/authentication identifiers, and the associated keys.

4. Based on the protection mode, and the contents of the SA, a MAPsec message is created and sent to NE-B. In case the protection mode is 0, an unprotected MAP message can also be sent. This differs from a MAPsec message with protection mode 0, as the latter has a header (containing SPI and original component identifier), which is not present in an unprotected MAP message.

5. At the receiving end, NE-B determines whether an unprotected MAP message or a MAPsec message is received.

6. In case an unprotected MAP message is received, and the SPD does not mandate the use of MAPsec, or if fallback to unprotected mode is allowed, the message is normally processed. However, if both the conditions are violated (i.e. SPD mandates use of MAPsec and fallback to unprotected mode is not allowed), the message is discarded. If required (e.g. when the MAP dialogue is still open), an error is returned to NE-A.

7. In case a protected MAP message is received, NE-B decomposes it into a header and payload. The SPI and original component Id is extracted from the security header

8. If the received SPI is not valid, or MAPsec is not to be used for this message, the message is discarded. If required, an error is returned to NE-A.

9. Else, NE-B uses the SPI to extract the information from SA. The SA specifies the protection mode applied by the sender of the message (i.e. 0, 1 or 2). In case the protection mode 0 was applied, the MAP message is simply processed. (Note that for protection mode 0, only the SPI and original component identifier is received. The other three parameters of the security header are not present).

10. For protection mode 1 and 2, the freshness of the message is tested using the Time Variant Parameter (TVP). In case the message is not fresh, it is discarded. If required, an error is returned to NE-A.

11. After the freshness of message is ensured, based on the protection mode applied, the authentication and integrity checks are carried out. In case of success, the cleartext payload is available for further processing. In case of failure of any of the checks, the message is discarded. If required, an error is returned to NE-A.

14.4.4   Key Distribution in MAPsec

Secured MAP (MAPsec) was first introduced in Rel4 specifications. To stagger the standardization of MAPsec protocol, it was assumed in Rel4 that the network entities established Security Association (SA) manually through operator intervention. No standardized mechanism was specified in Rel4 standards for distribution of SA information. In Rel5, it was proposed that this shortcoming be removed by introduction of a new MAPsec architecture. This new MAPsec architecture introduced a new network node and three well-defined interfaces, namely Zd, Ze and Zf.

Figure 14.13 shows the security architecture for SA distribution in MAPsec. As depicted, a new node called the Key Administration Center (KAC) is introduced Its role is to negotiate SA with KAC of other PLMNs and distribute the SA to all the network elements of the PLMN. Two KACs establish SA using the Internet Key Exchange (IKE) protocol. The IKE is explained later in this chapter in relation to IPSec protocol.

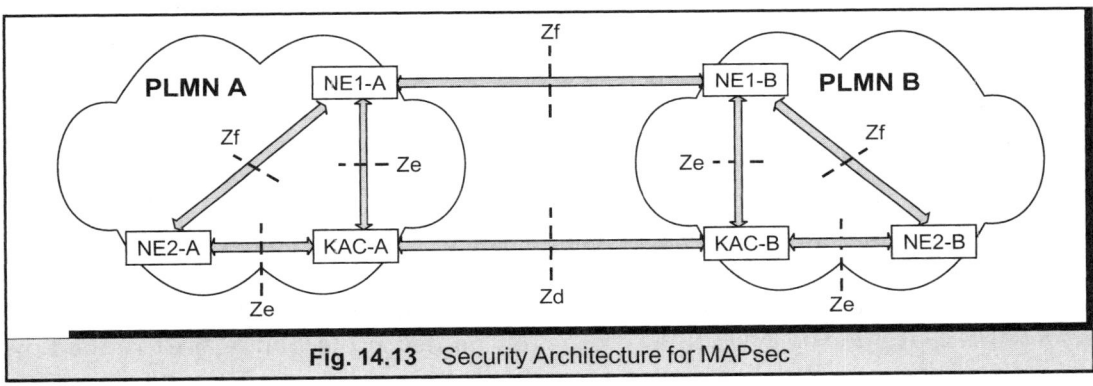

Fig. 14.13   Security Architecture for MAPsec

Figure 14.13 also shows the three interfaces defined by the MAPsec architecture. The Zd interface exists between two KACs. This interface is used to exchange IKE information and to establish SAs between PLMNs. Once the SAs are established, they are distributed by KAC to the network elements over the Ze interface. Once the network elements have obtained the SAs, they communicate with other MAPsec enabled entities using the MAPsec protocol over the Zf interface.

Unfortunately, there was disagreement on technical and non-technical issues related to the standardization of Ze interface. Thus, the Ze interface specifications could not be completed in time. As a result, the automatic key management for MAPsec was moved to Rel6 standardization. *Thus, MAPsec architecture depicted in Figure 14.13 was removed from Rel5 MAPsec specifications.* This architecture is supposed to be part of Rel6 specifications.

14.5   NETWORK DOMAIN SECURITY USING IP SECURITY

In the Core Network, the CS and PS domain use MAP for exchange of signaling information. From Release 4 onwards, security provisions have been added in MAP.

This security enabled MAP, called MAPsec, has been discussed in the previous section. In Release 5, a new domain, called the IP Multimedia Subsystem (IMS), has been introduced. The IMS domain uses the IP protocol at the network layer, the SCTP protocol at the transport layer, and SIP or Diameter at the application layer. Among these protocols, SCTP, Diameter and SIP protocols do not provide any security features. Thus, the onus of secured exchange is with the IP layer using the IP Security (IPSec) protocol. The IPSec, as defined in RFC 2401, provides the security architecture for IP-based transport. The applicability of IPSec for IMS domain and for other domains using IP-based transport (e.g. Sigtran for CS/PS domain) is defined in 3GPP TS 33.210 'Network Domain Security: IP network layer security (Release 5)'. This specification is also referred to as NDS/IP (meaning Network Domain Security for IP-based protocols).

Network Domain Security for IP-based protocols (NDS/IP), as defined in 3GPP TS 33.210, is not a new protocol specification. Rather, it merely refers to the IPSec protocol suite specified in RFC 2401 to RFC 2412 and defines how this protocol suite is applicable to UMTS. For example, while IPSec defines use to two security protocols, namely Authentication Header (AH) and Encapsulating Security Payload (ESP), 3GPP TS 33.210 specifies that only ESP is adequate to satisfy the requirements of NDS/IP.

The security features provided by NDS/IP are similar to those of MAPsec. The features include *data integrity, data origin authentication, anti-replay protection* and *confidentiality* (optional). NDS/IP also provides limited protection against traffic flow analysis when confidentiality is applied.

14.5.1 Architecture for NDS/IP

The architecture for NDS/IP uses two important concepts or notions. The first concept is that of a *security domain*. A security domain is defined as *the collection of networks that are managed by a single administrative authority*. Within a security domain the same level of security and usage of security services is applied. Thus, a network operated by a single operator typically constitutes one security domain although an operator may subsection its network into separate sub-networks if the need arises (The concept of security domain is somewhat analogous to the concept of Autonomous Systems (AS) used in relation to IP routing. In other words, the relation between security domain and security policies is similar to that between autonomous systems and routing policies).

The second concept is that of Security Gateway (SEG). The SEG is responsible *for enforcing the security policy of a security domain towards other SEGs in the destination security domain*. A security domain may have more than one SEG to avoid a single point of failure or to improve performance. In such a case, the SEG may provide connectivity to all reachable security domain destinations or to only a subset of reachable destinations.

Figure 14.14 shows the architecture for NDS/IP. In the figure, two security domains communicate with each other using SEGs. If this architecture is compared to the architecture of MAPsec depicted in Figure 14.13, it becomes clear that direct communication between two network elements (NEs) belonging to different security domains is not possible. In other words, all communication takes place using SEGs. This is different from MAPsec where two NEs of different PLMNs can communicate directly without an intervening gateway. The difference arises because MAP resides at the application level whereas IP is a hop-by-hop routing protocol. It is easier to use a gateway at the border of a security domain.

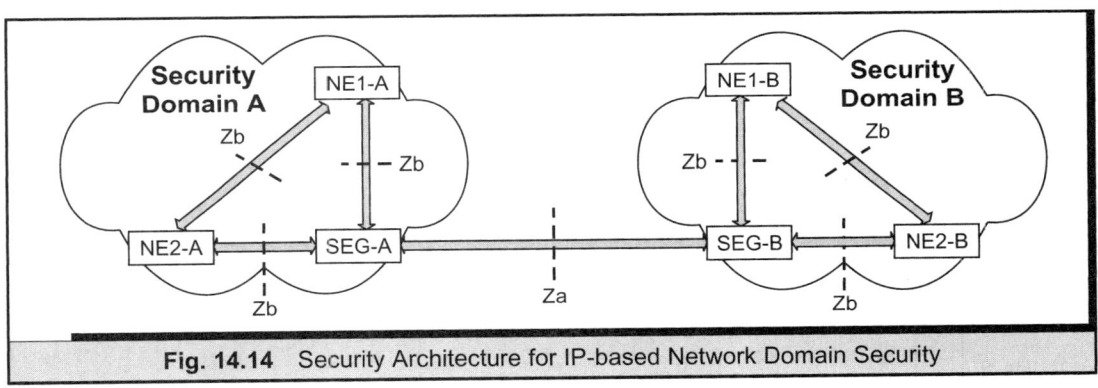

Fig. 14.14 Security Architecture for IP-based Network Domain Security

As shown in Figure 14.14, the interface between two SEGs (belonging to different security domains) is called the Za interface. The Za interface uses the Internet Key Exchange (IKE) protocol to negotiate, establish and maintain a secure tunnel between them. After the tunnel is established, the Encapsulating Security Payload (ESP) protocol is used to provide encryption and authentication. The networks complying with 3GPP TS 33.210 must provide the Za interface.

The interface between NE and a SEG or between two NEs is called the Zb-interface. The implementation of this interface is optional. However, if this interface is supported, it uses ESP and IKE. The following sub-sections elaborate upon ESP and IKE protocols.

14.5.2 Encapsulating Security Payload (ESP)

In the IPSec suite of protocols, there are two security protocols, namely Authentication Header (AH) and Encapsulating Security Payload (ESP). Among the two, the 3GPP TS 33.210 specification chooses the ESP protocol. The ESP protocol is defined in RFC 2406. The features provided by ESP include confidentiality, origin authentication, integrity protection, anti-replay protection and limited protection from analysis of traffic flow.

Before the ESP operates, a Security Association (SA) must first be established between the communicating entities. This may be through IKE or through other means. Whatever be the means, assuming that an SA exists, the ESP provides security services by adding a header and a trailer to the payload. The ESP header and trailer are shown in Figure 14.15. The header contains a Security Parameter Index (SPI), which in combination with the destination IP address uniquely identifies a SA. The header also contains a 32-bit sequence number, which is used to protect against replay of stored messages. At the start of communication, the sequence number is set to 0. To prevent any possibility of replay, the wrap-around of sequence number is not possible. Thus, the SA can only be used to send 2^{32} packets. After this, the sequence number cannot be used. In such cases, a new SA must be established and used.

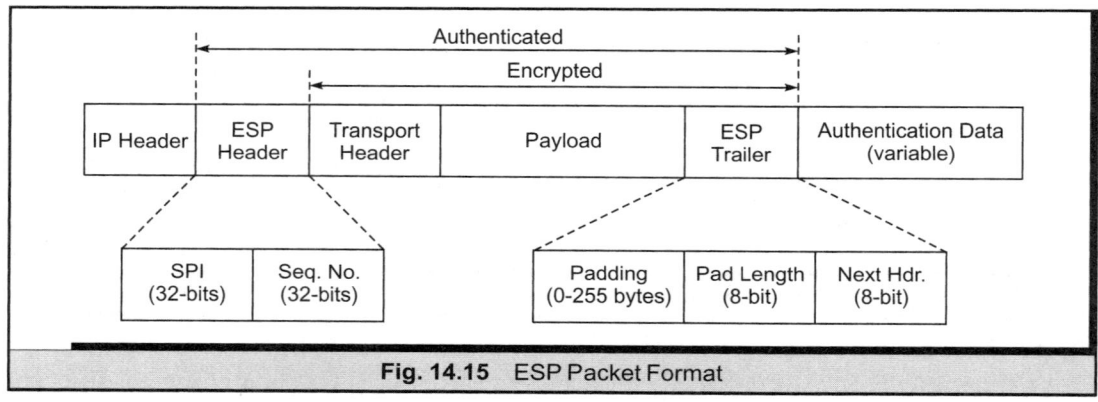

Fig. 14.15 ESP Packet Format

In the trailer, there are padding bytes, and a length field that gives the number of padding bytes. The padding is required when certain authentication/encryption algorithms require that the plaintext be a multiple of given number of bytes. The padding and length bytes are followed by the 8-bit 'next-header' field. This field is mandatory and identifies the higher layer PDU in the payload (e.g. TCP data).

The trailer is followed by the optional ESP authentication data with a variable length field containing the Integrity Check Value (ICV). The ICV is used for data integrity. The length of the ICV field depends upon the authentication function selected by ESP. This authentication data field is optional and is present only if it is specified in the security association.

The ESP protocol operates in two modes: the *transport mode* and the *tunnel mode*. The transport mode is typically used by IP hosts, as it does not protect the IP header. In the tunnel mode, the IP header is also protected. Figure 14.16 shows the modes of operation for ESP. As shown in the figure, the transport mode only protects the IP payload (comprising of TCP header and TCP payload in the example). The IP header is

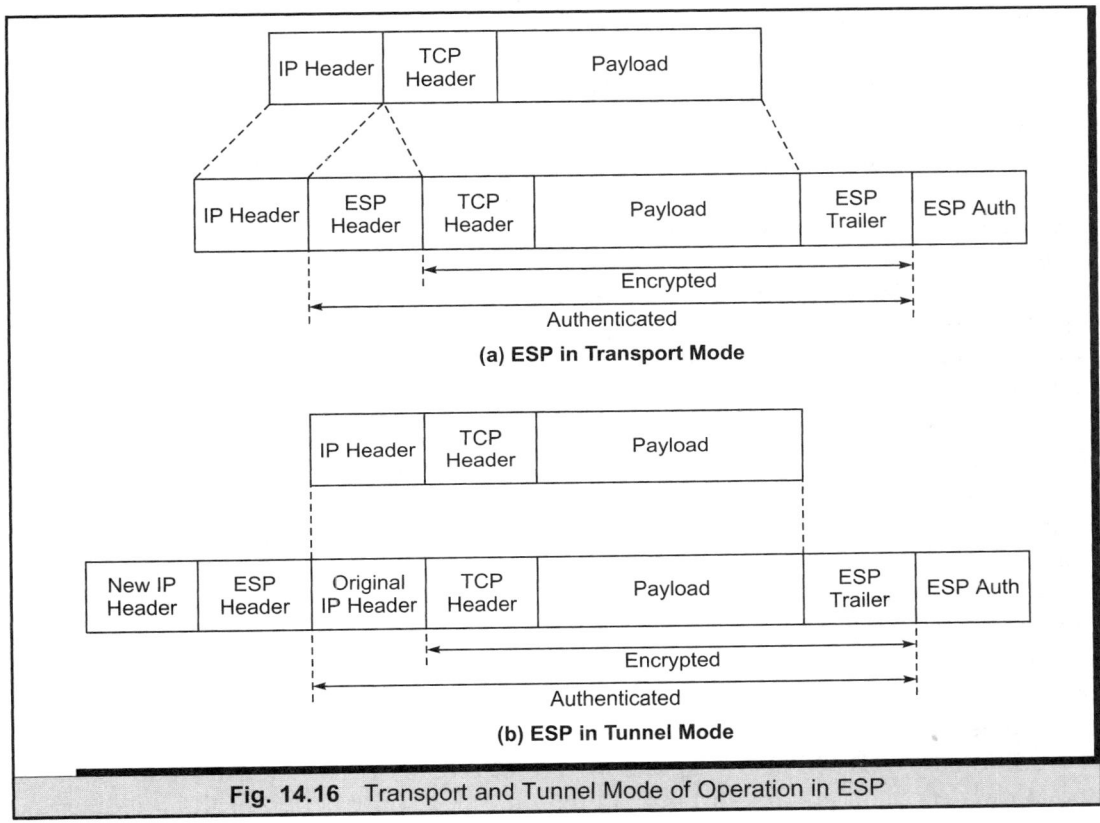

Fig. 14.16 Transport and Tunnel Mode of Operation in ESP

still vulnerable. The tunnel mode offers greater security as a new header is created from the original header. In the new header, the source and destination address is that of the tunnel end-points. This feature hides the original source and destination of the packet and makes traffic flow analysis difficult. Other fields of the original header are used in the new header without any change. The tunnel mode thus provides security to the original header. Changes in the original header are detected through integrity protection functionality.

14.5.2.1 Applicability of ESP to NDS/IP

The previous section briefly explained the ESP protocol as defined in RFC 2406. The RFC 2406 specification of ESP is generic in the sense that it caters to a wide set of requirements. Not all features defined in RFC 2406 are required for UMTS. To simplify implementation and to ease deployment—without compromising on the security obtained—3GPP TS 33.210 defines a subset of features of ESP that are applicable to

NDS/IP. An operator has the freedom to use the complete set of ESP features, although there seems no apparent need to do so (Here, it may be mentioned that in IPSec, both AH and ESP security protocols are defined, out of which only the use of ESP is specified in 3GPP TS 33.210).

As discussed earlier, two modes of operation are defined in ESP: the tunnel mode and the transport mode. In NDS/IP, only the use of the tunnel mode is specified. This is done because in inter-domain communication, all traffic flows through Security Gateways (SEGs).

Moreover, ESP supports a number of confidentiality transforms, among which the NULL transform (i.e. no transform) and DES transform (RFC 2405) are mandatory. However, DES algorithm is outdated and the new AES algorithm is all set to replace DES-based transforms. Therefore, in NDS/IP, the DES transform is not used. Instead, it is mandatory to support the ESP_AES transform. At the time of writing this book, the standardization for AES-based transforms was underway.

Like confidentiality transforms, support for a number of authentication transforms is mandatory for ESP. These transforms are the NULL transform, the HMAC_MD5 transform (RFC 2403), and the HMAC_SHA-1 transform (RFC 2404). However, for NDS/IP it is mandatory to provide integrity, data origin authentication, and anti-replay services. This implies that the NULL transform cannot be used. Between the remaining two, the use of only the HMAC_SHA-1 transform is permitted. This choice is made because the larger message digest size of HMAC_SHA-1 makes it less vulnerable to brute-force attacks, as compared to HMAC_MD5.

Additionally, NDS/IP also suggests the use of AES MAC transform. Apart from the above restrictions, there are additional guidelines for construction of the Initialization Vector (IV), for which the reader is referred to 3GPP TS 33.210.

14.5.3 Internet Key Exchange (IKE)

To use ESP or other IPSec protocols, a Security Association (SA) must first be established between communicating entities. The Internet Key Exchange (IKE) protocol provides the means for two entities to establish an SA so that data communication can take place using this SA. While the details of IKE are beyond the scope of this book, a short overview of IKE is provided here. For details of IKE, the reader is referred to RFC 2409.

In IPSec, there are two types of SA: *IKE SA* and *IPSec SA*. The IKE SA is used to *negotiate keys* in a secured manner. The IPSec SA is used to *exchange data* in a secured manner. Thus, the IKE protocol is first used to set up the IKE SA. This is referred to as the *first phase* of IKE. In order to ensure that exchanges in the first phase are authenticated, various mechanisms are possible. One of these is to use pre-shared secret keys. This is similar to the way AuC and USIM use a shared secret key for authentication. How the entities obtain the pre-shared secret key is outside the scope of IKE protocol. Apart from pre-shared secret keys, other schemes for authentication are Digital Signatures and Public Key Encryption. Whatever be the means, the essence of the matter is that the IKE peers establish an IKE SA in the first phase using some form of authentication.

Two modes of negotiations are possible in phase-1, namely, the 'Main Mode' and the 'Aggressive Mode'. In the main mode, three pairs of messages are exchanged between IKE peers. After these exchanges, an IKE SA is established and is ready for use. The aggressive mode, as the name suggests, uses fewer message exchanges. It is used when identity protection is not needed.

Once an IKE SA is established, the *second phase* of IKE is used to negotiate IPSec SA. Only 'Quick Mode' is supported in this phase. The parameters negotiated in this phase include:

- Authentication Mechanism
- Encryption Algorithm
- Hash Algorithm
- Key Values and Key Lifetimes

An important concept in IKE is *Perfect Forward Secrecy*. In simple terms, this implies that the key used to protect transmission of data must not be used to derive any additional keys. Moreover, if this key was derived from some other keying material, that material must not be used to derive any more keys.

14.5.3.1 Applicability of IKE to NDS/IP

The previous section briefly explained the IKE protocol, as defined in RFC 2409. The RFC 2409 specification of IKE is generic in the sense that it caters to a wide set of requirements. Not all features defined in RFC 2409 are required for UMTS. To simplify implementation and to ease deployment, without compromising on the security obtained, 3GPP TS 33.210 defines a subset of features of IKE that are applicable to NDS/IP. An operator has the freedom to use the complete set of IKE features, although there seems no apparent need to do so.

The following additional requirement on IKE is made mandatory for inter-security domain SA negotiations over the Za interface.

For IKE phase-1 (IKE SA), the following conditions apply:

- The use of pre-shared secret key for authentication shall be supported.
- Only Fully Qualified Domain Names (FQDN) shall be used.
- Only Main Mode shall be used.
- Support of AES in CBC mode shall be mandatory for confidentiality.
- Support of SHA-1 shall be mandatory for integrity/message authentication.

Note that phase-1 IKE SAs shall be persistent with respect to the IPsec SAs derived from these. That is, IKE SAs shall have a lifetime of at least the same duration as do the derived IPsec SAs.

For IKE phase-2 (IPsec SA), the following conditions apply:

- Perfect Forward Secrecy is optional.
- Only IP addresses or subnet identity types shall be mandatory address types.
- Support of notifications shall be mandatory. Notifications are used to exchange varied information, including error conditions between IKE peers.

SUMMARY

This chapter covered the security aspects of UMTS network. In particular, four functional blocks of security have been covered: User Domain Security, Access Security, Network Domain Security—Application layer and Network Domain Security—Network layer. Among these the user domain security is most simple; it ensures that the access to MS is provided only to the authenticated users (through the use of PINs).

Access security is a challenge—response protocol (called the UMTS AKA) and it provides mutual authentication. Once the mutual authentication is sucessful, the CK and IK are available. CK is used to encrypt user and signaling data over the air interface. IK is used to provide data integrity for signaling data over the air interface. The integrity protection is not applied to user data due to performance reasons.

The NDS, at the application layer, is performed using the MAPsec. Its applicability is limited to two entities that use the MAP for communication (e.g. VLR and SGSN). It is not applicable for other protocols (e.g. GTP protocol used between SGSN and GGSN).

The NDS, at the network layer, is performed using the IPsec. It is more generic and applicable for entities using the IP layer. Legacy entities using SS7 may use the MAPsec instead of the IPSec. The concept of NDS, both at the application layer as well as the network layer, is useful primarily for inter-domain communication. For intra-domain communication, it can be assumed that the network is secure and thus, the NDS principles may be ignored.

Part

4

IP Initiatives in UMTS Network

The last decade has witnessed a major transformation in the way people communicate. Till some time back, there were clear distinctions between Circuit-Switched and Packet-Switched networks. While the circuit-switched network was used to carry voice, the packet-switched network was primarily used to carry data. The latter was considered unsuitable for voice because of its high and unpredictable delay. However, various developments in the recent past have altered the landscape of the communication world.

First of all, the Internet Protocol (IP) has grown way beyond the most optimistic expectations from it. Its popularity is driven by killer applications like the World Wide Web (WWW) and Internet-based Email. The sudden surge in the popularity of IP has forced people to align their networking infrastructure with the already deployed Internet infrastructure. An important fallout of this development is that the Public Switched Telephone Network (PSTN) is converging with IP-based networks. This has led to standardization of the means to carry SS7 traffic over the IP network. These standardized mechanisms replace the SS7 protocols like Message Transfer Part (including MTP1, MTP2 and MTP3) by Adaptation protocols (like M3UA) and IP protocol suite. During the transition phase, when both the SS7 and IP

networks co-exist, there are Signaling Gateways (SGs) that provide the means to connect IP-based networks with legacy SS7 networks. Using SS7 over IP solutions allows the network administrators to substantially reduce operational costs.

Apart from reduction in operational costs, the introduction of IP as a transport network has led to a general consensus in the networking fraternity to move towards an all IP network. This step does not restrict the role of IP as a transport network for data carrier. Instead, the IP is now aimed at providing a whole gamut of services that include voice as well as enhanced multimedia services. This development has been fuelled by multiple factors, for example, the advent of new protocol suites like H.323 and SIP. These protocol suites provide a wide range of services that enable transfer of voice and multimedia over packet networks like the IP. Another factor that contributes to the packetization of voice/multimedia is the enhanced Quality of Service (QoS) features of IP network. Traditionally, the IP has been a best-effort protocol. However, a spate of technological advances has made QoS in IP networks a distinct possibility.

Both these trends (i.e. use of IP to carry signaling information and the advent of signaling protocols like SIP) are important, especially in the context of the developments in 3GPP community for UMTS networks. *This part of the book focuses on these trends. The next chapter, Chapter 15, explores the changes in the UMTS Network, where SS7 transport (including MTP1, MTP2, MTP3 and SCCP) is replaced by IP Signaling Transport, referred to as Signaling Transport (or simply SIGTRAN). Two alternatives for IP-based transport are discussed, namely M3UA and SUA. The 3GPP has chosen M3UA as its option. Different options for deployment of M3UA are explained.*

In Chapter 16, a new subsystem, called the IP Multimedia Subsystem (IMS), is discussed. The IMS has been developed to enable transfer of voice, video, data (e.g. shared white board) and other media-types over the UMTS networks. The IP Multimedia Subsystem uses the Session Initiation Protocol (SIP) as the protocol for communication. SIP is used for handling multimedia sessions in the IP Multimedia Subsystem. It is also used to provide various IMS services such as VoIP, Internet-based services, and Unified messaging. The role of SIP in the IMS domain is briefly explained in Chapter 16.

IP-BASED SIGNALING TRANSPORT

15.1 INTRODUCTION

In the fiercely competitive world of telecommunication, reducing operational costs is a very important business consideration. It is a known fact that packet-switched networks (like the IP-based Internet) offer greater utilization of network resources and thereby help in reducing costs. A good example to prove the point is the substantial reduction in the long-distance telephony cost following the introduction of Internet telephony. Voice over IP is just one of the many enhancements in the IP world. Another very important development is the introduction of IP-based signaling transport. This development is important not only for the telecommunication world at large but also for the UMTS networks. It is perceived that IP will play a significant role in carrying signaling traffic in the years to come.

Traditionally, the signaling traffic has been carried by SS7 networks, using protocols like Message Transfer Part (MTP). The use of SS7 implies that a network entity has to support two interfaces—one for signaling (based on SS7) and another for data (based on IP). If there were means to also carry signaling traffic on IP, then the two networks could be converged. To achieve this end, the Signaling Transport (SIGTRAN) working group was formed in the Internet Engineering Task Force (IETF). The work of SIGTRAN falls under two categories. The first part relates to the development of a *reliable transport protocol* that could remove some of the drawbacks of TCP (e.g. support for multiple IP addresses over the same connection). For this, the Stream Control Transmission Protocol (SCTP) was developed by SIGTRAN. The SCTP protocol ensures reliable signaling transmissions over IP and removes the shortcomings of TCP. The second part relates to the development of *adaptation protocols*. These protocols are designed in such a way that their users are unaware of the fact that SS7 protocols have been replaced with IP protocols. This design approach has the advantage that minimum changes are required in the users of adaptation protocols, while the SS7 protocols are systematically replaced by IP protocols.

Among the many adaptation layers introduced, the two important protocols developed by the SIGTRAN working group are MTP3 User Adaptation Layer (M3UA) and SCCP User Adaptation Layer (SUA). The 3GPP has chosen M3UA for IP signaling transport in the Access Network as well as in the Core Network. Given this, the M3UA protocol is discussed in detail, while some of the important aspects of SUA are also mentioned. The differences between M3UA and SUA are also discussed in this chapter.

15.2 IP-BASED SIGNALING TRANSPORT FROM SIGTRAN

IP-based signaling transport provides a means to transparently carry message-based signaling protocols (e.g. SCCP) over the IP networks. The IP-based signaling transport is being developed by the SIGTRAN working group of IETF. Initially, the SIGTRAN effort was to develop a protocol between VoIP architectural entities; for example, it was intended to reside between the Signaling Gateway (SG) and Media Gateway Controller (MGC). However, given the importance of SIGTRAN's work in diverse fields, the effort is now being made at a more general level.

15.2.1 Requirements

The primary goal of the SIGTRAN working group is to address the 'transport of packet-based PSTN signaling over the IP networks, taking into account *functional* and *performance* requirements of the PSTN signaling', which are detailed in the following sub-sections:

15.2.1.1 *Functional Requirements*

The functional requirements for SIGTRAN govern the functionality of various protocols developed by SIGTRAN. These requirements include:

- Provide the means to transport a variety of protocols (including MTP3, ISUP, SCCP, TCAP, MAP, etc.).
- Provide the means to identify the particular protocol being transported.
- Provide a common base protocol that defines header formats, security extensions and procedures for signaling transport, as well as support extensions as necessary to add individual protocols if and when required.
- In conjunction with the underlying network protocol (IP), provide the following functionality:
  - Flow control
  - In-sequence delivery of signaling messages within a control stream

  - Logical identification of entities on which signaling messages originate or terminate
  - Logical identification of the physical interface controlled by signaling message
  - Error detection
  - Recovery from failure of components in the transit path
  - Retransmission and other error correcting methods
  - Detection of non-availability of peer entities.

- Support the ability to multiplex several higher layer sessions on one underlying signaling transport session. In general, in-sequence delivery is required for signaling messages within a single control stream, but is not necessarily required for messages that belong to different control streams. The protocol should take advantage of this property to avoid blocking delivery of messages in one control stream due to errors within another control stream. It should also provide the means to send different control streams to different interfaces of the same host.
- Provide the means to transport complete messages of greater length than the segmentation/reassembly limitations of the underlying Switched Circuit Network (SCN). For example, signaling transport should not be constrained by the length limitations defined for the SS7 lower layer protocol (e.g. 272 bytes in the case of narrowband SS7) but should be capable of carrying longer messages without the need for segmentation.
- Allow for a range of suitably robust security schemes to protect signaling information being carried across networks. For example, signaling transport should be able to operate over proxiable sessions, and be transported through firewalls.
- Provide for guards against congestion in the Internet, by supporting appropriate controls on signaling traffic generation and reaction to network congestion.

15.2.1.2 Performance Requirements

Performance requirements for SIGTRAN govern the performance issues of various protocols developed by SIGTRAN. These requirements relate to message delay, message loss, sequencing loss and percentage availability. Inability to meet these requirements could result in adverse and undesirable signaling and call behaviour. The performance requirements include:

- **Message Delay:** The responses must be received within 500 to 1200 ms. The message delay includes round trip time and processing at the remote end.

- **Message Loss:** This is no more than 1 in 10^7 messages lost due to transport failure.

- **Sequence Error:** This includes out-of-sequence and duplicate messages received due to transport problems. The value must be less than 1 in 10^{10} messages.

- **Message Error:** This includes undetected errors by the transport protocol. The value must be less than 1 in 10^{10}.

- **Availability:** The availability of any signaling route set must be 99.9998% or better. This figure denotes a downtime of 10 minutes in a year or less. A signaling route set is the complete set of allowed signaling paths from a given signaling point towards a specific destination.

The above requirements apply to transport of MTP3 messages over SIGTRAN. Similar requirements are defined for the transport of messages of other protocols like ISUP, TCAP and Q.931 over SIGTRAN. For more on these requirements, the reader is referred to RFC 2719 'Framework Architecture for Signaling Transport'.

15.2.2 SIGTRAN Protocol Layering

The SIGTRAN protocol stack defines the use of a common transport layer protocol called the Stream Control Transmission Protocol (SCTP). The SCTP is defined in RFC 2960. It uses the services of IP layer. In turn, the SCTP layer provides services to various adaptation layers.

Figure 15.1 depicts the protocol layering in SIGTRAN. As shown in the figure, there is a common network layer (IP) and a common transport layer (SCTP). Over the SCTP, there are various adaptation layers. These adaptation layers are as follows:

- **ISDN Q.921 User Adaptation Layer (IUA):** This layer is used for providing services to the Q.931 signaling layer, which assumes Q.921 as the underlying layer. The IUA adaptation layer replaces the Q.921 protocol layer. The IUA is defined in RFC 3057.

- **SS7 MTP2 User Adaptation Layer (M2UA):** This layer is used for providing services to the MTP3 signaling layer. The M2UA adaptation layer replaces the MTP2 protocol layer. The M2UA is defined in RFC 3331.

- **SS7 MTP3 User Adaptation Layer (M3UA):** This layer is used for providing services to users of MTP3 layer (e.g. SCCP). The M3UA adaptation layer replaces the MTP3 protocol layer. The M3UA is defined in RFC 3332.

- **Signaling Connection Control Part User Adaptation Layer (SUA):** This layer is used for providing services to users of SCCP layer (e.g. TCAP). The SUA adaptation layer replaces the SCCP protocol layer. At the time of writing, the SUA was available as an Internet-draft.

Apart from these, there are few other adaptation layers defined by the SIGTRAN working group. For an exhaustive list, the reader is referred to the website of the SIGTRAN working group as provided in the References.

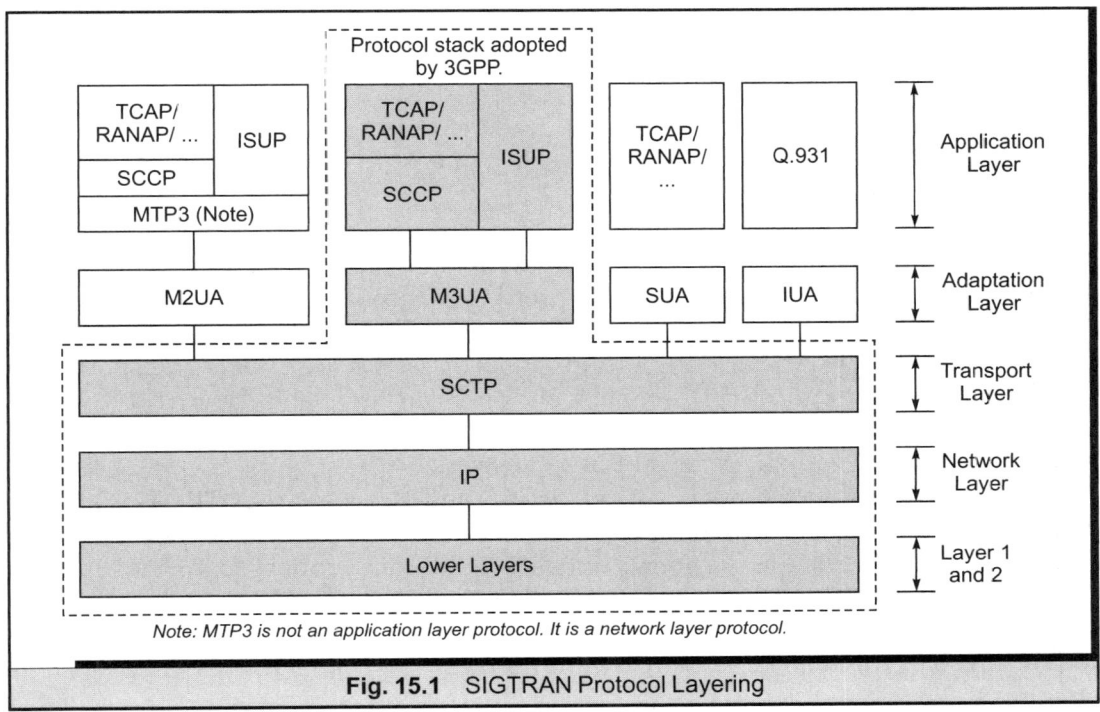

Fig. 15.1   SIGTRAN Protocol Layering

In the context of 3GPP, the adaptation protocols IUA and M2UA are not used. Hence, these are not discussed any further in this chapter.

15.3   STREAM CONTROL TRANSMISSION PROTOCOL (SCTP)

For many years, TCP and UDP had been used in the Internet as transport layer protocols. Among these protocols, TCP was used to provide reliable service over an unreliable layer (i.e. over IP). Developed in 1981, it has been more than two decades since TCP was first standardized. While some changes have been made in the way the TCP operates, the basic framework of the protocol remains unchanged. Given this, an increasing number of applications have found TCP too restricted and have developed their own reliable data transfer protocol on top of UDP.

Among the limitations of TCP, the most important one is the head-of-line blocking. This refers to the problem whereby some of the packets hold up the entire delivery of TCP packets. This happens because of the window-based flow control mechanisms of TCP in which an unacknowledged TCP segment holds up the transmission of new packets. It is found that some applications may do without sequence maintenance while others would be satisfied with partial ordering of the data. The TCP, with its head-of-line blocking problem, is unable to satisfy the above requirements.

Another problem with TCP is that it is stream-oriented. To make things more manageable, applications must add their own headers/trailers to delineate their messages and must make explicit use of the push facility to ensure that a complete message is transferred in a reasonable time.

Moreover, TCP is also found vulnerable to denial of service attacks.

For SIGTRAN applications, all the above limitations restrict the applicability of TCP. This prompted the IETF to develop a new protocol that could remove these restrictions. This new protocol is called the Stream Control Transmission Protocol (SCTP) and is defined by the SIGTRAN working group in RFC2960. The SCTP is a transport level datagram transfer protocol that operates on top of an unreliable layer (IP). Like the TCP, SCTP provides a reliable transport service, ensuring that data is transported across the network without error and in sequence.

The SCTP works on the notion of *associations* and *streams*. An SCTP association is similar to a TCP connection, just that it can support multiple IP addresses at either or both ends. This added functionality makes SCTP more robust to various failures (e.g. failure of an interface card can be handled by using multiple addresses attached to different interface cards). Further, the SCTP association is comprised of multiple logical streams. The sequenced delivery of user messages is guaranteed within a single stream.

SCTP establishes the reliable message transport service by retransmitting lost messages, like TCP. However, unlike TCP, the retransmission by SCTP of a lost message in one stream does not block the delivery of messages in other streams. The use of multiple streams within SCTP resolves the issue of head-of-line blocking associated with the use of TCP.

15.3.1   Functions of SCTP

The functions of SCTP are as follows:

- **Reliable Delivery:** SCTP provides an acknowledged, error-free and non-duplicated transfer of user data.

- **Fragmentation and Reassembly:** In case SCTP is not able to transfer a user message due to lower layer Maximum Transfer Unit (MTU) limitations, it fragments the user messages to conform to MTU size. On receipt of messages, fragments are reassembled into complete messages before being passed on to the SCTP user.

- **Sequenced Delivery within Streams:** As stated earlier, SCTP works on the notion of associations and streams. There can be multiple streams for an association between two SCTP peer entities. The number of streams to be supported by the association is defined at startup time of an association. Each message sent by SCTP contains a Stream Identifier and a Stream Sequence Number. These fields are used by the receiving side to ensure that messages within a given stream are delivered to the SCTP user in sequence. The benefit of using streams is that while one stream may be blocked waiting for the next in-sequence user message, delivery from other streams may proceed. SCTP also provides the flexibility to bypass the sequenced delivery. Hence, it is possible to deliver a message to the SCTP user as soon as it is received, bypassing the sequence numbering.

- **Chunk Bundling:** SCTP provides the facility to bundle multiple user messages into a single SCTP packet. To do so, an SCTP packet consists of a common header followed by one or more chunks. Further, each chunk uses a Type-Length-Value format. A chunk contains either signaling information or it contains user data.

- **Path Management:** For a multi-homed host with more than one IP address, SCTP provides the ability to form an association with multiple IP addresses. The path management function reports the eligible set of local transport addresses to the peer during association startup, and reports the transport addresses returned from the peer to the SCTP user. The path management function also monitors reachability through heartbeats. The support of multi-homing at either or both ends of an association provides network level fault tolerance.

15.4 SS7 MTP3 USER ADAPTATION LAYER (M3UA)

The SS7 MTP3 User Adaptation Layer (or M3UA in short) provides the means to carry messages of MTP3 users like SCCP and ISUP. In case of SCCP, the protocols that use its services (like RANAP or TCAP) are carried transparently in M3UA as SCCP payloads (i.e. SCCP is not aware of the protocol carried by its user protocol). M3UA uses the services of SCTP to provide MTP3-like services.

To provide interworking between nodes belonging to the IP domain (i.e. the IP node) and nodes belonging to the SS7 domain (i.e. SS7 node), a Signaling Gateway (SG) is used. This gateway provides the interworking functions between M3UA and MTP3 (refer to Section 15.4.2 for details).

15.4.1 Functions of M3UA

The set of primitives provided by M3UA to MTP3-users is equivalent to the one provided by MTP3 to its users at an SS7 end-point. Hence, the M3UA users are unaware of the fact that the MTP3 services are offered by M3UA and not by the MTP3 layer.

In general, M3UA provides the following functions to its users:

- **Support for Transport of MTP3-user messages:** This is the fundamental function of M3UA. It enables the transport of MTP3-user messages using the MTP-Transfer primitive. In case an IP node has interfaces with multiple Signaling Gateways (SGs), the M3UA layer chooses the one that is most appropriate to route the message to its destination. Moreover, by doing away with MTP2, M3UA removes the limitation that does not allow the information field to exceed 272 bytes. In other words, blocks larger than 272-octets can now be carried by M3UA without the user layers having to segment/reassemble these.

- **Native Management Functions:** This refers to basic operations like protocol error handling and error reporting.

- **Interworking with MTP3 Network Management Functions:** The M3UA provides this function at a signaling gateway in order to enable seamless interworking between SS7 and the IP domain. The interworking includes indicating whether an SS7 destination is reachable or not. Apart from this, information related to congestion is also exchanged.

- **Management of SCTP Associations:** Since an M3UA protocol entity can have more than one peer, it manages SCTP associations with each one of them and maintains the availability of peer nodes.

15.4.2   Scenarios for Deployment of M3UA in UMTS Network

There are two basic configurations for the deployment of M3UA in any IP-based network. In the first, a signaling gateway is used to connect two nodes, one belonging to the SS7 domain and other to the IP domain. The other configuration is used between two nodes, when both the nodes belong to the IP domain. In this second configuration, the nodes communicate with each other over the IP network. The signaling gateway is not required in this case.

The above configurations apply to the UMTS networks as well. As provided in 3GPP TR 29.903, the configuration for using the M3UA protocol with the signaling gateway is depicted in Figure 15.2. As shown in the figure, an IP node (say HLR) communicates with a signaling gateway over an IP network. The MAP/TCAP protocols are carried transparently in M3UA payload over the IP network.

The signaling gateway communicates with a SS7 node (say VLR) over the SS7 network. In this configuration, the signaling gateway performs the relay functions between MTP3 and M3UA.

Another configuration for using the M3UA protocol without the signaling gateway is depicted in Figure 15.3. In this case, the SS7 network does not come in the picture; only the IP network is used. This configuration is suitable when there is no requirement to interwork with legacy SS7 networks.

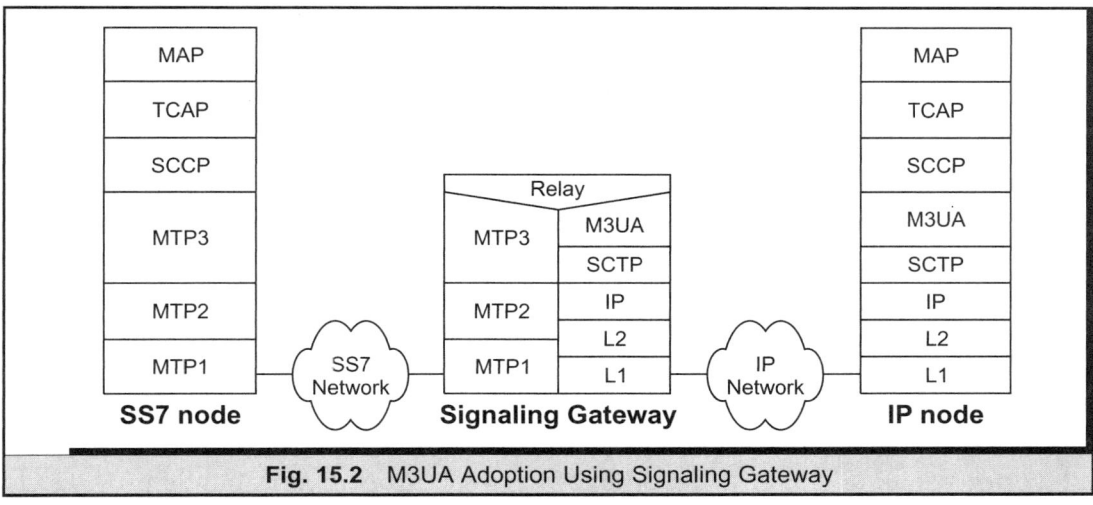

Fig. 15.2 M3UA Adoption Using Signaling Gateway

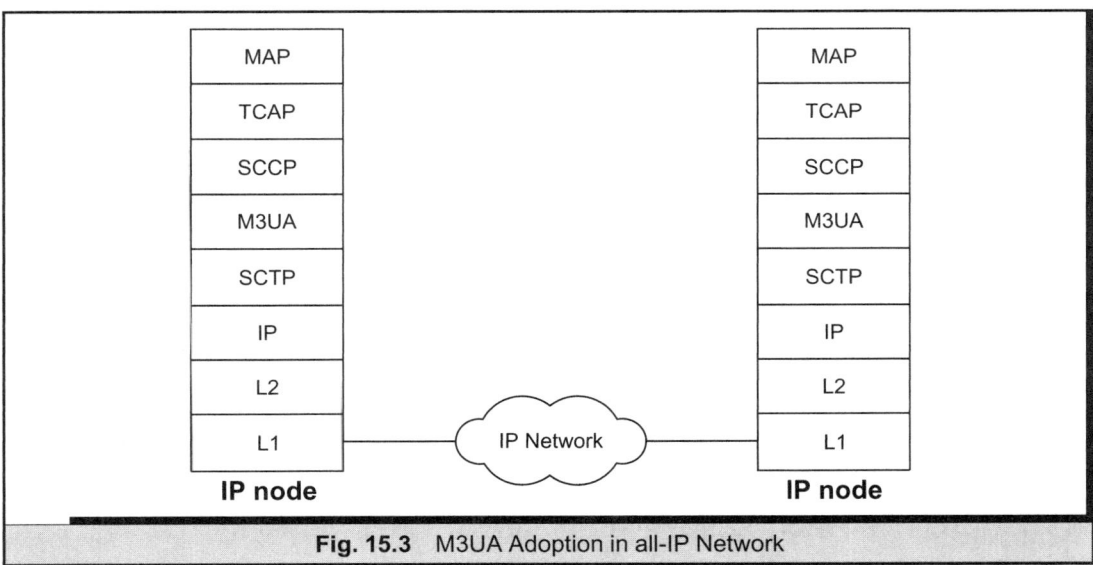

Fig. 15.3 M3UA Adoption in all-IP Network

15.5 SCCP USER ADAPTATION LAYER (SUA)

The SCCP User Adaptation Layer (or SUA in short) provides the means to carry messages of SCCP users like TCAP and RANAP. In case of TCAP, the MAP PDU is carried transparently in SUA as TCAP payload. SUA is not aware of the protocol carried by its user protocol. It uses the services of SCTP to provide SCCP-like services.

A Signaling Gateway (SG) is used to provide interworking between an IP and a SS7 node. This gateway provides the interworking functions between SUA and SCCP.

15.5.1   Functions of SUA

The set of primitives provided by SUA to SUA-users is equivalent to the one provided by SCCP to its users at an SS7 end-point. SUA provides the following functions to its users:

- **Support for Transport of SCCP-user messages:** This is the fundamental function of SUA. It provides the transport of SUA-user messages (e.g. TCAP and RANAP messages). For message transfer, SUA provides two types of service, namely *connectionless service* and *connection-oriented service*.

- **Native Management Functions:** This refers to basic operations like protocol error handling and error reporting. In case of errors, notification is provided to local management entity and/or to remote peer entity.

- **Interworking with SUA Network Management Functions:** The SUA provides this function at a signaling gateway, in order to enable seamless interworking between SS7 and the IP domain. The interworking includes indicating whether an SS7 destination is reachable or not. Apart from this, information related to congestion is also exchanged.

- **Management of SCTP Associations:** Since a SUA protocol entity can have more than one peer, it manages SCTP associations with each one of them and maintains the availability of peer nodes.

- **Address Mapping Function (AMF):** The SUA uses SCTP as its underlying layer. Thus, any message received by it must be sent on a particular SCTP association. For this, an Address Mapping Function (AMF) is required. The AMF uses routing information like Point Codes (PC) and Sub-System Numbers (SSN) to select a particular SCTP association.

15.5.2   Scenarios for Deployment of SUA in UMTS Network

The deployment configurations of SUA are similar to those of M3UA. Figure 15.4 depicts the configuration for using the SUA protocol with a signaling gateway. As shown in the figure, an IP node (say HLR) communicates with a signaling gateway over an IP network. The signaling gateway communicates with an SS7 node (say VLR) over the SS7 network. In this configuration, the signaling gateway performs the relay functions between the SCCP and SUA.

Another configuration for using the SUA protocol, this time without the signaling gateway, is depicted in Figure 15.5. Here, the SS7 network does not come in the

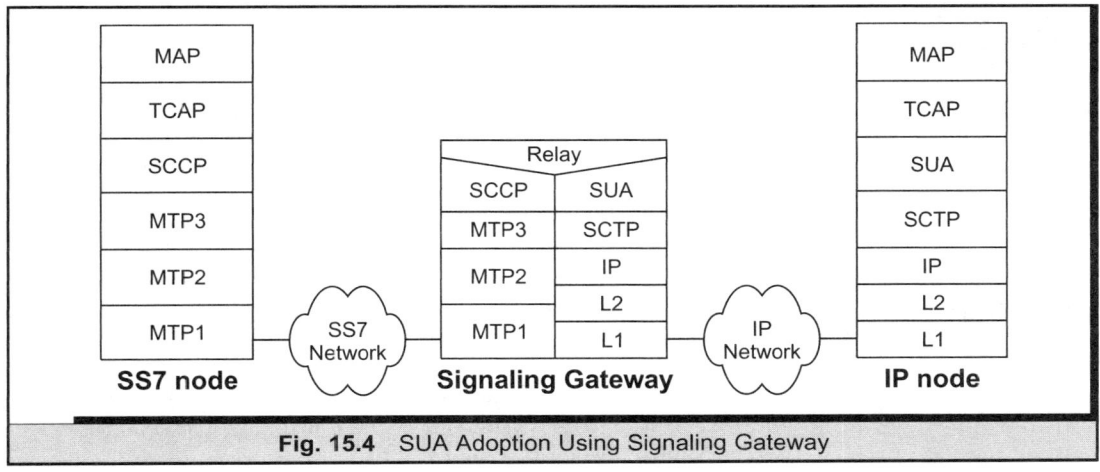

Fig. 15.4 SUA Adoption Using Signaling Gateway

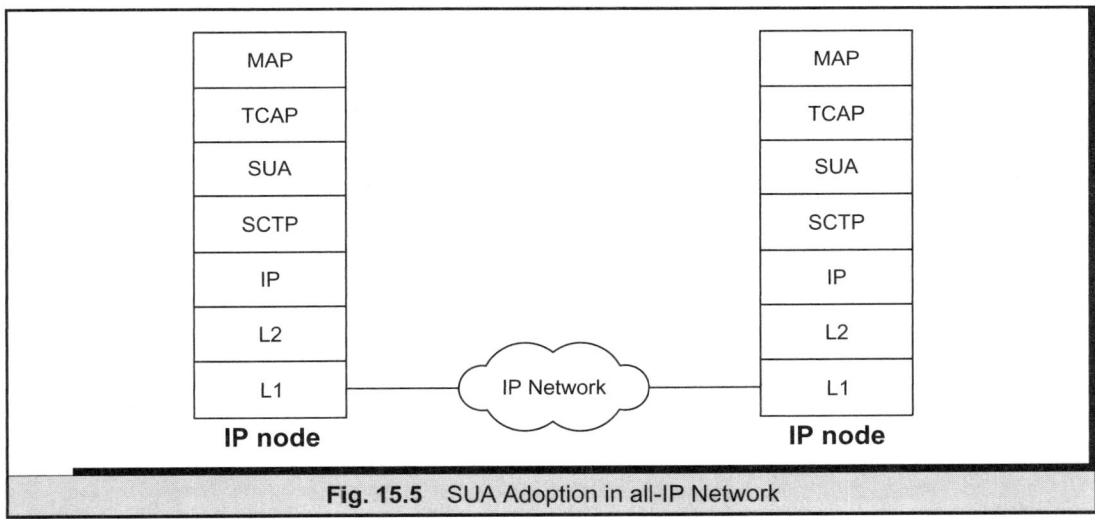

Fig. 15.5 SUA Adoption in all-IP Network

picture; only the IP network is used. This configuration is suitable when there is no requirement to interwork with legacy SS7 networks.

15.6 COMPARISON BETWEEN M3UA AND SUA

Both M3UA and SUA offer a means to employ IP to carry signaling information. There is considerable debate as to which protocol is better. Table 15.1 compares the two (refer to 3GPP TR 29.903).

Table 15.1 Comparison between M3UA and SUA

Aspect	M3UA	SUA
System Complexity	M3UA requires SCCP layer.	SUA does not require SCCP layer. This means one protocol less in the system resulting in reduced complexity and cost. Moreover, the capabilities of SUA make SCCP and M3UA unnecessary.
Point Codes	M3UA needs to be routed on Point Codes even in an all-IP scenario. This means managing two sets of addresses (Point Codes and IP addresses). Moreover, there are two sets of translation, namely Global Title to Point Code and Point Code to IP address.	In an all-IP scenario, the SUA can route messages using Global Titles without the involvement of point codes. Thus, there is only one translation from Global Title to IP address.
Interoperability issues	Different flavours of SCCP are needed to inter-operate with different national systems (e.g. ANSI SCCP and ITU-T SSCP).	Since Release 4 has already adopted M3UA, adopting SUA in Release 5 leads to upgradation costs and interoperability issues (with M3UA).
Support for MTP3 users	M3UA supports ISDN User Part (ISUP) and Bearer Independent Call Control (BICC).	MTP3 users like ISDN User Part (ISUP) and Bearer Independent Call Control (BICC) are not supported by SUA as they do not use SCCP.
Functional duplication	There are some functional redundancies in the SCCP/M3UA/SCTP stack. For example, message segmentation and reassembly mechanism are specified at both the SCTP and the SCCP layer.	SUA removes some of the functional redundancies, thus better utilizing the network and the processor.
Scalability	M3UA overlays a hop-by-hop, connectionless protocol mechanism over an end-to-end, connection-oriented protocol. This results in flexibility and scalability issues.	SUA provides much better scalability and flexibility for signaling network implementation in an all-IP network as compared to the SCCP/M3UA option.

SUMMARY

The SIGTRAN protocols (especially M3UA and SCTP) have found many uses in 3GPP. As already discussed in Chapter 5, the UTRAN uses M3UA/SCTP/IP in the Iu_PS and Iur interface. The Iu_PS interface is used to carry RANAP over SCCP over M3UA (see 3GPP TS 25.410). Similarly, the Iur interface is used to carry RNSAP over SCCP over M3UA (see 3GPP TS 25.420).

In the Core Network, the MTP-based transport for MAP-based nodes (like HLR, VLR or SGSN) is replaced by IP-based transport using M3UA. In this case, MAP messages are carried over SCCP, which in turn is carried over M3UA/SCTP/IP. The 3GPP TS 29.202 defines the use of IP Signaling Transport in the Core Network.

The M3UA-based signaling transport was introduced for the first time by 3GPP in Release 4 specifications. The reason for this was that IETF started work on M3UA before starting work on SUA. Hence, M3UA was adopted by 3GPP in Release 4 specifications. By the time Release 5 standards were being finalized, both M3UA and SUA were available as Internet-drafts. Thus, the option of SUA was also discussed. A Technical Report (TR) was also prepared by 3GPP—3GPP TR 29.903 'Feasibility Study on SS7 signaling transport in the Core Network with SCCP-User Adaptation Layer (SUA)'—that discussed the feasibility of adopting SUA in the CN. However, after hot debate and a voting among 3GPP members, the option of M3UA was retained by 3GPP. The SUA may be adopted in a future standard.

Further, in the IP Multimedia Subsystem (IMS), which will be discussed in the next chapter, the SCTP protocol is used at the transport layer. Since the SCCP or a similar protocol is not used in IMS, there is no requirement for any adaptation layer in IMS.

IP MULTIMEDIA SUBSYSTEM

16.1 INTRODUCTION

Till the Release 4 specifications of 3GPP, there were two distinct domains in the UMTS networks: the Circuit Switched (CS) domain and the Packet Switched (PS) domain. The first was used primarily for basic voice call, and the second for accessing the Internet and other packet-based services. With the growing demand for multimedia applications (like video conferencing, data collaboration, and network games), the services offered by the CS and PS domain seemed inadequate. To provide support for multimedia applications, the IP Multimedia Subsystem (IMS) was introduced for the first time in the Release 5 specifications of 3GPP. The driving force behind the introduction of IMS was to use the ubiquity of the IP protocol and its potential to provide the operators the freedom to develop totally new and value-added multimedia applications.

Certain key decisions/assumptions were made while designing the IP Multimedia Subsystem (IMS). First, unlike the CS and PS domain—where there are a number of standardized services (including supplementary and other services)—the IMS merely defines a framework, using which services/applications can be built by third party vendors. The standardization of IMS services is deliberately kept outside the purview of IMS standards. Only basic services are provided by the IMS specifications. This framework enables an operator to deploy IP multimedia applications in a network-independent manner. The operator does not have to depend on 3GPP to standardize services.

Further, it is assumed that the standard services available in the Internet fraternity (e.g. chat and instant messaging) will be re-used. This will help channel the vast experience available in developing HTML and JAVA applications towards developing applications for the IMS.

Another important design assumption is that the PS domain is used to provide underlying transport for both the signaling and bearer channels. The CS domain is not relevant to the functioning of IMS. Figure 16.1 depicts how the PS domain is used in the context of IMS. The Access Network (i.e. UTRAN) and the IP-based transport (i.e. PS domain) collectively provide the bearer for user and control plane. In future 3GPP releases, some other transport mechanism may be used (e.g. WLAN) for carrying IMS traffic.

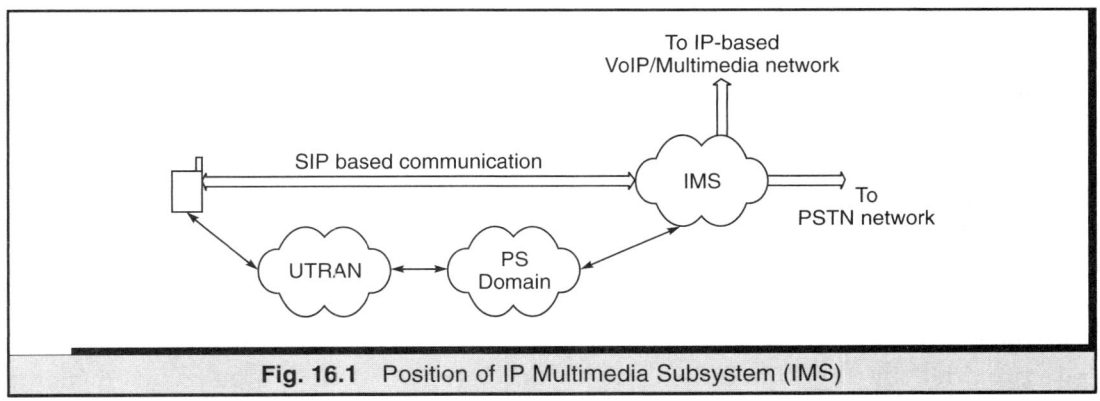

Fig. 16.1   Position of IP Multimedia Subsystem (IMS)

Among various protocols that exist for multimedia transport (e.g. H.323 and SIP), the 3GPP has chosen the Session Initiation Protocol (SIP) and the related protocols for session establishment and management. SIP is used for signaling between the UE and the IMS as well as between the entities within the IMS. The IMS also uses SIP to terminate voice and multimedia sessions in the Internet. SIP was chosen because of its simplicity, extensibility and its availability. Availability of a wide variety of commercial SIP phones (both in hardware and software) indicates the popularity of SIP. The 3GPP attempts to use this popularity to provide multimedia services in IMS.

To explain the IP Multimedia Subsystem (IMS), this chapter is organized as follows: it first details the various entities that form the IMS. These entities include the Home Subscriber Server (HSS), Call Session Control Function (CSCF), Application Server (AS) and interworking entities. The various interfaces between these entities are covered next. Thereafter, the IMS Addressing and the IMS Subscriber Data are explained. This is followed by details of various IMS procedures, which fall under two categories: *session-related* and *session-unrelated* procedure.

The IMS protocols are explained next. The most important IMS protocol is the SIP, details of which are beyond the scope of this book. Thus, only a brief description of SIP is provided towards the end of this chapter. Apart from SIP, the Diameter protocol, which is used between the HSS and CSCF, is explained.

Lastly, the security aspects of IMS are detailed; in particular, the access security. The Network Domain Security for IMS has already been covered in Chapter 14.

16.2 ENTITIES OF IP MULTIMEDIA SUBSYSTEM

Given that the IMS uses SIP for signaling, it requires the following functional entities in order to function:

- **SIP Components:** The requirements of IMS architecture are governed by the needs of the SIP protocol. The key components of any network using this protocol are: User Agents (UA), Proxy and Registrar. The IMS architecture provides these functional components; it is just that there is no one-to-one correspondence between these SIP components and the IMS entities. The UE acts as the end-point and provides the UA functionality. The Proxy/Registrar functionality is provided by the Call Session Control Function (CSCF), which is of three types: P-CSCF, I-CSCF, and S-CSCF. The P-CSCF and I-CSCF can be viewed as SIP proxy, which forward the SIP requests to the S-CSCF. The S-CSCF acts as the Registrar. To avail of IMS services, a UE has to first register with the S-CSCF.

- **Subscriber Database:** Apart from the SIP components, the IMS requires a central database for subscriber data storage and management. This functionality is provided by the Home Subscriber Server (HSS), which is an evolved Home Location Register (HLR). HSS not only includes interfaces for the CS and PS domain, but also the functions necessary for handing the database requirements of IMS. The I-CSCF/S-CSCF communicates with HSS to obtain subscriber data over the Cx interface, and the Application Server communicate with it to obtain this data over the Sh interface.

- **Service Platform:** The actual service execution logic resides on the service platform. The service platform is also called the Application Server (AS). In IMS, there are various application servers, namely the SIP Application Server (SIP AS), the Open Services Architecture Service Capability Server (OSA SCS) and the IP Multimedia Service Switching Function (IM-SSF). While SIP AS hosts the SIP-based service, OSA is a relatively new concept where various services can be provided by third-party applications using OSA APIs. The IM-SSF is used to host CAMEL network features.

- **Interworking Entities:** Besides the entities mentioned above, the IMS architecture also includes the Breakout Gateway Control Function (BGCF), Media Gateway (MGW) and Media Gateway Control Function (MGCF). These entities are required for interworking with PSTN and other networks.

Figure 16.2 expands the IMS depicted in Figure 16.1 and provides a high-level representation of IMS architecture. As shown in Figure 16.1, a SIP-enabled UE uses the UMTS radio interface in the access side. As in the UMTS, the UE has a USIM; similarly, in the IMS there is an ISIM. The ISIM holds the UE's identity and other necessary information. To expedite deployment of IMS networks, it is possible that a UE, which only has a USIM, can access the IMS services. For this, the IMS identities are derived from USIM identities, details of which are provided in Section 16.4.

The SGSN/GGSN of the PS domain provide packet-based transport for IMS. Before an IMS user can avail of IMS services, the PS-connectivity must be present. Thus, mobility aspects are handled by protocols of the PS domain. Note that the IMS is intended to be access independent in the sense that other access networks (e.g. Wireless LAN) can also be used to avail of IMS services. These options are, however, outside the scope of the ReL5 3GPP standards.

The P-CSCF is the first contact point in the IMS. If the user is roaming, the P-CSCF is in the visited network. If the user is in the home network, the P-CSCF is also in the home network. In the Figure 16.2, it is assumed that the user is roaming and the P-CSCF is in the visited network.

In the home network, the first contact point is the I-CSCF. The location of I-CSCF is obtained using the domain name of the user and DNS. The I-CSCF uses the information provided by HSS to contact an S-CSCF.

The S-CSCF, in association with the service platform, provides IMS services.

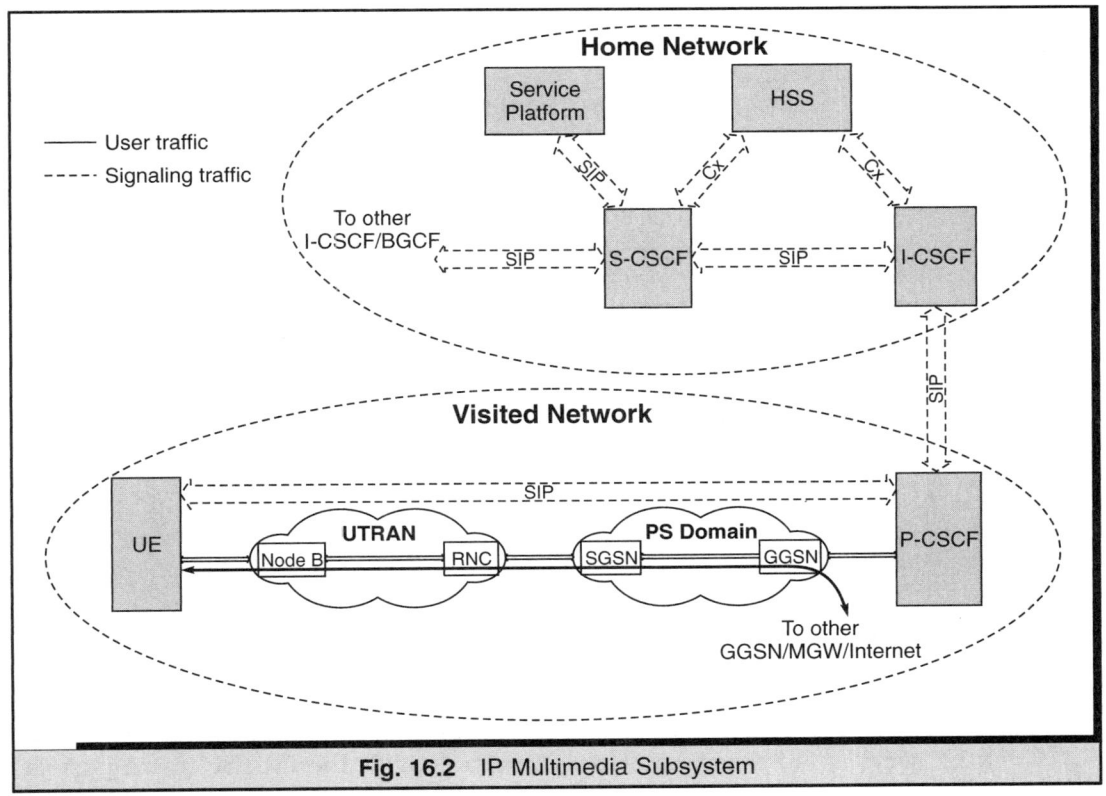

Fig. 16.2 IP Multimedia Subsystem

As is clear from Figure 16.2, the SIP protocol is used for signaling in IMS. Once the SIP sessions are established, the transport bearer does not follow the signaling path. The bearer path extends from the UE to the destination network via SGSN/GGSN. The destination network may be another GGSN, or an IP network, or an MGW. Typically, the Real Time Protocol (RTP) is used in the user plane.

The different IMS entities are explained in the following sub-sections.

16.2.1   Home Subscriber Server (HSS)

In Release5, the HLR functionality is included as part of a new entity referred to as the Home Subscriber Server (HSS). The HSS can be viewed as an entity that performs the functionality of HLR for the CS and PS domain, as well as the database functionality required for IMS. In this chapter, the reference is to the latter.

The HSS for IMS can be viewed as the database that hosts subscriber data for IMS subscribers and contains the subscription-related information required for handling SIP sessions. The following information is maintained at the HSS (also refer to Sections 16.4 and 16.5 for details of these terms):

- Private User Identity (IMPI)
- One or more Public User Identity (IMPU) associated with an IMPI
- Registration status of IMPUs
- The name of S-CSCF with which the IMPUs are registered
- Shared Secret Key and Authentication Data
- Server Capabilities required for S-CSCF selection
- Initial Filter Criteria and other subscriber profile information.

Based on this information, the HSS carries out the following functions in IMS:

- **Session Establishment support:** The HSS supports the session establishment procedures in the IMS. For mobile-terminating session, the HSS provides information as to which S-CSCF the user is registered with.

- **Authentication Information Generation:** This information, generated by the HSS, is used by S-CSCF for authentication.

- **Access Authorization:** Access authorization by HSS allows access only to authorized users. This is done by checking whether the user is allowed to roam in the visited network.

- **Service Profile Provisioning and Management:** HSS supports the provisioning of service profile data. This information is provided to S-CSCF and Application Servers (e.g. SIP AS and the OSA SCS).

- **Facilitating a Host of Services:** The HSS facilitates a host of services including the CAMEL service. For example, it communicates with the IM-SSF to support the CAMEL services related to the IMS.

16.2.2 Call Session Control Function (CSCF)

In the IMS, the SIP functionality is provided by the CSCF. There are three types of CSCF, as follows:

- **Proxy CSCF (P-CSCF):** The P-CSCF is the first contact point for the UE within the IMS.

- **Interrogating CSCF (I-CSCF):** The I-CSCF is primarily the contact point for all IMS connections assigned to a subscriber within an operator's network.

- **Serving CSCF (S-CSCF):** The S-CSCF actually handles the session signaling in the IMS.

The following subsections provide details of various CSCF.

16.2.2.1 Proxy-CSCF (P-CSCF)

The Proxy-CSCF or P-CSCF is the first contact point for the UE in the IMS. In case the UE is roaming in a network, the P-CSCF is a part of the visited network. Otherwise, it is a part of the home network. The P-CSCF is essentially a SIP proxy server. Its functions are as follows:

- It forwards the SIP Register request received from the UE to an I-CSCF. This is done using the home domain name as provided by the UE.
- It forwards the SIP messages received from the UE to the SIP Server (e.g. S-CSCF).
- It forwards messages received from I-CSCF/S-CSCF to the UE.
- It maintains a security association with the UE. This is required for secured communication in the Access Network.

16.2.2.2 Interrogating-CSCF (I-CSCF)

While the P-CSCF is the first contact point for a UE in the visited network, the Interrogating-CSCF or I-CSCF is the first contact point within an operator's network. By providing a single point of contact for entry into the network, I-CSCF hides the configuration of a network from other network operators.

Like the P-CSCF, I-CSCF is a SIP proxy. Its functions are:

- It selects an S-CSCF during the registration procedure. An S-CSCF is selected using the capabilities required for a subscriber and the capabilities offered by various S-CSCFs in the network.
- It obtains the address of the S-CSCF (from HSS) with which a subscriber is registered and forwards the SIP request/response to the appropriate S-CSCF.

16.2.2.3 *Serving-CSCF (S-CSCF)*

The Serving-CSCF or S-CSCF is the most important entity among all CSCFs. It provides the session control services to the UE. A network can have several S-CSCFs, each capable of providing different services. The I-CSCF chooses the S-CSCF that can serve a subscriber. The S-CSCF acts as a SIP Registrar. The functions of S-CSCF are as follows:

- Accepts registration requests from the UE and provides this information to HSS.
- Provides the session control services to the UE.
- Interacts with the services platforms to support various services.
- Serves the originating UE by providing various services. These services include obtaining the address of the I-CSCF for the network operator serving the destination UE; forwarding the request/response to the chosen I-CSCF; and forwarding the SIP request/response to the breakout gateway (BGCF) for call routing to the PSTN or the CS domain.
- Serves the destination UE by providing various services, which include forwarding the SIP request/response to a P-CSCF/I-CSCF for a mobile-terminating procedure, and forwarding this request/response to breakout gateway (BGCF) for call routing to the PSTN or the CS domain.
- Rejects communication to/from entities that are barred from using IMS.

16.2.3 Server Locator Function (SLF)

The Server Locator Function (SLF) is an optional entity in the IMS network. It is required if there are more than one HSS in a network. The SLF is accessed by the I-CSCF/S-CSCF over the Dx interface. It is used to find the address of HSS, which holds the subscriber data for a given IMS subscriber.

16.2.4 Application Server (AS)

In IMS, there are many application servers that offer value added services. They influence and impact the SIP sessions on behalf of the services supported by the operator's network. There are three types of AS:

- **SIP Application Server (SIP AS):** These are SIP-based AS that host and execute IMS services. The SIP AS SIP influence and impact the SIP sessions on behalf of the services supported by the operator's network.

- **Open Services Architecture Service Capability Server (OSA SCS):** This application server interfaces with the OSA framework application server using OSA API to execute service logic.

- **IP Multimedia Service Switching Function (IM-SSF):** This is a CAMEL-based application server. The IM-SSF hosts the network features that include trigger detection points and CAMEL Service Switching Finite state machine.

The role of application servers in providing SIP services is explained in Section 16.7.1.

16.2.5 Entities used for Interworking

For IMS to interwork with PSTN and other circuit-switched networks, three different entities are used. These entities are as follows:

- **Breakout Gateway Control Function (BGCF):** The BGCF selects the network in which PSTN breakout is to occur. Here, two scenarios are possible. In the first, the breakout occurs in the same network in which BGCF is present. In this case, the BGCF selects the MGCF for further progress of the call. In the second scenario, the breakout occurs in another network, in which case the BGCF selects another BGCF that can facilitate breakout. The call flow diagram depicting the role of BGCF is explained in Section 16.7.3.

- **Media Gateway Control Function (MGCF):** The MGCF communicates with S-CSCF over SIP and with MGW using the MEGACO protocol. The MEGACO protocol is used by MGCF to establish bearer connections through MGW. The MGCF also selects a CSCF depending on the number for calls originated at the legacy networks. The MGCF manages one or more MGWs.

- **Media Gateway Function (MGW):** The primary function of MGW is to convert media from one format to another. This conversion usually takes place between the IP-based packet format to PCM-based voice format and vice-versa. The MGW interacts with MGCF using MEGACO or ISUP. These interactions are used for terminating bearer channels from switched circuit network, and media streams from a packet network. The media gateway function may also involve payload processing (e.g. echo cancellation, codec support, etc.).

16.2.6 Signaling Gateway Function (SGW)

The SGW carries out the signaling conversion at the transport level between the SS7-based transport signaling used in pre-Rel4 networks, and the IP-based transport signaling, possibly used in post-R99 networks (i.e. between Sigtran SCTP/IP and SS7 MTP). The SGW does not interpret the application layer (e.g. MAP, CAP, BICC, ISUP) messages but may have to interpret the underlying SCCP or SCTP layer message to ensure proper routing of signaling message.

16.3 NETWORK INTERFACES OF IP MULTIMEDIA SUBSYSTEM

In the previous section, various network entities of IMS were explained. This section details the interfaces that exist between these network entities. The important interfaces of IMS are shown in Figure 16.3.

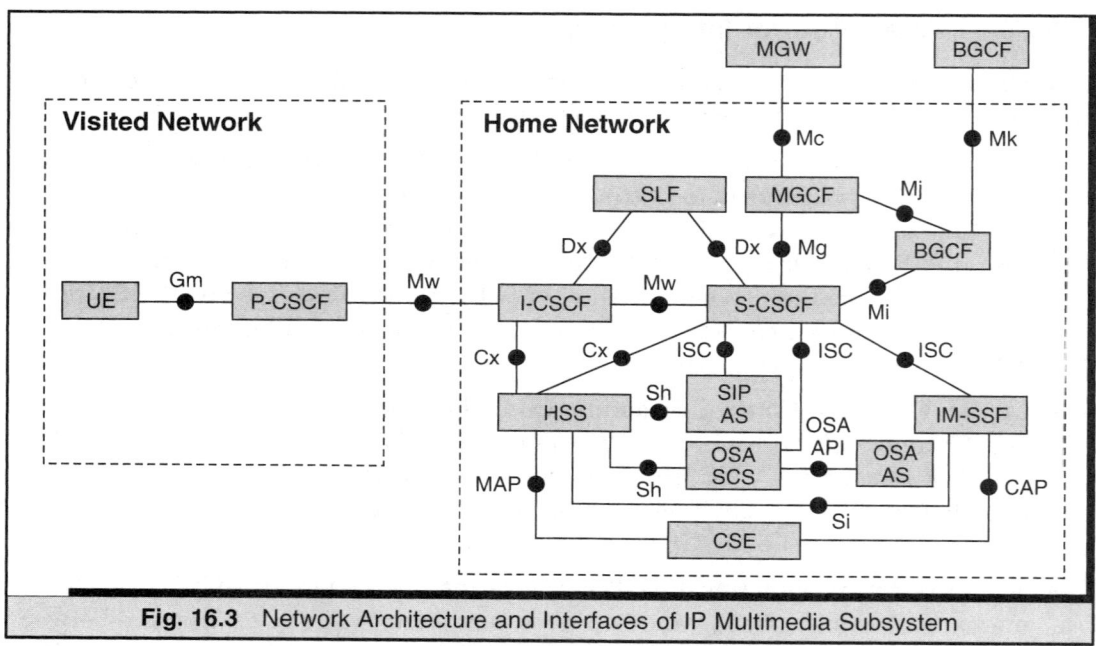

Fig. 16.3 Network Architecture and Interfaces of IP Multimedia Subsystem

The CSCF are the key component of the IMS. The I-CSCF and S-CSCF communicate with HSS over the Diameter-based Cx interface. In case the address of HSS is not known to CSCF, the I-CSCF/S-CSCF communicate with SLF over the Dx interface to obtain the same.

The S-CSCF interfaces with I-CSCF over the SIP-based Mw interface. Then, S-CSCF uses the Service Control Interface (ISC) to interact with various application servers (i.e. SIP AS, OSA SCS and IM-SSF). The OSA SCS uses OSA APIs to communicate with OSA AS. The IM-SSF uses CAP to communicate with CSE. S-CSCF also uses the SIP protocol to interact with MGCF and BGCF over the Mg and Mi interface respectively. The MGCF and BGCF communicate with each other using the Mj interface. The MGCF interacts with MGW over the Mc interface. The BGCF interacts with another BGCF using the Mk interface.

Apart from the Cx interface, HSS interfaces with various applications servers (e.g. SIP AS and OSA SCS) over the Diameter-based Sh interface. The Si interface is used between the HSS and IM-SSF. For CAMEL, the HSS uses MAP for communicating with the CSE.

Table 16.1 Lists the IMS interfaces, giving a brief description of the interface, the protocol applicable to the interface and the relevant specification(s).

Table 16.1 IMS Interfaces

Interface	Between	Description	Protocol	Specification
Cx	S-CSCF – HSS I-CSCF – HSS	Used by the I-CSCF and S-CSCF to obtain subscriber information from HSS.	Diameter-based Cx protocol	29.228/29.229
Dx	S-CSCF – SLF I-CSCF – SLF	Enables the I-CSCF and S-CSCF to find the HSS address, which holds the data of a particular subscriber.	Diameter-based Dx protocol	29.228/29.229
Sh	HSS – AS	Provides AS facility to download data from HSS, update data maintained at HSS, subscribe to notifications, and receive notifications.	Diameter-based Sh protocol	29.328/29.329
Si	HSS – IM-SSF	Used to exchange CAMEL information.	MAP	29.002
ISC	S-CSCF – AS S-CSCF – IM-SSF S-CSCF – OSA SCS	Used by S-CSCF to interact with service platforms, which host and execute SIP services.	SIP (Note 1)	24.229
Gm	UE – CSCF	Used by UE for registering with a CSCF and to take part in originating/terminating sessions.	SIP (Note 1)	24.229
Mc	MGCF – MGW	Used by MGCF to control the MGW.	MEGACO	H.248
Mg	S-CSCF – MGCF	Used for calls originated at PSTN.	SIP (Note 1)	24.229
Mw	x-CSCF – y-CSCF	Used for communication between different CSCFs.	SIP (Note 1)	24.229
Mi	S-CSCF – BGCF	Used for interworking with PSTN.	SIP (Note 1)	24.229
Mj	MGCF – BGCF	Used by BGCF to forward the session signaling to the MGCF for interworking with PSTN networks.	SIP (Note 1)	24.229
Mk	BGCF – BGCF	Used when breakout has to occur in another network.	SIP (Note 1)	24.229

Note 1: The SIP protocol is defined in RFC 3261. However, 3GPP TS 24.229 defines the applicability of SIP and its extensions in the IMS. For a list of SIP extensions applicable to IMS, refer to 3GPP TS 24.229.

16.3.1 Cx Interface between HSS – CSCF

The Cx interface is used by the I-CSCF and S-CSCF to obtain subscriber information from HSS. The information generally flows in the direction of HSS to I-CSCF/S-CSCF.
The main procedures applicable to this interface are as follows:

- Transfer of server capability information from HSS to I-CSCF so that the I-CSCF can select the S-CSCF, that can in turn serve the IMS subscriber.
- Verification of access permissions and roaming agreements by HSS so that only authorized subscribers get access to network services.
- Obtaining the name of the S-CSCF where the IMS subscriber is registered (if registered at all).
- Transfer of authentication data from HSS to S-CSCF.
- Notifying the HSS of the registration/de-registration of an IMS subscriber by the S-CSCF.
- Transfer of user profile information (e.g. initial filter criteria, implicitly registered IMPUs, etc.) from HSS to S-CSCF.
- Updating of user profile information at S-CSCF in case it is modified at the HSS.

The Cx interface is defined in 3GPP TS 29.228 and TS 29.229. These specifications in turn use Diameter Base Protocol (DBP), which is defined in RFC 3588.

The messages used for the Cx interface are listed in Table 16.2, wherein, each row indicates a transaction comprising of a request and an answer. The request and answer collectively define a complete transaction. For example, the request message UAR is sent by I-CSCF to HSS and the answer message UAA is sent by HSS to I-CSCF. The UAR-UAA transaction is used by I-CSCF to request the HSS to authorize user registration, check the user's roaming agreements and to obtain the S-CSCF name where the user is registered. In case the user is not registered, the HSS provides I-CSCF the capability information to select an S-CSCF.

16.3.2 Dx Interface between CSCF and SLF

The Dx interface enables the I-CSCF and S-CSCF to find the address of HSS, which holds the subscriber data for a given IMS subscriber. In case there is only one HSS in the IMS network, the Dx functionality is not required.

The Dx interface is based on the redirection functionality offered by the Diameter protocol. When the address of the HSS is not known, the I-CSCF/S-CSCF sends a request directed to HSS to an SLF. In such a case, the SLF acts as a redirect agent and sends the request back to the I-CSCF/S-CSCF along with the address of HSS. The returned address is then used to send the message to HSS.

The Dx interface is defined along with the Cx interface in RFC 3588, 3GPP TS 29.228 and TS 29.229.

Table 16.2 Cx Messages

Message	Abbr	From	To	Description
User-Authorization-Request/Answer	UAR/ UAA	I-CSCF/ HSS	HSS/ I-CSCF	Used by I-CSCF to request the HSS to authorize user's registration, check user's roaming agreements and to obtain capability information for S-CSCF selection.
Multimedia-Authentication-Request/Answer	MAR/ MAA	S-CSCF/ HSS	HSS/ S-CSCF	Used by S-CSCF to obtain authentication information from HSS. This information is used to authenticate a user.
Server-Assignment-Request/Answer	SAR/ SAA	S-CSCF/ HSS	HSS/ S-CSCF	Used by S-CSCF to notify HSS of IMPU's registration/de-registration. The message is also used to download user profile information.
Location-Info-Request/Answer	LIR/ LIA	I-CSCF/ HSS	HSS/ I-CSCF	Used by I-CSCF to obtain the name of S-CSCF where the user is registered.
Push-Profile-Request/Answer	PPR/ PPA	HSS/ S-CSCF	S-CSCF/ HSS	Used by HSS to update user profile information maintained at S-CSCF.
Registration-Termination-Request/Answer	RTR/ RTA	HSS/ S-CSCF	S-CSCF/ HSS	Used by HSS to inform S-CSCF of HSS-initiated user de-registration.

16.3.3 Sh Interface between HSS and AS

The IMS network architecture is such that the HSS is the master database that maintains subscriber data of an IMS subscriber. There may be occasions when application servers (like SIP AS or OSA SCS) require downloading data from the HSS. This may be done to download the subscriber profile information from the HSS. The Sh interface may also be used by application servers to use HSS as a temporary data store for service data. To provide the aforementioned functionality, the Sh interface is specified between the HSS the and the Application Server. The main procedures applicable to this interface are as follows:

- Transfer of data (e.g. repository data, subscriber profile and registration status) from HSS to AS using the pull mechanism.
- Update by AS of data maintained by HSS using the update mechanism.
- AS subscribing to notifications from HSS whenever certain data (e.g. repository data and registration status) maintained at HSS changes.
- The notifications from HSS to AS when data actually changes.

The list of messages used for the Sh interface is given in Table 16.3.

Table 16.3   Sh Messages

Message	Abbr	From	To	Description
User-Data-Request/Answer	UDR/ UDA	AS/HSS	HSS/AS	The transaction is used by AS to obtain transparent and/or non-transparent data from HSS (Note 1).
Profile-Update-Request/Answer	PUR/ PUA	AS/HSS	HSS/AS	This is used by AS to update data maintained at HSS.
Subscribe-Notification-Request/Answer	SNR/ SNA	AS/HSS	HSS/AS	Used by AS to subscribe to notifications from the HSS for changes in data maintained by the latter.
Push-Notification-Request/Answer	PNR/ PNA	HSS/AS	AS/HSS	PNR is sent by the HSS to an AS notifying the latter of the changes made by HSS. These changes had earlier been subscribed to by the AS.

Note 1: Transparent data is the data whose semantics is not understood by HSS (e.g. the repository data, stored in the HSS by AS). In such cases, the HSS merely acts as a data store. Non-transparent data on the other hand, is the data whose contents can be read and understood (e.g. subscriber profile information).

16.3.4   Si interface between HSS – CAMEL

The CAMEL Application Server (IM-SSF) communicates with the HSS using the Si interface. The MAP protocol defined in 3GPP TS 29.002 is used for this interface.

16.3.5   ISC Interface between S-CSCF and AS

This interface between S-CSCF and the application servers (i.e., SIP AS, OSA SCS, or CAMEL IM-SSF) is used to provide SIP services for the IMS.

The SIP AS influences and impacts the SIP session based on the services desired. When a SIP session request is received by the S-CSCF, it decides whether some processing is required by AS or not. This decision is based on the filter information received from HSS (using SAR Cx message). The filter information contains the list of filter criteria, the name of AS associated with the filter, the priority of the filter, and the Service Point Trigger (SPT). Based on this information, if a request is to be sent to an AS, the S-CSCF sends it using the ISC interface. The ISC information is also used for the exchange of charging information.

16.3.6   Gm Interface between UE and CSCF

Communication between the UE and CSCF in IMS takes place using the SIP protocol over the Gm interface. For the UE, this is the most important interface. The main procedures applicable to the Gm interface are as follows:

- UE registering with an S-CSCF.

- Authentication between UE and S-CSCF.
- Originating a SIP session (MO session).
- Terminating a SIP session (MT session).

16.3.7 Mc Interface between MGCF and MGW

The MGCF communicates with MGW using MEGACO over Mc interface. The MEGACO messages are used by MGCF to establish bearer connections through MGW.

16.3.8 Mg Interface between MGCF and S-CSCF

This interface provides the means whereby a call originated in PSTN can be routed to the IMS endpoint via the MGCF and the S-CSCF. The Mg interface is based on SIP.

16.3.9 Mw Interface between x-CSCF and y-CSCF

This interface allows the Interrogating CSCF to direct mobile-terminated calls to the Serving CSCF.

16.3.10 Mi Interface between S-CSCF and BGCF

In case a call is intended for the PSTN network, the S-CSCF uses the Mi interface, which is based on SIP, to forward the session signaling to BGCF.

16.3.11 Mj Interface between BGCF and MGCF

This interface allows the BGCF to forward the session signaling to the MGCF for interworking with PSTN networks. The Mj interface is based on SIP.

16.3.12 Mk Interface between BGCF and BGCF

In case a BGCF determines that the breakout for a session signaling has to occur in another network, it forwards the session to another BGCF over the Mk interface. The next BGCF then takes further action. The Mk interface is also based on SIP.

16.4 IMS ADDRESSING

IMS users are addressed by two types of entity: the private identity (IMPI) and the public identity (IMPU). The following sections explain the nature of these two identities.

16.4.1 IMS Private User Identity (IMPI)

An IMS subscription is identified by the IMS Private User Identity (IMPI). The IMPI is used within the network for key functions like registration, authentication, authorization and accounting. The IMPI does not play any role in routing (i.e. it is not used for establishing contact with the IMS subscriber).

IMPI is assigned by the home network operator and is valid for the duration of the user's subscription with the home network. The IMPI is secured within the ISIM and it cannot be modified by the user. The IMPI is used for authentication during registration and re-registration, and is present in all registration, re-registration and de-registration requests.

The IMPI takes the form of Network Access Identifier (NAI), which is defined in RFC 2486. In simple terms, IMPI is of the form user@realm. For example, 'john@imsdomain.com' is a valid IMPI, where 'john' is the user and 'imsdomain.com' the realm.

It is possible that the UE does not have an ISIM. In this scenario, the IMSI of the USIM is used to derive the IMPI. The realm is made using the MCC and MNC of the IMSI and is of the form <mnc>.<mcc>.IMSI.3gppnetwork.org. For example, if IMSI is 243160999999999, with MCC=243, and MNC=16, the IMPI is '243160999999999@16.243.IMSI.3gppnetwork.org'.

16.4.2 IMS Public User Identity (IMPU)

Unlike the IMPI, the IMS Public User Identity (IMPU) is used to communicate with an IMS subscriber (i.e. an IMPU is used for SIP routing). Before an IMPU is used to originate/terminate IMS sessions, it must be registered with an S-CSCF. Exceptions are made in the case of services related to unregistered users (for example, call forwarding). Barring these, an IMPU must be registered before availing of services. The IMPUs are not authenticated during registration procedures.

The IMPU is a SIP URI (RFC3261) and is of the form sip:user@domain. Example of a valid SIP URI is 'sip:alice@atlanta.com'. There is at least one IMPU stored in the ISIM. Other IMPUs may or may not be stored in the ISIM. In case there is no ISIM application, the IMPU is of the form 'sip:<IMPI>'. For example, if the IMPI is '243160999999999@16.243.IMSI.3gppnetwork.org', then the IMPU for this is 'sip:243160999999999@16.243.IMSI.3gppnetwork.org'.

The IMPU can also be a E.164 number (a TEL URL). For example tel:+358-555-1234567.

16.4.3 Relationship of IMPI and IMPU

An IMS subscription, for which there can be one or more IMPUs, is identified by an IMPI. Each IMPU is associated with one, and only one *service profile*. This service profile has information like filter criteria, which determines how a SIP request is handled. A service profile can have more than one IMPU associated with it. However, all service profiles are associated with one and the same IMPI.

Further, all IMPUs of an IMPI register with the same S-CSCF. This S-CSCF uses the service profiles maintained for the IMPUs to provide various services. This restriction applies to Release 5 specifications. It can be relaxed for future release of IMS specifications.

Apart from these notions, the IMS defines another term—the *Implicit Registration Set*. This set contains one or more IMPUs associated with the same or a different profile. The IMPUs of an implicit registration set are registered or de-registered together. The HSS maintains the list of IMPUs belonging to the implicit registration set. When one of these is registered, all the other IMPUs belonging to the set are also registered at the same time. Similarly, when one of the IMPUs of the set is de-registered, the others belonging to the set are de-registered at the same time.

The relationship between the IMS subscription, IMPI, IMPU, Service Profile, Implicit Registration Set and S-CSCF is shown in Figure 16.4. This relationship is as per Rel5 of 3GPP specifications. In Rel6, there is a proposal to modify this relationship.

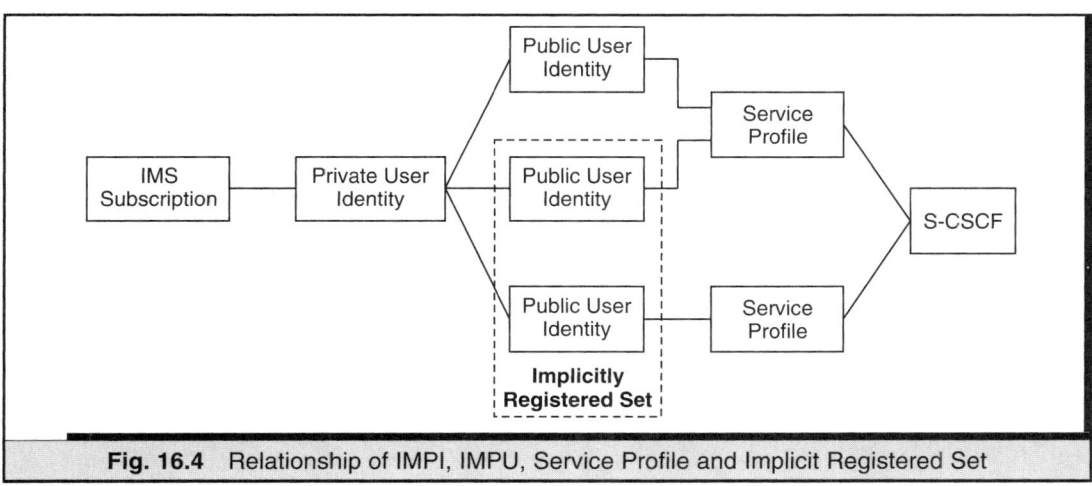

Fig. 16.4   Relationship of IMPI, IMPU, Service Profile and Implicit Registered Set

16.5   SUBSCRIBER DATA

To provide the various services to a subscriber, the subscriber is associated with various data elements, referred to as Subscriber Data. Subscriber data is required for various purposes, including subscriber identification, authentication, session handling, charging, operation and maintenance. Elements of subscriber data fall under two categories: Permanent Subscriber Data and Temporary Subscriber Data.

Permanent Subscriber Data is provisioned by the network operator and cannot be changed dynamically (e.g. the IMPI is a permanent data). Temporary Subscriber Data, on the other hand, may change dynamically. For example, the S-CSCF name is a temporary data as it may change when a user registers with another S-CSCF.

In the IMS, different network entities maintain different subsets of subscriber data required for their functioning. Table 16.4 lists various elements of subscriber data and the applicability of this data at HSS, CSCF, IM-SSF, and the AS.

Table 16.4 Subscriber Data maintained in the IMS

Type	Element	Relevant at	Description
Address/ Identifiers	IMPI	HSS, S-CSCF, P-CSCF	Uniquely identifies a subscription. Used for functions like registration, authentication, authorization and accounting.
	IMPU	HSS, S-CSCF, P-CSCF	Used for communicating with an IMS subscriber.
	Implicit Registration Set	HSS, S-CSCF, P-CSCF	Set that contains one or more associated IMPUs that are registered or de-registered together.
	Barring indication	HSS, S-CSCF	Indicates that the IMPU is barred from any IMS communication (except registrations and re-registrations).
Authentication	K	HSS	Long-term secret key, used to generate Authentication Vectors.
	SQN	HSS	Sequence number, used for synchronization.
	Authentication Vector	S-CSCF	Consists of: 1) Random Challenge (RAND); 2) Expected Response (XRES); 3) Cipher Key (CK); 4) Integrity Key (IK); and 5) Authentication Token (AUTN). These are used for authentication, confidentiality and data integrity.
Registration	Registration Status	HSS	Status of registration of an IMPU (e.g. registered, not registered, pending authentication or unregistered).
	S-CSCF Name	HSS	Identifies the S-CSCF allocated to the subscriber when the subscriber is registered. It is used during the mobile terminated sessions setup, registration, re-registration and de-registration.
	Diameter Client Address of S-CSCF	HSS	Used by HSS to communicate with S-CSCF at the Diameter level.
	Diameter Server Address of HSS	S-CSCF	Used by S-CSCF to communicate with HSS at the Diameter level.

Contd.

Table 16.4	Contd.

Type	Element	Relevant at	Description
Profile Information	Initial Filter Criteria	HSS, S-CSCF	Each set of filter criteria includes the Application Server Address, AS priority, Default Handling, Trigger Points and optional Service Information.
	Service Indication	HSS, AS	Identifies exactly one set of service related transparent data, which is stored in the HSS in an operator network.
	Subscribed Media Profile Identifier	HSS, S-CSCF	Identifies a set of session description parameters that the subscriber is authorized to request.
Others	Capability Information	HSS, I-CSCF	Used by the I-CSCF for the selection of an S-CSCF. Includes mandatory capabilities, optional capabilities, and preferred S-CSCFs for a service profile.
	Charging Information	HSS, S-CSCF	Contains addresses of entities providing the charging function.
	CAMEL-related Information	HSS and/or IM-SSF	Data related to support CAMEL functionality (e.g. CSI information and gsmSCF address).

16.6   SESSION-UNRELATED PROCEDURES

The IMS procedures can be classified into two distinct categories. In the first category are the procedures that are not directly used for session origination or termination. These are called *session-unrelated* procedures, which include the following:

1. Establishing IMS Transport
2. Registration
3. De-registration
4. Profile Update

16.6.1   Establishing IMS Transport

The IMS procedures assume the presence of an IP bearer. Thus, prior to communicating with the IMS, the transport is established. This involves three important steps. The first step in the establishment of IMS Transport is to execute the 'GPRS-attach' procedure. This procedure enables a UE to avail of packet services. For this, the UE sends the 'Attach' message to the SGSN. The SGSN then executes the necessary steps (including location update with HLR). Upon completion of these steps, the SGSN sends an

'Attach Complete' message to the UE. The details of GPRS-attach procedure were provided in Chapter 9.

On completion of Attach procedure, the UE activates the PDP context. Through this procedure, the UE obtains an IPv6 address for communicating with IP-based entities. The PDP context activation also establishes an association between SGSN and GGSN. For IMS, there are certain enhancements in the PDP context activation procedure. To support SIP, the UE may indicate, through the Access Point Name (APN), its desire to set the PDP context for SIP signaling. The SGSN uses the information provided through the APN to select a GGSN that can provide SIP services. The PDP context thus provided could be a dedicated one, in which case it is used exclusively for SIP signaling. Alternatively, a general purpose PDP context could be activated, which could then be used to carry SIP signaling as well as the bearer.

The third and the last step is to obtain the P-CSCF address. Since P-CSCF is the first contact point in the IMS, the UE needs to know its address in order to avail of IMS services. The P-CSCF address can be obtained through two mechanisms. It can either be sent by GGSN during the context activation procedure (Figure 16.5), or the UE could use the Dynamic Host Configuration Protocol (DHCP) for IPv6. Either of these means can be used to obtain the P-CSCF address.

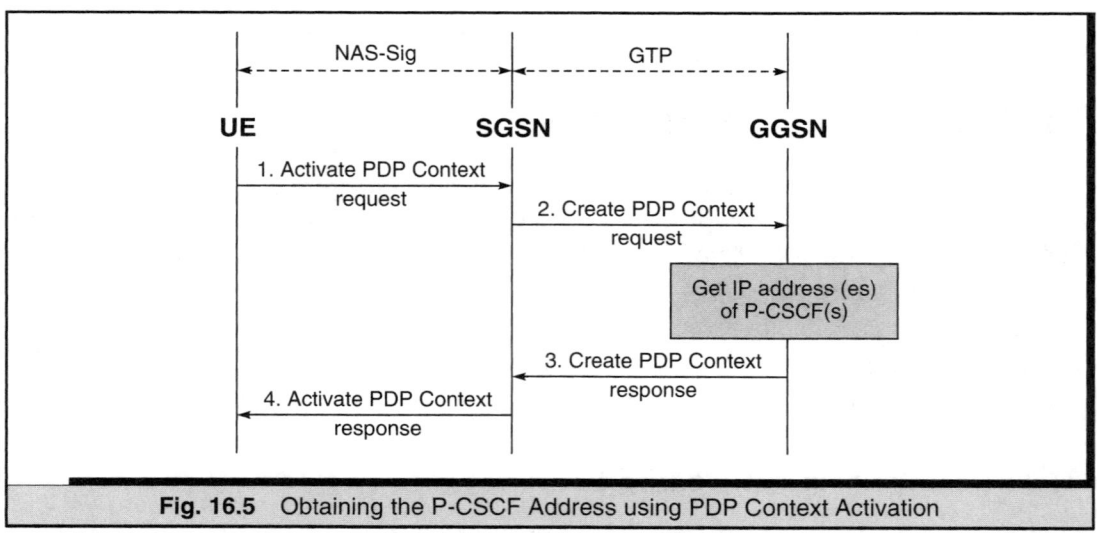

Fig. 16.5 Obtaining the P-CSCF Address using PDP Context Activation

16.6.2 Registration

After obtaining the P-CSCF address, the UE performs the registration procedure. It is mandatory for an IMS subscriber to register itself with the S-CSCF of the home network in order to avail of IMS services. This registration is required for several reasons. First, it allows the home network to authenticate and authorize the user before services are granted. The authorization check is based on the profile of the subscriber and operator limitations (if any). The second and more important reason is to choose an S-CSCF that will provide IMS services to the subscriber. The S-CSCF acts as a registrar and maintains the contact information of the user (e.g. the current IP address). This information is then used for various services, including routing of mobile-terminated calls.

Figure 16.6 shows the messages exchanged in the registration procedure. Here, it is assumed that the PDP context has been activated, the P-CSCF is discovered and the UE has the IP connectivity before the procedure starts. The steps involved in the registration procedure are summarized as follows:

1. The UE sends a 'Register' message to the proxy server (i.e. P-CSCF) with appropriate information (i.e. IMPI, IMPU, home domain name and the UE IP address).

2. The P-CSCF uses the 'home domain name' and DNS to forward the message to the I-CSCF of the home domain.

3. The I-CSCF, upon receipt of the message, requests the HSS to authorize the request of the user. This is done using the 'User-Authorization-Request (UAR)' message. If the user is authorized to register, the HSS sends capability information needed for I-CSCF to select an S-CSCF to which the registration request can be forwarded.

4. Upon receipt of 'User-Authorization-Answer (UAA)' message, the I-CSCF uses the received capability information to carry out the 'S-CSCF selection' procedure. The I-CSCF maintains as configuration information the capabilities of various S-CSCF in the network. The S-CSCF selection procedure is carried out using the capability information received from HSS and the configuration information maintained by I-CSCF. This procedure gives the I-CSCF the address of the S-CSCF to which the request is forwarded.

5. The 'Register' message is forwarded to the selected S-CSCF.

6. Upon receiving the 'Register' message, S-CSCF requests the HSS to send authentication vectors for authentication of the user. The request is made using the 'Multimedia-Authorization-Request (MAR)'.

7. HSS sends authentication vectors to S-CSCF using the 'Multimedia-Authorization-Answer (MAA)' message. It also stores the name of S-CSCF with which the user is currently registering. This information is subsequently sent to I-CSCF in the 'UAA' message so that I-CSCF chooses the same S-CSCF and another S-CSCF selection is not required.

8. Upon receiving the 'MAA' message, the S-CSCF uses the first authentication vector to authenticate the UE. For this, the parameters RAND, AUTN, CK and IK

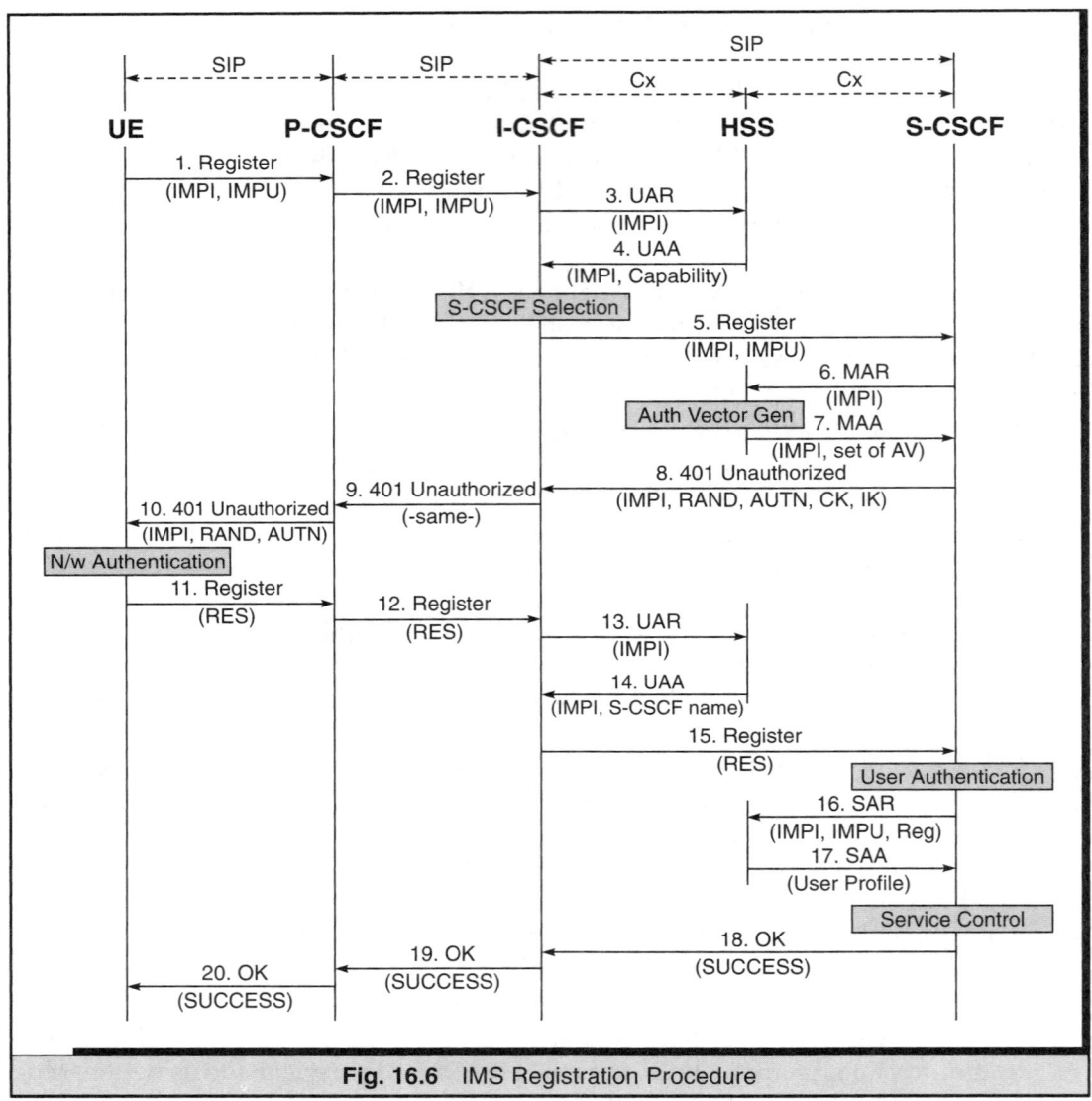

Fig. 16.6 IMS Registration Procedure

are sent to P-CSCF in the SIP '401 Unauthorized' message. The parameter XRES is not sent to P-CSCF (The XRES is used by S-CSCF to authenticate the user).

9. The I-CSCF forwards the '401 Unauthorized' message received from S-CSCF to P-CSCF.

10. P-CSCF keeps the cipher and integrity key (CK and IK), and sends the '401 Unauthorized' message containing the RAND and AUTN to UE.

11. Using the received information, the UE authenticates the serving network. In case of success, the UE computes the value of RES and sends it in the 'Register' message to P-CSCF.
12. The P-CSCF forwards the 'Register' message to I-CSCF.
13. I-CSCF uses the 'UAR' message to obtain the S-CSCF name.
14. The 'S-CSCF name' is returned to I-CSCF using the 'UAA' message.
15. The 'Register' message is then forwarded to S-CSCF.
16. Next, the S-CSCF authenticates the user. In case the received RES matches the stored XRES, the user authentication is complete. The S-CSCF then conveys the success of authentication to HSS using the 'Server-Assignment-Request (SAR)' message.
17. HSS responds by returning the user profile using the 'Server-Assignment-Answer (SAA)' message. The SAA message contains the list of implicitly registered public identities along with the subscriber profile information.
18. Using the user profile, the S-CSCF performs service control. This entails sending the registration information to service control platform (i.e. AS) and perform whatever service control procedures that are appropriate. S-CSCF also sends a success message to UE using SIP protocol.
19. The I-CSCF forwards the success message to P-CSCF.
20. The P-CSCF forwards this message to UE. The receipt of the message by UE completes the user-initiated registration procedure.

When registration is complete, the UE can avail of the IMS services. At this stage, a security association is formed between the UE and the P-CSCF which enables these two entities to communicate securely over the access network. Following registration, the S-CSCF has the UE IP Address along with the Proxy name/address and the P-CSCF network identifier. This is used for contacting the UE (e.g. for mobile-terminated sessions).

Apart from registration procedures, there are re-registration procedures as well. The re-registration is required by the UE to refresh an existing registration. This is done because a registration is applicable for a finite duration only. The procedures of re-registration are similar to those of registration except that some of the message flow may not be required (for example, S-CSCF may choose not to authenticate the user for re-registration).

16.6.3 De-registration

De-registration is the cancellation of registration of a user. The de-registration procedure may be initiated by a user when he/she does not wish to avail any services from the network. This procedure may also be initiated by the network. The network-initiated de-registration procedure can be initiated in the following scenarios:

- When the registration timer maintained for a user expires
- For network maintenance purposes (e.g. in case of a lost SIM card)

- Change of service profile
- Change in roaming restrictions
- Non-payment of charges

Both user-initiated and network-initiated de-registration procedures are explained in the following sub-sections.

16.6.3.1 *User-Initiated De-registration*

After a UE has availed the services of IMS and wants to de-register from it, it follows the de-registration procedure. De-registration involves sending a Register request with expiration time of zero. The data flow is similar to the Registration flow (Figure 16.7). The steps involved in the de-registration procedure can be summarized as follows:

1. The UE sends a 'Register' message to the P-CSCF with expiration time set to zero.
2. The P-CSCF uses the 'home domain name' to forward the 'Register' message to I-CSCF of the home domain.
3. The I-CSCF, upon receipt of the message, requests HSS to send the name of the S-CSCF with which the user is registered. This is done using the 'User-Authorization-Request (UAR)' message.
4. The HSS responds with the S-CSCF name (if maintained by it) using 'User-Authorization-Answer (UAA)'.

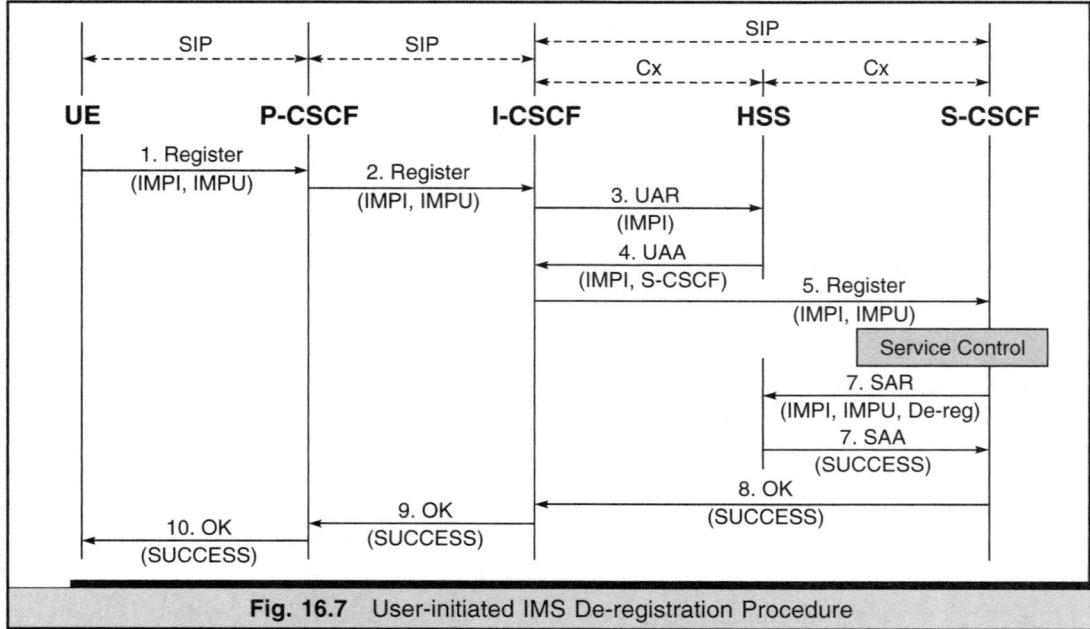

Fig. 16.7 User-initiated IMS De-registration Procedure

5. Upon receipt of 'UAA', I-CSCF forwards the request to the S-CSCF.
6. S-CSCF processes the de-registration request and sends the de-registration information to the service control platform. The filter criteria is used for sending this information. The service control platform removes all subscription information for the subscriber. The S-CSCF then sends the 'Server Assignment Request (SAR)' message to inform HSS about the de-registration.
7. HSS acknowledges the message by sending 'Server-Assignment-Answer (SAA)'. Moreover, depending upon implementation, the S-CSCF may or may not remove the profile information stored by it for a registered user. In case the S-CSCF wishes to store the profile information for a subscriber and wants the HSS not to remove the S-CSCF name associated with the subscriber, it indicates this to HSS in 'SAR' message. If HSS accepts the request, it informs S-CSCF through the 'SAA' message, in which case the profile information can be used for future registrations. In case S-CSCF does not want HSS to keep the S-CSCF name, it indicates so in the 'SAR' message. The HSS has the option to keep or remove the S-CSCF name. Whatever be the case, HSS informs S-CSCF about its decision in the 'SAA' message. This functionality is similar to the *Super Charger* functionality discussed in Chapter 9.
8. The S-CSCF sends a success message to UE.
9. The I-CSCF forwards the message to P-CSCF.
10. P-CSCF forwards the message to UE. The P-CSCF also removes the registration information for this specific registration of the user. The receipt of the message by UE completes the user-initiated de-registration procedure.

16.6.3.2 *Network-Initiated De-registration*

As mentioned earlier, the home network can also initiate de-registration procedure for various reasons (e.g. due to network maintenance, loss of SIM, authentication failure, contract expiry, etc). The de-registration procedure can either be initiated by the HSS or by S-CSCF. In this section, the de-registration procedure initiated by HSS is explained. The de-registration procedure initiated by S-CSCF is on similar lines. Its details are available in 3GPP TS 23.228 and TS 29.228.

The HSS-initiated de-registration procedure is shown in Figure 16.8. In this scenario, the I-CSCF does not come in the picture. The S-CSCF has information about the P-CSCF to which the de-registration request is to be sent. This information is obtained during the course of the registration procedure. Note that the Cx interface is only between the I-CSCF and HSS and between S-CSCF and HSS; the P-CSCF has no Cx interface with the HSS.

The steps involved in the HSS-initiated de-registration are summarized as follows:

1. For administrative or other reasons, the HSS initiates a de-registration procedure by sending a 'Registration-Termination-Request (RTR)' message to S-CSCF. The

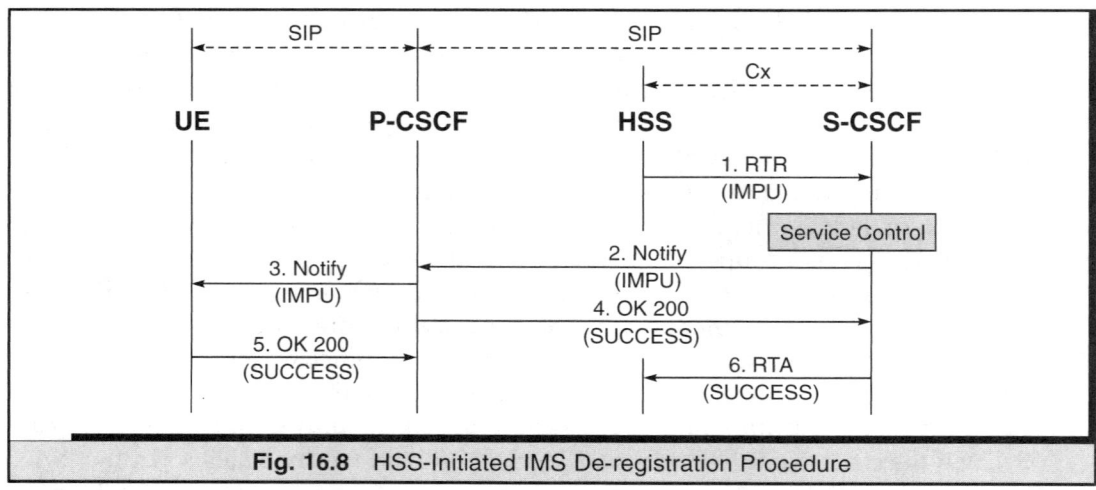

Fig. 16.8   HSS-Initiated IMS De-registration Procedure

'RTR' message contains the reason for de-registration, as well as an IMPI or an IMPU, or a set of IMPUs that are to be de-registered. In the figure, only an IMPU is shown. The detailed procedure at S-CSCF and P-CSCF differs when other choices are exercised.

2. On receipt of 'RTR', the service control procedure is carried out by S-CSCF by sending the de-registration information to the service control platform. The filter criteria is used for sending this information. The S-CSCF also informs the UE of the de-registration event, provided that the UE has subscribed to this service. For this, the S-CSCF sends a 'NOTIFY' message containing one or more IMPUs to the P-CSCF.

3. The P-CSCF then forwards this message to the UE, and also deletes the stored information for the IMPUs that are de-registered. In case there are no more registered IMPUs of an IMPI, the P-CSCF removes any security association that it maintains with the UE.

4. The P-CSCF then acknowledge the receipt of the message to S-CSCF.

5. The UE also acknowledges the receipt of message to P-CSCF.

6. The S-CSCF sends the 'Registration-Termination-Answer (RTA)' message to HSS to complete the HSS-initiated de-registration procedure.

16.6.4   Profile Update

Apart from the registration, re-registration and de-registration procedure, there is the Profile Update procedure, applicable to the Cx interface. Whenever the profile information for a subscriber is modified at HSS, it sends the complete profile to the S-CSCF (provided that at least one IMPU is registered with the S-CSCF). The Profile Update procedure is carried out using the Push-Profile-Request (PPR) and Push-Profile-Answer (PPA) Cx messages.

16.7 SESSION-RELATED PROCEDURES

The session-related procedures use the Session Initiation Protocol (SIP) and Session Description Protocol (SDP) to establish/release SIP sessions. While the SIP is used to manage SIP sessions, the SDP is used to describe the nature of media required for communication.

A typical call flow in IMS (for mobile-originated session) is shown in Figure 16.9. The session establishment procedure is initiated by sending an INVITE to destination end-point. This is followed by negotiation as to the media required for communication, using the SDP protocol. When the negotiation is over, the resources are reserved and the session establishment procedure is confirmed (using OK and ACK). Thereafter, the session is in progress.

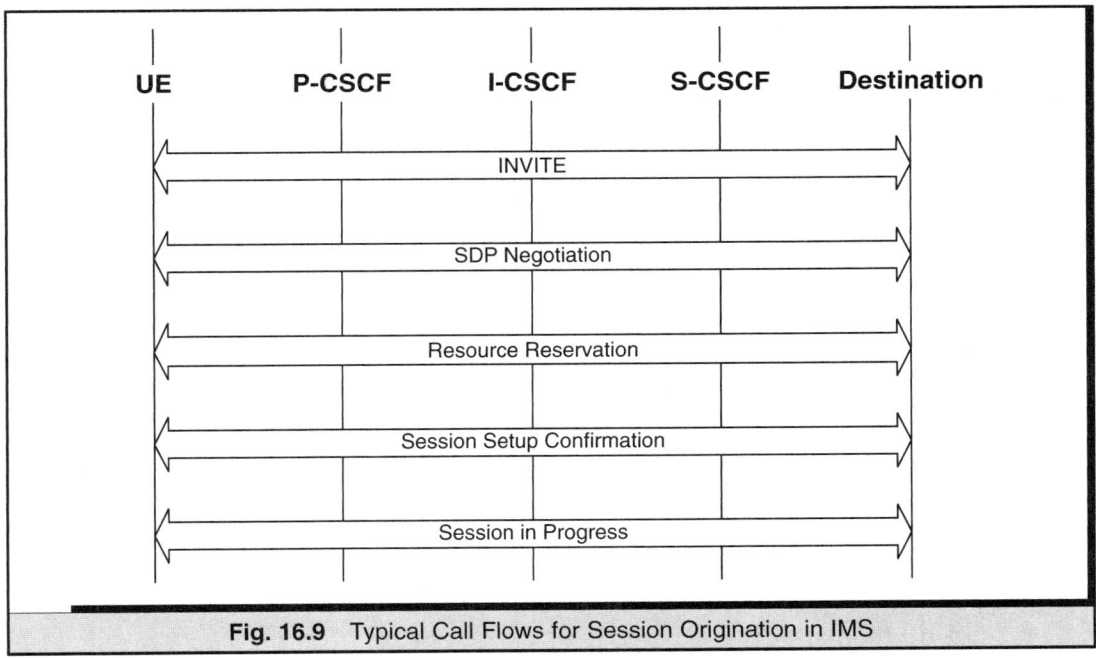

Fig. 16.9 Typical Call Flows for Session Origination in IMS

A complete session establishment procedure can be viewed as a concatenation of three distinct procedures, namely

1. Session origination procedure

2. Interworking procedure between two S-CSCFs (Note that interworking may also take place between an S-CSCF and BGCF),
3. Session Termination procedure.

Each of these procedures is explained in the following sub-sections. Before delving into specific call flows of session-related procedures, a brief description of service control is provided. Service control describes the interaction between the S-CSCF and AS.

16.7.1 Service Control

In the IMS, Service Control is carried out by the S-CSCF and the application servers. The following steps summarize how services are implemented in the IMS:

1. During registration, the S-CSCF downloads profile information stored at the HSS. This information contains, besides other things, the Initial Filter Criteria (IFC).
   Note: 3GPP TS 23.218 also introduces the concept of subsequent Filter Criteria (sFC). But this term is not clarified in the specifications.
2. The two main components of an IFC are the Service Point Triggers (SPTs) and address of AS.
3. The SPT can be set on various elements as follows:
   - Any known or unknown SIP method (e.g. REGISTER, INVITE, SUBSCRIBE, MESSAGE).
   - Presence or absence of any header field.
   - Content of any header field or Request-URI.
   - Direction of the request with respect to the served user (i.e. mobile-originated or mobile-terminated session).

   When any of these SPT criteria is fulfilled, the corresponding AS is contacted for further processing.
4. In order to allow the S-CSCF to handle the different filter criteria in the right sequence, a priority is assigned to each IFC. The same priority is not assigned to more than one IFC for a given user.
5. In case more than one IFC is sent from the HSS to the S-CSCF, the S-CSCF handles a received SIP message by checking the IFC one by one according to their indicated priority. The steps followed by the S-CSCF are:
   (a) Set up the list of IFC according to their priority.
   (b) Parse the received request in order to find out the Service Point Triggers (SPTs) that are contained in the message.
   (c) Check whether the trigger points of the next highest priority filter criteria are matched by the SPTs of the request. If the two do not match, the S-CSCF

shall immediately proceed further. If the two match, the S-CSCF takes the following actions:

  – Adds an indication to the request which allows the S-CSCF to identify the message on the incoming side, even if its dialog identification has been changed, e.g. due to the AS executing third party call control.
  – Forward the request via the ISC interface to the AS indicated in the current IFC. The AS then performs the service logic, may modify the request and may send the request back to the S-CSCF via the ISC interface.
  – Proceed further if the request was received again by the S-CSCF from the AS via the ISC interface.

  (d) Repeat the above steps for every IFC that was initially set up, until the last IFC has been checked.
  (e) Route the request based on the normal SIP routing behaviour.

6. The AS may act in various ways to influence a SIP session when a SIP message is received by it from S-CSCF over the ISC interface. In particular, the AS may act in one of the following manners:

  • It may act as a terminating user agent,
  • as a originating user agent,
  • as a redirect server,
  • or it may act as a Back-to-Back User Agent (B2BUA).

For more details on service control in the IMS, refer to 3GPP TS 23.228 and TS 23.218.

16.7.2 Session Origination

There are various scenarios related to session origination. Two important ones are as follows:

  1. SIP session originating in the IMS network and terminating in any network.
  2. Call originating in PSTN network and terminating in the IMS network.

16.7.2.1 Session Origination in IMS

The IMS session origination is carried out through the SIP INVITE message. The exact message flow depends upon whether the user is roaming or is in the home network. In case the user is roaming, another aspect determines the message flows. This aspect is related to Topology Hiding and determines whether a network operator is willing to expose the network topology information or mandates that all messages flow through an I-CSCF. In the example presented, it is assumed that the user is roaming and all external message flows are through an I-CSCF.

Figure 16.10 shows the messages exchanged during an MO session. The steps involved in the registration procedure are summarized as follows:

1. UE sends a SIP 'INVITE' message to the proxy server (i.e. P-CSCF). The 'INVITE' message contains the originating party's SIP address, the destination address and an initial SDP. The initial SDP may represent one or more media for a multimedia session.
2. The P-CSCF remembers (using the registration procedure) the next hop for UE. In this example, the P-CSCF forwards the message to I-CSCF.
3. The I-CSCF forwards the message to S-CSCF.
4. The S-CSCF executes the service control by validating the service profile and carrying out any 'origination service control' required for this subscriber. This includes authorization of the requested SDP based on the user's subscription for the multimedia service. For example, the S-CSCF may reject the session if the originating user is using an identity that is barred from communication. The S-CSCF then forwards the message. This message can be sent to another S-CSCF or a BGCF in case of interworking. The various scenarios are explained in Section 16.7.3 where the handling of a session at S-CSCF is discussed.
5. The terminating party responds with its own SDP capabilities.
6. The S-CSCF forwards the 'SDP answer' message to I-CSCF.
7. The I-CSCF forwards the 'SDP answer' message to P-CSCF.
8. The P-CSCF authorizes the resource necessary for this session. It then forwards the 'SDP answer' message to the originating UE.
9. Based on the SDP answer, the originating UE determines the SDP to be offered. The offered SDP is labeled as 'SDP off' in the figure.
10. The P-CSCF forwards the 'SDP offered' message to I-CSCF.
11. The I-CSCF forwards the 'SDP offered' message to S-CSCF.
12. The S-CSCF forwards the 'SDP offered' message to the terminating endpoint.
13. The terminating endpoint responds with an answer.
14. The S-CSCF forwards the 'SDP answer' message to I-CSCF.
15. The I-CSCF forwards the 'SDP answer' message to P-CSCF.
16. The P-CSCF forwards the 'SDP answer' message to the originating UE.
17. When the originating UE receives the 'SDP answer' message, the resource reservation process is successfully complete. The UE indicates this by sending the 'SUCCESS' message.
18. The P-CSCF forwards the 'SUCCESS' message to I-CSCF.
19. The I-CSCF forwards the 'SUCCESS' message to S-CSCF.
20. The S-CSCF forwards the 'SUCCESS' message to the terminating endpoint.
21. The terminating endpoint responds with an answer.
22. The S-CSCF forwards the 'SDP answer' message to I-CSCF.
23. The I-CSCF forwards the 'SDP answer' message to P-CSCF.

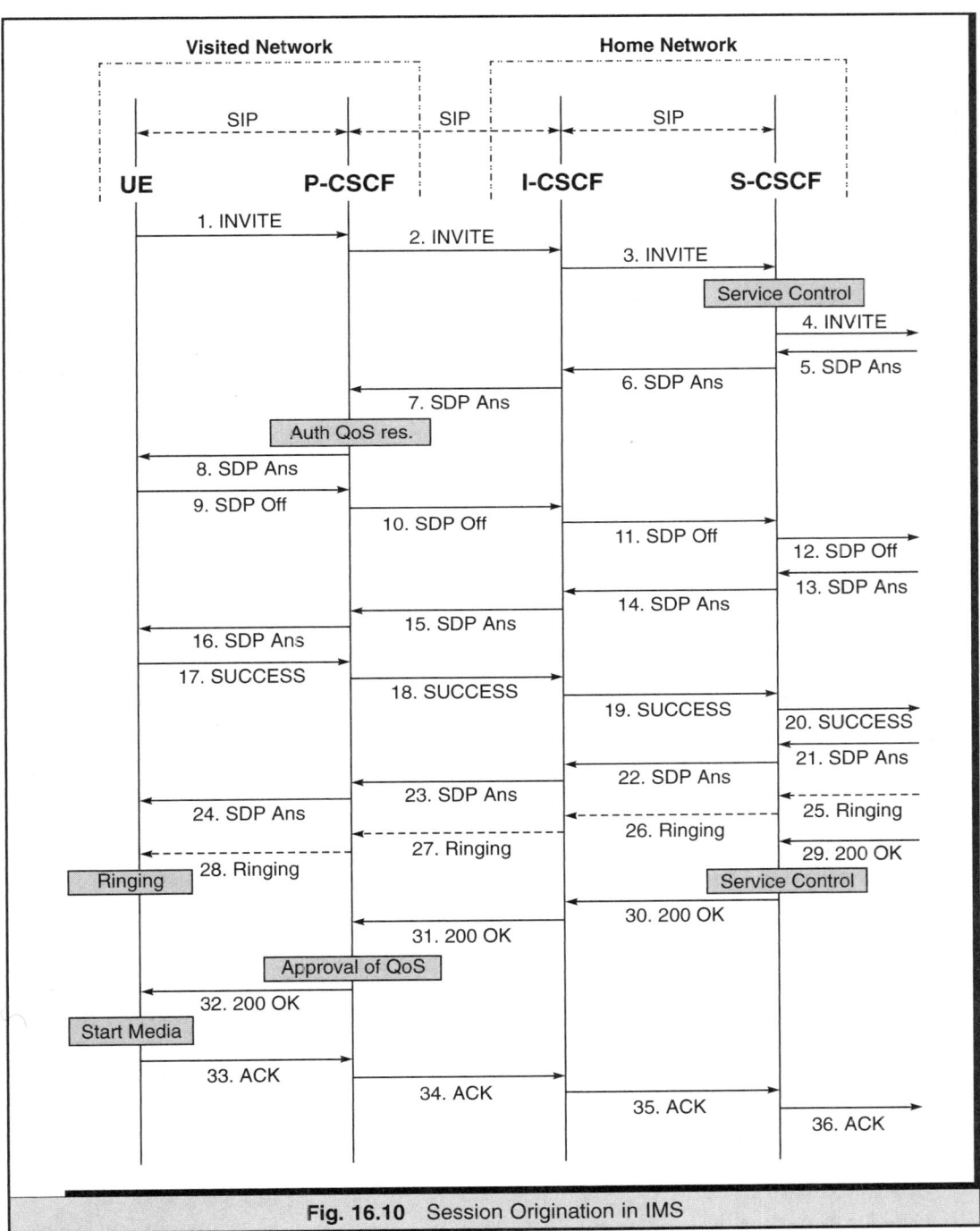

Fig. 16.10 Session Origination in IMS

24. The P-CSCF forwards the 'SDP answer' message to the originating UE.
25. Optionally, the terminating endpoint may respond with the 'Ringing' message.
26. If a 'Ringing' message is received, the S-CSCF forwards the message to I-CSCF.
27. The I-CSCF forwards the 'Ringing' message to P-CSCF.
28. The P-CSCF forwards the 'Ringing' message to the originating UE. If a 'Ringing' message is received, the UE indicates the event to the originating endpoint.
29. When the destination party answers, the terminating endpoint sends a '200 OK' message to S-CSCF.
30. The S-CSCF processes the message and conducts service control. It then forwards the '200 OK' message to I-CSCF.
31. The I-CSCF forwards the '200 OK' message to P-CSCF.
32. On receipt of the '200 OK' message, the P-CSCF indicates its approval regarding the use of the resources reserved for the session. The P-CSCF also forwards the '200 OK' message to the originating UE.
33. The UE starts the media flow (i.e. actual data transfer). The UE also sends an 'ACK' message to P-CSCF.
34. The P-CSCF forwards the 'ACK' message to I-CSCF.
35. The I-CSCF forwards the 'ACK' message to S-CSCF.
36. The S-CSCF forwards the 'ACK' message to the terminating endpoint.

16.7.2.2 Call Origination in PSTN network

The previous section depicted the scenario in which the session originated in the IMS. It is possible that a call originates in the PSTN network and terminates in the IMS. For this, the ISUP signaling is used between the originating endpoint and the MGCF. The MGCF uses the MEGACO protocol with MGW to establish bearer channels. The MGCF also uses the SIP protocol to communicate with S-CSCF.

Figure 16.11 shows the messages exchanged during the origination of call in the PSTN. The steps involved in this procedure are summarized as follows:

1. A PSTN signaling node sends the 'Initial Address Message (IAM)' ISUP message to the MGCF. Upon receipt of this message, the MGCF initiates a H.248 interaction to seize a trunk and an IP port.
2. The MGCF then sends the SIP 'INVITE' message to initiate the session establishment procedure. This message contains an initial SDP.
3. The terminating IMS endpoint party responds with its own SDP capabilities using the 'SDP answer' message.
4. On receipt of the 'SDP answer' message, the MGCF initiates a H.248 command to modify the connection parameters and instructs the MGW to reserve the resources for the session. The MGCF then responds with the 'SDP offered' message.
5. The terminating IMS endpoint party responds with the 'SDP answer' message.

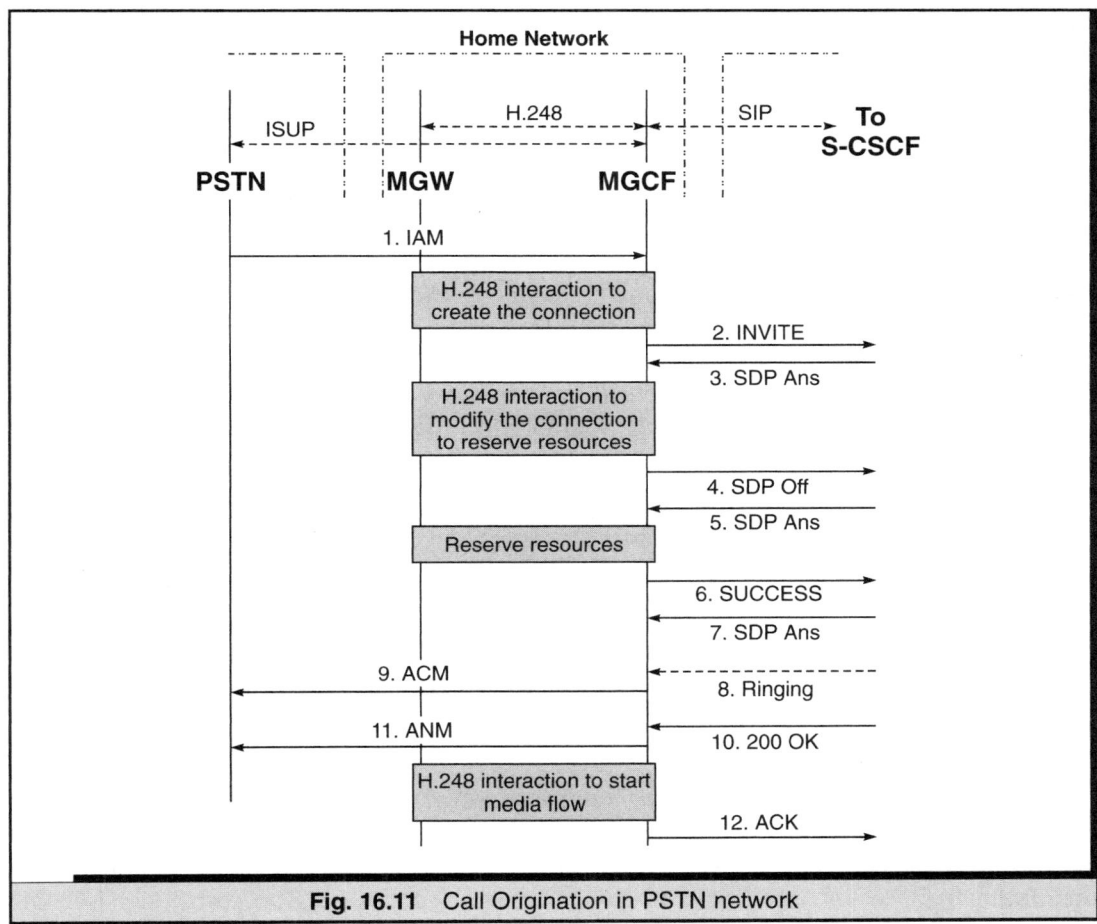

Fig. 16.11 Call Origination in PSTN network

6. The MGW then reserves the resources. After resource reservation, the MGCF sends the 'SUCCESS' message.
7. The terminating IMS endpoint party responds with the 'SDP answer' message.
8. Optionally, the IMS endpoint sends a 'Ringing' message.
9. If the terminating IMS endpoint sends a 'Ringing' message, the MGCF sends 'Address Complete Message (ACM)' ISUP message to the originating node.
10. The terminating IMS endpoint responds with '200 OK' message.
11. The MGCF sends the 'Answer Message (ANM)' ISUP message to the originating node.
12. MGCF initiates a H.248 command to alter the connection at MGW to make it bi-directional. The MGCF also sends the 'ACK' message to the terminating IMS endpoint.

16.7.3 Interworking Procedure

The previous sections detail the procedures necessary for session initiation. This section details the procedures for interworking between two S-CSCFs or between an S-CSCF and a BGCF.

Interworking between two S-CSCFs is required when both the originating and terminating endpoints are in the IMS. This scenario is depicted in Figure 16.12. In this scenario, the originating and terminating endpoints belong to the same network. Hence, both the S-CSCFs are shown in the same home network. Had this not been the case, the S-CSCF would have been reached by an I-CSCF of another network. In such a scenario, it is possible that there were two I-CSCFs in the path (instead of one). The first I-CSCF is used by the first operator to conceal the position of its S-CSCF. The second I-CSCF is used by the second operator to hide its own S-CSCF.

In Figure 16.12, the basic message exchanges are the same as those in the Session Origination procedure. One important addition in the LIR/LIA message exchange is the interface between the I-CSCF and HSS. The Location Information Request (LIR) is used by I-CSCF to obtain the address of the S-CSCF with which the terminating UE is registered. This is required because the I-CSCF does not have any information about the S-CSCF. The HSS obtains this information when the user is registered. This scenario can contrast with the session origination scenario (of Section 16.7.2.1) where the I-CSCF can derive the S-CSCF address from the home network contact point information sent to P-CSCF during registration. This information is not a available in terminating sessions.

In another scenario, the S-CSCF of the originating endpoint may determine that the session is not meant for an IMS user. In the scenario depicted in Figure 16.12, the S-CSCF used the terminating endpoint's address to determine that the session was to terminate in its own network, and hence forwarded the session to a local I-CSCF. However, if the S-CSCF determines that the session needs to terminate in the PSTN, it chooses a breakout gateway BGCF to forward the session further. This scenario is depicted in Figure 16.13. In this scenario, two sub-scenarios are possible. In the first, the BGCF determines that the breakout has to occur locally, in which case it forwards the session signaling to an MGCF. Alternatively, the BGCF forwards the session signaling to another BGCF. This corresponds to the scenario where the session breaks out in another network.

16.7.4 Session Termination

The Session Termination scenario is almost identical to the Session Origination scenario explained in Section 16.7.2. Only the direction of the message is reversed. Hence, the details of session termination are not explained any further.

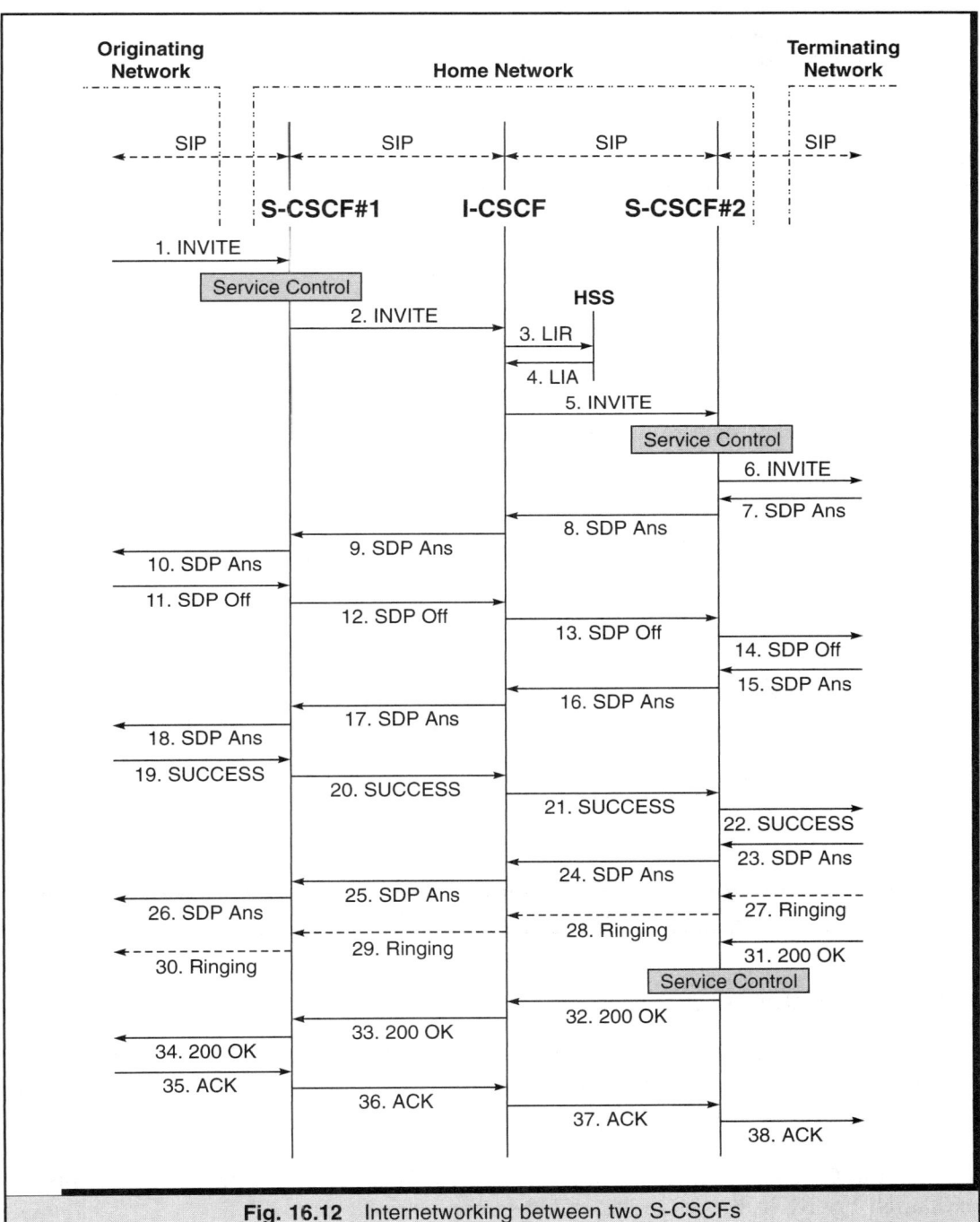

Fig. 16.12 Internetworking between two S-CSCFs

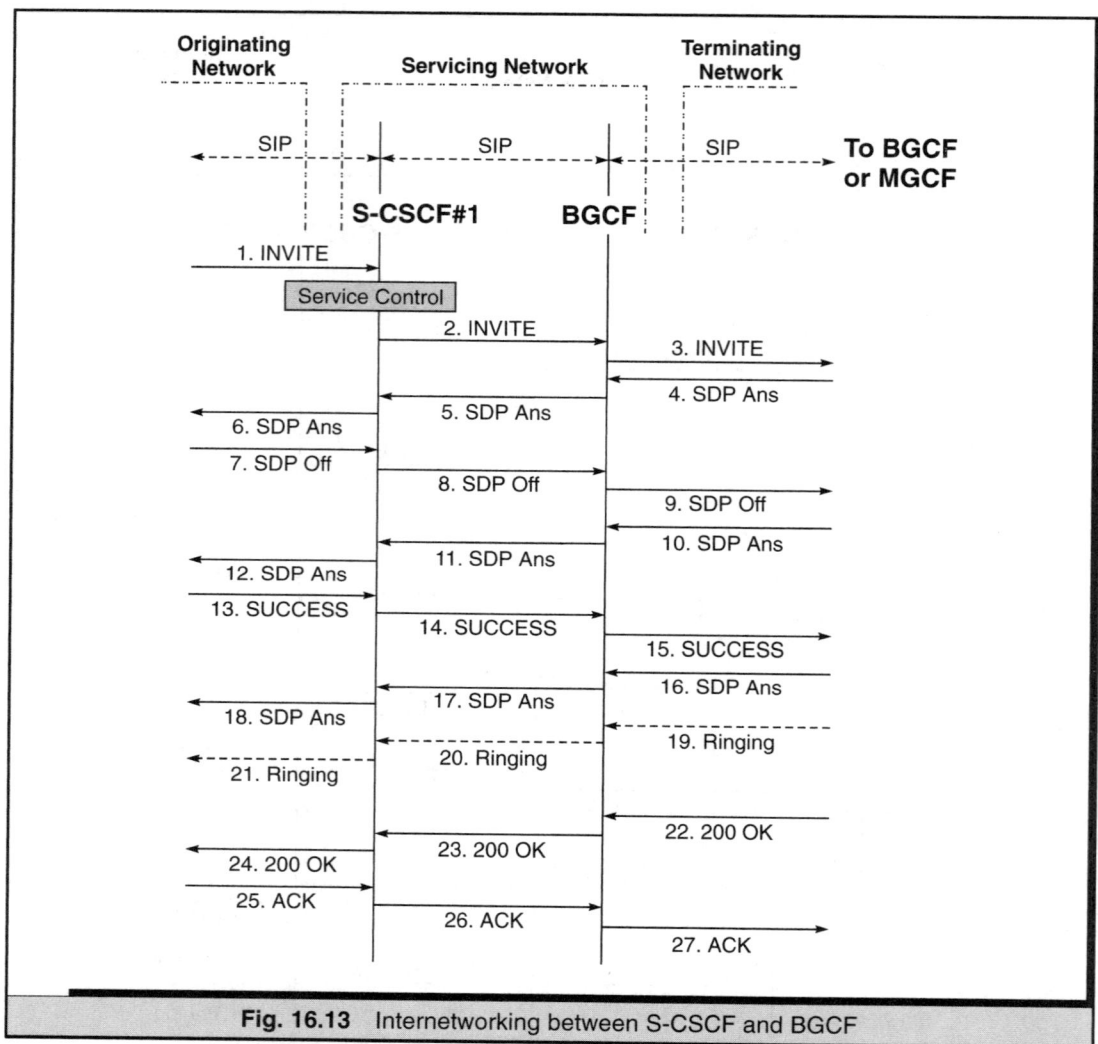

Fig. 16.13 Internetworking between S-CSCF and BGCF

16.8 IMS PROTOCOLS

In the discussion so far, the protocols used in the IMS have briefly been mentioned. The most important among these is the Session Initiation Protocol (SIP), which carries a brief explanation in the next sub-section. Apart from SIP, a relatively new protocol 'Diameter' is used for communication with the HSS. This protocol has two variants, one for the Cx/Dx interface and another for the Sh interface. Both these variants are explained briefly in the following sub-sections.

Apart from SIP and Diameter, there is the MAP protocol for the Si interface. The MAP protocol and the ISUP protocol were discussed in Chapter 6.

Then, there are the MEGACO and SDP protocols, details of which are not within the scope of this book. For more information on these, the reader is referred to the relevant specifications as given in the Reference section of this book.

16.8.1 Session Initiation Protocol (SIP)

Session Initiation Protocol (SIP) is used to establish, modify and tear down SIP sessions between one or more endpoints in an IP network. SIP is a textual client-server protocol and is based on the Simple Mail Transfer Protocol (SMTP) and Hypertext Transfer Protocol (HTTP). In particular, the SIP reuses a lot of syntax and semantics of the HTTP protocol. SIP resides on TCP or the UDP.

The details of SIP are beyond the scope of this book. The reader may refer to RFC 3261 and 3GPP TS 24.229 for more on the subject.

16.8.2 Diameter

The diameter protocol derived its name from Radius, which has been used for Authentication, Authorization and Accounting (AAA). As Diameter is more comprehensive than Radius, this protocol was so named.

The Diameter protocol is unique in the sense that it cannot be used alone. Hence, the protocol specified in RFC 3588 is called the Diameter Base Protocol (DBP). The DBP can only be used in conjunction with another protocol; it is then referred to as the Diameter Extension protocol. The DBP provides generic functionality, while the extension protocol provides specific functionality. The generic functionality of the DBP includes the following:

- Exchange of Attribute Value Pairs (AVP). AVP is the basic unit of information exchange in Diameter. The AVP is a tuple consisting of an AVP-type, AVP-length and AVP-data. There can be one or more AVPs in a single message. A set of AVPs can be grouped together to form a Grouped AVP. The exact definition of a Diameter is provided using the Augmented Backus-Naur Form (ABNF) specified in RFC 2234.
- Exchanging and negotiating protocol capabilities so that only supported messages are exchanged between peers.
- Establishing and terminating user sessions.
- Monitoring the state of peer connections.
- Exchanging accounting information.

The DBP is not developed by 3GPP. It is an IETF specification and is used for a diverse set of users including the Mobile IP and NASREQ. The 3GPP also uses the DBP for the

Cx/Dx and Sh interfaces of the IMS. There is an extension of Diameter for Cx/Dx (TS 29.229) and another extension for the Sh interface (TS 29.329).

Another important aspect of the DBP in relation to IMS is that only a subset of the protocol is applicable to 3GPP extensions (i.e. for Cx/Dx and Sh interface). The complete DBP is not used, as it is not required for the 3GPP specific extensions. The 3GPP TS 29.229, which specifies the Diameter extension for the Cx and Dx interface, lists the restrictions of the DBP when used in the IMS. These restrictions include:

- Support of only the SCTP protocol as the transport layer. Diameter allows use of both the SCTP and TCP transport layer protocols.
- No support for Accounting features.
- No use of sessions.

The above restrictions are applicable to the Sh interface as well. Given these restrictions, only a few messages, defined in the DBP, are applicable in the IMS. They are Capability-Exchange-Request (CER) and Capability-Exchange-Answer (CEA). These are the initial messages exchanged as soon as the SCTP transport layer connection is established between the communicating peers. Then there is the Device-Watchdog-Request (DWR) and Device-Watchdog-Answer (DWA), which are exchanged between peers to monitor the health of the Diameter application. Further, when the Diameter node does not wish to continue communication with a peer, it sends a Disconnect-Peer-Request (DPR). This is acknowledged by the peer using the Disconnect-Peer-Answer (DPA). The extension messages are exchanged following the exchange of the CER and CEA. Figure 16.14 shows one such scenario, where the UAR and UAA message is exchanged between the I-CSCF and HSS over the Cx interface. In the figure, the exchange of DBP messages is also highlighted.

Figure 16.14 assumes that there is direct communication between the I-CSCF and HSS. This may not always be true. To make the Diameter-based communication scalable and extensible, the Diameter defines three intermediate nodes or agents, which are *relays*, *proxies* and *redirects*. The relays and proxies forward a received Diameter message to the next hop using the Destination host address or the Destination realm. The redirect, on the other hand sends a message back to the sender with information that helps the sender in forwarding the message to the destination. The concept of redirection is used for the Dx interface, which exists between the I-CSCF/S-CSCF and the Subscriber Locator Function (SLF). In case the I-CSCF/S-CSCF do not know the HSS address, they forward the message to SLF, which redirects it back with the address of the HSS. This address is then used for sending the message to HSS.

16.9 SECURITY IN IP MULTIMEDIA SUBSYSTEM

In the IMS, the security features can be broadly categorized under, *access security* and *network domain security*. Access security refers to corroborating the authenticity of a

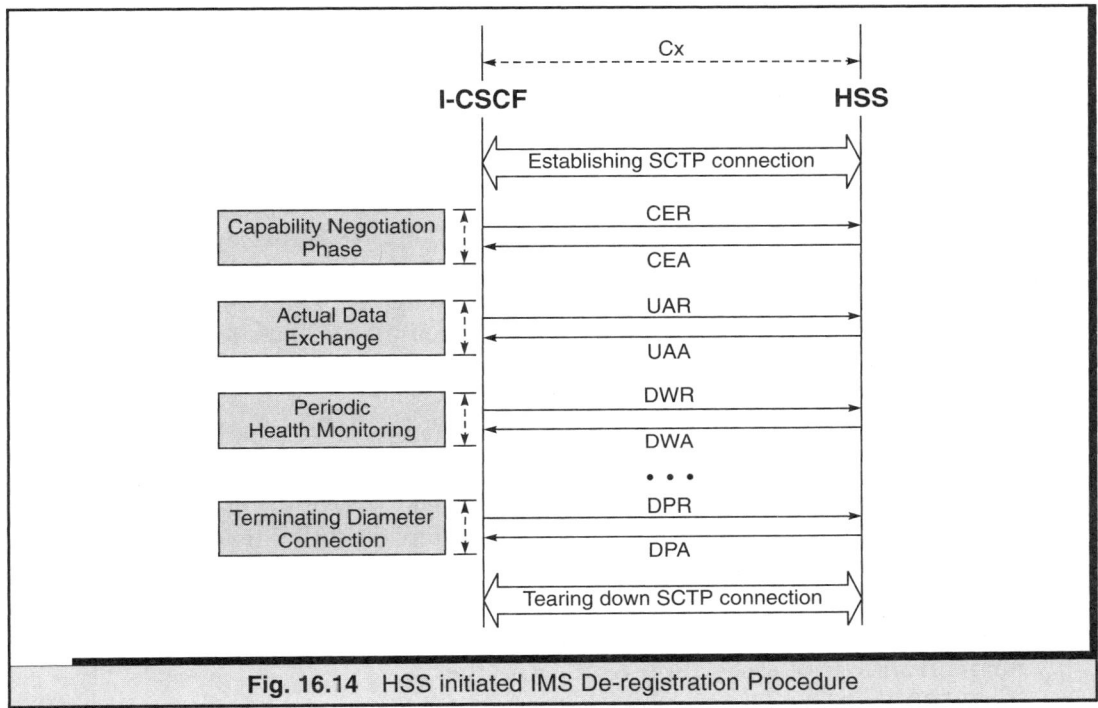

Fig. 16.14 HSS initiated IMS De-registration Procedure

user before granting it access to the IMS services. The authentication procedure is executed at the time of registration/re-registration. Besides authentication, other security features like *confidentiality* and *integrity protection* also come under access security.

Apart from access security, the IMS also has security features for communication between two network nodes (e.g. between P-CSCF and S-CSCF). This secured network communication comes under the purview of Network Domain Security (NDS). The NDS is applied at the IP layer using the IPSec protocol.

The NDS for IP was discussed in Chapter 14. The access security in IMS is discussed in the following subsection.

16.9.1 Access Security

Before an IMS subscriber can avail of the services provided by the home network, it must first register at the S-CSCF and prove its authenticity. The procedures to achieve this come under the purview of IMS access security. In the UMTS, access security is achieved using the UMTS AKA procedures (that were discussed in Chapter 14). The IMS access security is based on UMTS AKA, and is known as the IMS AKA procedure.

The features of IMS AKA can be summarized as follows:

- The authentication is based on the IMPI. The IMPI in IMS is equivalent to IMSI used in the CS/PS domain.
- The IMS authentication keys and functions at the user side reside on the IM Services Identity Module (ISIM). The ISIM may or may not be the same as the USIM.
- The authentication is done using a long-term secret key shared between ISIM and the Home Subscriber Server (HSS). Apart from the secret key, the entities also maintain the sequence number SQN, also called SQN_{ISIM} and SQN_{HSS} at the ISIM and HSS respectively. The handling of SQN is similar to that in UMTS AKA.
- The generation and use of authentication vectors (RAND, XRES, CK, IK and AUTN) is similar to that of the UMTS.
- The transport of the authentication vectors between S-CSCF and HSS takes place over the Cx interface using the MAR/MAA message. The exchange of authentication information between the S-CSCF and UE takes place using the SIP protocol.
- While the HSS is responsible for the generation of authentication vectors, the S-CSCF actually authenticates the IMS user.
- After the IMS AKA procedure is executed, the CK/IK at the UE and P-CSCF are available for confidentiality and integrity protection. For this, the UE and P-CSCF maintain a security association with each other.
- The 3GPP TS 33.203 defines the scope of confidentiality and integrity protection in IMS. According to this specification, the following are applicable:
  - Confidentiality is not applied to SIP signaling messages between the UE and the P-CSCF.
  - Integrity protection applies between the UE and the P-CSCF for protecting the data integrity of SIP signaling messages.
- Between the P-CSCF and S-CSCF the security mechanisms are applied using the IP Security.

SUMMARY

This chapter provided an overview of the IMS entities (i.e. CSCF, HSS, MGC, MGCF and BGCF), IMS addressing scheme (i.e. IMPI, IMPU and Implicit Registration Set), IMS procedures (session-unrelated and session-related) and IMS protocols (SIP and Diameter).

The IMS subsystem has been introduced by 3GPP in Rel5 specifications. At the time of its introduction, IMS was supposed to be an overlay over PS domain. As there were pressures on 3GPP to complete the specifications in time, only a basic infrastructure was ready by June 2003, when further work on Rel5 was frozen it may be noted that work on Rel5 specifications continued even after its deadline for completion (i.e. March, 2002) expired. At the time of writing, the work on IMS was underway in Rel6. In this release, the notion of IP-based Connectivity Access Network (IP-CAN) is introduced. The IMS utilizes the IP-CAN to transport the signaling and bearer traffic for IMS. In this model, the mobility of the terminal is handled by IP-CAN and thus such moves are hidden from the IMS. The 3GPP specifications define how PS domain can be used as IP-CAN. Another option can be WLAN.

Apart from this, various IMS services are also being standardized in Rel6. Some of the important services include presence, messaging, conferencing, and group service capabilities.

DEPLOYMENT OF 3G NETWORKS

The 3G networks have taken reasonably long time to get standardized. After long periods of standardization, and subsequent network development and trials, the mobile operators worldwide are now deploying 3G networks. The two contending technologies are the *WCDMA UMTS* as standardized by 3GPP and the *cdma2000* as standardized by 3GPP2. Worldwide, the WCDMA UMTS has gained more popularity than *cdma2000*. Given that GSM is the pre-dominant among different 2G technologies and migrating to UMTS from GSM is easier, there are more followers for UMTS WCDMA than *cdma2000*.

This section looks at 3G deployments based on UMTS WCDMA technology in three selected countries: Japan, UK and Italy. Among various operators, Hutchison Whampoa and NTT DoCoMo spearhead the deployment of UMTS WCDMA networks in Europe and Asia.

Note: The authors claim that data is supplied for information only. The information has been collected from various sources. The authors or the publishers are not responsible for its accuracy or completeness of the information provided. The reader is referred to the web-link [UMTS Deployment] provided in the reference section to have an up to date information of UMTS deployment across the world.

A.1 JAPAN

Japan has been the front-runners in the deployment of 3G networks. Japanese operator NTT DoCoMo launched the world's first commercial 3G service Freedom Of Mobile multimedia Access (FOMA) on 1st of October, 2001. Through this launch, the idea was to replicate the runaway success of I-mode on the 3G platform. To align with the global standards, as provided by 3GPP, the FOMA was based on the WCDMA UMTS technology. FOMA provides high-speed packet transmission at speeds of up to 384 Kbps for receiving side and up to 64Kbps for sending side. Apart from this, FOMA provides various applications that includes entertaining video games, video clips through email, automatically updated stock quotes and weather reports. Trials are underway for live video communication.

The progress of FOMA has been slow as even after two years of launch FOMA has barely on million users till October, 2003. However, the last few months have seen an impressive growth and the subscriber base has nearly doubled from 1 million to 2 million (see Table A.1). The success of FOMA is attributed to better and cheaper phones, with lower rates and a wider coverage. However, this fares poorly as compared to NTT DoCoMo's I-mode service that has a staggering forty million subscribers.

Table A.1 | 3G Deployment in Japan

Operator/Brand	Technology	Date of Launch	Subscriber Base (Note 1)
NTT DoCoMo/ FOMA	WCDMA UMTS	Oct, 2001	1.88 million
KDDI/Au	cdma2000	Apr, 2002	11.764 million
Vodafone KK	WCDMA UMTS	Dec, 2002	0.111 million

Note 1: Source http://mobilemediajapan.com/ dated 12th January, 2004.

In contrast to FOMA, KDDI's Au brand based on CDMA2000 1xEV-DO standard has become very popular with a subscriber base of nearly 12 million. Some contend that while FOMA subscribers were acquired afresh, KDDI's Au subscribers were those who upgraded to 3G platform. The Au provides Ezweb-based services like eznavigation (location-based services), ezmovie (video distribution), EZweb@email (multimedia messaging) and ezplus (high-speed web access). One of the most popular serivice is sending movie clips up to 15 seconds through MMS.

The third operator Vodafone KK has been a late entrant and has a low subscriber base of nearly 100 thousand.

The future of 3G market in Japan looks bright and some predict that year 2004 could be the year of 3G. By the growth of FOMA in recent months, some expect that FOMA could attract about a million subscribers every month. The success of 3G in Japan is critical for the world-wide success of 3G.

A.2 UK

The UMTS WCDMA based 3G deployments in UK is spearheaded by Hutchison 3G UK. The Hutchison 3G UK is a joint venture between Hutchison Whampoa Limited (HWL), NTT DoCoMo and KPN Mobile. Its brand 3 has roped in more than two hundred thousand subscribers by October, 2003 (see Table A.2). While many operators (e.g. Vodafone) are treading cautiously with regards the 3G deployments, Hutchison 3G is going all out and trying to get a first-mover advantage. The important services offered by Hutchison3G includes two-way mobile video call, location-based services (including route map between two points and 'Quick Map' of current location), sports information, news, video clips and games (including action games, puzzles, sports and classic games) among others.

Table A.2 | 3G Deployment in UK

Operator/Brand	Technology	Date of Launch	Subscriber Base
Hutchison3G/3	WCDMA UMTS	May, 2003	216,900 (Note 1)
Vodafone	WCDMA UMTS	n/a	–

Note 1: Source http://www.3gnewsroom.com/html/stats/index.shtml as on October, 2003

Apart from Hutchison3G and Vodafone, a Btspin-off MM02 is also planning to launch 3G network in UK.

A.3 ITALY

Like in UK, the H3G Italy (owned by the Hutchison Whampoa group) has gone ahead aggressively and launched the 3G service named 3. Till October, 2003, H3G had acquired more than three hundred thousand subscribers (see Table A.3) and it is expected that it will have a million subscribers by March, 2004, a year after its launch.

Table A.3 3G Deployment in Italy

Operator/Brand	Technology	Date of Launch	Subscriber Base
Hutchison3G/3	WCDMA UMTS	Mar, 2003	332,900 (Note 1)
Omnitel Vodafone	WCDMA UMTS	Early 2004	–
Telecom Italia Mobile	WCDMA UMTS	2004	–

Note 1: Source http://www.3gnewsroom.com/html/stats/index.shtml as on October, 2003.

Other operators like Vodafone and Telecom Italia Mobile are treading cautiously. Both are expected to launch commercial services sometime in 2004. Vodafone is planning a soft launch in early 2004 to be followed by a bigger launch in mid-2004. Telecom Italia Mobile is testing its network and is planning to begin commercial operations in 2004.

By end of 2004, It is expected that there will be three to four million 3G subscribers in Italy.

SUMMARY

Apart from deployments in Japan, UK and Italy, 3G deployments are going in other parts of the world. Hutchison 3G has launched commercial 3G services based on WCDMA UMTS in Austria, Australia and Sweden and trial services in Hong Kong and Singapore. Commercial deployments are also taking place in Korea, Germany, Russia and many other parts of the world. It is expected that 3G deployments will peak in 2004.

*While there has been steady progress, the rapid growth of 3G deployment is restricted by many factors including **interoperability issues, low battery life, limited coverage, dropped calls on the network,** and a **lack of affordable and attractive handsets.** Another major factor that delayed 3G deployments was the **expensive auction** of spectrum licenses. In Europe alone, 3G operators have shelled out 125 billion dollars to get 3G licenses. This has meant that after paying huge costs for spectrum licenses, network operators are left with little fund to invest towards 3G deployments. Further, the licensing requirements in Europe are very stringent, making the life of 3G operators very difficult.*

*Another reason for the lack of popularity of 3G has been the **lack of a killer application.** While the SMS proved to be a killer application for 2G mobile networks, operators are still looking out for similar killer applications for 3G. The lack of a killer application has meant that network operators have not been able to earn revenues commensurate with the investments.*

B

FOURTH GENERATION (4G) MOBILE NETWORKS

After a long period of standardization, development and network trials, mobile operators worldwide are now deploying 3G networks. However, the path to 3G deployments hasn't been smooth. As noted in Appendix A, the growth of 3G deployment is restricted by many factors which include *interoperability issues, limited coverage, lack of affordable and attractive handsets, lack of funds* with network operators due to *expensive auctions* of spectrum licenses, and a *lack of killer application* for 3G. While 3G deployments continue slowly and steadily, In the meanwhile, research has been ongoing towards the Fourth Generation (4G) Mobile Networks. While it is not clearly defined what 4G exactly is, the term 4G is being used quite liberally in the research industry. In most simple terms, 4G is being developed as a successor to 3G, not only to overcome the limitations of 3G, but also to make use of the latest developments in wireless technology domain. The following sections try to address the four queries related to 4G: the 'Why', 'What', 'How' and 'When' of 4G.

B.1 WHY 4G?

The 4G mobile networks are being developed with two main objectives. One of these objectives is to overcome the shortcomings and limitations of 3G, prime amongst which is the issue of *available bandwidth*. The 3G specifications have been defined to offer maximum bandwidths of 2 Mbps. However, this figure of 2 Mbps can be highly misleading. While 2 Mbps is the maximum achievable bandwidth in 3G, in practical scenarios, the bandwidth available to users will be quite less than 2Mbps. It is expected that in most realistic scenarios, users of 3G networks will be able to receive bandwidths of around 384 Kbps. Bandwidths near the 2 Mbps value will only be achieved in restricted cases, involving low mobility and in Pico cells. Even though the bandwidths offered by 3G networks is an order of magnitude better than their 2G counterparts, it is still not sufficient to support all types of multimedia communication. 4G mobile networks are being envisioned to offer higher bandwidths, up to a value of 100 Mbps.

Besides the bandwidth limitation of 3G, other shortcomings of 3G are related to the issue of *global roaming* and *network scalability*. 3G technologies were originally proposed to provide global roaming. Global roaming implies that a subscriber roaming across the globe would always remain connected. The original goal as envisaged by ITU was to have a single radio interface that provided global roaming. However, in actuality, a set of five standards for the radio interface were adopted for Third Generation (3G) networks, since it was felt that having multiple standards fostered competition and also catered to the migration of the installed base of Second Generation networks. Hence, current implementations of 3G have failed to achieve the requirement of global roaming. Therefore, global roaming has been added as one of the requirements from the future 4G networks. Further, 3G specifications define three different Core Network CN domains, where each domain provides a different set of services. The CS domain is defined to provide circuit-switched services and, the PS domain the packet switched services, and the IMS domain, the IP multimedia services. Many researchers feel that the present 3G network architecture is not scalable and is not suitable for the next generation mobile networks. Hence, for the 4G networks, enhanced network architecture is being proposed which is expected to be entirely a packet switched network.

The second main objective behind 4G developments is to make good use of the achievements in the area of wireless technology. The sluggish pace of 3G deployments has provided an opportunity for other wireless technologies to capture a sizeable portion of the market. Prominent amongst these other wireless technologies is the Wireless LAN (WLAN) and Bluetooth. As the name suggests, WLAN provides a wireless solution to networking in LANs, while Bluetooth is used to communicate between smart devices in a Personal Area Network (PAN). Satellite-based mobile networks (e.g. Thuraya) are also popular, and these provide an efficient means of offering wireless services in sparsely populated areas. Thus, the next generation of mobile networks would have to accommodate such other prominent wireless networks as well. The 4G network architecture is expected to consist of a collection of such wireless networks, the 3G cellular network being just one of these.

B.2 WHAT IS 4G?

Unlike the predecessor networks of 4G (i.e. 2G and 3G), which consisted of well-defined cellular network components, 4G networks are expected to consist of a collection of wireless networks. These would include the Personal Area Networks using, for example, Bluetooth, the local area networks using WLAN, the satellite-based mobile networks, and enhanced 3G cellular networks, besides others. The vision of 4G mobile networks is to bind these different wireless technologies together in such a manner so as to provision broadband access and global roaming using the most appropriate of these technologies.

B.2.1 4G Network Hierarchy

The vision of 4G can be better appreciated by understanding the expected 4G network hierarchy, as depicted in Figure B.1. The 4G network hierarchy is expected to consist of four broad levels of networks: the Personal Networks, the Local Networks, the Cellular Networks, and the Satellite-based Networks.

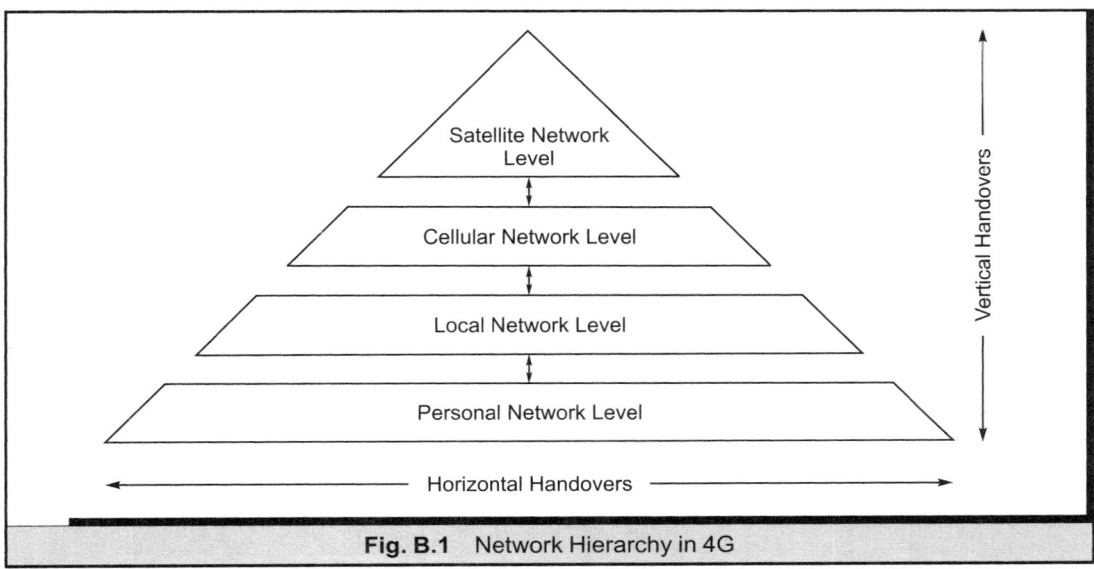

Fig. B.1 Network Hierarchy in 4G

At the lowest layer are the personal networks, which constitute of smart devices communicating with each other over wireless links. An example of a personal network can be a set of devices consisting of one or more of computer desktop, laptop, printer, modem, etc., of a residential user, that are communicating with each other using Bluetooth. The smart devices could even consist of the coffee machine, the temperature control device or the washing machine for that matter, all communicating via Bluetooth.

At the next level are the local networks. These could consist of Local Area Networks (LANs) using the Wireless LAN technology. Local area networks based on WLANs would normally have a greater coverage area than personal networks based on Bluetooth, and would be used in office complexes or in hot spots like cafes, hotels and airports.

The cellular network level comes next, which will consist of existing 2G and 3G cellular networks, as well as enhanced 3G cellular networks.

At the top-most level will be the Satellite-based mobile networks, which have a much greater coverage area than the cellular networks, or any of the networks at the lower levels.

As a result of a multi-tier hierarchy, 4G user devices would be expected to perform vertical handovers, besides horizontal handovers. Horizontal handovers will take place within one network level. Horizontal handovers at cellular network level include Soft Handovers and Hard Handovers, which were discussed in Chapter 2 of this book. Vertical handovers, on the other hand, will be performed between different network levels. As an example, a user device using WLAN to communicate with servers within an office premises would perform a handover to the cellular network when the employee moves out of the office complex with the user device. Thus, 4G user devices would have to be intelligent devices that cannot only work with different wireless technologies, but also select the most appropriate technology to avail a particular service by performing horizontal and vertical handovers.

B.2.2   Features of 4G Networks

While it is not clearly defined as to what networks can be categorized as 4G networks, there are some features that are expected to be supported by most 4G networks. These features include:

- **Higher Bandwidths:** It is expected that 4G networks would provide higher bandwidths to support multimedia services. Bandwidths up to 100 Mbps will be possible to achieve in 4G networks.

- **Packet-switched Network:** While 3G networks consisted of both circuit-switched and packet-switched domains, 4G networks are expected to be entirely based on packet-switched networks. IP is expected to be used as the packet-switched network in 4G.

- **Stringent Network Security:** Network security in 4G networks is expected to be further improvised. Security mechanisms in 3G networks may be enhanced to provide better and tighter security.

- **Global mobility and network scalability:** While 3G networks will not be able to provide true global mobility, this is one of the requirements from 4G networks. Also, the network architecture for 4G networks, discussed in the previous section, provides mechanisms for easy scalability.

A broad comparison of the 3G and 4G mobile networks is provided in Table B.1.

Table B.1   Comparison of 3G and 4G

3G Mobile Networks	*4G Mobile Networks*
Evolved from the 2G Mobile Networks providing backward compatibility.	Consists of a collection of different wireless networks. Network architecture at cellular network level is an enhancement over 3G network architecture.
Consists of both circuit-switched and packet-switched networks.	Consists of entirely packet-switched networks.
Data Rates upto 2 Mbps.	Data Rates upto 100 Mbps.
Practical data rates of ~384 Kbps mean that applications like live high-definition video, HDTV, etc. are difficult to support.	Higher practical data rates (~ 2Mbps) mean that bandwidth hungry applications like HDTV can easily be supported.
Network security not adequate.	Enhancements to network security.

B.3   HOW TO ACHIEVE 4G?

A lot of research and technological advancements have taken place in the wireless arena, and 4G networks are expected to incorporate some of these advancements to support broadband access and seamless global roaming. This section discusses some of the important technological advances that would help in the evolution towards 4G. These technological advances include:

- **Evolution of the Physical Layer:** Support of multimedia traffic effectively requires higher bandwidths than what is supported by current 2.5G and 3G networks. One of the means towards achieving higher bandwidths is by evolving the radio interface physical layer. A lot of research is currently happening on this front. This includes research on Multiple Input Multiple Output

(MIMO) systems using advanced antenna configurations and space-time processing. Besides, research is ongoing in the area of signal processing, modulation schemes and coding algorithms, which will all contribute towards increasing the efficiency of the physical layer.

- **Enhancements in TDD-CDMA:** While 3G specifications define both FDD and TDD mode of operation over the air interface, initial deployments of 3G will mostly use the FDD mode. However, from the viewpoint of asymmetric services (e.g. Internet Access), it is expected that the TDD mode of operation would be more suitable. There are many reasons for such consideration. Amongst the most important reasons is the fact that TDD mode can operate over unpaired frequency spectrums, unlike FDD mode, where a paired spectrum of equal uplink and downlink bandwidth is required. Using FDD mode of operation would lead to inefficient utilization of the uplink bandwidth in cases where services are asymmetrical in nature, with more traffic downlink. This is characteristic of the Internet Access service, where more data flows downstream from the server to the client, rather than vice-versa. Asymmetrical services are expected to constitute the bulk of the services in future mobile networks. Hence, a lot of research is currently directed towards TDD-CDMA as the multiple-access strategy for the radio interface.

- **Enhancements to Mobile Network Architecture:** Besides the research initiatives on evolution of the physical layer, and the use of TDD-CDMA, a lot of focus is also on evolving the mobile network architecture for 4G networks. One of the key areas towards evolution of the network architecture is to move towards an All-IP Network, which will use the inherent flexibility of the IP protocol to transparently integrate diverse wireless networks. In evolving the network architecture, the following are some of the focus areas:

  - **Migration towards OpenRAN Architecture:** In 4G, the access network also expected to evolve further. The paper [OpenRAN J. Kempf] describes a new architecture for the RAN, the 'OpenRAN', which is based on a distributed processing model. Existing RAN architectures are based on centralized RNCs connected to Node Bs via point-to-point links. Such architecture is prone to a single point of failure in case the RNC fails. The OpenRAN architecture proposes a distributed processing model with a routed IP network as the underlying transport fabric. The architecture applies principles borrowed from the data communication networks, which aim towards reducing network deployment costs, and towards improving the reliability of the networks.

  - **Use of Cellular IP:** This area concerns with incorporating the principles of mobility within the IP protocol. The Mobile IP protocol proposed by the IETF was the first step towards supporting mobility in IP networks. However, Mobile IP cannot be used in cellular networks, since it only provides *macro-mobility and not micro-mobility*. This means that while Mobile IP does allows a user to migrate to another geographical region, attach with the visited IP network, and avail services without having to change to a new IP address, it does not provide means for a user to avail services while on the move. Cellular IP provides mechanisms to support micro-mobility, by including principles such as IP paging and seamless handover control.

  - **Development of SNRM platform:** 4G networks are expected to be a collection of multiple disparate networks, ranging from personal area networks, WLANs, cellular networks to satellite-based networks. Management of network resources and services in such an environment is expected to be an uphill task. The Service and Network Resource

Management (SNRM) platform provides means to manage network resources and services in a composite wireless network. The service providers can make use of the SNRM platform to provide services with appropriate-QoS levels, by using the most-appropriate wireless network technology.

B.4 WHEN SHOULD WE EXPECT 4G?

Mobile networks are normally expected to have a 10 years lifetime (see Figure B.2). The 1990's saw the deployment and use of 2G mobile networks, prominently the GSM technology. In fact, 2G mobile networks are even currently being used in the year 2004. Trials for 3G mobile networks started with the coming of the 21st century, and it is expected that they will last at least till the year 2010. In the meanwhile, research has already started towards 4G networks, and these are expected to be deployed in the years post-2010.

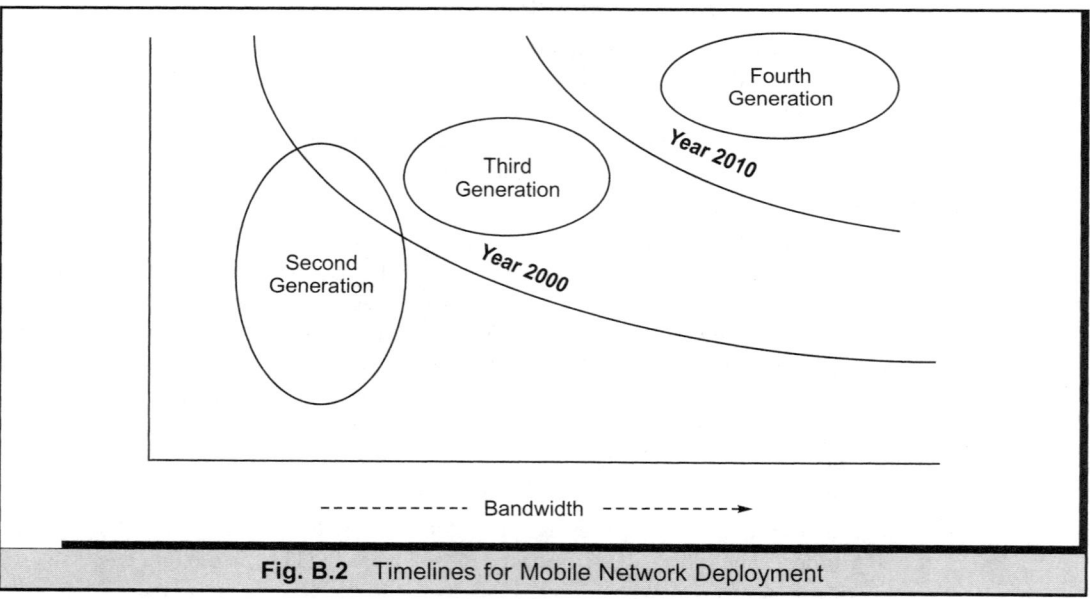

Fig. B.2 Timelines for Mobile Network Deployment

Amongst some of the leading names in the area of 4G-research has been the name of NTT DoCoMo. In June 2003, NTT DoCoMo, which has been conducting research in the area of 4G since 1998, announced plans for a field trial of 4G mobile communications systems. As per this announcement, the field trial will take place in Yokosuka, Kanagawa which houses an R&D center belonging to DoCoMo. The planned field trial is expected to employ Variable Spreading Factor Orthogonal Frequency and Code Division Multiplexing (VSF-OFCDM) and Variable Spreading Factor Code Division Multiple Access (VSF-CDMA) technologies. With regards to 4G deployments, a lot will depend on the success of this field trial by NTT DoCoMo. A successful field trial might lead to an early deployment of 4G networks, maybe even earlier than the year 2010.

REFERENCES

I SPECIFICATIONS

This section lists the specifications from 3GPP, IETF and ITU-T that were used in the book. The important specifications are **boldfaced**. For 3GPP specifications, unless otherwise mentioned, the Release 5 version as available on December 2002 were referred.

I.1 Introduction

Specification Number	Specification Title
3GPP TR 21.900	**Technical Specification Group Working Methods**

I.2 Principles of WCDMA

Specification Number	Specification Title
3GPP TS 25.214	Physical Layer Procedures (FDD)
3GPP TS 25.401	**UTRAN Overall Description**
3GPP TR 25.922	Radio Resource Management Strategies

I.3 UMTS Network Architecture

Specification Number	Specification Title
3GPP TS 21.103	3rd Generation Mobile System Release 5 Specifications
3GPP TR 21.905	Vocabulary for 3GPP Specifications
3GPP TS 22.002	Circuit Bearer Services supported by a PLMN
3GPP TS 22.003	Circuit Tele-services Supported by a PLMN
3GPP TS 22.060	General Packet Radio Service (GPRS): Service Description (Stage 1)

3GPP TS 22.101	Service Aspects; Service Principles
3GPP TS 23.002	**Network Architecture**
3GPP TS 23.003	**Numbering, Addressing and Identification**
3GPP TS 23.101	**General UMTS Architecture**
3GPP TS 23.107	**Quality of Service (QoS) Concept and Architecture**
3GPP TS 23.060	General Packet Radio Service (GPRS): Service Description (Stage 2)
3GPP TS 24.002	GSM–UMTS Public Land Mobile Network (PLMN) access Reference Configuration
3GPP TS 25.401	UTRAN Overall Description
3GPP TS 43.068	Voice Group Call Service (VGCS) (Stage 2)
3GPP TS 43.069	Voice Broadcast Service (VBS) (Stage 2)

I.4 User Equipment

Specification Number	Specification Title
3GPP TS 24.002	GSM–UMTS Public Land Mobile Network (PLMN) access Reference Configuration
3GPP TS 27.001	**General on Terminal Adaptation Functions (TAF) for Mobile Stations (MS)**
3GPP TS 27.002	Terminal Adaptation Functions (TAF) for Services using Asynchronous Bearer Capabilities
3GPP TS 27.003	Terminal Adaptation Functions (TAF) for Services using Synchronous Bearer Capabilities
3GPP TS 27.005	Use of Data Terminal Equipment–Data Circuit terminating Equipment (DTE-DCE) interface for Short Message Service (SMS) and Cell Broadcast Service (CBS)
3GPP TS 27.007	AT command set for 3G User Equipment (UE)
3GPP TS 31.101	**UICC-terminal interface; Physical and logical characteristics**
3GPP TS 31.102	Characteristics of the USIM Application
3GPP TS 31.111	USIM Application Toolkit (USAT)

I.5 Access Network

Specification Number	Specification Title
3GPP TS 23.002	**Network Architecture**
3GPP TS 25.215	Physical layer: Measurements (FDD)
3GPP TS 25.225	Physical layer: Measurements (TDD)
3GPP TS 25.301	**Radio Interface Protocol Architecture**
3GPP TS 25.308	UTRA High Speed Downlink Packet Access (HSPDA): Overall description (Stage 2)
3GPP TS 25.321	Medium Access Control (MAC) protocol specification

3GPP TS 25.322	Radio Link Control (RLC) protocol specification
3GPP TS 25.323	Packet Data Convergence Protocol (PDCP) specification
3GPP TS 25.324	Broadcast/Multicast Control (BMC)
3GPP TS 25.331	**Radio Resource Control (RRC) protocol specification**
3GPP TS 25.401	**UTRAN overall description**
3GPP TS 25.410	UTRAN Iu interface: General Aspects and Principles
3GPP TS 25.411	UTRAN Iu interface Layer 1
3GPP TS 25.412	UTRAN Iu interface signaling transport
3GPP TS 25.413	**UTRAN Iu interface RANAP signaling**
3GPP TS 25.414	UTRAN Iu interface data transport & transport signaling
3GPP TS 25.415	UTRAN Iu interface user plane protocols
3GPP TS 25.419	UTRAN Iu-BC interface: Service Area Broadcast Protocol (SABP)
3GPP TS 25.420	UTRAN Iur Interface General Aspects and Principles
3GPP TS 25.421	UTRAN Iur interface Layer 1
3GPP TS 25.422	UTRAN Iur interface signaling transport
3GPP TS 25.423	**UTRAN Iur interface RNSAP signaling**
3GPP TS 25.424	UTRAN Iur interface data transport and transport signaling for CCH data streams
3GPP TS 25.425	UTRAN Iur interface user plane protocols for CCH data streams
3GPP TS 25.427	UTRAN Iur and Iub interface user plane protocols for DCH data streams
3GPP TS 25.430	UTRAN Iub interface: General Aspects and Principles
3GPP TS 25.431	UTRAN Iub interface Layer 1
3GPP TS 25.432	UTRAN Iub interface: signaling transport
3GPP TS 25.433	**UTRAN Iub interface NBAP signaling**
3GPP TS 25.434	UTRAN Iub interface data transport and transport signaling for CCH data streams
3GPP TS 25.435	UTRAN Iub interface user plane protocols for CCH data streams
3GPP TS 48.004	BSS – MSC interface: Layer 1 specification
3GPP TS 48.006	Signaling transport mechanism specification for the BSS – MSC interface
3GPP TS 48.008	MSC – BSS interface: Layer 3 specification
3GPP TS 48.014	BSS – SGSN interface: Gb interface Layer 1
3GPP TS 48.016	BSS – SGSN interface: Network service
3GPP TS 48.018	BSS – SGSN interface: BSS GPRS Protocol (BSSGP)
3GPP TS 48.051	BSC – BTS interface: General aspects
3GPP TS 48.052	BSC – BTS interface: Interface principles
3GPP TS 48.054	BSC – BTS interface: Layer 1 structure of physical circuits
3GPP TS 48.056	BSC – BTS interface: Layer 2 specification
3GPP TS 48.058	BSC – BTS interface: Layer 3 specification
ITU-T I.361	B-ISDN ATM Layer specification
ITU-T I.363	B-ISDN ATM Adaptation Layer specification

ITU-T I.363.2	B-ISDN ATM Adaptation Layer specification Type 2 AAL
ITU-T I.363.5	B-ISDN ATM Adaptation Layer specification Type 5 AAL
ITU-T I.366.1	Segmentation and Reassembly Service Specific Convergence Sub-layer for AAL2
ITU-T Q.2100	B-ISDN Signaling ATM Adaptation Layer (SAAL) overview description
ITU-T Q.2110	B-ISDN ATM adaptation layer – Service Specific Connection Oriented Protocol (SSCOP)
ITU-T Q.2130	B-ISDN ATM Adaptation Layer – Service Specific Coordination Function for Signaling at the User-Network Interface (SSCF at UNI)
ITU-T Q.2140	B-ISDN ATM Adaptation Layer – Service Specific Coordination Function for Signaling at the Network-Node Interface (SSCF at NNI)
ITU-T Q.2150.1	Signaling Transport Converter on MTP3 and MTP3b
ITU-T Q.2150.2	Signaling Transport Converter on SSCOP
ITU-T Q.2210	Message transfer part level 3 functions and messages using the services of ITU-T Recommendation Q.2140
ITU-T Q.2630.1	AAL type 2 signaling protocol (ALCAP) – Capability set 1
ITU-T Q.2630.2	AAL type 2 signaling protocol (ALCAP) – Capability set 2
ITU-T Q.704	Signaling network functions and messages
RFC 2507	IP Header Compression
RFC 3095	Robust Header Compression (ROHC)

I.6 Core Network

Specification Number	Specification Title
3GPP TS 23.002	**Network Architecture**
3GPP TS 23.003	Numbering, Addressing and Identification
3GPP TS 23.008	Organization of subscriber data
3GPP TS 23.012	Location management procedures
3GPP TS 23.060	**General Packet Radio Service (GPRS) Service description**
3GPP TS 24.007	Mobile radio interface signaling Layer 3: General Aspects
3GPP TS 24.008	**Mobile radio interface Layer 3 specification: Core Network protocols (Stage 3)**
3GPP TS 24.010	Mobile radio interface Layer 3 – Supplementary Service specification
3GPP TS 24.011	Point-to-Point (PP) Short Message Service Support on mobile radio interface
3GPP TS 29.002	**Mobile Application Part (MAP) specification**
3GPP TS 29.007	General requirements on interworking between the PLMN and the ISDN or PSTN
3GPP TS 29.016	SGSN – VLR Gs interface: Network Service Specification
3GPP TS 29.018	SGSN – VLR Gs interface: Layer 3 specification

3GPP TS 29.060	**GPRS Tunnelling Protocol (GTP) across the Gn and Gp interface**
3GPP TS 29.061	Interworking between the PLMN supporting Packet Based services and Packet Data Networks (PDN)
ITU-T E.164	Numbering plan for the ISDN era
ITU-T E.212	Identification plan for land mobile stations
ITU-T E.214	Structuring of the land mobile global title for the signaling connection control part
ITU-T I.363.2	B-ISDN ATM Adaptation Layer specification: Type 2 AAL
ITU-T Q.700	Introduction to ITU-T Signaling System No. 7
ITU-T Q.701	**Functional Description of the Message Transfer Part (MTP) of SS7**
ITU-T Q.703	SS7 MTP: Signaling link
ITU-T Q.704	SS7 MTP: Signaling network functions and messages
ITU-T Q.711	Functional description of the Signaling Connection Control Part
ITU-T Q.712	Definition and function of Signaling Connection Control Part messages
ITU-T Q.713	Signaling Connection Control Part formats and codes
ITU-T Q.714	**Signaling Connection Control Part procedures**
ITU-T Q.761	ISDN User Part functional description
ITU-T Q.762	ISDN User Part general functions of messages and signals
ITU-T Q.763	ISDN User Part formats and codes
ITU-T Q.764	**ISDN User Part signaling procedures**
ITU-T Q.771	Functional description of Transaction Capabilities
ITU-T Q.772	Transaction Capabilities information element definitions
ITU-T Q.773	Transaction Capabilities formats and encoding
ITU-T Q.774	**Transaction Capabilities procedures**
ITU-T Q.775	Guidelines for using transaction capabilities
RFC 1889	RTP: A Transport Protocol for Real-Time Applications

I.7 Radio Resource Control Procedures

Specification Number	Specification Title
3GPP TS 25.331	**Radio Resource Control (RRC) protocol specification**

I.8 UTRAN Signaling Procedures

Specification Number	Specification Title
3GPP TS 25.324	Broadcast/Multicast Control (BMC)
3GPP TS 25.331	Radio Resource Control (RRC) protocol specification
3GPP TS 25.401	**UTRAN overall description**
3GPP TS 25.413	UTRAN Iu Interface RANAP Signaling
3GPP TS 25.419	UTRAN Iu-BC interface: Service Area Broadcast Protocol (SABP)

3GPP TS 25.423	UTRAN Iur Interface RNSAP Signaling
3GPP TS 25.433	UTRAN Iub Interface NBAP Signaling
3GPP TR 25.931	**UTRAN Functions, Examples on Signaling Procedures**
ITU-T Q.2630.1	AAL type 2 signaling protocol (ALCAP) – Capability set 1
ITU-T Q.2630.2	AAL type 2 signaling protocol (ALCAP) – Capability set 2

I.9 Mobility Management

Specification Number	Specification Title
3GPP TS 23.012	**Location management procedures**
3GPP TS 23.060	**General Packet Radio Service (GPRS) Service description**
3GPP TS 23.116	Super-Charger
3GPP TS 23.122	Non-Access-Stratum functions related to Mobile Station (MS) in idle mode
3GPP TS 24.007	Mobile radio interface signaling layer 3; General Aspects
3GPP TS 24.008	**Mobile radio interface Layer 3 specification; Core network protocols**
3GPP TS 25.304	UE Procedures in Idle Mode and Procedures for Cell Reselection in Connected Mode
3GPP TS 29.002	Mobile Application Part (MAP) specification (Note 1)

Note 1: Refer Sections 8 and 19 for Mobility Management messages and procedures.

I.10 Call Handling

Specification Number	Specification Title
3GPP TS 22.032	Immediate Service Termination (IST)
3GPP TS 23.018	**Basic Call Handling**
3GPP TS 23.082	Call Forwarding Supplementary Services
3GPP TS 23.088	Call Barring (CB) Supplementary Services
3GPP TS 24.008	**Mobile radio interface Layer 3 specification; Core network protocols**
3GPP TS 24.079	Support of Optimal Routing (SOR)
3GPP TS 29.002	Mobile Application Part (MAP) specification (Note 1)
ITU-T Q.763	ISDN User Part (ISUP) Formats and Codes

Note 1: Refer Sections 10 and 21 for Call Handling messages and procedures.

I.11 Session Management

Specification Number	Specification Title
3GPP TS 23.060	**General Packet Radio Service (GPRS) Service description**

3GPP TS 24.008	Mobile radio interface Layer 3 specification; Core network protocols
3GPP TS 29.060	GPRS Tunnelling Protocol (GTP) across the Gn and Gp interface

I.12 Supplementary Services

Specification Number	Specification Title
3GPP TS 22.004	**General on Supplementary Services**
3GPP TS 22.030	Man-Machine Interface (MMI) of the User Equipment (UE)
3GPP TS 23.011	**Technical Realization of Supplementary services – General Aspects**
3GPP TS 29.011	Signaling Interworking for Supplementary Services
3GPP TS 23.067	Enhanced Multi-Level Precedence and Preemption Service (EMLPP)
3GPP TS 23.072	Call Deflection Supplementary Service
3GPP TS 23.081	Line Identification Supplementary Services
3GPP TS 23.082	Call Forwarding Supplementary Services
3GPP TS 23.083	Call Waiting and Call Hold Supplementary Services
3GPP TS 23.084	Multi-Party Supplementary Services
3GPP TS 23.085	Closed User Group Supplementary Services
3GPP TS 23.086	Advice of Charge Supplementary Services
3GPP TS 23.087	User to User Signaling
3GPP TS 23.088	Call Barring (CB) Supplementary Services
3GPP TS 23.090	Unstructured Supplementary Service Data (USSD)
3GPP TS 23.091	Explicit Call Transfer (ECT) Supplementary Service
3GPP TS 23.093	Completion of Calls to Busy Subscriber (CCBS)
3GPP TS 23.094	Follow Me Supplementary Service
3GPP TS 23.096	Name Identification Supplementary Service
3GPP TS 23.097	Multiple Subscriber Profile (MSP)
3GPP TS 23.135	Multi Call
3GPP TS 24.080	Mobile radio Layer 3 supplementary service specification; Formats and coding

I.13 Value-Added Services

Specification Number	Specification Title
3GPP TS 23.040	**Technical realization of Short Message Service (SMS)**
3GPP TS 24.011	Point-to-Point (PP) Short Message Service (SMS) Support on Mobile Radio Interface
3GPP TS 29.002	Mobile Application Part (MAP) specification (Note 1)
3GPP TS 23.041	**Technical realization of Cell Broadcast Service (CBS)**
3GPP TS 25.324	Broadcast/Multicast Control (BMC)

3GPP TS 25.419	UTRAN Iu-BC interface: Service Area Broadcast Protocol (SABP)
3GPP TS 23.140	**Multimedia Messaging Service (MMS); Functional description**
3GPP TS 26.140	Multimedia Messaging Service (MMS); Media formats and codes
3GPP TS 23.271	**Location Services (LCS)**
3GPP TS 24.030	Location Services (LCS); Supplementary service operations
3GPP TS 22.038	USIM/SIM Application Toolkit (USAT/SAT) Service description
3GPP TS 22.057	Mobile Execution Environment (MExE) Service description
3GPP TS 22.078	Customized Applications for Mobile network Enhanced Logic (CAMEL) Service description
3GPP TS 22.105	Services and service capabilities
3GPP TS 22.121	Service aspects; The Virtual Home Environment
3GPP TS 22.127	Service Requirement for the Open Services Access (OSA)

Note 1: Refer Sections 12 and 23 for Short Message Service messages and procedures.

I.14 Security Management

Specification Number	Specification Title
3GPP TS 33.102	**3G Security; Security Architecture**
3GPP TS 33.120	3G Security; Security Principles and Objectives
3GPP TS 33.200	**Network Domain Security; MAP application layer security**
3GPP TS 33.203	**Access security for IP-based services**
3GPP TS 33.210	**Network Domain Security: IP network layer security**
3GPP TS 35.201	Specification of the 3GPP Confidentiality and Integrity Algorithms; Document 1: f8 and f9 Specification
3GPP TS 35.202	Specification of the 3GPP Confidentiality and Integrity Algorithms; Document 2: KASUMI Specification
3GPP TS 35.203	Specification of the 3GPP Confidentiality and Integrity Algorithms; Document 3: Implementors' Test Data
3GPP TS 35.204	Specification of the 3GPP Confidentiality and Integrity Algorithms; Document 4: Design Conformance Test Data
3GPP TS 35.205	Specification of the MILENAGE Algorithm Set: An example algorithm set for the 3GPP authentication and key generation functions f1, f1*, f2, f3, f4, f5 and f5*; Document 1: General
3GPP TS 35.206	Specification of the MILENAGE Algorithm Set: An example algorithm set for the 3GPP authentication and key generation functions f1, f1*, f2, f3, f4, f5 and f5*; Document 2: Algorithm Specification
3GPP TS 35.207	Specification of the MILENAGE Algorithm Set: An example algorithm set for the 3GPP authentication and key generation functions f1, f1*, f2, f3, f4, f5 and f5*; Document 3: Implementors' Test Data
3GPP TS 35.208	Specification of the MILENAGE Algorithm Set: An example algorithm set for the 3GPP authentication and key generation functions f1, f1*, f2, f3, f4, f5 and f5*; Document 4: Design Conformance Test Data

3GPP TS 35.209	Specification of the MILENAGE Algorithm Set: An example algorithm set for the 3GPP authentication and key generation functions f1, f1*, f2, f3, f4, f5 and f5*; Document 5: Summary and results of design and evaluation
RFC 2401	Security Architecture for the Internet Protocol
RFC 2403	The Use of HMAC-MD5-96 within ESP and AH
RFC 2404	The Use of HMAC-SHA-1-96 within ESP and AH
RFC 2405	The ESP DES-CBC Cipher Algorithm With Explicit IV
RFC 2406	IP Encapsulating Security Payload
RFC 2409	The Internet Key Exchange (IKE)

I.15 IP-Based Signaling Transport

Specification Number	*Specification Title*
3GPP TS 25.410	UTRAN Iu Interface: general aspects and principles
3GPP TS 25.420	UTRAN Iur Interface General Aspects and Principles
3GPP TS 29.202	**SS7 Signaling Transport in Core Network (Stage 3)**
3GPP TR 29.903	Feasibility Study on SS7 signaling transport in the core network with SUA
RFC 2719	Architectural Framework for Signaling Transport
RFC 2960	Stream Control Transmission Protocol
RFC 3057	ISDN Q.921 User Adaptation Layer (IUA)
RFC 3331	SS7 MTP2 User Adaptation Layer (M2UA)
RFC 3332	**SS7 MTP3 User Adaptation Layer (M3UA)**
Internet Draft	**Signaling Connection Control Part User Adaptation Layer (SUA)**

I.16 IP Multimedia Subsystem

Specification Number	*Specification Title*
3GPP TS 22.228	Service requirements for the IP Multimedia Core Network Sub-system (Stage 1)
3GPP TS 22.941	IP Based Multimedia Services Framework
3GPP TS 23.002	**Network Architecture**
3GPP TS 23.003	Numbering , Addressing and Identification
3GPP TS 23.008	Organization of Subscriber Data
3GPP TS 23.060	General Packet Radio Service (GPRS): Service description (Stage 2)
3GPP TS 23.218	IP Multimedia Session handling and Call Model (Stage 2)
3GPP TS 23.221	Architectural requirements
3GPP TS 23.228	**IP Multimedia Subsystem (Stage 2)**
3GPP TS 24.228	Signaling Flows for IM Call Control based on SIP and SDP
3GPP TS 24.229	**IP Multimedia Call Control based on SIP and SDP**

3GPP TS 29.002	Mobile Application Part (MAP) specification (Note 1)
3GPP TS 29.228	IMS Cx and Dx Interfaces: Signaling Flows and message contents
3GPP TS 29.229	Cx and Dx Interfaces based on the Diameter protocol: Protocol details
3GPP TS 29.328	IMS Sh Interface: Signaling Flows and message contents
3GPP TS 29.329	Sh Interface based on the Diameter protocol: Protocol details
3GPP TS 33.102	Security Architecture
3GPP TS 33.203	Access Security for IP based services
3GPP TS 33.210	Network Domain Security: IP network layer security
ITU-T E.164	Numbering plan for the ISDN era
ITU-T H.248	Gateway control protocol
ITU-T Q.763	ISDN User Part (ISUP) Signaling Procedures
RFC 1889	RTP: A Transport Protocol for Real-Time Applications
RFC 2327	SDP: Session Description Protocol
RFC 2396	Uniform Resource Identifiers (URI): Generic Syntax
RFC 2486	The Network Access Identifier
RFC 2960	Stream Control Transmission Protocol (SCTP)
RFC 3261	**SIP: Session Initiation Protocol**
RFC 3539	Authentication, Authorization and Accounting (AAA) Transport Profile
RFC 3588	Diameter Base Protocol
RFC 3589	Diameter Command Codes for 3GPP Release 5

Note 1: Refer Section 24A for messages exchanged over Si interface.

II BOOKS AND PAPERS

The following books and papers have been arranged using keywords (e.g. 3G, 4G, ATM, etc.).

[3G A. Roos] Anders Roos et al., "Critical Issues for Roaming in 3G", *IEEE Wireless Communications*, Feb 2003, pp. 29–35.

[3G B. Landfeldt] Bjorn Landfeldt et al., "Providing Scalable and Deployable Addressing in Third-Generation Cellular Networks", *IEEE Wireless Communications*, Feb 2003, pp. 36–42.

[3G E. Rousseau] E. Rousseau and T. Sagar, "Third Generation Terminals", *Alcatel Telecommunications Review*, First Quarter, 2001, (url: *http://www.alcatel.com/atr*).

[3G G. Patel] G. Patel and S. Dennett, "The 3GPP and 3GPP2 Movements Toward an All-IP Mobile Network", *IEEE Personal Communication*, Aug 2000, Vol. 7, No. 4, pp. 62–64.

[3G Hughes] Hughes White Paper, "*GSM TO 3G: Evolution or Revolution*", (url: *http://www.hssworld.com/whitepapers/overview.htm*).

[3G J. Korhonen] Juha Korhonen, "*Introduction to 3G Mobile Communications*", Artech House.

[3G M. Oliphant] Malcolm W. Oliphant, "Radio Interfaces Make the Difference in 3G Cellular Systems", *IEEE Spectrum*, Oct 2000, pp. 53–58.

[3G Nokia] Nokia White Paper, "*Introducing Mobile IPv6 in 2G and 3G Mobile Networks*", (url: *http://www.nokia.com/downloads/solutions/operators/intro_to_mipv6.pdf*).

[3G Northstream] Northstream Report, *"3G Rollout Status"*, (url: *http://www.3gnewsroom.com/html/whitepapers/year_2003.shtml*).

[3G P. Sehier] P. Sehier et al., "Standardization of 3G Mobile Systems", *Alcatel Telecommunications Review*, First Quarter, 2001, (url: *http://www.alcatel.com/atr*).

[3G R. Prasad] R.Prasad, W. Mohr and W. Konhauser, *"Third Generation Mobile Communication Systems"*, Artech House.

[3G S. Tabbane] Sami Tabbane, *"3G Wireless Networks Developments"*, (url: *http://www.wmrc.com/businessbriefing/pdf/wireless2002/reference/12.pdf*).

[3G Trillium] Trillium White Paper, *"Third Generation (3G) Wireless White Paper"*, (url: *http://www.trillium.com/news-events/white-papers/index.html*).

[3G UMTS Forum] UMTS Forum White Paper, *"Evolution to 3G/UMTS Services"*, White Paper No. 1, UMTS Forum, Aug 2002, (url: *http://www.3gnewsroom.com/html/whitepapers/year_2002.shtml*).

[4G A. Campbell] Andrew T. Campbell et al., Design, Implementation, and Evaluation of Cellular IP, *IEEE Personal Communications*, Aug 2000, pp. 42–49.

[4G J. Kempf] James Kempf and Parviz Yegani, "OpenRAN: A New Architecture for Mobile Wireless Internet Radio Access Networks", *IEEE Communications Magazine*, May 2002, pp. 118–123.

[4G O. Haase] Oliver Haase and Kazutaka Murakami, "Unified Mobility Manager: Enabling Efficient SIP/UMTS Mobile Network Control", *IEEE Wireless Communications*, Aug 2003, pp. 66–75.

[4G P. Demestichas] Panagiotis Demestichas et al., Management of Networks and Services in a Composite Radio Context, *IEEE Wireless Communications*, Aug 2003, pp. 44–51

[4G R. Esmailzadeh] Riaz Esmailzadeh and Masao Nakagawa, "TDD-CDMA for the 4th Gene-ration of Wireless Communications", *IEEE Wireless Communications*, Aug 2003, pp. 8–15.

[4G S. Uskela] Sami Uskela, "Key Concepts for Evolution Toward Beyond 3G Networks", *IEEE Wireless Communications*, Feb 2003, pp. 43–48.

[4G V. Kumar] V. Kumar, "Wireless Communications Beyond 3G", *Alcatel Telecommunications Review*, First Quarter, 2001, (url: *http://www.alcatel.com/atr*).

[ATM S. Kasera] Sumit Kasera, *"ATM Networks: Concepts and Protocols"*, Tata McGraw-Hill.

[CDMA A. Morrison] A. Morrison et al., "An Iterative DOA Algorithm for a Space-Time DS-CDMA Rake Receiver", *3G Mobile Communication Technologies Conference*, London, March 2000.

[CDMA A. Sampath] A. Sampath, P. Kumar and J. Holtzman, "On Setting Reverse Link Target SIR in a CDMA System", *Proceedings of VTC'97*, Arizona, May 1997.

[CDMA A. Viterbi] A. J. Viterbi, *"Principles of Spread Spectrum Communication"*, Addison Wesley.

[CDMA F. Adachi] F. Adachi, M. Sawahashi and K. Okawa, "Tree-structured Generation of Orthogonal Spreading Codes with Different Lengths for Forward Link of DS-CDMA Mobile", *Electronics Letters*, 1997, Vol. 33, No. 1, pp. 27–28.

[CDMA Z. Liu] Zhao Liu and Magda El Zarki, "SIR-based Call Admission Control for DS-CDMA Cellular Systems"*IEEE Journal on Selected Areas in Communications*, May 1994, Vol. 12, No. 4, pp. 638–644.

[GPRS A.Salkintzis] Apostolis K. Salkintzis et al., "WLAN-GPRS Integration for Next-Generation Mobile Data Networks", *IEEE Wireless Communications*, Oct 2002, pp. 112–124.

[GPRS B. Ghribi] Brahim Ghribi and Luigi Logrippo, *"Understanding GPRS: The GSM Packet Radio Service"*, (url: *http://lotos.site.uottawa.ca/ftp/pub/Lotos/Papers/GPRS_Tutorial.pdf*).

[GPRS C. Bettstetter] C. Bettstetter, H. J. Vogel, J. Eberspacher, "General Packet Radio Service GPRS: Architecture, Protocols, and Air Interface", *IEEE Communications Survey*, Third Quarter 1999, Vol. 2, No. 3.

[GPRS Cisco] Cisco White Paper, "*GPRS White Paper*", (url: *http://www.cisco.com/warp/public/cc/so/neso/gprs/gprs_wp.htm*).

[GPRS G. Heine] G. Heine, "*GPRS from A – Z*", Artech House.

[GSM A. Mehrotra] A. Mehrotra, "*GSM System Engineering*", Artech House.

[GSM G. Heine] G. Heine, "*GSM Networks: Protocols, Terminology, and Implementation* ", Artech House.

[GSM J. Eberspacher] J. Eberspacher and H. J. Vogel, "*GSM: Switching, Services and Protocols*", John Wiley & Sons.

[GSM M. Mouly] M. Mouly and M. B. Pautet, "*The GSM System for Mobile Communications*", Published by the Authors.

[GSM M. Rahnema] M. Rahnema, "Overview of the GSM System and Protocol Architecture" *IEEE Communications Magazine*, April 1993, Vol. 31, No. 4, pp. 92–100.

[GSM S. Redl] S. Redl, M. Weber and M. Oliphant, "*An Introduction to GSM*", Artech House.

[IMS H. Montes] Hector Montes et al., "Deployment of IP Multimedia Streaming Services in Third-Generation Mobile Networks", *IEEE Wireless Communications*, Oct 2002, pp. 84–92.

[IMS N. Parameshwar 1] Narayan Parameshwar et al., "*Advanced SIP Series: SIP and 3GPP Operations*", (url: *http://www.awardsolutions.com/research/white_papers.shtm*).

[IMS N. Parameshwar 2] Narayan Parameshwar and Chris Reece, "*Advanced SIP Series: SIP and 3GPP*", (url: *http://www.awardsolutions.com/research/white_papers.shtm*)

[IMS P. Frene] P. Frene et al., "Mobile Evolution Towards Full IP Multimedia", *Alcatel Telecommunications Review*, First Quarter, 2001, (url: *http://www.alcatel.com/atr*).

[IMS UMTS Forum] Report from UMTS Forum, "*IMS Service Vision for 3G Markets*", Report No. 20, UMTS Forum, April 2002. (url: *http://www.3gnewsroom.com/html/whitepapers/year_2002.shtml*).

[IMT-2000 H. Yumiba] Hideaki Yumiba et al., "The Design Policy for a GSM-based IMT-2000 Network", *IEEE Wireless Communications*, Feb 2003, pp. 7–14.

[IMT-2000 T. Tamura] Toshiyuki Tamura et al., "IMT-2000 Core Network Node Systems", *IEEE Wireless Communications*, Feb 2003, pp. 15–21.

[LCS M. Dru] M. A. Dru and S. Saada, "Location-based Mobile Services: The Essentials", *Alcatel Telecommunications Review*, First Quarter, 2001, (url: *http://www.alcatel.com/atr*)

[OSA R. Stretch] R. M. Stretch, "The OSA API and Other Related Issues", *BT Technol J*, Jan 2001, Vol. 19, No. 1.

[SIP G. Bonnet] G. Bonnet and Y. Shen, "Next Generation Telecommunication Services Based on SIP", *Alcatel Telecommunications Review*, Second Quarter, 2002, (url: *http://www.alcatel.com/atr.*

[SS7 Verisign] Verisign Tutorial, "*Signaling System 7 (SS7)*", (url: *http://www.iec.org/online/tutorials/ss7*).

[UMTS E. Nikula] E. Nikula et al., " FRAMES Multiple Access for UMTS and IMT-2000", *IEEE Personal Communications Magazine*, April 1998, pp. 16–24.

[UMTS J. Park] Jeong-Hyun Park, "Wireless Internet Access for Mobile Subscribers Based on the GPRS/UMTS Network", *IEEE Communications Magazine*, April 2002, pp. 38–49.

[UMTS J. Wiljakka] Juha Wiljakka, "Transition to IPv6 in GPRS and WCDMA Mobile Networks", *IEEE Communications Magazine*, April 2002, pp. 134-140.

[UMTS S. Baudet] S. Baudet et al., "QoS Implementation in UMTS Networks" *Alcatel Telecommunications Review*, First Quarter, 2001, (url: *http://www.alcatel.com/atr*).

[UMTS S. Breyer] S. Breyer et al., "UMTS Node B Architecture in a Multi-Standard Environment", *Alcatel Telecommunications Review*, First Quarter, 2001, (url: *http://www.alcatel.com/atr*).

[UMTS S. Kim] Soojin Kim et al., "Interoperability Between UMTS and CDMA2000 Networks", *IEEE Wireless Communications*, Feb 2003, pp. 22–28.

[UMTS Tektronix] Tektronix Tutorial, *"UMTS Protocols and Protocol Testing"*, (url: *http://www.iec.org/ online/tutorials/umts*).

[UMTS Y. Lin] Yi-Bing Lin et al., "An All-IP Approach for UMTS Third-Generation Mobile Networks", *IEEE Network*, Sep/Oct 2002, pp. 8–19.

[UMTS Y. Lin] Yi-Bing Lin et al., "Mobility Management: From GPRS to UMTS", *Wireless Communications and Mobile Computing*, 2001, Vol. 1, pp. 339–359.

[UTRAN A. Toskala] A. Toskala, O. Lehtinen and P. Kinnunen, "UTRA GSM Handover from Physical Layer Perspective", *Proceedings of ACTS Summit*, Italy, June 1999.

[UTRAN M. Kiril] Mitrevski Kiril and Zgonjanin Dus Ko, "Principles of the UMTS Terrestrial Radio Access Network – UTRAN", *IX Telecommunications Forum'2001*, Beograd, Nov 2001.

[WCDMA E. Dahlman] E. Dahlman et al., "WCDMA—The Radio Interface for Future Mobile Multimedia Communications", *IEEE Transactions on Vehicular Technology*, Nov 1998, Vol. 47, No. 4, pp. 1105–1118.

[WCDMA H. Holma] H. Holma et al., "Asynchronous Wideband CDMA for IMT-2000", *SK Telecom Journal*, South Korea, 1998, Vol. 8, No. 6, pp. 1007–1021.

[WCDMA H. Holma] Harri Holma and Antti Toskala, *"WCDMA for UMTS: Radio Access for Third Generation Mobile Communications"*, John Wiley & Sons.

[Wireless B. Walke] B. Walke , *"Mobile Radio Networks"*, John Wiley & Sons.

[Wireless J. Vriendt] Johan De Vriendt et al., "Mobile Network Evolution: A Revolution on the Move", *IEEE Communications Magazine*, April 2002, pp. 104–111.

III WEB RESOURCES

For the benefit of readers and to avoid typing mistakes, all the web-links are available at the book's website http://3gbook.tripod.com

III.1 Standardization Bodies

Topic	Link
Third Generation Partnership Project (3GPP)	*http://www.3gpp.org/*
Third Generation Partnership Project 2 (3GPP2)	*http://www.3gpp2.org/*
Association of Radio Industries and Businesses (ARIB), Japan	*http://www.arib.or.jp/english/*

Australian Communications Industry Forum (ACIF)	*http://www.acif.org.au/*
China Wireless Telecommunication Standards Group (CWTS)	*http://www.cwts.org/cwts/index_eng.html*
Digital Enhanced Cordless Telecommunications (DECT)	*http://www.dectweb.com/DECTForum/ overview/Evolution.htm*
European Telecommunication Standards Institute (ETSI)	*http://www.etsi.org/*
International Telecommunication Union (ITU)	*http://www.itu.int/*
Internet Engineering Task Force (IETF)	*http://www.ietf.org/*
T1, USA	*http://www.t1.org/*
Telecommunication Technology Committee (TTC), Japan	*http://www.ttc.or.jp/e/*
Telecommunications Industry Association (TIA), USA	*http://www.tiaonline.org/*
Telecommunications Standards Advisory Council of Canada (TSACC)	*http://www.tsacc.ic.gc.ca/*
Telecommunications Technology Association (TTA), Korea	*http://www.tta.or.kr/English/main/*
Universal Wireless Communications Consortium (UWCC)	*http://www.uwcc.org/*

III.2 White Papers and Tutorials

Topic	Link
3G Americas white papers	*http://www.3gamericas.com/English/Technology_Center/ WhitePapers/*
3G white papers	*http://www.3gnewsroom.com/html/whitepapers/ index.shtml*
Alcatel Telecommunications Review	*http://www.alcatel.com/atr/*
Award Solution white papers	*http://www.awardsolutions.com/research/white_papers. shtm*
CDMA white papers	*http://www.cdg.org/resources/white_papers.asp*
Hughes Software white papers	*http://www.hssworld.com/whitepapers/overview.htm*
IEC tutorials	*http://www.iec.org/online/tutorials/*
IT papers	*http://www.itpapers.com/*
Trillium white papers	*http://www.trillium.com/news-events/white-papers/ index.html*

UMTS Forum papers	*http://www.umts-forum.org/servlet/dycon/ztumts/ umts/Live/en/umts/Resources_Papers_index/*

III.3 | Information and News

Topic	Link
3G Americas	*http://www.3gamericas.com/*
3G Information	*http://www.3g.co.uk/*
3G IP	*http://www.3gip.org/*
3G News	*http://www.3gnewsroom.com/3g_news/*
3G Subscribers	*http://www.3gnewsroom.com/html/stats/index.shtml*
4G Sun Research	*http://research.sun.com/features/4g_wireless/*
CDMA Development Group	*http://www.cdg.org/*
Cellular coverage	*http://www.cellular-news.com/coverage/*
Cellular news	*http://www.cellular-news.com/3G/*
Global mobile Suppliers Association (GSA)	*http://www.gsacom.com/*
GSM Association	*http://www.gsmworld.com/*
IETF SIGTRAN charter	*http://www.ietf.org/html.charters/sigtran-charter.html*
IPv6 Forum	*http://www.ipv6forum.com/*
Japanese Mobile Market	*http://mobilemediajapan.com/*
Open Mobile Alliance	*http://www.openmobilealliance.org/*
SIP Information	*http://www.sipcenter.com/*
UMTS Forum	*http://www.umts-forum.org/*
UMTS World	*http://www.umtsworld.com*
UMTS Deployment	*http://www.umts-forum.org/servlet/dycon/ztumts/ umts/Live/en/umts/Resources_Deployment_index*

III.4 | Select 3G Operators

Topic	Link
Hutchison 3G (Australia)	*http://www.three.com.au/*
Hutchison 3G (Italy)	*http://www.tre.it/*
Hutchison 3G (UK)	*http://www.hutchison3g.com/*
Hutchison Whampoa	*http://www.hutchison-whampoa.com/eng/telco/ telecoms.htm*
KDDI (Japan)	*http://www.au.kddi.com/english/index.html*
NTT DoCoMo (Japan)	*http://foma.nttdocomo.co.jp/english/*
Vodafone KK (Japan)	*http://www.vodafone.jp/scripts/english/top.jsp*

ABBREVIATIONS*

1G	First Generation	**ALCAP**	Access Link Control Application Part
2G	Second Generation	**AM**	Acknowledged Mode
3G	Third Generation	**AMF**	Authentication Management Field/Address Management Function
3GPP	Third Generation Partnership Project		
3GPP2	Third Generation Partnership Project 2	**AMR**	Adaptive Multi Rate
		AN	Access Network
ABNF	Augmented Backus-Naur Form	**ANM**	Answer Message
AAA	Authentication, Authorization and Accounting	**ANSI**	American National Standards Institute
AAL	ATM Adaptation Layer	**AoC**	Advice of Charge
AAL2	ATM Adaptation Layer type 2	**APDU**	Application Protocol Data Unit
AAL5	ATM Adaptation Layer type 5	**API**	Application Programming Interface
ACIF	Australian Communications Industry Forum	**APN**	Access Point Name
ACM	Address Complete Message	**ARIB**	Association of Radio Industries and Businesses
AES	Advanced Encryption Standard		
AH	Authentication Header	**AS**	Access Stratum/Application Server/Autonomous System
AK	Anonymity Key		
AKA	Authentication and Key Agreement	**ASE**	Application Service Entity
		AT Command	ATtention Command

---

* The 3GPP TR 21.905 defines the abbreviations used in 3GPP specifications. However, the definitions of many abbreviations are not consistent across different 3GPP specifications. For example, NMSI is referred to as National Mobile Station Identifier in 3GPP TR 21.905 and National Mobile Subscriber Identity in 3GPP TR 23.003. There are more such inconsistencies with regards to the definition of URA, UICC, CSCF, and few others. In the above abbreviation list, the definition provided in the most relevant specification is used, which may not necessary match with that provided in 3GPP TR 21.905. At some places, multiple definitions are provided for same term when the exact definition cannot be ascertained.

ATM	Asynchronous Transfer Mode	**CAMEL**	Customized Application for Mobile network Enhanced Logic
AuC	Authentication Centre		
AUTN	Authentication Token	**CAP**	Camel Application Part
AV	Authentication Vector	**CB**	Call Barring/Cell Broadcast
AVP	Attribute Value Pairs	**CBC**	Cell Broadcast Centre
		CBE	Cell Broadcast Entity
B2BUA	Back-to-Back User Agent	**CBR**	Constant Bit Rate
BAIC	Barring of All Incoming Calls	**CBS**	Cell Broadcast Service
BAOC	Barring of All Outgoing Calls	**CC**	Call Control/Country Code
BCCH	Broadcast Control Channel	**CCBS**	Completion of Calls to Busy Subscriber
BCFE	Broadcast Control Functional Entity		
		CCCH	Common Control Channel
BCH	Broadcast Channel	**CCTrCH**	Coded Composite Transport Channel
BER	Bit Error Rate/Bit Error Ratio		
BG	Border Gateway	**CD**	Call Deflection
BGCF	Breakout Gateway Control Function	**CDMA**	Code Division Multiple Access
		CDR	Charging Data Record
BIC	Barring of Incoming Calls	**CF**	Call Forwarding
BICC	Bearer Independent Call Control	**CEA**	Capability Exchange Answer
B-ISDN	Broadband ISDN	**CER**	Capability Exchange Request
BLER	Block Error Rate/Block Error Ratio	**CFB**	Call Forwarding on mobile subscriber Busy
BMC	Broadcast/Multicast Control	**CFN**	Connection Frame Number
BOIC	Barring of Outgoing International Calls	**CFNRc**	Call Forwarding on mobile subscriber Not Reachable
BOIC-exHC	Barring of Outgoing International Calls except those directed to the Home PLMN Country	**CFNRy**	Call Forwarding on No Reply
		CFU	Call Forwarding Unconditional
		CGI	Cell Global Identity
		CH	Call Handling/Call Hold
BS	Base Station/Basic Service	**CI**	Cell Identity
BSC	Base Station Controller	**CK**	Cipher Key
BSG	Basic Service Group	**CLIP**	Calling Line Identification Presentation
BSS	Base Station Sub-system		
BSSAP	Base Station Sub-system Application Part	**CLIR**	Calling Line Identification Restriction
BSSGP	Base Station Sub-system GPRS Protocol	**CM**	Connection Management
		CN	Core Network
BSSMAP	Base Station Sub-system Management Application Part	**COLP**	Connected Line Identification Presentation
BTS	Base Transceiver Station	**COLR**	Connected Line Identification Restriction
C-RNTI	Cell-Radio Network Temporary Identity	**CPCH**	Common Packet Channel

CPCS	Common Part Convergence Sublayer	**DRNC**	Drift Radio Network Controller
CPS	Common Part Sublayer	**DRNS**	Drift Radio Network Sub-system
CRC	Cyclic Redundancy Check	**D-RNTI**	Drift RNC Radio Network
CRNC	Controlling Radio Network Controller		Temporary Identity
CS	Circuit Switched/Convergence Sublayer	**DRX**	Discontinuous Reception
		DS-CDMA	Direct-Sequence Code Division Multiple Access
CSCF	Call Session Control Function	**DSCH**	Downlink Shared Channel
CSE	Camel Service Environment	**DSSS**	Direct Sequence Spread Sequence
CSI	Camel Subscription Information	**DTCH**	Dedicated Traffic Channel
CTCH	Common Traffic Channel	**DTMF**	Dual Tone Multiple Frequency
CUG	Closed User Group	**DTN**	Deflected-To-Number
CW	Call Waiting	**DTX**	Discontinuous Transmission
CWTS	China Wireless Telecommunication Standards Group	**DWA**	Device Watchdog Answer
		DWR	Device Watchdog Request
		ECR	Enhanced Call Routing
D-AMPS	Digital-Advanced Mobile Phone Services	**ECT**	Explicit Call Transfer
		EDGE	Enhanced Data Rates for Global Evolution
DBP	Diameter Base Protocol		
DC	Dedicated Control	**EGPRS**	Enhanced GPRS
DCA	Dynamic Channel Allocation	**EIR**	Equipment Identity Register
DCFE	Dedicated Control Function Entity	**eMLPP**	Enhanced Multilevel Precedence and Pre-emption
DCCH	Dedicated Control Channel	**ESP**	Encapsulating Security Payload
DCH	Dedicated Channel	**ETSI**	European Telecommunications Standards Institute
DCH-FP	Dedicated Channel-Frame Protocol		
DCN	Data Communication Network	**FACH**	Forward Access Channel
DES	Data Encryption Standard	**FAX**	Facsimile
DECT	Digital Enhanced Cordless Telecommunications	**FCS**	Frame Check Sequence
		FDD	Frequency Division Duplex
DHCP	Dynamic Host Configuration Protocol	**FDM**	Frequency Division Multiplex
		FDMA	Frequency Division Multiple Access
DNS	Domain Name System		
DPA	Disconnect Peer Answer	**FEC**	Forward Error Correction
DPC	Destination Point Code	**FHSS**	Frequency Hopping Spread Spectrum
DPCCH	Dedicated Physical Control Channel		
		FM	Fault Management
DPCH	Dedicated Physical Channel	**FN**	Frame Number
DPDCH	Dedicated Physical Data Channel	**FOMA**	Freedom of Mobile Multimedia Access
DPR	Disconnect Peer Request		

FP	Frame Protocol	**HN**	Home Network
FQDN	Fully Qualified Domain Name	**HO**	Handover
FSM	Finite State Machine	**HPLMN**	Home Public Land Mobile
FTN	Forwarded-To-Number		Network
FTNW	Forwarded-To-Network	**HSCSD**	High-Speed Circuit Switched
FTNW-LEC	Local Exchange in Forwarded-		Data
	To-Network	**HS-DSCH**	High-Speed Downlink Shared
FTP	File Transfer Protocol		Channel
		HSDPA	High-Speed Downlink Packet
GC	General Control		Access
GERAN	GSM/EDGE Radio Access	**HS-PDSCH**	High-Speed Physical Downlink
	Network		Shared Channel
GGSN	Gateway GPRS Support Node	**HSS**	Home Subscriber Server
GMLC	Gateway Mobile Location	**HTTP**	Hyper Text Transfer Protocol
	Centre	**IAM**	Initial Address Message
GMM	GPRS Mobility Management	**IC**	Integrated Circuit
GMSC	Gateway Mobile Switching	**ICC**	Integrated Circuit Card
	Center	**ICMP**	Internet Control Message
GPRS	General Packet Radio Service		Protocol
gprsSSF	GPRS Service Switching Function	**I-CSCF**	Interrogative-Call Session
GPS	Global Positioning System		Control Function
GSA	Global mobile Suppliers	**ICV**	Integrity Check Value
	Association	**ID**	Identifier
GSM	Global System for Mobile	**IE**	Information Element
	communications	**IETF**	Internet Engineering Task Force
gsmSCF	GSM Service Control Function	**IK**	Integrity Key
gsmSSF	GSM Service Switching Function	**IKE**	Internet Key Exchange
gsmSRF	GSM Specialized Resource	**IMEI**	International Mobile Equipment
	Function		Identity
GSN	GPRS Support Nodes	**IMPI**	IMS Private User Identity
GT	Global Title	**IMPU**	IMS Public User Identity
GTP	GPRS Tunneling Protocol	**IMS**	IP Multimedia Subsystem
GTP-C	GPRS Tunnelling Protocol for	**IMS AKA**	IMS Authentication and Key
	Control Plane		Agreement
GTP-U	GPRS Tunnelling Protocol for	**IMSI**	International Mobile Subscriber
	User Plane		Identity
GTT	Global Title Translation/Global	**IM-SSF**	IP Multimedia Service Switching
	Text Telephony		Function
		IMT-2000	International Mobile
HE	Home Environment		Telecommunications 2000
HFN	Hyper Frame Number	**IN**	Intelligent Network
HHO	Hard Handover	**IP**	Internet Protocol
HLR	Home Location Register		

IPv4	Internet Protocol version 4	**MAP**	Mobile Application Part
IPv6	Internet Protocol version 6	**MAPsec**	MAP security
IPLMN	Interrogating Public Land Mobile Network	**MAA**	Multimedia Authorization Answer
ISC	Interface Service Control	**MAR**	Multimedia Authorization Request
ISDN	Integrated Services Digital Network	**MBMS**	Multimedia Broadcast/Multicast Service
ISIM	IM Services Identity Module		
ISO	International Organization for Standardization	**MC**	Multi-Call
		MCC	Mobile Country Code/Mobile Competence Centre
ISP	Internet Service Provider		
IST	Immediate Service Termination	**MCEF**	Memory Capacity Exceeded Flag
ISUP	ISDN User Part	**M-CSI**	Mobility Management Camel Subscription Information
ITU	International Telecommunication Union		
		ME	Mobile Equipment
IUA	ISDN Q.921 User Adaptation	**MExE**	Mobile Execution Environment
IV	Initialization Vector	**MGC**	Media Gateway Controller
IWF	Inter-Working Function	**MGCF**	Media Gateway Control Function
IWMSC	Inter-Working MSC		
		MGW	Media GateWay
KAC	Key Administration Center	**MIB**	Management Information Base
KSI	Key Set Identifier	**MM**	Mobility Management
		MMI	Man–Machine Interface
LA	Location Area	**MMS**	Multimedia Messaging Service
LAC	Location Area Code	**MNC**	Mobile Network Code
LAI	Location Area Identity	**MNP**	Mobile Number Portability
LAN	Local Area Network	**MNRF**	MS Station Not Reachable Flag
LAU	Location Area Update	**MNRG**	MS Not Reachable for GPRS
LCS	Location Services	**MNRR**	MS Not Reachable Reason
LEC	Local Exchange	**MO**	Mobile Originated
LI	Length Indicator	**MPTY**	Multiparty
LIA	Location Information Answer	**MRF**	Media Resource Function
LIR	Location Information Request	**MS**	Mobile Station
LLC	Logical Link Control	**MSC**	Mobile Switching Centre
LMSI	Local Mobile Station Identity	**MSIN**	Mobile Subscriber Identification Number
LR	Location Registration		
LSP	Locally Significant Part	**MSISDN**	Mobile Subscriber ISDN Number
LU	Location Update		
		MSP	Multiple Subscriber Profile
M2UA	MTP2 User Adaptation	**MSRN**	Mobile Station Roaming Number
M3UA	MTP3 User Adaptation		
MAC	Medium Access Control/ Message Authentication Code	**MT**	Mobile Terminated/Mobile Termination/Mobile Terminal

MTP	Message Transfer Part	**OSA**	Open Service Access/Open Service Architecture
MTP1	Message Transfer Part layer 1		
MTP2	Message Transfer Part layer 2	**OSA SCS**	Open Service Architecture Service Capability Server
MTP3	Message Transfer Part layer 3		
MTP3-B	Message Transfer Part 3 for Broadband	**OSI**	Open System Interconnection
		OSI RM	OSI Reference Model
MTU	Maximum Transfer Unit	**OSS**	Operator Specific Service
MWD	Messages Waiting Data	**OTDOA**	Observed Time Difference Of Arrival
MWIF	Mobile Wireless Internet Forum		
NAI	Network Access Identifier	**PC**	Point Code
NAM	Network Access Mode	**PCCH**	Paging Control Channel
NAS	Non-Access Stratum	**PCCPCH**	Primary Common Control Physical Channel
NBAP	Node B Application Part		
NDC	National Destination Code	**PCH**	Paging Channel
NDS	Network Domain Security	**PCM**	Pulse Code Modulation
NE	Network Element/Network Entity	**PCPCH**	Physical Common Packet Channel
NMSI	National Mobile Subscriber Identity	**PCS**	Personal Communication System
NNI	Network–Node Interface/ Network–Network Interface	**P-CSCF**	Proxy-Call Session Control Function
NSAP	Network Service Access Point	**PCU**	Packet Control Unit
NSAPI	Network Service Access Point Identifier	**PDC**	Personal Digital Communications
NSS	Network Sub System	**PDCP**	Packet Data Convergence Protocol
Nt	Notification		
NT	Network Termination/Non-Transparent	**PDH**	Plesiochronous Digital Hierarchy
		PDN	Public Data Network/Packet Data Network
O&M	Operations and Maintenance	**PDP**	Packet Data Protocol
OAM	Operation, Administration and Maintenance	**PDSCH**	Physical Downlink Shared Channel
O-CSI	Originating-Camel Subscription Information	**PDU**	Protocol Data Unit
		PLMN	Public Land Mobile Network
ODB	Operator Determined Barring	**PNFCE**	Paging and Notification Control Function Entity
ODMA	Opportunity Driven Multiple Access		
		PPA	Push Profile Answer
OFDMA	Orthogonal Frequency Division Multiple Access	**PPI**	Protection Profile Indicator
		PPP	Point-to-Point Protocol
OMAP	Operations Maintenance and Administration Part	**PPR**	Push Profile Request
		PRACH	Physical Random Access Channel
OR	Optimal Routing		

PRN	Provide Roaming Number	**ROHC**	Robust Header Compression
PROP	Proprietary field	**RRC**	Radio Resource Control
PS	Packet Switched	**RRM**	Radio Resource Management
PSAP	Public Safety Answering Point	**RSZI**	Regional Subscription Zone Identity
PSTN	Public Switched Telephone Network	**RT**	Radio Termination
P-TMSI	Packet-TMSI	**RTA**	Registration Termination Answer
PUSCH	Physical Uplink Shared Channel	**RTP**	Real Time Protocol
PVC	Permanent Virtual Circuit	**RTR**	Registration Termination Request
QoS	Quality of Service	**RTT**	Round Trip Time
R99	Release 1999 of 3GPP	**S-RNTI**	Serving RNC-Radio Network Temporary Identity
RA	Routing Area	**SA**	Security Association
RAB	Radio Access Bearer	**SABP**	Service Area Broadcast Protocol
RAC	Routing Area Code	**SAA**	Server Assignment Answer
RACH	Random Access Channel	**SAAL**	Signaling ATM Adaptation Layer
RADIUS	Remote Authentication Dial In User Service	**SAD**	Security Association Database
RAI	Routing Area Identity	**SAGE**	Security Algorithms Group of Experts
RAN	Radio Access Network	**SAM**	Subsequent Address Message
RANAP	Radio Access Network Application Part	**SAP**	Service Access Point
RAND	Random number	**SAPI**	Service Access Point Identifier
RAT	Radio Access Technology	**SAR**	Segmentation and Reassembly/ Server Assignment Request
RAU	Routing Area Update	**SC**	Service Center/Service Code
RB	Radio Bearer	**SCCP**	Signaling Connection Control Part
REL	Release Message	**SCCPCH**	Secondary Common Control Physical Channel
Rel4 (or R4)	Release 4 of 3GPP	**SCF**	Services Capability Feature
Rel5 (or R5)	Release 5 of 3GPP	**SCFE**	Shared Control Function Entity
Rel6 (or R6)	Release 6 of 3GPP	**SCN**	Switched Circuit Number
RF	Radio Frequency	**S-CSCF**	Serving-Call Session Control Function
RFC	Request For Comments	**SCTP**	Stream Control Transmission Protocol
RFE	Routing Functional Identity	**SDH**	Synchronous Digital Hierarchy
RLC	Radio Link Control/Release Complete Message	**SDP**	Session Description Protocol
RLCP	Radio Link Control Protocol	**SDU**	Service Data Unit
RNC	Radio Network Controller		
RNS	Radio Network Sub-system		
RNSAP	Radio Network Sub-system Application Part		
RNTI	Radio Network Temporary Identity		

SEAL	Simple and Efficient Adaptation Layer	**SNRM**	Service and Network Resource Management
SEG	Security Gateway	**SPC**	Signaling Point Code
SG	Signaling Gateway	**SPD**	Security Policy Database
SGSN	Serving GPRS Support Node	**SPI**	Security Parameter Index
SGW	Signalling Gateway	**SPT**	Service Point Trigger
SHCCH	Shared Channel Control Channel	**SQN**	Sequence Number
		SRB	Signaling Radio Bearer
SHO	Soft Handover	**SRI**	Send Routing Information
SI	Supplementary Information	**SRNC**	Serving Radio Network Controller
SIB	System Information Block		
SIFIC	Send Information For Incoming Call	**SRNS**	Serving Radio Network Sub-system
SIFOC	Send Information For Outgoing Call	**SS**	Supplementary Service
		SS7	Signaling System No. 7
SIGTRAN	Signaling Transport	**SSADT**	Service Specific Assured Data Transfer Sublayer
SIM	Subscriber Identity Module		
SIP	Session Initiated Protocol	**SSCF**	Service Specific Co-ordination Function
SIR	Signal-to-Interference Ratio		
SLF	Server Locator Function	**SSCF-NNI**	Service Specific Co-ordination Function for NNI
SLA	Service Level Agreement		
SM	Session Management/Short Message/Support Mode	**SSCF-UNI**	Service Specific Co-ordination Function for UNI
SME	Short Message Entity	**SSCOP**	Service Specific Connection Oriented Protocol
SMG	Special Mobile Group		
SMLC	Serving Mobile Location Centre	**SSCS**	Service Specific Convergence Sublayer
SMS	Short Message Service	**SS-CSI**	Supplementary Service-Camel Subscription Information
SMS-CB	SMS-Cell Broadcast		
SMS-CSI	SMS-Camel Subscription Information	**SSDT**	Site-Selection Diversity Transmit
		SSF	Service Switching Function
SMS-GMSC	SMS-Gateway MSC	**SSN**	Sub-System Number
SMS-IWMSC	SMS-Inter-Working MSC	**SSSAR**	Service Specific Segmentation and Re-assembly Sublayer
SMS-PTP	Short Message Service-Point to Point		
		SSTED	Service Specific Transmission Error Detection Sublayer
SMSC	Short Message Service Centre		
SN	Serving Network/Subscriber Number	**STC**	Signaling Transport Converter
		STP	Signaling Transfer Point
SNDCP	Sub-Network Dependent Convergence Protocol	**SUA**	SCCP User Adaptation Layer
		SVC	Switched Virtual Circuit
SNMP	Simple Network Management Protocol	**T**	Transparent
SNR	Serial Number	**TA**	Terminal Adaptation

TAC	Type Allocation Code	**UDP**	User Datagram Protocol
TCAP	Transaction Capabilities/	**UE**	User Equipment
	Transaction Capability	**UICC**	Universal Integrated Circuit
	Application Part		Card or
T-CSI	Terminating-Camel Subscription	**UMTS**	Integrated Circuit Card
	Information	**UM**	Unacknowledged Mode
TCP	Transmission Control Protocol	**UMTS**	Universal Mobile
TDD	Time Division Duplex		Telecommunications System
TDMA	Time Division Multiple Access	**UMTS AKA**	UMTS Authentication and Key
TE	Terminal Equipment		Agreement
TEID	Tunnel End Point Identifier	**UNI**	User–Network Interface
TFT	Traffic Flow Template	**UP**	User Plane
TIA	Telecommunications Industry	**UPC**	Usage Parameter Control
	Association	**URA**	UTRAN Registration
TLLI	Temporary Logical Link		Area
	Identity	**URL**	Uniform Resource Locator
TLV	Type Length Value	**U-RNTI**	UTRAN-Radio Network
TM	Transparent Mode		Temporary Identity
TME	Transfer Mode Entity	**USAT**	USIM Application Toolkit
TMN	Telecom Management	**USCH**	Uplink Shared Channel
	Network	**USIM**	Universal Subscriber Identity
TMSI	Temporary Mobile Subscriber		Module
	Identity	**USP**	Unique Selling Point
TOS	Type of Service	**USSD**	Unstructured Supplementary
TPC	Transmit Power Control		Service Data
TR	Technical Report	**UTRA**	Universal Terrestrial Radio
TS	Technical Specification		Access
TSACC	Telecommunications Standards	**UTRAN**	Universal Terrestrial Radio
	Advisory Council of Canada		Access Network
TSG	Technical Specification Group	**UWCC**	Universal Wireless
TTA	Telecommunications Technology		Communications Consortium
	Association	**UUS**	User-to-User Signaling
TTC	Telecommunication Technology	**UUI**	User-to-User Information
	Committee		
TTI	Transmission Timing Interval	**VBR**	Variable Bit Rate
TVP	Time Variant Parameter	**VC**	Virtual Circuit
TUP	Telephone User Part	**VCI**	Virtual Channel Identifier
		VGCS	Voice Group Call Service
UA	User Agent	**VHE**	Virtual Home Environment
UAA	User Authorization Answer	**VLR**	Visitor Location Register
UAR	User Authorization Request	**VMSC**	Visited MSC
U-CSI	USSD-Camel Subscription	**VoIP**	Voice Over IP
	Information	**VPI**	Virtual Path Identifier

VPLMN	Visited Public Land Mobile Network	**WAP**	Wireless Application Protocol
VPN	Virtual Private Network	**WCDMA**	Wideband Code Division Multiple Access
VSF-CDMA	Variable Spreading Factor CDMA	**WG**	Working Group
		WLAN	Wireless LAN
VSF-OFCDM	Variable Spreading Factor Orthogonal Frequency and Code Division Multiplexing	**WTDMA**	Wideband Time Division Multiple Access
		WWW	World Wide Web
VT-CSI	VMSC Terminating-Camel Subscription Information	**XRES**	Expected Response

INDEX

2.5 Generation Networks, 5

A Interface, 94
AAL2 Signaling
 functions, 146
 layer architecture, 145
 messages, 147
 overview, 144
Abis Interface, 92
Access Link Control Application Part. *See*
 AAL2 Signaling
Access Network
 architecture, 91
 entities, 89
 interfaces, 91
 overview, 48
Access Point Name, 357
Access Security
 data confidentiality. *See* Data
 Confidentiality
 data integrity. *See* Data Integrity
 mutual authentication. *See* Mutual
 Authentication
 overview, 439, 441

 user identity confidentiality. *See* User
 Identity Confidentiality
Access Stratum, 54, 220
Advice of Charge, 402
Application Server, 488, 492
Asynchronous Transfer Mode
 functions, 134
 layer architecture, 133
 overview, 131
ATM Adaptation Layer 2
 functions, 136
 layer architecture, 135
 overview, 135
ATM Adaptation Layer 5
 functions, 138
 layer architecture, 137
 overview, 137
Authentication, 438
Authentication and Key Agreement, 442
Authentication Center, 51, 166
Authentication Token, 442
Authentication Vector, 442, 444, 445
Autocorrelation, 26

---

* The index is made for full forms only. For indexing an abbreviation, first find the full form using the Abbreviation section and then refer the full form in this index. For example, to find 2G, find full form of 2G in Abbreviations section (which is Second Generation), and then refer this index for Second Generation. For very long phrases, a part of abbreviations is used as index key word (for e.g. 'GPRS Tunneling Protocol' is used instead of 'General Packet Radio Services Tunneling Protocol').

B Interface, 172
Background Class, 75
Barring of All Incoming Calls. *See* Call
    Barring
Barring of All Incoming Calls when Roaming
    outside the HPLMN country. *See* Call
    Barring
Barring of All Outgoing Calls. *See* Call
    Barring
Barring of Outgoing International Calls. *See*
    Call Barring
Barring of Outgoing International Calls
    except to HPLMN Country. *See* Call
    Barring
Base Station Controller, 48, 89
Base Station Sub-system, 48, 89
Base Station Sub-system Application Part+
    functions, 228
    layer architecture, 226
    messages, 228
    overview, 226
Base Transceiver Station, 48, 89
Basic Services, 373
Bearer Services, 68
Black list, 167
Border Gateway, 170, 358
Breakout Gateway Control Function, 488, 493
Broadcast/Multicast Control
    functions, 123
    layer architecture, 123
    overview, 123

C Interface173
Call Barring
    barring of all incoming calls, 394
    barring of all incoming calls when
    roaming outside the HPLMN country, 394
    barring of all outgoing calls, 393
    barring of outgoing international calls,
        394
    barring of outgoing international calls
        except those directed to the Home
        PLMN country, 394
    interaction with call control procedures,
        338
    overview, 393

Call Control
    concepts, 193, 324
    interaction with call barring, 338
    interaction with call forwarding, 338
    procedures, 330, 333
Call Deflection, 388
Call Forwarding
    call forwarding on mobile subscriber busy,
        392
    call forwarding on mobile subscriber not
        reachable, 393
    call forwarding on no reply, 392
    call forwarding unconditional, 392
    interaction with call control procedures,
        338
    overview, 392
Call Forwarding on mobile subscriber Busy.
    See Call Forwarding
Call Forwarding on mobile subscriber Not
    Reachable. *See* Call Forwarding
Call Forwarding on No Reply. *See* Call
    Forwarding
Call Forwarding Unconditional. *See* Call
    Forwarding
Call Handling. *See* Call Control
Call Hold, 394
Call Session Control Function, 488, 491
Call Waiting, 394
Calling Line Identification Presentation, 389
Calling Line Identification Restriction, 389
Calling Name Presentation, 410
Cell Broadcast Area, 421
Cell Broadcast Center,171, 421
Cell Broadcast Entity, 421
Cell Broadcast Service,
CBS pages, 421
    interfaces, 421
    network architecture, 421
    overview, 171, 420
    procedures, 422
Cell Global Identity, 59
Cell Identity, 59
Cell Update, 255, 287
Challenge-Response Protocol, 330
Channelization Codes, 24
Cipher Key, 441, 442

Circuit-Switched Connection, 50
Circuit-Switched Domain
  architecture, 171
  control plane, 181
  entities, 167
  interfaces, 171
  overview, 50
  protocols, 180
  user plane, 180
Circuit-Switching, 50
Closed Loop Power Control
  concepts, 35
  inner loop, 36
  outer loop, 36
Closed User Group, 398
CM Sublayer, 298
Code Division Multiple Access
  concepts, 21
  direct-sequence. *See* Direct-Sequence
  CDMA
Completion of Calls to Busy Subscriber, 411
Confidentiality, 438
Connected Line Identification Presentation,
  390
Connected Line Identification Restriction,
  390
Control Channels, 114
Controlling RNC, 91
Conversational Class, 73
Core Network
  circuit-switched domain. *See* Circuit-
  Switched Domain
  entities, 165
  functions, 49, 189
  home network, 53
  overview, 50
  packet-switched domain. *See* Packet-
  Switched Domain
  serving network, 52
  subscriber data, 196
  transit network, 53
Country Code, 62
Cross Correlation, 26
Cu Interface, 83

Customized Applications for Mobile network
  Enhanced Logic, 171, 436
Cx Interface, 496

D Interface, 174
Data Confidentiality, 451
Data Integrity, 438, 453
Deflected-To-Number, 388
Deployment of 3G Networks,
  Italy, 528
  Japan, 527
  overview, 526
  UK, 527
Despreading, 24
Diameter, 521
Digital Enhanced Cordless
  Telecommunications, 7
Digital-Advanced Mobile Phone Services, 4
Direct Sequence Spread Spectrum, 22
Direct-Sequence CDMA, 22
Drift RNC, 40, 91
Dx Interface, 496
Dynamic Host Configuration Protocol, 504

E Interface, 174
Early Call Forwarding, 346
Encapsulating Security Payload, 465
Encapsulation, 359
Enhanced Call Routing, 434
Enhanced Data Rates for Global Evolution, 5
Enhanced GPRS, 6
Enhanced Multilevel Precedence and
  Preemption, 387
Equipment Identity Register, 51, 167
Explicit Call Transfer, 406

F Interface, 174
Forwarded-To-Number, 392, 393
Fourth Generation Networks
  features, 532
  network hierarchy, 530
  overview, 529
Framing Protocols
  functions, 162
  layer architecture, 161
  overview, 161

Frequency Division Duplex, 6, 24
Frequency Division Multiple Access, 21
Frequency Hopping Spread Spectrum, 23

G Interface, 174
Gateway GPRS Support Node, 51, 170
Gateway Mobile Location Center, 171
Gateway Mobile Switching Center, 51, 169
Gb Interface, 94
Gc Interface, 180
Gd Interface, 180
General Packet Radio Services, 5
Gf Interface, 179, 186
Gi Interface, 178
Global Positioning System, 432
Global System for Mobile communications, 4
Global Title, 202, 218
Global Title Translation, 202
Gm Interface, 498
Gn Interface, 176, 186
Gp Interface, 176, 186
GPRS Attach, 192, 314, 503
GPRS Detach, 192, 314
GPRS Tunneling Protocol
    functions, 225
GTP tunnel, 360
    layer architecture, 225
    messages, 226
    overview, 224
Gr Interface, 178, 186
Grey list, 167
Gs Interface, 179, 189
GTP-MAP Protocol Converter, 188

Handover, 32, 36, 107
Hard Handover, 38, 258
High-Speed Circuit Switched Data, 5
Home Location Register, 51, 165, 488
Home Public Land Mobile Network, 324
Home Subscriber Server, 488, 490
Hybrid FDMA/TDMA, 21

Immediate Service Termination
    alert service, 350
    command service, 352
    concepts, 350
IMSI Attach, 191, 306, 311

IMSI Detach, 190, 312
Initial Filter Criteria, 503, 512
Initial UE Identity, 242
Integrity Key, 441, 442
Interactive Class, 75
International Mobile Equipment Identity, 64, 167
International Mobile Subscriber Identity, 60
International Mobile Telecommunication-2000, 6
Internet Key Exchange, 468
Interrogating CSCF. *See* Call Session Control Function
Interrogating Public Land Mobile Network, 324, 344
Inter-system Handover, 38, 259
Interworking Function, 169
IP Multimedia Subsystem
    access security, 523
    addressing, 499
    application server. *See* Application Server
    de-registration, 507
    entities, 488
    implicit registration set, 501
    interfaces, 494
    interworking entities, 488
ISIM, 488
    network domain security, 523
    overview, 486
    private identity, 500
    profile update, 510
    protocols, 520
    public identity, 500
    registration, 504
    role of SIP, 486
    security, 522
    service control, 512
    service platform, 488
    service profile, 500
    session interworking, 518
    session origination, 513
    session termination, 518
    session-related procedures, 511
    session-unrelated procedures, 503
    subscriber data, 502
    underlying transport, 503

IP Security
  architecture, 464
  key exchange, 468
  overview, 464
  protocols, 465
  security domain, 464
  security gateway, 464
  transport mode, 466
  tunnel mode, 466
IS-95, 4
ISC Interface, 498
ISDN User Part
  functions, 204
  layer architecture, 204
  messages, 204
  overview, 203
Iu Bearer, 73
Iu Interface, 94, 180, 182, 184, 186
Iu User Plane Protocol
  layer architecture, 159
  messages, 160
  overview, 158
  support mode, 159
  transport mode, 159
Iu_BC Interface, 105
Iu_CS Interface, 98
Iu_PS Interface, 100
Iub Interface, 92, 103
Iur Interface, 93, 101

KASUMI, 453

Late Call Forwarding, 346
LCS Client, 428
LCS Server, 428
Le Interface, 430
Lg Interface, 431
Lh Interface, 431
Local Mobile Station Identity, 63
Location Area, 57, 295
Location Area Code, 58
Location Area Identity, 58
Location Number, 65
Location Registration, 303
Location Services
  control procedures, 429
  interfaces, 430
  network architecture, 430
  overview, 428
  positioning methods, 431
  reference model, 428
  types of services, 433
Location Update, 191, 306
Location-based Charging, 433
Logical Channels, 113

Macrodiversity, 30
Man-Machine Interface
  MMI codes, 384
  MMI codes for standardized supplementary
    services, 385
  overview, 382
MAP Security
  fallback to unprotected mode, 459
  key administration center, 463
  key distribution, 462
  message format, 457
  operations, 461
  overview, 457
  protection mode, 458
  protection profile identifier, 461
  security association, 457
  security association database, 459
  security policy database, 459
Master Information Block, 237
Mc Interface, 499
Media Gateway, 488, 493
Media Gateway Control Function, 488, 493
Medium Access Control
  functions, 114
  layer architecture, 114
  logical channels. *See* Logical Channels
  overview, 95, 113
Message Transfer Part
  functions, 201
  layer architecture, 200
  message transfer part 1, 200
  message transfer part 2, 200
  message transfer part 3, 200
  overview, 200
  users, 200

Message Transfer Part 3b
  functions, 143
  layer architecture, 142
  overview, 142
Mg Interface, 499
Mi Interface, 499
MILENAGE, 450
Mj Interface, 499
Mk Interface, 499
MM Connection, 191, 297
MM1 Transfer Protocol, 425
MM3 Transfer Protocol, 426
MM4 Transfer Protocol, 426
MMS Relay/Server, 423
MMS User Agent, 422
MMS VAS Applications, 424
Mobile Application Part
  example scenario, 216
  functions, 210
  interfaces, 183
  layer architecture, 209
  message routing, 216
  messages, 212
  overview, 209
Mobile Competence Centre, 9
Mobile Country Code, 56
Mobile Equipment, 47, 78, 79
Mobile Execution Environment, 437
Mobile Network Code, 56
Mobile Station Roaming Number, 63, 327, 329
Mobile Subscriber ISDN, 60
Mobile Switching Center, 51, 168
Mobile Termination
  functional groups, 79
  functions, 84
  overview, 79
Mobile-Originated Call
  architecture, 325
  overview, 193, 324
  priority handling, 387
Mobile-Terminated Call
  architecture, 326
  overview, 193, 324
  priority handling, 387

Mobility Management
  cell selection, 301
  concepts, 189, 292
  connection management procedures, 297
  GMM common procedure, 192
  GMM procedure, 191
  GMM specific procedure, 192
  location registration procedures. *See* Location Registration
MM common procedure, 190
MM connection. *See* MM Connection
MM procedure, 190
MM specific procedure, 190
  PLMN selection, 300
  procedures overview, 296
  state model, 293
MTP3 User Adaptation Layer
  deployment in UMTS, 480
  functions, 480
  overview, 479
Multi Party, 397
Multi-Call, 408
Multimedia Messaging Service
  overview, 422
  procedures, 427
  protocol framework, 424
  reference architecture, 422
Multipath Diversity, 30
Multipath propagation, 26
Multiple Subscriber Profile, 410
Mutual Authentication
  authentication vector. *See* Authentication Vector
  challenge-response protocol, 447, 449
  flow diagram, 455
  in UMTS, 442
  overview, 439, 441, 442
  synchronization failure, 449
Mw Interface, 499

National Destination Code, 62
Nb Interface, 175
Nc Interface, 175, 183
Network Domain Security
  applicability of ESP, 467

applicability of IKE, 469
IP security. *See* IP Security
MAP security. *See* MAP Security
    overview, 439
Network Service Access Point Identifier, 356
Network Termination, 80
Node B, 48, 91
Node B Application Part
    functions, 155
    layer architecture, 153
    messages, 156
    overview, 153
Non-Access Stratum, 54, 220
Non-Access Stratum Signaling
    connection establishment, 267
    functions, 221
    layer architecture, 220
    messages, 221
    overview, 220

Observed Time Difference of Arrival, 432
Open Loop power control, 34
Open Service Access, 436
Optimal Routing. *See* Support for Optimal
    Routing
Origin Authentication. *See* authentication

Packet Data Convergence Protocol
    functions, 122
    layer architecture, 122
    overview, 121
Packet Data Protocol
    addressing, 64, 355
    context. *See* PDP Context
    dynamic addressing, 64, 355
    protocol states, 361
    static addressing, 64, 355
    types, 356
Packet-Switched Connection, 51
Packet-Switched Domain
    architecture, 175
    control plane, 185
    entities, 169
    interfaces, 175
    overview, 50
    protocols, 184
    user plane, 184

Packet-Switching, 51
Paging
    overview, 127, 239, 296, 329
    pre-paging, 329, 333, 337
    procedure, 266, 334, 335
    UE dedicated paging, 240
Paired Spectrum, 6
PDP Context
    activation, 64, 193, 356, 358, 362, 504
    deactivation, 194, 356, 358, 370
    information elements, 356
    modification, 193, 367
    MS-initiated activation, 363
    MS-initiated deactivation, 370
    MS-initiated modification, 368
    network-initiated modification, 370
    network-requested activation, 365
    overview, 356
    secondary PDP Context, 362, 364
    SGSN-initiated deactivation, 371
    SGSN-initiated modification, 369
    unsuccessful network-requested
        activation, 366
Personal Digital Communications, 4
Physical Layer
    channel coding, 110
    functions, 110
    overview, 95, 110
    transport channels. *See* Transport
    Channels
    Point Code, 201, 218
Power Control
    closed loop. *See* Closed Loop Power
    Control
    mechanisms, 34
    open loop. *See* Open Loop Power Control
    requirement, 33
Proxy CSCF. *See* Call Session Control
    Function
PS Attach. *See* GPRS Attach
PS Detach. *See* GPRS Detach
Pseudo-Noise sequence, 23
Public Land Mobile Network, 56
Public Safety Services, 433

QoS Architecture, 72
QoS Classes, 73

R Interface, 84
Radio Access Bearer,
    control procedures, 273, 278
    definition, 72, 149
Radio Access Network Application Part
    functions, 149
    layer architecture, 148
    messages, 150
    overview, 148
Radio Access Schemes, 21
Radio Bearer
    control procedures, 250
    definition, 72
Radio Interface, 95
Radio Link Control
    acknowledged mode, 119
    functions, 120
    layer architecture, 118
    modes of operation, 118
    overview, 95, 117
    transparent mode, 118
    unacknowledged mode, 119
Radio Network Controller, 48, 90
Radio Network Sub-system, 48, 89
Radio Network Sub-system Application Part
    functions, 152
    layer architecture, 150
    messages, 153
    overview, 150
Radio Network Temporary Identity, 63
Radio Resource Control
    connection, 240
    connection management procedures, 236,
        268, 272
    connection mobility procedures, 254
    functions, 126
    layer architecture, 124
    measurement procedures, 260
    messages, 128
    overview, 95, 124
    protocol states, 233
    radio bearer control procedures, 249
    security functions, 247
    service access points, 124
Radio Resource Management, 107
Radio Termination, 79

RADIUS, 521
RAKE receiver, 30, 37
Release 4, 5, 6, 15, 16, 99
Replay Protection, 438
Routing Area, 58, 295
Routing Area Code, 59
Routing Area Identity, 59
Routing Area Update, 192, 316

SCCP User Adaptation Layer
    deployment in UMTS, 482
    functions, 482
    overview, 481
Scrambling, 28
Scrambling Codes, 29
Second Generation
    concepts, 3
    limitations, 4
    technologies, 4
Security Management
    2G security, 438
    3G security, 439
    access security. *See* Access Security
    network domain security. *See* Network
    Domain Security
    overview, 438
    user domain security. *See* User Domain
    Security
Server Locator Function, 492
Service Area Broadcast, 265
Service Area Broadcast Protocol, 157
Service Capability Features
    overview, 435
    toolkits, 436
    types of, 436
Service Point Trigger, 512
Service Specific Connection Oriented Part
    functions, 139
    layer architecture, 139
    overview, 139
Service Specific Co-ordination Function for
    NNI, 140
Service Specific Co-ordination Function for
    UNI, 141
Serving CSCF. *See* Call Session Control
    Function

Serving GPRS Support Node, 51, 169
Serving RNC, 39, 91
Session Initiation Protocol
    components, 488
    overview, 521
    proxy, 488, 491
    registrar, 488, 492
    role in IMS, 487
    session origination, 513
    session termination, 518
    user agents, 488
Session Management
    concepts, 193, 355
    packet filtering, 361
    packet routing, 358
Sh Interface, 497
Short Message Service
    entities, 167, 417
    interfaces, 417
    network architecture, 416
    overview, 69, 416
    procedures, 418, 419
Short Message Service Center, 167, 417
Si Interface, 498
Signaling Connection Control Part
    functions, 202
    Global Title Translation. *See* Global Title
    Translation
    layer architecture, 202
    overview, 202
    sub-system numbering. *See* Sub-system
    Number
Signaling Gateway, 493
Signaling Radio Bearers, 242
Signaling Transport
    comparison between M3UA and SUA, 483
    ISDN Q.921 User Adaptation Layer, 476
    MTP2 User Adaptation Layer, 476
    MTP3 User Adaptation Layer. *See* MTP3
    User Adaptation Layer
    overview, 474
    protocols, 476
    requirements, 474
SCCP User Adaptation Layer. *See* SCCP User
    Adaptation Layer

Stream Control Transmission Protocol. *See*
    Stream Control Transmission Protocol
    uses, 484
Signaling Transport Converter
    functions, 144
    layer architecture, 143
    overview, 143
Signal-to-Interference Ratio, 35
SIGTRAN. *See* Signaling Transport
Site-Selection Diversity Transmit, 32
SMS Gateway Mobile Switching Center, 417
SMS Inter-Working Mobile Switching Center,
    417
Soft Handover
    concepts, 36
    procedures, 257, 282
Softer Handover, 38
Spreading
    benefits, 27
    concepts, 24
Spreading Codes, 24
Spreading Factor, 25
SRNS Relocation
    concepts, 39
    procedure, 285
Stream Control Transmission Protocol, 478
Streaming Class, 75
Subscriber Tracing, 307, 309
Sub-system Number, 202
Super-Charger, 320
Supplementary Services
    call independent management, 379
    call related management, 379
    concepts, 194, 374
    MMI Codes. *See* Man-Machine Interface
    operations, 194
    overview, 71, 373
Support for Optimal Routing, 341
System Information Block
    master information block. *See* Master
    Information Block
    organization, 237
    overview, 126, s237
System Information Broadcast, 126, 264

Technical Specifications Group
  CN, 10
  GERAN, 10
  overview, 10
  RAN, 10
  SA, 11
Terminal, 11
Teleservices, 70
Temporary Mobile Subscriber Identity, 62, 454
Terminal Adaptation
  functions, 85
  overview, 80
Terminal Equipment
  functions, 85
  overview, 80
Third Generation
  air interface requirements, 19
  overview, 7
  radio interface systems, 7
Third Generation Partnership Program
  evolution, 17
  individual members, 9
  interaction, 8
  market representation partners, 8
  objectives, 9
  observers, 9
  organizational partners, 8
  overview, 8
  release 4. *See* Release 4
  release 5. *See* Release 5
  release 6. *See* Release 6
  release 99. *See* Release 99
  specification releases, 15
  specification versions, 13
  specifications, 13
  standardization process, 12
  technical specification group. *See* Technical Specification Group
Time Division Duplex, 7, 24
Time Division Multiple Access, 21
TMSI Reallocation, 332, 337
Tracing. *See* Subscriber Tracing
Tracking Services, 434
Traffic Channels, 114
Traffic Flow Template, 361

Transaction Capabilities
  component sub-layer, 206
  layer architecture, 205
  overview, 204
  transaction sub-layer, 207
Transport Channels, 111
Transport Format, 117, 253
Transport Format Set, 253
Tu Interface, 84
Tunnel Endpoint Identifier, 357
Tunneling, 360

Um Interface, 82
UMTS Attach. *See* GPRS Attach
UMTS Detach. *See* GPRS Detach
Universal Integrated Circuit Card, 48, 78
Universal Mobile Telecommunications System
  access network. *See* Access Network
  addresses and identifiers, 60
  core network. *See* Core Network
  deployment. *See* Deployment of 3G Networks
  definition, 7
  hierarchical structure, 56
  network architecture, 45
  QoS architecture. *See* QoS Architecture
  user equipment. *See* User Equipment
Universal Subscriber Identity Module
  authentication, 447
  functions, 85
  overview, 47, 78, 79
Universal Terrestrial Radio Access Network
  broadcase/multicast functions, 110
  definition, 90
  functions, 105
  global signaling procedures, 264
  mobility management, 107
  overview. *See* Access Network
  protocol architecture, 97
  protocols, 110
  radio network layer, 97
  radio resource management. *See* Radio Resource Management
  security functions, 106
  signaling procedures for specific UE, 265

system access control, 106
transport network layer, 97
Universal Wireless Communications
  Consortium, 7
Unpaired Spectrum, 7
Unstructured Supplementary Service Data
  application, 412
  architecture, 411
  handler, 412
  message flows, 413
  overview, 69, 411
Unstructured Supplementary Services. *See*
  Unstructured Supplementary Service Data
URA Update, 256, 289
User Domain Security, 440
User Equipment
  classification, 87
  components, 47, 78
  functions, 84
  interfaces, 81
  MT-TE functionality split, 47, 79
  multi-network mode, 88
  multi-radio mode, 87
  overview, 47, 78
  protocols, 85
  single network mode, 88

single radio mode, 87
user equipment combination, 80
User Identity Confidentiality, 454
User-to-User Information, 404, 406
User-to-User Signaling, 69, 404
USIM Application Toolkit, 436
UTRAN Registration Area
  overview, 59, 295
  URA update, *See* URA Update
Uu Interface, 83

Virtual Home Environment, 437
Visited Public Land Mobile Network, 324
Visitor Location Register, 51, 168

White list, 167
Wideband Code Division Multiple Access
  concepts, 22
  modes of operation, 23
  synchronization aspects, 23

Za interface, 465
Zb interface, 465
Zd interface, 463
Ze interface, 463
Zf interface, 463

Nishit and Sumit (left to right)

Sumit Kasera is Senior Technical Leader at Hughes Software Systems, India. He has a B Tech. degree in computer science and engineering from IIT, Kharagpur, India. His current areas of interests include software development for mobile networks for both access and core network, software development for networking protocols (like ATM and TCP/IP), network modeling and simulation and routing protocols over satellite.

Sumit is the author of the book *ATM Networks: Concepts and Protocols.* He is an active participant in various technical forums like ITU-T, ATM and 3GPP. He has presented papers and conducted seminars in these forums.

Nishit Narang is Senior Technical Leader at Hughes Software Systems, India. He has a B Tech. degree in computer science and engineering from IIT, Delhi, India. An expert in developing high-availability high-performance 3G HLR, his areas of interest include software development for Node B and RNC, software development for networking protocols (like ATM and TCP/IP), network convergence, and routing protocols over satellite-based networks.

Nishit is an active participant in various technical forums like IEEE ACM, ATM and 3GPP. He has presented papers and conducted seminars in these forums.